BASIC TECHNICAL COLLEGE MATHEMATICS

Third Edition

Len Mrachek

with contributions by
Charles Komschlies

Upper Saddle River, New Jersey
Columbus, Ohio

Library of Congress Cataloging-in-Publication Data
Mrachek, Len
Basic technical college mathematics / Len Mrachek — 3rd ed.
p. cm.
Includes index.
ISBN 0-13-091751-6
1. Mathematics. II. Title.
QA39.2 .M687 2004
513'.14—dc21

2002043171

Editor in Chief: Stephen Helba
Executive Editor: Gary Bauer
Editorial Assistant: Natasha Holden
Media Development Editor: Michelle Churma
Production Editor: Louise N. Sette
Production Supervision: Carlisle Publishers Services
Design Coordinator: Diane Ernsberger
Cover Designer: Bryan Huber
Production Manager: Brian Fox
Marketing Manager: Leigh Ann Sims

This book was set in Times by Carlisle Communications, Ltd. It was printed and bound by Banta Book Group. The cover was printed by Phoenix Color Corp.

Pearson Education Ltd.
Pearson Education Singapore Pte. Ltd.
Pearson Education Canada, Ltd.
Pearson Education—Japan
Pearson Education Australia Pty. Limited
Pearson Education North Asia Ltd.
Pearson Educación de Mexico, S.A. de C.V.
Pearson Education Malaysia Pte. Ltd.

PEARSON
Prentice
Hall

10 9 8 7 6 5 4 3 2 1
ISBN 0-13-091751-6

PREFACE

Since the beginning of recorded history, mathematics has broadened the outreach of every civilization and generation. Today, with our knowledge doubling every five years, the role of mathematics in our everyday life cannot be overestimated. Mathematics has become essential to life itself. Every vocation requires some practical knowledge of mathematics. The major objective of the third edition of *Basic Technical College Mathematics* is to help you gain the mathematical skills you need in your chosen profession.

To help you attain these skills, I have been careful to present mathematical concepts in words and examples that are simple to understand and easy to remember. Frequent use is made of mathematical problems which daily confront the technician and craftsman. Consequently, this book is written for you, the student, to learn mathematics in the most efficient manner.

To provide additional essential mathematics, the third edition of *Basic Technical College Mathematics* includes three new chapters: *Chapter 6—Basic Measurement, Chapter 7—Data, Graphs and Basic Statistics, and Chapter 13—Graphs and Linear Equations.* Former *Chapter 14—Construction of Simple Geometric Figures* has been moved to the Appendix. These new chapters provide mathematics content that is becoming more important in solving daily practical problems. An extensive Mathematical Glossary assists learning.

It is important to use this text properly to optimize your learning. Read the following description of the features in each chapter so you will understand how to use the text.

1. *Chapter Title:* The chapter title gives you a glimpse of the contents of the chapter and enables you to rapidly select your area of study.
2. *Objectives:* The objectives tell you what you will learn and what you will be able to do when you complete the chapter.
3. *Self-Test:* Use the self-test to measure your need to study the chapter. If you do well with the self-test, you may already have a good comprehension of the material presented in the chapter. Thus, you may want to move on to the next self-test and the beginning of the next chapter.
4. *Definitions and Basics:* A brief introduction of the topics and a glimpse of the concepts that are used in the chapter. Careful study of the definitions of mathematical terms is essential to fully understand and learn the topics covered in the chapter.

5. *Examples and Applications:* This is the meat of the chapters. You are introduced to the mathematical skills and knowledge you need to master in order to meet the chapter objectives. The numerous examples and illustrations help you gain these skills and knowledge.
6. *Exercises including Applications:* These problems give you practice in skills you have learned. Most often, the exercises are drawn from realworld problems and everyday experiences.
7. *Think Time:* This feature gives you the opportunity to think of ways you can apply what you have learned. Perhaps you can think of problems you face in your profession that can be solved more effectively with what you have learned.
8. *Procedures to Remember:* Mathematical procedures and operations you need to remember are written concisely and to the point. You will find these procedures to be of great help to you in solving problems quickly and efficiently.
9. *Chapter Summary:* The chapter summary serves as a composite of all the information that has been presented within the chapter. It includes those concepts that must be learned for further development of your mathematical skills.
10. *Chapter Test:* The test at the end of each chapter evaluates your ability to do the mathematical operations taught in the chapter. Wrong answers and/or procedures for solutions indicate the precise area where mathematical skills have not been properly or adequately developed. Correct answers and procedures indicate properly developed skills and your readiness to begin work on the skill taught in the next chapter.

Calculators are necessary tools for everyone in everyday life. I recommend that you use the calculator at the appropriate time—once the basic arithmetic skills are mastered. However, remember that the calculator is only a computational aid; you must have the mathematical understanding to know which buttons to push. The use of the hand-held calculator is encouraged to allow more time to investigate additional problems.

ACKNOWLEDGMENTS

I would like to express my sincere appreciation to Charles Komschlies for his contributions to the first edition of this text. Charles contributed some very good ideas to the manuscript, including the Think Time and Self-Test features, and assisted in refining the presentation of the original material. My thanks to Delisa Whiting for her suggestions and the word processing of the StudyWizard manuscript. I would also like to acknowledge the reviewers of this text: Dr. Harry Hoffman, TCI—College for Technology (CT); Larry Kropp—Rich Mountain Community College (AR); and Daniel L. Timmons—Alamance Community College (NC). And I wish to acknowledge my children, grandchildren, and my students—past, present, and future—who inspire me to continue to attempt to provide good mathematical instruction.

Len Mrachek
University of Minnesota
Formerly at Hennepin Technical College

CONTENTS

Section Four: Trigonometry

FUNDAMENTAL OPERATIONS OF ARITHMETIC

OBJECTIVES

After studying this chapter, you will be able to:

1.1. Understand the properties of whole numbers, and perform the basic operations with whole numbers, including addition, subtraction, multiplication, division, rounding and estimating.

1.2. Solve applied problems using whole numbers.

SELF-TEST

S–1–1.
$$\begin{array}{r} 474 \\ 439 \\ 56 \\ +822 \\ \hline \end{array}$$

S–1–2.
$$\begin{array}{r} \$1598 \\ 39 \\ +\ 199 \\ \hline \end{array}$$

S–1–3.
$$\begin{array}{r} 789 \\ -454 \\ \hline \end{array}$$
(round difference to the nearest ten)

S–1–4.
$$\begin{array}{r} \$1239 \\ -\ 444 \\ \hline \end{array}$$

S–1–5.
$$\begin{array}{r} 47{,}463 \\ -16{,}587 \\ \hline \end{array}$$

S–1–6.
$$\begin{array}{r} \$1289 \\ \times\ \ 12 \\ \hline \end{array}$$

S–1–7.
$$\begin{array}{r} 539 \\ \times\ 321 \\ \hline \end{array}$$
(round product to the nearest hundred)

S–1–8. $7\overline{)287}$

S–1–9. $4\overline{)\$556}$

S–1–10. $408{,}060 \div 20 =$

S–1–11. An electrician worked 48, 42, 44, and 50 hours. Find the total number of hours he worked.

S–1–12. Find the total points scored in the following NBA games: 83, 79, 124, 146, and 116.

S–1–13. Kay Cobb's brother, Ruferthana, received checks for $586, $1429, $2246, $429, and $1842 as the deli manager in the supermarket. How much did he earn?

S–1–14. A painter used 28 gallons of paint from a 55-gallon drum. How much paint was left?

S–1–15. A water meter read 2412 cubic feet on September 1 and 2901 cubic feet on October 1. How many cubic feet of water were used during the month of September?

S–1–16. A boat and trailer together weigh 1235 pounds. The trailer weighs 359 pounds. How much does the boat weigh?

S–1–17. Al Eisele's car odometer read 93,842 at the beginning of the summer and 99,326 at the end of the summer. How many miles did he drive during the summer?

S–1–18. If there are 5280 feet in 1 mile, how many feet are there in 12 miles? (Round result to the nearest ten thousand.)

S–1–19. If gasoline costs 199 cents per gallon, find the cost of 125 gallons.

S–1–20. A test tube weighs 14 grams. Find the weight of a gross of test tubes. (144 = 1 gross.)

S–1–21. If a ham weighs 17 pounds, how many 1-ounce slices will you be able to obtain from this ham? (16 oz = 1 lb.)

S–1–22. How long would a steel bar need to be to make 18 pieces 17 inches long? (Disregard the amount of material lost in cutting.)

S–1–23. How much wire would be needed to wrap 14 coils if 2528 inches of wire is needed for each coil?

S–1–24. The "Front Four" of a football team are listed in the program as weighing 280, 265, 260, and 275 pounds. What is their average weight?

S–1–25. Find the value of D in Figure 1–1.

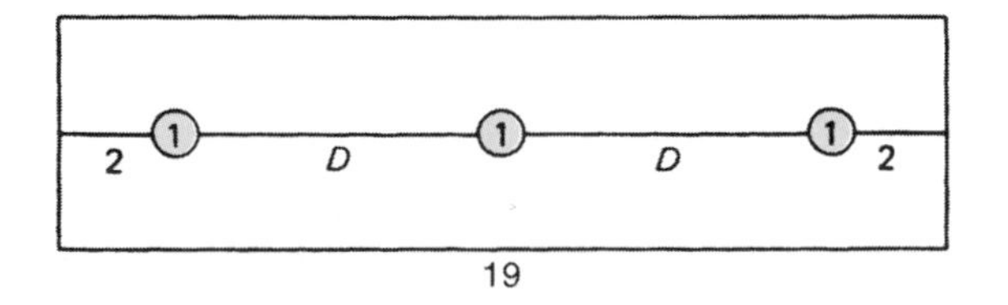

Figure 1–1

1.1 UNDERSTANDING WHOLE NUMBER PROPERTIES AND BASIC OPERATIONS

Numbers play an important part in our everyday lives. Most occupations require the development of technical skills based on the ability to read, write and solve problems.

You need to know how to read and write numbers. When writing a check, for example, it is necessary to write the numbers in words. You need to know the definitions and properties of whole numbers to be able to converse with fellow workers. You need to know the fundamental operations of addition, subtraction, multiplication, division and rounding to perform basic problem solving.

Although we are living in an age of calculators and computers, much of the basic operations can be done more efficiently by hand or mentally. Many trade and technical jobs require you to demonstrate that you can do basic calculations before you get the job. Thus, it is essential to perform the basic operations of whole numbers as a foundation for problem solving.

The numbers used in counting are called integers. An *integer* is a whole or natural number used in counting. Each integer is made up of one or more *digits.* Thus the number 8 has one digit, the number 23 has two digits (2 and 3), and the number 4146 has four digits (4, 1, 4, and 6). Each digit has a place value. An *even integer* is a number whose last digit (reading from left to right) is divisible by 2. Examples of even numbers are 2, 4, 6, 8, 12, and 96. An *odd integer* is a number whose last digit is not divisible by 2. Thus 3, 5, 7, 9, 17, and 361 are examples of odd numbers. A *prime number* is a number that can be divided only by itself and the number 1. The first eight prime numbers are: 2, 3, 5, 7, 11, 13, 17, and 19. The number 1 may or may not be considered a prime number.

A *composite number* is a whole number greater than 1 that can be divided by whole numbers other than itself. For example, 8 is a composite number $8 = 2 \times 4$ or $8 = 2 \times 2 \times 2$.

Rounding means giving an approximate value for a number. We round numbers for various reasons, generally to give an estimated amount. For example, we may say the highway distance from Minnesota to Florida is about 1,800 miles, when actually it would vary according to the location in the state and the route. Rounding is helpful when attempting to estimate an answer.

The place value chart names the value of the digits in a number. For example, 5,280 is read "five thousand two hundred eighty," since the 5 is in the thousands place and the other digits follow the chart.

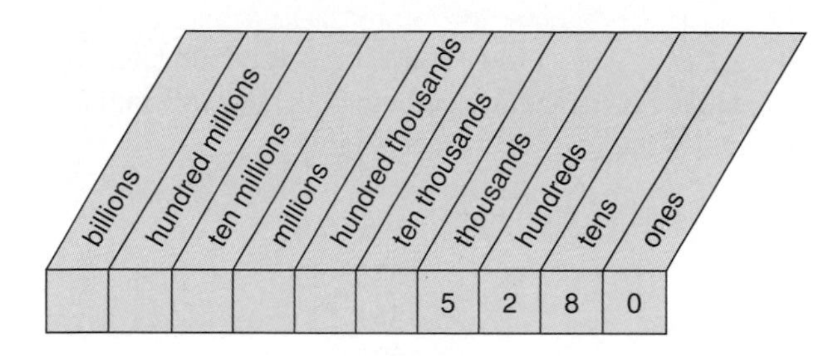

When numbers are written in *standard form,* they are separated into 3 digit groups, called *periods,* by *commas.*

The procedure for rounding whole numbers:

- If the digit to the right of the designated value is greater or equal to 5, increase the value of the given place value by 1 digit and replace all the digits to the right to zeros.
  Thus $38{,}764 \overrightarrow{(\text{hundreds})} 38{,}800$
  Note the arrowhead indicates to round to the given value.
- If the digit to the right of the designated value is less than 5, leave the digit as it is and replace all digits to the right with zeros.
  Thus $78{,}544 \overrightarrow{(\text{hundreds})} 78{,}500$
  $625{,}760 \overrightarrow{(\text{thousands})} 626{,}000$
  $1896 \overrightarrow{(\text{tens})} 1{,}900$

There are several symbols or signs used with whole numbers: for example, in $7 + 2 = 9$. The equal sign (=) indicates the quantities are the same or equal. The symbol ($\neq$) means that one quantity is not equal to the other. Thus $7 \neq 5$. Other signs are $<$ and $>$: these are inequality symbols. The arrow always points to the smaller quantity: $5 > 3$ (this is read "5 is greater than 3") or $8 < 14$ (this is read "8 is less than 14"). The ($\approx$) symbol indicates that the quantities or numbers are approximately equal. Thus $19 \approx 20$ is read "19 is approximately equal to 20" and $98 \approx 100$ is read "98 is approximately equal to 100."

Numbers that are combined or added to give a total or sum are called *addends.* Addition is indicated by the + symbol, called the *plus sign* of operation. The plus sign of operation tells you to add or combine the numbers to obtain a total. The result of combining the addends is called the *sum.*

$$\begin{array}{rl} 12 & \text{(addend)} \\ +25 & \text{(addend)} \\ \hline 37 & \text{(sum)} \end{array}$$

The purpose of *subtraction* is to find out how much larger one number is than another. The larger number is placed on top of the smaller number. The larger number is called the *minuend,* and the smaller number is called the *subtrahend.* The result is called the *difference.* The sign of operation of subtraction is the *minus sign* (−). Subtraction is the opposite operation of addition.

$$\begin{array}{rl} 25 & \text{(minuend)} \\ -12 & \text{(subtrahend)} \\ \hline 13 & \text{(difference)} \end{array}$$

You may wish to add the same number several times. For example, if you wanted to find out how many inches there are in 7 feet, you would add seven 12's together to arrive at the answer of 84.

$$\begin{array}{r} 12 \\ 12 \\ 12 \\ 12 \\ 12 \\ 12 \\ +12 \\ \hline 84 \end{array} \quad \text{or} \quad \begin{array}{rl} 12 & \text{(multiplicand)} \\ \times\ 7 & \text{(multiplier)} \\ \hline 84 & \text{(product)} \end{array}$$

You could multiply 12 by 7 and obtain the same answer of 84. This process is called *multiplication* and is considered a rapid form of addition. It saves time and decreases the chances of making mistakes. When two numbers are multiplied, the resulting answer is called the *product.* The top number is called the *multiplicand,* and the bottom number is called the *multiplier.* The sign of operation of multiplication is the *times sign* (×).

You may wish to divide a quantity into two or more parts. For example, if you wanted to find out how many dozen rolls you would need for 72 people, you could keep on subtracting 12's from 72 until you reach a number that was less than 12. You would then count the number of 12's you had subtracted to find out how many dozen rolls you would need.

$$\begin{array}{r} 72 \\ -12 \\ \hline 60 \end{array} \quad \begin{array}{r} 60 \\ -12 \\ \hline 48 \end{array} \quad \begin{array}{r} 48 \\ -12 \\ \hline 36 \end{array} \quad \begin{array}{r} 36 \\ -12 \\ \hline 24 \end{array} \quad \begin{array}{r} 24 \\ -12 \\ \hline 12 \end{array} \quad \begin{array}{r} 12 \\ -12 \\ \hline 0 \end{array}$$

You could also solve this problem by dividing 72 by 12 to arrive at the same answer, 6.

$$\begin{array}{rrl} & 6 & \text{(quotient)} \\ \text{(divisor)} & 12\overline{)72} & \text{(dividend)} \end{array}$$

This process, called *division,* may be considered as rapid subtraction. When one number is divided by another, the resulting answer is called the *quotient.* The number that is used to divide is called the *divisor,* and the number that is divided is called the *dividend.* The sign of operation of division is the *divide sign* (÷). When this sign appears in a problem, it tells you to divide the second number into the first number. For example, 72 ÷ 12 tells you to divide 72 by 12. When you attempt to solve the problem by division, you rewrite the problem so you can more easily arrive at the correct answer. Thus you could rewrite 72 ÷ 12 as $12\overline{)72}$.

The four basic operations of mathematics are addition, subtraction, multiplication, and division. They are the essential skills to solve applied problems with whole numbers.

EXERCISE 1–1 OPERATIONS WITH WHOLE NUMBERS

Do the following problems using the basic arithmetic operations. Be sure to check your results.

1–1.
$$\begin{array}{r} \$9.98 \\ 4.49 \\ +\ 3.29 \\ \hline \end{array}$$

1–2.
$$\begin{array}{rl} 10.8 & \text{gal} \\ 5.7 & \text{gal} \\ 9.4 & \text{gal} \\ +\ 8.0 & \text{gal} \\ \hline \end{array}$$

1–3.
$$\begin{array}{rl} 1673 & \text{lb} \\ 1841 & \text{lb} \\ 1397 & \text{lb} \\ +\ 224 & \text{lb} \\ \hline \end{array}$$

1–4.
$$\begin{array}{rl} 386 & \text{mi} \\ 513 & \text{mi} \\ 121 & \text{mi} \\ +\ 86 & \text{mi} \\ \hline \end{array}$$

1–5.
$$\begin{array}{rl} 4.25 & \text{hr} \\ 8.75 & \text{hr} \\ 10.25 & \text{hr} \\ +\ 4.25 & \text{hr} \\ \hline \end{array}$$

1–6.
$$\begin{array}{r} 5008 \\ -\ 946 \\ \hline \end{array}$$

1–7.
$$\begin{array}{r} 78{,}462 \\ -\ 3{,}008 \\ \hline \end{array}$$

1–8.
$$\begin{array}{rl} 28.8 & \text{gal} \\ -\ 8.9 & \text{gal} \\ \hline \end{array}$$

1–9.
$$\begin{array}{rl} 17{,}001 & \text{mi} \\ -\ 9{,}314 & \text{mi} \\ \hline \end{array}$$

1–10.
$$\begin{array}{rl} 50{,}000 & \text{mi} \\ -19{,}876 & \text{mi} \\ \hline \end{array}$$

1–11.
$$\begin{array}{r} 98 \\ \times\ 8 \\ \hline \end{array}$$

1–12.
$$\begin{array}{r} 2406 \\ \times\ 39 \\ \hline \end{array}$$

1–13.
$$\begin{array}{r} 5280 \\ \times\ 707 \\ \hline \end{array}$$

1–14.
$$\begin{array}{r} \$1498 \\ \times\ 9 \\ \hline \end{array}$$

1–15.
$$\begin{array}{r} \$3.89 \\ \times\ 14 \\ \hline \end{array}$$

1–16. $6\overline{)528}$ 1–17. $17\overline{)663}$ 1–18. \$22.50 ÷ 6 =

1–19. 5412 ÷ 451 = 1–20. $16\overline{)\$223{,}968}$

1.2 SOLVE APPLIED PROBLEMS USING WHOLE NUMBERS

Application problems are the reason we learn the basic skills and study mathematics. To solve applied problems, read the problem carefully to determine what is to be found. Note the given quantities, and estimate the value of the answer. Use common sense. Then translate the printed words into a form of the basic skills or into an equation, and solve. Test your answer, and attach the proper units. The following examples illustrate how to use the basic operations of mathematics

Example 1–1: Addition of Whole Numbers

PROBLEM: Jack drove 169 miles the first week, 96 miles the second week, 174 miles the third week, and 191 miles the fourth week, using super unleaded gasoline. How many miles did he drive using super unleaded gasoline?

SOLUTION: Add (from top to bottom).

$$\begin{array}{rl} 169 & \text{(addend)} \\ 96 & \text{(addend)} \\ 174 & \text{(addend)} \\ +191 & \text{(addend)} \\ \hline 630 & \text{(sum)} \end{array}$$

CHECK: Re-add from bottom to top.

Example 1–2: Subtraction of Whole Numbers

PROBLEM: How many miles did Jack drive on a trip if the odometer read 79,386 at the start and 81,234 at the end of the trip?

SOLUTION: Subtract 79,386 from 81,234; borrow as necessary.

$$\begin{array}{rl} 81{,}234 & \text{(minuend)} \\ -79{,}386 & \text{(subtrahend)} \\ \hline 1{,}848 & \text{(difference)} \end{array}$$

CHECK: Add.

$$\begin{array}{rl} 79{,}386 & \text{(subtrahend)} \\ +\ 1{,}848 & \text{(difference)} \\ \hline 81{,}234 & \text{(minuend)} \end{array}$$

Example 1–3: Multiplication of Whole Numbers

PROBLEM: Jack paid 134 cents for each gallon of super unleaded gasoline. How much did he pay for the 8 gallons of super unleaded gasoline?

SOLUTION: Multiply the gallons by the price per gallon.

$$\begin{array}{rl} 134 & \text{price (cents) per gallon (multiplicand)} \\ \times\ 18 & \text{gallons of unleaded (multiplier)} \\ \hline 1072 & \\ 134 & \\ \hline 2412 & \text{cents (product)} \end{array}$$

CHECK: Divide 18 into 2412.

$$\begin{array}{l} \qquad\qquad\quad\ \ 134\ \text{(quotient)} \\ \text{(divisor)}\ 18\overline{)2412}\ \text{(dividend)} \\ \qquad\qquad\quad\ \underline{18} \\ \qquad\qquad\quad\ \ 61 \\ \qquad\qquad\quad\ \ \underline{54} \\ \qquad\qquad\quad\ \ \ 72 \\ \qquad\qquad\quad\ \ \ \underline{72} \end{array}$$

Example 1–4: Division of Whole Numbers

PROBLEM: Jack used 21 gallons of unleaded gasoline to drive 630 miles. How many miles did he travel per gallon of gasoline?

SOLUTION: Divide the miles driven (630) by the total number of gallons of leaded gasoline (21).

$$\begin{array}{l} \qquad\qquad\quad\ 30\ \text{miles per gallon (quotient)} \\ \text{(divisor)}\ 21\overline{)630\ \text{miles driven (dividend)}} \\ \qquad\qquad\quad\ \underline{63} \\ \qquad\qquad\qquad 0 \end{array}$$

CHECK: Multiply 30 times 21.

$$\begin{array}{rl} 30 & \text{(multiplicand)} \\ \times 21 & \text{(multiplier)} \\ \hline 30 & \\ 60 & \\ \hline 630 & \text{(product)} \end{array}$$

Example 1–5: Multiplication and Addition of Whole Numbers

PROBLEM: Jack filled his car with gasoline five times. It took 14 gallons to fill it the sixth time. If the capacity of the fuel tank is 23 gallons, how much gasoline has he purchased?

SOLUTION: Multiply 5 times 23 gallons and add 14 gallons.

$$\begin{array}{rl} 23 & \text{gal} \\ \times\ 5 & \\ \hline 115 & \text{gal} \end{array} \qquad \begin{array}{rl} 115 & \text{gal} \\ +\ 14 & \\ \hline 129 & \text{gal} \end{array}$$

CHECK: Re-do the multiplication and then re-add the addition.

Example 1–6: Subtraction of Whole Numbers

PROBLEM: Jack wanted to find out which cost him less to use, unleaded or super unleaded gasoline. It cost Jack 2604 cents for unleaded gasoline to drive the 630 miles and 2412 cents for super unleaded to drive the same distance. When driving the 630 miles, how much less did it cost Jack for the super unleaded than for the unleaded?

SOLUTION: Subtract the cost of unleaded (2412 cents) from the cost of superunleaded (2604 cents).

$$\begin{array}{rl} & \text{cost (cents) for unleaded gasoline} \\ 2604 & \text{(minuend)} \\ & \text{cost (cents) for superunleaded gasoline} \\ -2412 & \text{(subtrahend)} \\ \hline 192 & \text{less using unleaded gasoline (difference)} \end{array}$$

CHECK: Add 2412 and 192.

$$\begin{array}{rl} 2412 & \text{(addend)} \\ +\ 192 & \text{(addend)} \\ \hline 2604 & \text{(sum)} \end{array}$$

Example 1–7: Addition and Division of Whole Numbers

PROBLEM: Jack wondered how many miles he drove on average each week. The miles driven during the nine weeks were: 169, 96, 174, 191, 154, 117, 137, 211, and 128. Find the average miles driven per week.

SOLUTION: Add the total number of miles driven each week.

$$\begin{array}{rl} 169 & \text{(addend)} \\ 96 & \text{(addend)} \\ 174 & \text{(addend)} \\ 191 & \text{(addend)} \\ 154 & \text{(addend)} \\ 117 & \text{(addend)} \\ 137 & \text{(addend)} \\ 211 & \text{(addend)} \\ +\ 128 & \text{(addend)} \\ \hline 1377 & \end{array}$$

Then divide the sum by the total number of weeks.

$$\begin{array}{rl} & \text{153 average miles (quotient)} \\ \text{(divisor) 9 weeks} & \overline{)\text{1377 miles driven (dividend)}} \\ & \underline{9} \\ & 47 \\ & \underline{45} \\ & \ \ 27 \\ & \ \ \underline{27} \end{array}$$

CHECK: First, re-add from bottom to top. Then multiply the average mileage by the number of weeks.

$$\begin{array}{rl} 153 & \text{(multiplicand)} \\ \times\ \ 9 & \text{(multiplier)} \\ \hline 1377 & \text{(product)} \end{array}$$

Example 1–8: Applied Whole Number Problem

PROBLEM: Given the drawing of a 53 by 28-foot foundation of warehouse, find the width of the foundation on the shorter and longer sides.

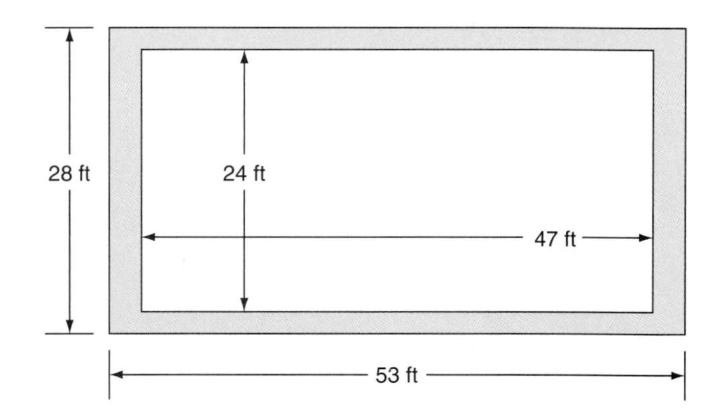

SOLUTION: Subtract the outside length (53 ft) from the inside length (47 ft) and divide by 2. This will give the foundation width of the length side.

Thus

$$\begin{array}{r} 53\text{ ft} \\ -47\text{ ft} \\ \hline 6\text{ ft} \end{array} \qquad \frac{6\text{ ft}}{2} = 3\text{ ft wide}$$

Then for the width of the foundation, subtract the outside width (28 ft) from the inside width (24 ft) and divide by 2.

$$\begin{array}{r} 28\text{ ft} \\ -24\text{ ft} \\ \hline 4\text{ ft} \end{array} \qquad \frac{4\text{ ft}}{2} = 2\text{ ft wide}$$

Therefore, the foundation is 3 feet wide on the long side, and 2 feet wide on the short side.

EXERCISE 1–2 WHOLE NUMBER APPLIED PROBLEMS

Do the following problems using the basic arithmetic operations. Be sure to check your results.

1–21. The air distance from Atlanta to Boston is about 837 miles. From Atlanta to New Orleans it is about 393 miles. How far is it from Boston to New Orleans via Atlanta?

1–22. A wholesaler sells 504 fishing reels to one dealer, 103 to another, 97 to another, and has 321 remaining. How many did he have at the beginning?

1–23. The area in square miles of the following states is: Alaska, 586,400; Alabama, 54,609; and Arkansas, 53,104. What is the total area of these states?

1–24. The total area in square miles of the five largest countries in the world is: the former Soviet Union, 8,599,300; Canada, 3,851,809; China, 3,691,500; United States, 3,675,633; and Brazil, 3,286,478. What is the total area of these countries?

1–25. Marshall Fields's service department did the following business for labor and parts in 1 hour:

	Labor	*Parts*
Hocker Hanson	$22.50	$19.50
Ducky Dunbar	22.75	27.00
Bird Benson	25.00	22.00
Phred Freeberg	20.50	19.85

(a) What is the total income from labor?
(b) What is the total income from parts?
(c) What is the total income from labor and parts for the 1 hour?

1–26. If a video camera costs $1495 at one store and $999 at a second store, how much would you save buying the camera at the second store?

1–27. The fuel capacity of a jet plane is 35,000 pounds. After a flight, it takes 27,673 pounds of fuel to refuel the plane. How much fuel was used on the flight?

1–28. How much greater is 12,000 than 9249?

1–29. The fuel capacity of a 1986 Chevrolet van is listed at 36 gallons. If it takes 19 gallons of gasoline to fill the tank, how much fuel was left in the tank?

1–30. A doctor recommended a dosage of 450 cubic centimeters per day and then changed it to 750 cubic centimeters. How much did he increase the dosage?

1–31. The warranty on a new Chrysler car is 70,000 miles. The odometer shows 39,875 miles. How many miles are left on the warranty?

1–32. A television set was priced at $579. Then it was marked down to $519. Later it was marked down further, to $495. What was the total discount?

1–33. Office space at a new office complex rents for $32 per square foot. What would the rent be for an office space of 1800 square feet?

1–34. If a truck averages 54 miles per hour, how many miles will the truck cover in 112 hours?

1–35. If it takes 3 hours of flying time between Seattle and Minneapolis, how much flying time would it take to make three round trips?

1–36. You signed a $13,500, 5-year lease for a Dodge Neon. If the lease payment is $225 per month and you have paid $8550, how many months are left on the lease?

1–37. Tony Beldon bought three compact discs (CDs) at $15 each and two CDs at $18 each. How much did he spend for the CDs?

1–38. What is the cost of 6 Saturns at $19,995 each?

1–39. The total for a catering bill for a wedding was $7209. If there were 267 guests, what was the cost for each guest?

1–40. The starting salary for a nurse at Med Center One is $23,340 per year. How much is this per month?

1–41. If you traveled 496 miles on 16 gallons of gas, what was your average mileage per gallon?

1–42. A lot in a posh suburb sold for $169,884. If the lot contained 4356 square feet, what was the cost per square foot?

1–43. Tom Melchior spent $3,588 for groceries for his family during the year. What was his average weekly cost for groceries? (There are 52 weeks in 1 year.)

1–44. What is the average age of a family whose father is 52; mother, 48; daughter, 25; daughter, 24; son, 21; daughter, 19; and daughter, 14?

1–45. The weekly salaries at the supermarket are as follows: Velda, $304; Jake, $289; Thelma, $273; Phred, $310; Harold, $265; and Brucy, $293. What is the average *annual* salary? (52 weeks/year)

1–46. Find the missing distance in Figure 1–2.

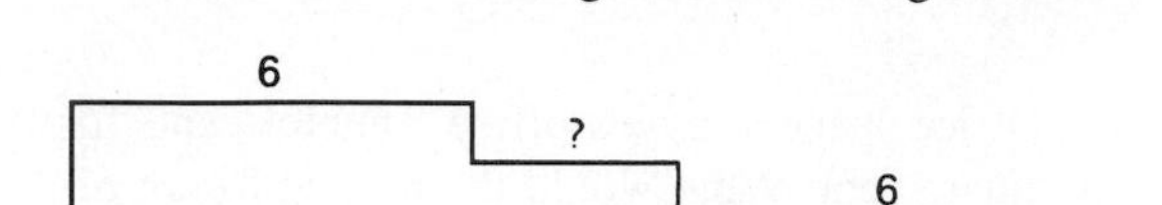

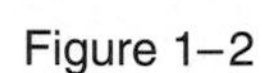
Figure 1–2

1–47. Find the missing dimension in Figure 1–3.

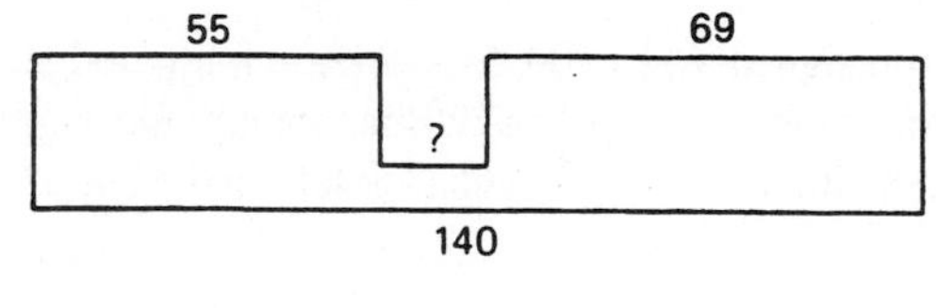

Figure 1–3

1–48. What is the thickness of the pipe shown in Figure 1–4?

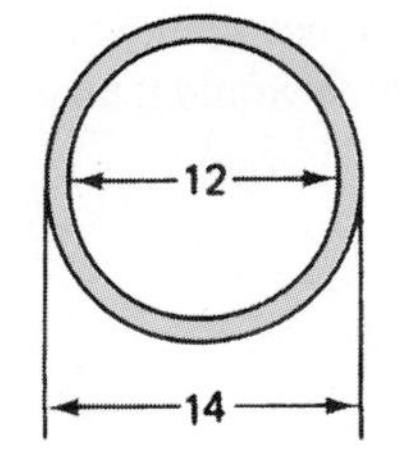

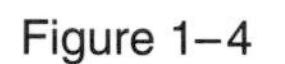
Figure 1–4

1–49. Find the missing dimension in Figure 1–5.

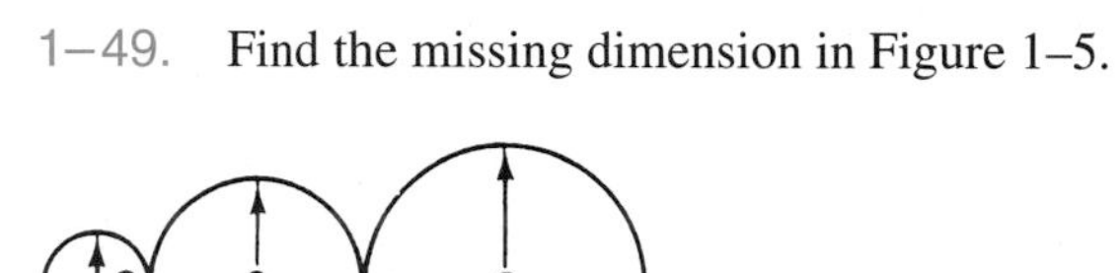

Figure 1–5

1–50. Find the missing dimension in Figure 1–6.

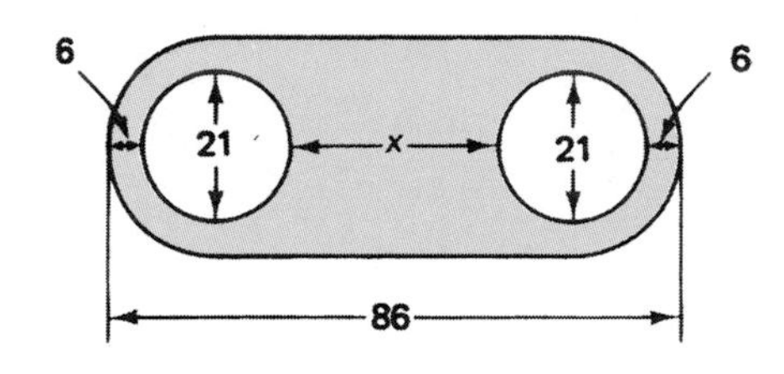

Figure 1–6

THINK TIME

As you worked the examples and exercises in this chapter, you used the four basic operations (addition, subtraction, multiplication, and division). Think of the things you have learned in this chapter that will help you on the job. If you have an opportunity to discuss and review some of these with a teacher or a friend, it would be to your advantage. The more capable you are in basic skills, the more successful you will be in many things.

PROCEDURES TO REMEMBER

The basic operations of arithmetic are:

1. Reading the signs of operation, equality, and inequality.
2. Performing the basic operations by the most efficient method and showing all necessary work.
3. Checking your work to make sure the answers are reasonable and correct. Methods of checking are as follows:

(a) *Addition:* Re-add from bottom to top.

(b) *Subtraction:* Add the difference to the subtrahend to obtain the minuend.

(c) *Multiplication:* Divide the product by the multiplier to obtain the multiplicand.

(d) *Division:* Multiply the quotient by the divisor to obtain the dividend.

4. *Rounding* procedures:
   - If the digit to the right of the desired value is greater than 5, increase the digit by one and replace all the digits to the right with zeros.
   - If the digit to the right of the desired value is less than 5, leave the digit as written and replace all digits to the right with zeros.

   Note: 76,564 $\overrightarrow{\text{(hundreds)}}$ 76,600

   224 $\overrightarrow{\text{(tens)}}$ 220

   Recall that the long arrow means to round; the quantity in parentheses indicates what value to round to.
5. When solving an *applied problem, read* the problem carefully, decide *what* is to be found, *estimate* the answer, *solve* the problem, and *check* the reasonableness of the answer.

CHAPTER SUMMARY

1. An *integer* is a whole or natural number used in counting.
2. An *even number* is a number that is divisible by 2.
3. An *odd number* is a number that is *not* divisible by 2.
4. A *prime number* is a number that can be divided only by itself and 1.
5. A *composite number* is a whole number greater than 1 that can be divided by a whole number other than itself.
6. The *place value chart* names the value of each digit in a number.
7. Numbers written in *standard form* are separated into 3 digit groups, called *periods,* that are separated by *commas.*
8. Symbols that indicate properties of whole numbers:
   "=" is equal to
   "≠" is not equal to
   ">" is greater than
   "<" is less than (The arrow always points to the smaller quantity.)
   "≈" is approximately equal to
9. The numbers that are added to give a *sum* or *total* are called *addends.* The plus sign (+) indicates addition. This is one of the *signs of operation* of mathematics.
10. *Subtraction* is the arithmetic operation of finding the *difference* between the *minuend* (top number) and the *subtrahend* (bottom number). The sign of operation is the minus or negative sign (−).
11. The *product* is obtained when you multiply the *multiplicand* (top number) by the *multiplier* (bottom number). This is called *multiplication.* The sign of operation is ×.
12. The *quotient* is obtained when you divide one number, the *divisor,* into another number, the *dividend.* The sign of operation of division is ÷. Divide the second number into the first. Thus we may show 20 ÷ 4 as $4\overline{)20}$. The *average* is the sum of all the numbers divided by the number of addends.

CHAPTER TEST

Solve the following problems using the fundamental operations of arithmetic. Show all necessary work.

T–1–1. $$\begin{array}{r} \$1529 \\ +\ 164 \\ \hline \end{array}$$

T–1–2. $$\begin{array}{r} \$18,398 \\ -\ 8,453 \\ \hline \end{array}$$

T–1–3. $3829 (round the sum to the nearest ten)

$$\begin{array}{r} \$3829 \\ 546 \\ 29 \\ 498 \\ +\ 39 \\ \hline \end{array}$$

T–1–4.

$$\begin{array}{r} \$139 \\ \times\ 15 \\ \hline \end{array}$$

T–1–5. $9\overline{)2016}$

T–1–6. If a master carpenter earns $41,288 in a year, what are his average weekly earnings?

T–1–7. Lawn mowers for cutting grass on a golf course cost $3,295 each. How much would three of them cost?

T–1–8. If a truck driver earns $170 a day, how much would he make in 250 working days (round to the nearest thousand)?

T–1–9. If water boils at 212° F at sea level and at 193° F in the McKinley range of mountains, what is the difference in boiling points?

T–1–10. An electrician used the following lengths of number 14 wire: 63, 132, 86, 168, and 1024 inches. How much wire did he use?

T–1–11.

$$\begin{array}{r} 8877 \\ 2345 \\ 618 \\ 6824 \\ +\ 36 \\ \hline \end{array}$$

T–1–12.

$$\begin{array}{r} \$20,382 \\ 527 \\ 322 \\ +\ 1,296 \\ \hline \end{array}$$

T–1–13.

$$\begin{array}{r} 23,273 \\ -\ 16,439 \\ \hline \end{array}$$

T–1–14.

$$\begin{array}{r} 999 \\ \times\ 99 \\ \hline \end{array}$$

T–1–15. $408\overline{)34,272}$

T–1–16. A washing machine in a bottling plant washes 3024 bottles in 12 hours. How many bottles does it wash per hour (round to the nearest ten)?

T–1–17. A running back carried the ball 22 times in one game for a total yardage of 198. What was the average yardage he gained per carry?

T–1–18. A garment worker makes 368 golf shirts in an 8-hour day. How many shirts does he make per hour?

T–1–19. An electrical generator produces 2700 watts of power. How many 75-watt light bulbs will this light?

T–1–20. The shipping weight of each of 15 electronic testing scopes is 219 pounds, including the carton. What is the total shipping weight?

T–1–21. A farmer wants to paint his barn, which has 14,950 square feet. How many gallons of paint will be needed if each gallon covers 650 square feet (round up to the next gallon)?

T–1–22. Vinny Vangogo, a painter, buys 22 gallons of paint at $13 per gallon and 11 gallons at $17 per gallon. How much did the paint cost?

T–1–23. Some duck hunters purchased a raft for $253 and a tent for $195. How many hunters were involved if each paid an equal amount of $112?

T–1–24. Members of the Jog for Joy club were comparing the distances they jogged each month. The distances stated were: 193, 124, 86, 113, and 59 kilometers. What is the average distance each member jogged per month?

T–1–25. Find T in Figure 1–7.

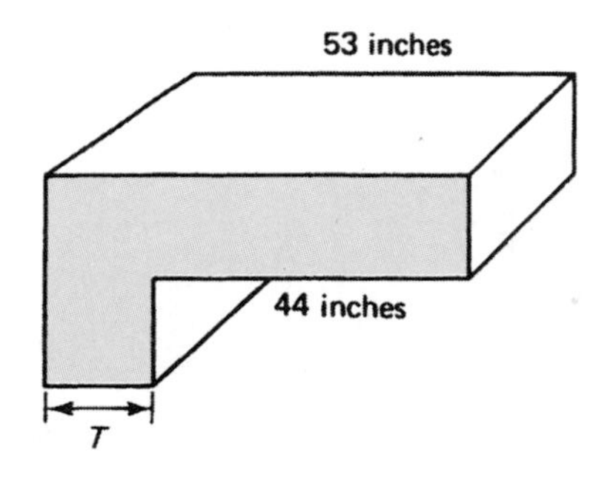

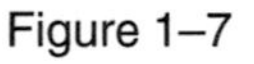

Figure 1–7

2

COMMON FRACTIONS

OBJECTIVES

After studying this chapter, you will be able to:

2.1. Understand the definitions and basic operations of fractions.
2.2. Multiply and divide fractions.
2.3. Add and subtract fractions.
2.4. Simplify complex fractions and perform the order of operations agreement.
2.5. Solve applied problems using fractions.

SELF-TEST

This test will determine your need to develop your skills with fractions. If you are able to work these problems, move on to decimals. Show your work; this will indicate areas where you may need to improve.

S–2–1. $\frac{5}{6} + \frac{3}{4} + \frac{1}{3}$

S–2–2. $24\frac{1}{3} + 10\frac{5}{6}$

S–2–3. $6\frac{3}{4} + 7\frac{1}{2} + 2\frac{2}{3}$

S–2–4. $\frac{4}{5} - \frac{1}{3} =$

S–2–5. $15\frac{2}{5} - 5\frac{2}{3}$

S–2–6. $34\frac{1}{3} - 29\frac{5}{6}$

S–2–7. $\frac{2}{3} \times \frac{6}{18} =$

S–2–8. $3\frac{1}{2} \times 1\frac{3}{4} =$

S–2–9. $24\frac{1}{2} \div 18\frac{5}{6} =$

S–2–10. $12\frac{1}{2} \div 18\frac{5}{6} =$

S–2–11. $\frac{3}{4} \div 12 =$

S–2–12. $\frac{7}{8} \times \frac{7}{16} \div \frac{21}{32} =$

S–2–13. $12\frac{1}{2} \div 4\frac{2}{3} \div 8\frac{3}{8} =$

S–2–14. $3\frac{2}{3} \times 1\frac{3}{5} \times 2\frac{1}{12} =$

S–2–15. $\dfrac{\frac{7}{8} - \frac{1}{2}}{4} =$

S–2–16. $\dfrac{3\frac{1}{2} + \frac{3}{4}}{3\frac{1}{2} \times 2\frac{1}{4}} =$

S–2–17. $\left(\frac{1}{3}\right)^2 + \left(\frac{3}{2} \div \frac{6}{4}\right)5 =$

S–2–18. Find length A in Figure 2–1.

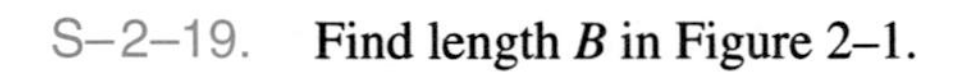

S–2–19. Find length B in Figure 2–1.

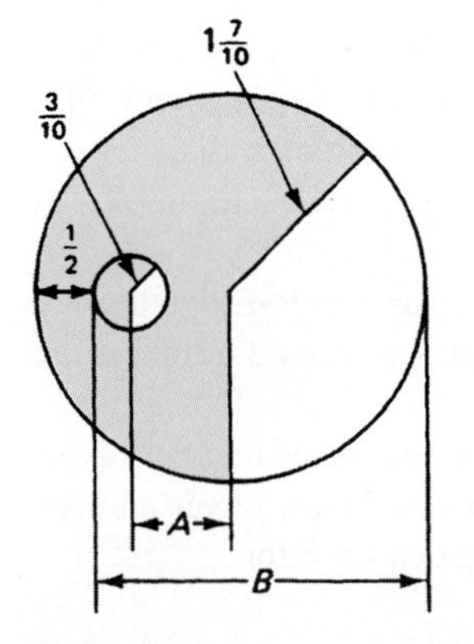

Figure 2–1

S–2–20. An oil stock sold for $11\frac{1}{2}$ a share when the stock market opened. The stock closed at $12\frac{1}{8}$. How much did the stock gain?

S–2–21. Brian drove 48 minutes to the Timberwolves game. If 18 minutes were spent in bumper-to-bumper traffic, what fractional part of his time was spent in traffic?

S–2–22. A Lear jet has 180 gallons of fuel remaining in its tanks. If the plane averages $4\frac{1}{3}$ miles per gallon, how far can the plane travel?

S–2–23. The gas tank of a car shows $\frac{3}{4}$ full. If the full tank contains 24 gallons and gasoline costs $2 per gallon, how much is the gas worth that is in the tank?

S–2–24. A truck travels $12\frac{4}{10}$ miles on a round trip from a plant to a construction site. On the eighth trip, the truck stops at the construction site. How many miles has the truck been driven?

S–2–25. A master painter can paint a garage in 6 hours. An apprentice can paint the same garage in 9 hours. Working together, how much of the garage can they paint in 1 hour?

2.1 UNDERSTANDING DEFINITIONS AND BASIC OPERATIONS OF FRACTIONS

Common fractions are an important part of our day-to-day living. We use fractions to indicate time—"I will meet you in a half hour"; quantity—"You have only a quarter of a tank of gasoline left"; and distance—"The horses have passed the three-quarter pole." Fractions are a part of the language of industry and technology. Terms such as "$\frac{3}{32}$-inch welding rods," "$\frac{5}{8}$-inch plywood," "three-quarters of a cup of flour," "$\frac{9}{64}$-inch drill," "three-quarter-ton truck," and "one-third off" are used many times. Therefore, if we are to survive and contribute to society, it is essential for us to have fraction skills.

A *fraction* is the division of one number by another number. Examples of fractions are $\frac{1}{8}$, $\frac{2}{3}$, and $\frac{6}{7}$. The number above the line is called the *numerator* and the number below the line is called the *denominator.* The numerator and denominator are the terms for a fraction. Thinking of the great state of North Dakota—ND—will help you remember that the numerator precedes (is above) the denominator. A fraction whose denominator is larger than its numerator is called a *proper fraction.* Examples of proper fractions are $\frac{3}{4}$, $\frac{7}{8}$, and $\frac{3}{16}$.

A fraction whose numerator is greater than its denominator is called an *improper fraction.* Examples of improper fractions are $\frac{5}{4}$, $\frac{32}{9}$, and $\frac{8}{5}$.

A whole number and a fraction written together are called *mixed numbers.* Examples of mixed numbers are $4\frac{3}{4}$, $5\frac{3}{16}$, and $6\frac{7}{8}$.

To change an improper fraction to a mixed number, divide the numerator by the denominator, write the whole number, and place the remainder over the denominator.

Thus $$\frac{19}{8} = 2\frac{3}{8}$$

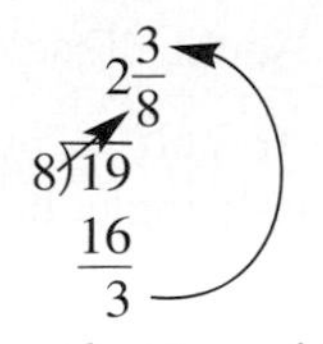

To change a mixed number to an improper fraction, multiply the whole number by the denominator and add the numerator; place that number over the original denominator.

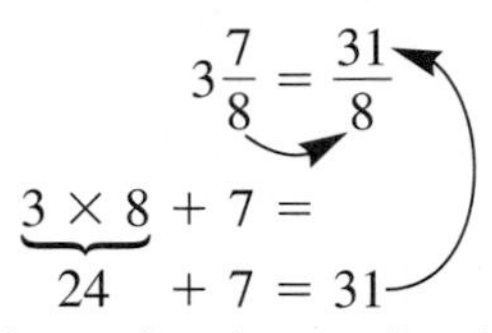

A fraction that has a fraction or fractions in the numerator and/or denominator is called a *complex fraction.* Examples of complex fractions are

$$\frac{\frac{6}{7}}{\frac{3}{5}}, \frac{\frac{2}{3}}{\frac{1}{2}}, \text{ and } \frac{\frac{3}{1}}{\frac{7}{8}}$$

To *reduce* a fraction to lowest terms, divide the numerator and the denominator by the same number.

A competent technician or skilled craftsperson needs to perform operations with fractions. These operations are addition, subtraction, multiplication, and division. We start with the multiplication of fractions, which is the easiest operation.

2.2 MULTIPLYING AND DIVIDING FRACTIONS

The procedure for multiplying fractions is to multiply the numerators and multiply the denominators. Wherever possible, reduce by dividing common numbers into the numerators and denominators.

Example 2–1: Multiplying Fractions

PROBLEM: $\frac{2}{3} \times \frac{1}{5} =$

SOLUTION: Multiply the numerator (2) by the numerator (1) and the denominator (3) by the denominator (5).

$$\frac{2}{3} \times \frac{1}{5} = \frac{2 \times 1}{3 \times 5} = \frac{2}{15}$$

Example 2–2: Multiplying Fractions

PROBLEM: $\frac{3}{4} \times \frac{7}{15} =$

SOLUTION: Before multiplying fractions, reduce the fractions to the lowest terms. Reducing before multiplying will make the multiplication process easier. The reducing may be done by different methods.

METHOD 1: Divide the numerator (3) and the denominator (15) by 3. The number to be divided evenly into the numerator and the denominator is done by observation. Practice will refine your selection ability.

$$\frac{\overset{1}{\cancel{3}}}{4} \times \frac{7}{\underset{5}{\cancel{15}}} = \frac{1 \times 7}{4 \times 5} = \frac{7}{20}$$

METHOD 2: Divide the top (numerator) of one fraction (3) into the bottom (denominator) of another fraction (15). Thus the numerator (3) divides into the denominator (15) five times and into itself (3) once. A simple rule to remember is "*top* into *bottom*" and "*bottom* into *top*." After reducing the fractions to lowest terms, multiply the numerators 1 and 7 and the denominators 4 and 5.

$$\frac{\overset{1}{\cancel{3}}}{4} \times \frac{7}{\underset{5}{\cancel{15}}} = \frac{1 \times 7}{4 \times 5} = \frac{7}{20}$$

Example 2–3: Multiplying Fractions

PROBLEM: $\frac{39}{9} \times \frac{45}{13} \times \frac{1}{10} =$

SOLUTION: When reducing fractions to lowest terms, there may be several possible reductions. Make certain you have made all possible reductions before you multiply the fractions.

METHOD 1: Divide the numerator (39) and the denominator (13) by 13. Divide the denominator (9) and the numerator (45) by 9. Then divide the new numerator (4) and the denominator (10) by 5.

$$\frac{\overset{3}{\cancel{39}}}{\underset{1}{\cancel{9}}} \times \frac{\overset{\overset{1}{\cancel{5}}}{\cancel{45}}}{\underset{1}{\cancel{13}}} \times \frac{1}{\underset{2}{\cancel{10}}} = \frac{3 \times 1 \times 1}{1 \times 1 \times 2} = \frac{3}{2}$$

METHOD 2: Divide the *bottom* (denominator) (9) into the *top* (numerator) (45) and into 9. Divide the *bottom* (denominator) (13) into the *top* (numerator) (39) and into itself (13). Divide the new *top* (numerator) (5) into the *bottom* (denominator) (10) and into itself (5). After reducing fractions, multiply the numerators (3, 1, 1) together and the denominators (1, 2, 2) together.

$$\frac{\overset{3}{\cancel{39}}}{\underset{1}{\cancel{9}}} \times \frac{\overset{\overset{1}{\cancel{5}}}{\cancel{45}}}{\underset{1}{\cancel{13}}} \times \frac{1}{\underset{2}{\cancel{10}}} = \frac{3 \times 1 \times 1}{1 \times 1 \times 2} = \frac{3}{2}$$

Then convert the improper fraction to a mixed number. Divide the numerator (3) by the denominator (2). The remainder (1) is the numerator.

$$\frac{3}{2} = 1\frac{1}{2}$$

The procedure for dividing fractions is similar to the procedure for multiplying fractions, with one major change. The first step in dividing fractions is to invert the second fraction or the fraction on the right. To *invert* means to interchange or reverse the positions of the denominator and the numerator. Thus the second fraction is literally turned upside down. Then proceed as with the multiplication of fractions.

Example 2–4: Dividing Fractions

PROBLEM: $\frac{4}{5} \div \frac{1}{3} =$

SOLUTION: After inverting the second fraction, reduce the fractions to lowest terms, multiply and convert to a mixed number. Remember, multiplication is done *after* the inverting, not before.

$$\frac{4}{5} \div \frac{1}{3} = \frac{4}{5} \times \frac{3}{1} = \frac{12}{5} = 2\frac{2}{5}$$

Example 2–5: Dividing Fractions

PROBLEM: $\frac{5}{8} \div \frac{3}{16} =$

SOLUTION: After inverting the second fraction, reduce and multiply, and then convert to a mixed number. Remember, multiplication is done *after* the inverting, not before.

$$\frac{5}{8} \div \frac{3}{16} = \frac{5}{\underset{1}{\cancel{8}}} \times \frac{\overset{2}{\cancel{16}}}{3} = \frac{10}{3} = 3\frac{1}{3}$$

Example 2–6: Dividing a Whole Number and a Fraction

PROBLEM: $14 \div \frac{7}{8} =$

SOLUTION: Note there is no denominator for 14. When a number is shown without a denominator, it is understood the denominator is 1. After inverting the fraction on the right, reduce to lowest terms and multiply.

$$14 \div \frac{7}{8} = \frac{\overset{2}{\cancel{14}}}{1} \times \frac{8}{\underset{1}{\cancel{7}}} = \frac{2 \times 8}{1 \times 1} = \frac{16}{1} = 16$$

Example 2–7: Applied Fraction Problem

PROBLEM: If half a pizza is to be divided equally among four people, how much will each person receive?

SOLUTION:

$$\frac{1}{2}\text{ pizza} \div 4\text{ people} = \frac{1}{4} \div \frac{4}{1} = \frac{1}{2} \times \frac{1}{4} = \frac{1}{8}$$

Note: Each person will receive $^1/_8$ of the *whole* pizza. See Figure 2–2.

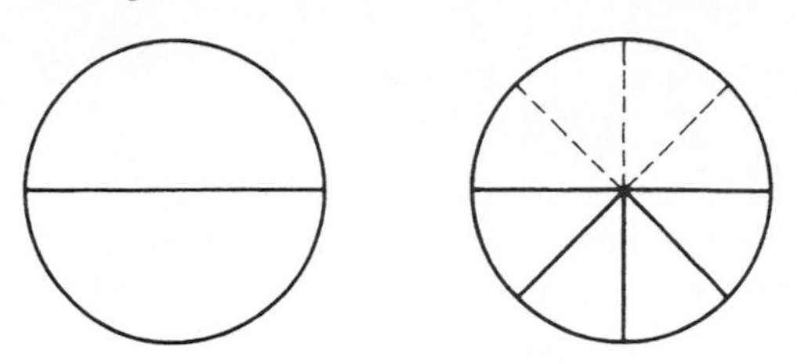

Figure 2–2

Example 2–8: Multiply a Fraction and a Whole Number

PROBLEM: $\frac{3}{8} \times 24$

SOLUTION: Write 24 with a denominator of 1, reduce by dividing 8 into 24 and multiply 3 × 3.

$$\frac{3}{\underset{1}{\cancel{8}}} \times \frac{\overset{3}{\cancel{24}}}{1} = 9$$

Example 2–9: Multiply a Fraction by a Mixed Number

PROBLEM: $\frac{3}{4} \times 5\frac{1}{2} =$

SOLUTION: To multiply a fraction times a mixed number first convert the mixed number to an improper fraction by multiplying 2 × 5, adding 1, and writing the result over 2.

Thus $5\frac{1}{2} = 2 \times 5 + 1 = \frac{11}{2}$

Now multiply the numerators and the denominators, and convert the improper fraction to a mixed numeral.

$$\frac{3}{4} \times \frac{11}{2} = \frac{33}{8} = 4\frac{1}{8}$$

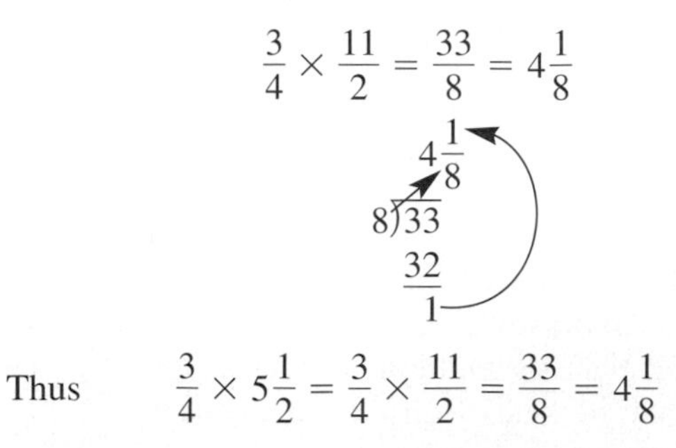

Thus $\frac{3}{4} \times 5\frac{1}{2} = \frac{3}{4} \times \frac{11}{2} = \frac{33}{8} = 4\frac{1}{8}$

Example 2–10: Multiply a Mixed Number Times a Mixed Number

PROBLEM: $5\frac{2}{3} \times 4\frac{3}{4} =$

SOLUTION: To multiply a mixed number times a mixed number, first change each mixed number to an improper fraction by multiplying the denominator by the whole number and adding the numerator. Multiply the numerators and denominators, and connect to a mixed number.

$$5\frac{2}{3} = 3 \times 5 + 2 = \frac{17}{3}$$

$$4\frac{3}{4} = 4 \times 4 + 3 = \frac{19}{4}$$

Thus $\quad \frac{17}{3} \times \frac{19}{4} = \frac{323}{12} = 26\frac{11}{12}$

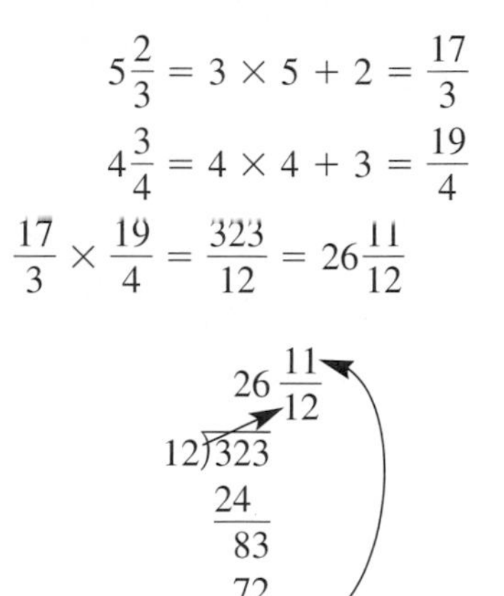

Example 2–11: Divide a Mixed Number by a Fraction

PROBLEM: $3\frac{5}{6} \div \frac{3}{4} =$

SOLUTION: To divide a mixed number by a fraction, first convert the mixed number to an improper fraction by multiplying the denominator by the whole number and adding the numerator. Write the mixed number as an improper fraction, invert the fractions that follows the division sign, reduce if possible, multiply the numerators and denominators, and convert to a mixed number.

Thus $\quad 3\frac{5}{6} = 6 \times 3 + 5 = \frac{23}{6}$

$$3\frac{5}{6} \div \frac{3}{4} =$$

$$\frac{23}{\cancel{6}_{3}} \times \frac{\cancel{4}^{2}}{3} = \frac{46}{9} = 5\frac{1}{9}$$

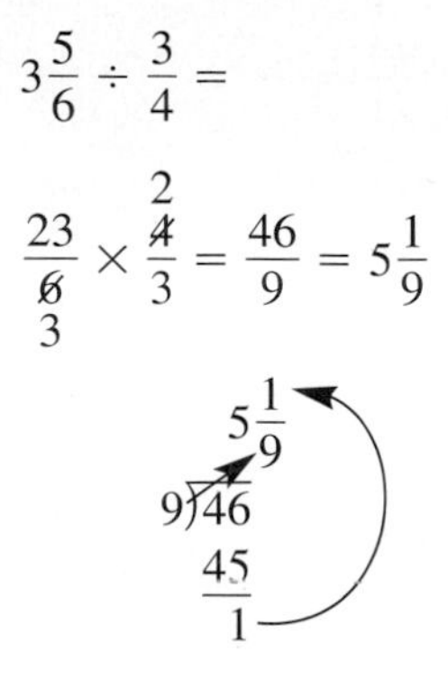

Example 2–12: Divide a Mixed Number by a Mixed Number

PROBLEM: $5\frac{3}{8} \div 3\frac{1}{4} =$

SOLUTION: To divide a mixed number by a mixed numbers, first convert each mixed number to an improper fraction, write the mixed numbers as improper fractions, invert the improper fraction that follows the division sign, reduce if possible, multiply the numerators and denominators, and convert to a mixed number.

Thus $\quad 5\frac{3}{8} = 8 \times 5 + 3 = 43 \text{ so } 5\frac{3}{8} = \frac{43}{8}$

and $\quad 3\frac{1}{4} = 4 \times 3 + 1 = 13 \text{ so } 3\frac{1}{4} = \frac{13}{4}$

Then $\quad \frac{43}{8} \div \frac{13}{4} =$

$$\frac{43}{\cancel{8}_{2}} \times \frac{\cancel{4}^{1}}{13} = \frac{43}{26} = 1\frac{7}{26}$$

$$\begin{array}{r} 1\frac{7}{26} \\ 26\overline{)43} \\ \underline{26} \\ 7 \end{array}$$

EXERCISE 2–1 MULTIPLYING AND DIVIDING FRACTIONS

Do the following problems to practice multiplying and dividing fractions.

2–1. $\frac{5}{9} \times \frac{2}{7} =$

2–2. $\frac{3}{4} \times \frac{7}{8} =$

2–3. $\frac{4}{1} \times \frac{3}{2} =$

2–4. $\frac{4}{7} \times \frac{2}{3} =$

2–5. $\frac{3}{4} \times \frac{1}{2} =$

2–6. $\frac{5}{6} \times \frac{12}{11} =$

2–7. $\frac{3}{4} \times \frac{8}{7} \times \frac{14}{21} =$

2–8. $\frac{3}{4} \times \frac{5}{12} \times \frac{7}{8} =$

2–9. $\frac{3}{4} \div \frac{2}{3} =$

2–10. $\frac{1}{2} \div \frac{1}{3} =$

2–11. $6 \div \frac{2}{3} =$

2–12. $\frac{1}{10} \div \frac{3}{5} =$

2–13. $15\frac{3}{5} \times \frac{15}{16}$

2–14. $10\frac{2}{3} \times 2\frac{1}{4}$

2–15. $10\frac{1}{5} \div 1\frac{7}{10}$

2.3 ADD AND SUBTRACT FRACTIONS

Example 2–13: Addition of Fractions

PROBLEM: $\frac{1}{2} + \frac{1}{3} =$

SOLUTION: To be able to add fractions, each fraction must have the same denominator. Thus the first step when adding fractions is to find the lowest common denominator. This means finding a number that is a multiple of each denominator. In this problem, find a number that is a multiple of 2 and 3. One way is to multiply the denominators together. However, this method does not always give the *lowest* common denominator. In our problem we can multiply the two denominators (2 × 3) and arrive at the common denominator of 6. Once a common denominator has been found, it is important to check if it is the lowest common denominator. In other words, is there a number less than 6 which is a multiple of 2 and 3? There is none, so 6 is the lowest common denominator for this problem.

The next step is to multiply each fraction by the lowest common denominator. Do this by using a fraction that is equal to 1. In this problem  will be used. Note the outline of the number 1, representing 1 or unity, is superimposed over the fraction for illustration. This is used only to emphasize the importance of making certain that the fraction is equal to 1.

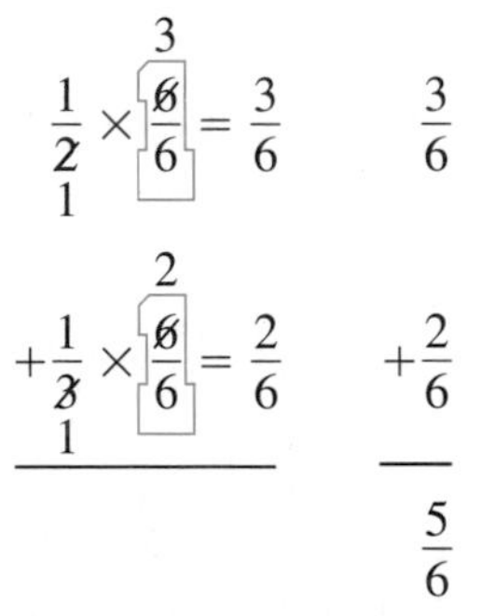

Then add the numerators. Note that *only* the numerators are added; the denominators remain the same.

Example 2–14: Addition of Fractions

PROBLEM:

$$\begin{array}{r} \frac{2}{5} \\ \frac{5}{8} \\ +\frac{3}{10} \\ \hline \end{array}$$

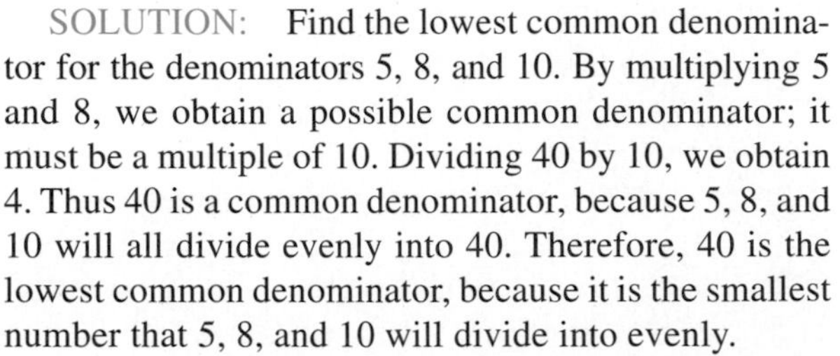

SOLUTION: Find the lowest common denominator for the denominators 5, 8, and 10. By multiplying 5 and 8, we obtain a possible common denominator; it must be a multiple of 10. Dividing 40 by 10, we obtain 4. Thus 40 is a common denominator, because 5, 8, and 10 will all divide evenly into 40. Therefore, 40 is the lowest common denominator, because it is the smallest number that 5, 8, and 10 will divide into evenly.

Then multiply each fraction by 1 or $\frac{40}{40}$:

$$\frac{2}{\underset{1}{\cancel{5}}} \times \frac{\overset{5}{\cancel{40}}}{40} = \frac{16}{40}$$

$$\frac{5}{\underset{1}{\cancel{8}}} \times \frac{\overset{5}{\cancel{40}}}{40} = \frac{25}{40}$$

$$+\frac{3}{\underset{1}{\cancel{10}}} \times \frac{\overset{4}{\cancel{40}}}{40} = \frac{12}{40}$$

An *alternative* second step is to multiply each fraction by a "special 1" that will give a common denomi-nator of 40. To do this, consider the first fraction, $^2/_5$. Ask yourself "5 times what number equals 40?" With the answer 8, we can now multiply by 1 or $\frac{8}{8}$ Following the same steps for the fraction $^5/_8$, we obtain 1 or $\frac{5}{5}$ and $\frac{4}{4}$ for $^3/_{10}$. This method eliminates cancellation.

$$\frac{2}{5} \times \frac{8}{8} = \frac{16}{40}$$

$$\frac{5}{8} \times \frac{5}{5} = \frac{25}{40}$$

$$+\frac{3}{10} \times \frac{4}{4} = \frac{12}{40}$$

Then add the fractions. Remember, add only the numerators and write the sum over the common denominator, 40.

$$\begin{array}{r} \frac{16}{40} \\ \frac{25}{40} \\ +\frac{12}{40} \\ \hline \frac{53}{40} \end{array}$$

Convert the improper fraction $^{53}/_{40}$ to a mixed fraction by dividing the denominator (40) into the numerator (53) and writing the remainder as a fraction.

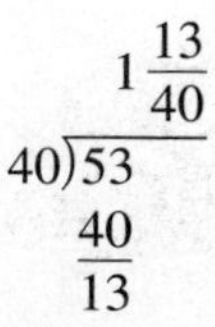

Example 2–15: Addition of Mixed Numbers

PROBLEM:

$$\begin{array}{r} 6\frac{1}{9} \\ 1\frac{9}{14} \\ +4\frac{5}{21} \\ \hline \end{array}$$

SOLUTION: Find the lowest common denominator for 9, 14, and 21. This could be done by multiplying $9 \times 14 \times 21$. The result (2646) could be reduced to a lowest common denominator through trial-and-error division. However, this procedure is time consuming. Another way to find the lowest common denominator is to identify the *prime factors* or *prime numbers* of each denominator. Prime factors are numbers that when multiplied together will be equal to each denominator. A *prime number* is a whole number greater than 1 that cannot be evenly divided except by itself and 1. A *composite number* is a whole number greater than 1 that can be divided by a whole number other than itself. In this problem, the prime factors of the denominator are

$$9 = 3 \times 3$$
$$14 = 2 \times 7$$
$$21 = 3 \times 7$$

The next step is to select those prime factors that will determine the lowest common denominator. Select the prime factors of the first fraction, 3×3, and then select additional prime factors that are different from 3×3. Use these additional prime factors only once. Thus the prime factors $3 \times 3 \times 7 \times 2$ are selected and the lowest common denominator is 126.

Then multiply each fraction by a "special 1" number. Method 1 uses $^{126}/_{126}$ Method 2 eliminates cancellation by selecting a number "1" that will produce the common denominator 126 through multiplication. Study both examples. To prevent mistakes, show all your work.

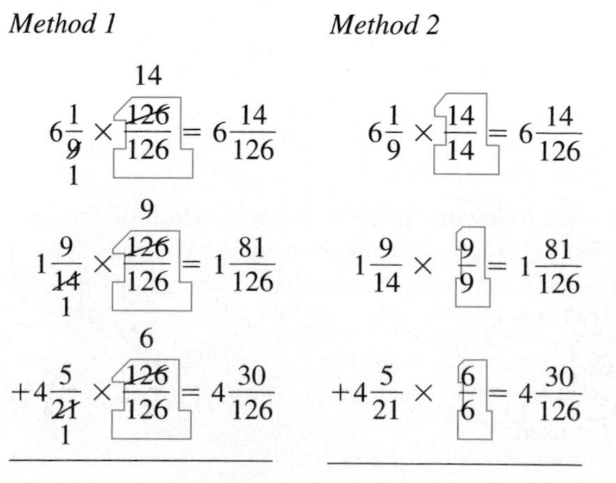

Then add the whole numbers and the denominators.

$$\begin{array}{r} 6\frac{14}{126} \\ 1\frac{81}{126} \\ +\ 4\frac{30}{126} \\ \hline 11\frac{125}{126} \end{array}$$

Example 2–16: Subtraction of Fractions

PROBLEM:
$$\begin{array}{r} 12\frac{7}{8} \\ -\ 4\frac{3}{16} \\ \hline \end{array}$$

SOLUTION: First, find the lowest common denominator for 8 and 16. By observation, this is easily determined to be 16. Then multiply each fraction by 1 or a "special 1."

Method 1

$$12\frac{7}{\not{8}_1} \times \frac{\not{16}^2}{16} = 12\frac{14}{16}$$

$$-4\frac{3}{\not{16}_1} \times \frac{\not{16}^1}{16} = 4\frac{3}{16}$$

Method 2

$$12\frac{7}{8} \times \frac{2}{2} = 12\frac{14}{16}$$

$$-4\frac{3}{16} \times \frac{1}{1} = 4\frac{3}{16}$$

Subtract the numerators and the whole numbers. Remember to subtract the subtrahend (bottom) numerator from the minuend (top) numerator. Denominators are *not* subtracted.

$$\begin{array}{rcl} 12\frac{7}{8} & = & \frac{14}{16} \\ -\ 4\frac{3}{16} & = & \frac{3}{16} \\ \hline & & 8\frac{11}{16} \end{array}$$

Example 2–17: Subtraction of Mixed Numbers

PROBLEM:
$$\begin{array}{r} 24\frac{1}{6} \\ -16\frac{5}{12} \\ \hline \end{array}$$

SOLUTION: Find the lowest common denominator for 6 and 12; by observation, it is 12. Then multiply by 1 or a "special 1" as shown.

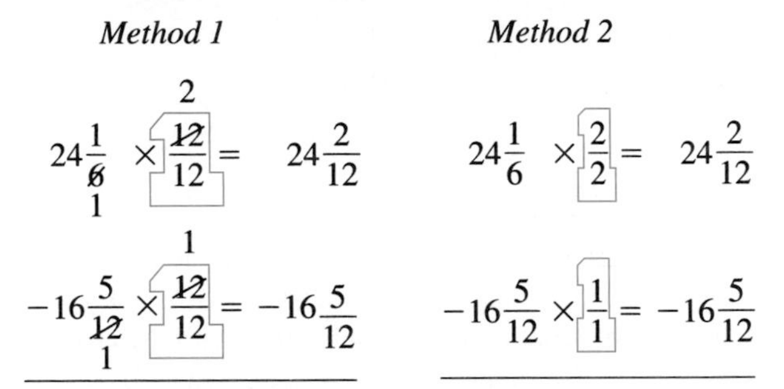

Method 1

$$24\frac{1}{\not{6}_1} \times \frac{\not{12}^2}{12} = 24\frac{2}{12}$$

$$-16\frac{5}{\not{12}_1} \times \frac{\not{12}^1}{12} = -16\frac{5}{12}$$

Method 2

$$24\frac{1}{6} \times \frac{2}{2} = 24\frac{2}{12}$$

$$-16\frac{5}{12} \times \frac{1}{1} = -16\frac{5}{12}$$

However, you cannot subtract 5 from 2; therefore, you must borrow "1," or $^{12}/_{12}$, from 24. Add $^{12}/_{12} + {^2}/_{12} = {^{14}}/_{16}$, then subtract the numerator 5 from the numerator 14 and subtract the whole number 16 from 23. Reduce $^9/_{12}$ by dividing the numerator and denominator by 3.

$$\begin{array}{rcl} \overset{23}{\not{24}}\frac{2}{12} + \frac{12}{12} & = & \frac{14}{12} \\ -16\frac{5}{12} & = & \frac{5}{12} \\ \hline & & 7\frac{9}{12} \end{array}$$

Note: When "borrowing," rather than adding $^{12}/_{12}$ to the top fraction, merely add the numerator and denominator, place the sum above the numerator, and reduce.

Thus $$7\frac{9}{12} = 7\frac{3}{4}$$

EXERCISE 2–2 ADDING AND SUBTRACTING FRACTIONS

Solve the following addition and subtraction of fractions and reduce to lowest terms. Remember, you must have a common denominator when adding or subtracting fractions.

2–16. $\begin{array}{r} \frac{2}{7} \\ +\frac{3}{7} \\ \hline \end{array}$

2–17. $\begin{array}{r} \frac{3}{5} \\ +\frac{4}{10} \\ \hline \end{array}$

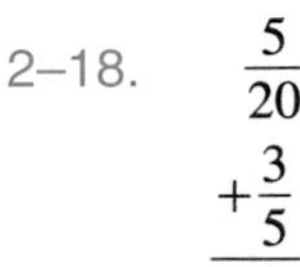

2–18. $\begin{array}{r} \frac{5}{20} \\ +\frac{3}{5} \\ \hline \end{array}$

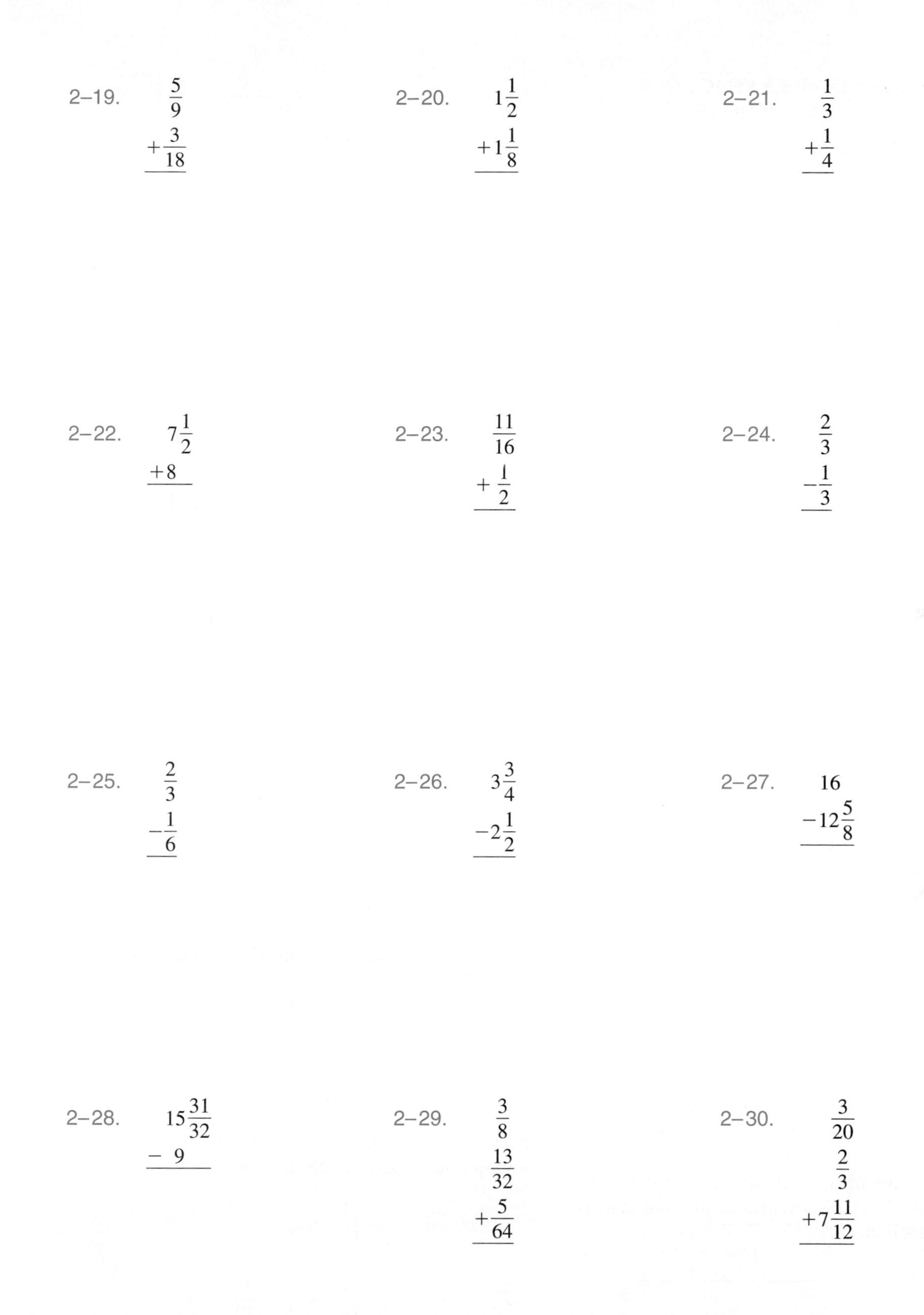

2–19. $\frac{5}{9} + \frac{3}{18}$

2–20. $1\frac{1}{2} + 1\frac{1}{8}$

2–21. $\frac{1}{3} + \frac{1}{4}$

2–22. $7\frac{1}{2} + 8$

2–23. $\frac{11}{16} + \frac{1}{2}$

2–24. $\frac{2}{3} - \frac{1}{3}$

2–25. $\frac{2}{3} - \frac{1}{6}$

2–26. $3\frac{3}{4} - 2\frac{1}{2}$

2–27. $16 - 12\frac{5}{8}$

2–28. $15\frac{31}{32} - 9$

2–29. $\frac{3}{8} + \frac{13}{32} + \frac{5}{64}$

2–30. $\frac{3}{20} + \frac{2}{3} + 7\frac{11}{12}$

2.4 SIMPLIFYING COMPLEX FRACTIONS AND THE ORDER OF OPERATIONS

So far, you have learned how to change fractions from one form to another; perform the fundamental operations of multiplication, division, addition, and subtraction; and simplify proper and improper fractions by reducing them to lowest terms. Simplifying *complex fractions* will require all of these skills.

The same rules for the order of operations that are used for whole numbers are used for fractions and mixed numbers.

Recall the rules are:

1. Do all operations inside the parentheses.
2. Evaluate the exponential expressions.
3. Do all the multiplications and divisions from left to right.
4. Do all the additions and subtractions from left to right.

Example 2–18: Simplifying a Complex Fraction

PROBLEM: Simplify

$$\frac{\frac{3}{4}}{\frac{1}{3}}$$

SOLUTION: The line between the two fractions indicates division; thus rewrite the fractions in regular division form and solve.

$$\frac{\frac{3}{4}}{\frac{1}{3}} = \frac{3}{4} \div \frac{1}{3} = \frac{3}{4} \times \frac{3}{1} = \frac{9}{4} = 2\frac{1}{4}$$

Example 2–19: Simplifying a Complex Fraction

PROBLEM: $$\frac{1}{\frac{1}{2}+\frac{1}{3}+\frac{1}{5}} =$$

SOLUTION: This complex fraction contains three fractions in the denominator. When simplifying complex fractions, first perform the operations indicated in the numerator and denominator. The lowest common denominator for the fractions in the denominator is 30. Thus a "special 1," $^{30}/_{30}$ is used to determine the equivalent for each fraction.

$$\frac{1}{\frac{1}{2}+\frac{1}{3}+\frac{1}{5}} = \frac{1}{\frac{1}{\cancel{2}_1}\times\frac{\cancel{30}^{15}}{30}+\frac{1}{\cancel{3}_1}\times\frac{\cancel{30}^{10}}{30}+\frac{1}{\cancel{5}_1}\times\frac{\cancel{30}^{6}}{30}}$$

$$= \frac{1}{\frac{15}{30}+\frac{10}{30}+\frac{6}{30}}$$

Then add the fractions in the denominator.

$$\frac{1}{\frac{15}{30}+\frac{10}{30}+\frac{6}{30}} = \frac{1}{\frac{31}{30}}$$

Simplify by writing the fraction in normal division of fraction terms and divide.

$$\frac{1}{\frac{31}{30}} = \frac{1}{1} \div \frac{31}{30} = \frac{1}{1} \times \frac{30}{31} = \frac{30}{31}$$

Example 2–20: Performing the Order of Operations

PROBLEM: Simplify $\left(\frac{3}{5}-\frac{1}{2}\right)^2 \div \left(\frac{3}{4}\times\frac{8}{15}\right) =$

SOLUTION: Perform the operations inside the parentheses. In the first parentheses, subtract; in the second, multiply.

$$\left(\frac{3}{5}-\frac{1}{2}\right)^2 \div \left(\frac{3}{4}\times\frac{8}{15}\right) =$$

$$\left(\frac{6}{10}-\frac{5}{10}\right)^2 \div \left(\frac{\cancel{3}^1}{\cancel{4}_1}\times\frac{\cancel{8}^2}{\cancel{15}_5}\right) =$$

$$\left(\frac{1}{10}\right)^2 \div \left(\frac{2}{5}\right) =$$

Evaluate the exponential expression by multiplying.

$$\left(\frac{1}{10}\times\frac{1}{10}\right) \div \left(\frac{2}{5}\right) =$$

$$\frac{1}{100} \div \frac{2}{5} =$$

Multiply and divide from left to right. Since there is no multiplication, perform the division.

$$\frac{1}{100} \times \frac{5}{2} =$$

$$\frac{1}{\cancel{100}_{20}} \times \frac{\cancel{5}^1}{2} = \frac{1}{40}$$

EXERCISE 2–3 SIMPLIFYING COMPLEX FRACTIONS AND ORDER OF OPERATIONS

Simplify the following complex fractions. Remember to use all the skills you have learned to simplify the complex fractions.

2–31. $\dfrac{\frac{3}{16}}{\frac{1}{4}}$

2–32. $\dfrac{\frac{1}{2}}{\frac{7}{8}} =$

2–33. $\dfrac{5}{\frac{3}{8}} =$

2–34. $\dfrac{\frac{7}{8}}{28} =$

2–35. $\dfrac{1\frac{1}{16}}{\frac{3}{16}} =$

2–36. $\dfrac{\frac{1}{7}}{3\frac{1}{7}} =$

2–37. $\dfrac{\frac{1}{2} + \frac{5}{8}}{\frac{5}{8} - \frac{1}{2}} =$

2–38. $\dfrac{\frac{5}{7} + \frac{3}{2}}{\frac{5}{3} \times 4} =$

2–39. $\dfrac{12\frac{1}{2} \times 4}{\frac{3}{4} \div \frac{6}{16}} =$

2–40. $\dfrac{\frac{23}{8} \div \frac{1}{8}}{69 \times \frac{1}{5}} =$

2–41. $\dfrac{1}{\frac{1}{2} + \frac{3}{4} + \frac{5}{8} + \frac{7}{16}} =$

2–42. $\frac{3}{4} \times 24 - 10\frac{1}{2} =$

2–43. $\left(\frac{3}{4}\right)^2\left(\frac{2}{3}\right)^3 =$

2–44. $\frac{1}{3} - \left(\frac{1}{3}\right)^2 + \left(\frac{2}{3}\right)^3 =$

2.5 SOLVE APPLIED PROBLEMS USING FRACTIONS

Example 2–21: Applied Blueprint Problem

PROBLEM: Find *A* and *B* given the blueprint shown in Figure 2–3.

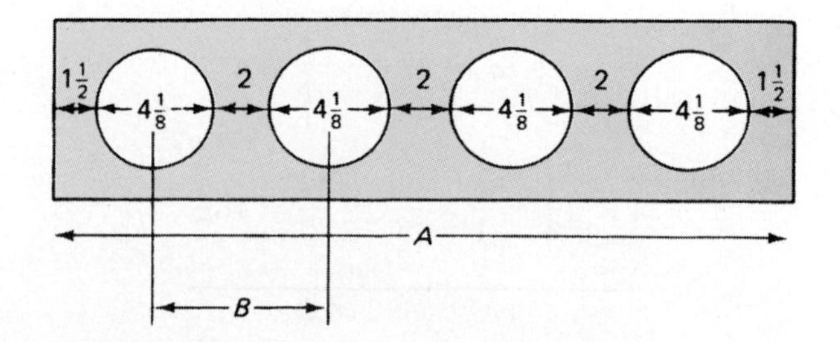

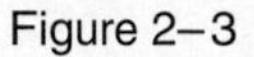

Figure 2–3

SOLUTION: Find A. A is equal to the sum of all the distances. Thus all the fractions should be added together.

$$A = 1\frac{1}{2} + 4\frac{1}{8} + 2 + 4\frac{1}{8} + 2$$
$$+ 4\frac{1}{8} + 2 + 4\frac{1}{8} + 1\frac{1}{2}$$

METHOD 1: Find the lowest common denominator. By observation, it is 8. Multiply each fraction, changing to the lowest common denominator (8) or $^8/_8$. Then add the fractions and reduce to lowest terms.

$$\begin{aligned}
1\frac{1}{\cancel{2}_1} \times \frac{\cancel{8}^4}{8} &= 1\frac{4}{8} \\
4\frac{1}{8} &= 4\frac{1}{8} \\
2 &= 2 \\
4\frac{1}{8} &= 4\frac{1}{8} \\
2 &= 2 \\
4\frac{1}{8} &= 4\frac{1}{8} \\
2 &= 2 \\
4\frac{1}{8} &= 4\frac{1}{8} \\
+\ 1\frac{1}{\cancel{2}_1} \times \frac{\cancel{8}^4}{8} &= 1\frac{4}{8} \\
\hline
&= 24\frac{12}{8} \\
&= 24 + 1\frac{4}{8} \\
&= 24 + 1\frac{1}{2} \\
&= 25\frac{1}{2}
\end{aligned}$$

METHOD 2: Another way to find A is to combine like dimensions by multiplying. Reduce to lowest terms and then add.

$$2 \times 1\frac{1}{2} = \frac{\cancel{2}^1}{1} \times \frac{3}{\cancel{2}_1} = 3$$
$$4 \times 4\frac{1}{8} = \frac{\cancel{4}^1}{1} \times \frac{33}{\cancel{8}_2} = \frac{33}{2} = 16\frac{1}{2}$$
$$\underline{\qquad\qquad\qquad\qquad\qquad\qquad}$$
$$+\ 3 \times 2 = 6$$
$$A = 25\frac{1}{2}$$

PROBLEM: Find B. B is a center-to-center length. B is equal to $^1/_2$ of $4\,^1/_8 + 2 + {}^1/_2$ of $4\,^1/_8$. Remember that "of" means multiply.

Thus $$B = \frac{1}{2} \times 4\frac{1}{8} + 2 + \frac{1}{2} \times 4\frac{1}{8}$$

Multiply by converting the mixed numbers, then add and reduce to lowest terms.

$$\begin{aligned}
B &= \frac{1}{2} \times \frac{33}{8} + 2 + \frac{1}{2} \times \frac{33}{8} \\
&= \frac{33}{16} + 2 + \frac{33}{16} \\
&= \frac{33}{16} + \frac{32}{16} + \frac{33}{16}
\end{aligned}$$

Thus $$B = \frac{98}{16} \text{ or } 6\frac{2}{16} = 6\frac{1}{8}$$

Example 2–22: Applied Fraction Problem

PROBLEM: How many square feet are there in a $^1/_3$-acre lot? (1 acre = 43,560 square feet.)

SOLUTION: Multiply $^1/_3$ times 43,560, since we need to find $^1/_3$ of 43,560.

$$43{,}560 \text{ sq ft} \times \frac{1}{3} =$$
$$\frac{\cancel{43{,}560}^{14{,}520}}{1} \times \frac{1}{\cancel{3}_1} = 14{,}520 \text{ sq ft}$$

Example 2–23: Applied Fraction Problem

PROBLEM: A production machinist can complete a drilling job every 3 minutes and 45 seconds. How many pieces can be completed in 1 hour?

SOLUTION: Convert the 45 seconds to a fractional part of a minute. Since there are 60 seconds in a minute, write $^{45}/_{60}$ and reduce $^{45}/_{60} = {}^3/_4$ minutes. Thus each piece could be completed in $3\,^3/_4$ minutes. Then divide $3\,^3/_4$ minutes into 60 minutes: this will indicate how many pieces can be completed in 1 hour.

Therefore,

$$\frac{45 \text{ sec}}{60 \text{ sec}} = \frac{3}{4} \text{ min} + 3 \text{ min} = 3\frac{3}{4} \text{ min/piece}$$
$$60 \div 3\frac{3}{4} =$$
$$\frac{60}{1} \div \frac{15}{4} =$$
$$\frac{\cancel{60}^4}{1} \times \frac{4}{\cancel{15}_1} = 16 \text{ pieces/hr}$$

Example 2–24: Applied Fraction Problem

PROBLEM: A land developer assigned $1\,^1/_2$ acres of a 20-acre plot for a park. How many $^1/_2$-acre lots can be developed from the remaining land?

SOLUTION: Subtract the $1\frac{1}{2}$-acre park portion from the 20 acres. Then find the available lots by dividing by $\frac{1}{2}$.

Thus
$$\frac{20 - 1\frac{1}{2}}{\frac{1}{2}} = \frac{19\frac{2}{2} - 1\frac{1}{2}}{\frac{1}{2}} = \frac{18\frac{1}{2}}{\frac{1}{2}}$$
$$= 18\frac{1}{2} \div \frac{1}{2}$$
$$= \frac{37}{2} \times \frac{2}{1} = 37 \text{ lots}$$

EXERCISE 2–4 SOLVING APPLIED PROBLEMS USING FRACTIONS

Solve the following fraction and applied problems that involve operations with fractions. Remember to use the procedure you have learned.

2–45. Find A given the flat washer shown in Figure 2–4.

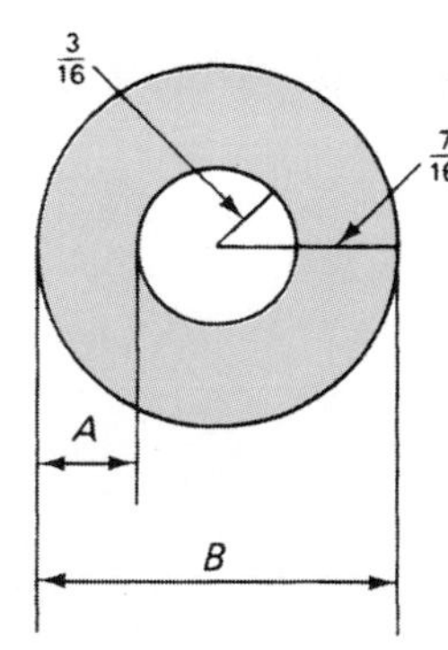

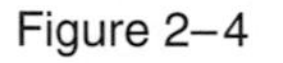
Figure 2–4

2–46. Find B given the flat washer shown in Figure 2–4.

2–47. Find C given the cone pulleys shown in Figure 2–5.

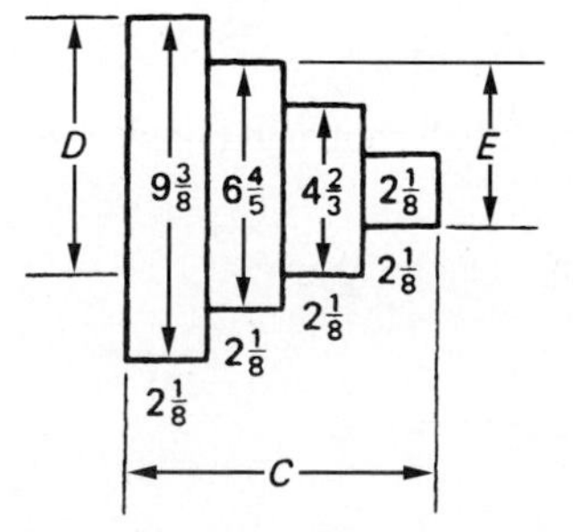

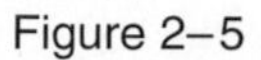
Figure 2–5

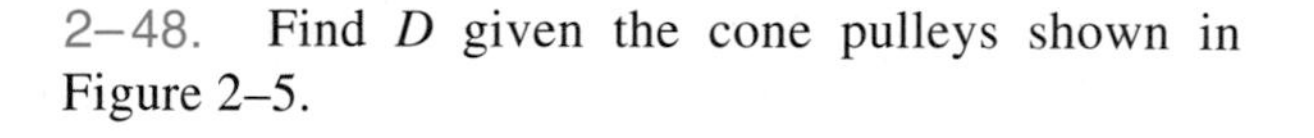

2–48. Find D given the cone pulleys shown in Figure 2–5.

2–49. Find E given the cone pulleys shown in Figure 2–5.

2–50. Find F given the concentric circles shown in Figure 2–6.

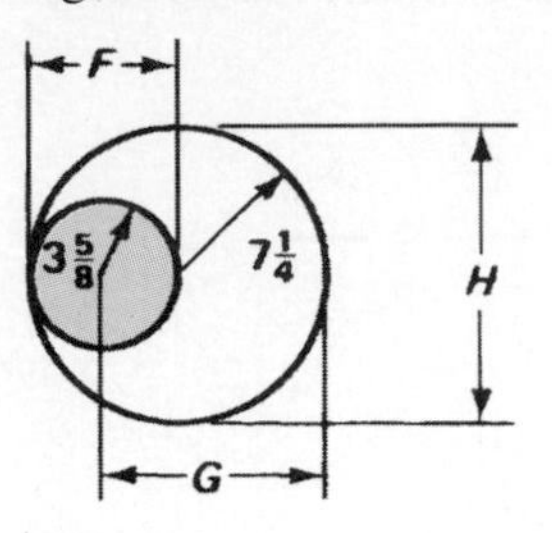

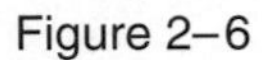
Figure 2–6

2–51. Find G given the concentric circles shown in Figure 2–6.

2–52. Find H given the concentric circles shown in Figure 2–6.

2–53. Find I given the eyebolt shown in Figure 2–7.

2–54. Find J given the eyebolt shown in Figure 2–7.

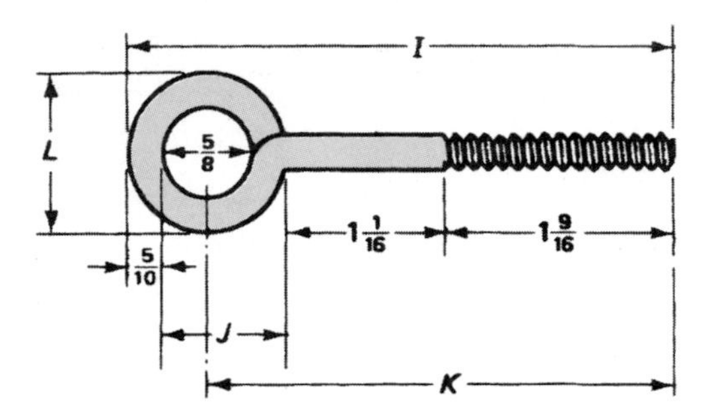

Figure 2–7

2–55. Find K given the eyebolt shown in Figure 2–7.

2–56. Find L given the eyebolt shown in Figure 2–7.

2–57. Find M given the crankshaft shown in Figure 2–8.

2–58. Find N given the crankshaft shown in Figure 2–8.

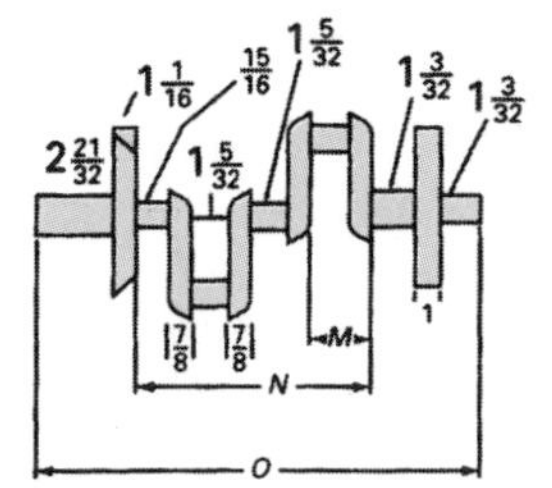

Figure 2–8

2–59. Find O given the crankshaft shown in Figure 2–8.

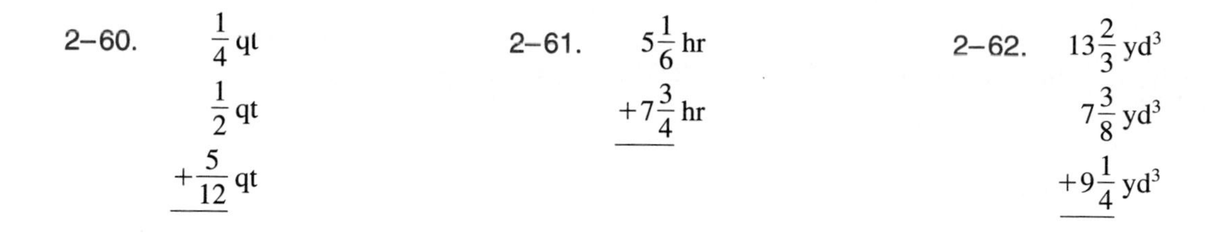

2–60. $\frac{1}{4}$ qt, $\frac{1}{2}$ qt, $+\frac{5}{12}$ qt

2–61. $5\frac{1}{6}$ hr, $+7\frac{3}{4}$ hr

2–62. $13\frac{2}{3}$ yd^3, $7\frac{3}{8}$ yd^3, $+9\frac{1}{4}$ yd^3

2–63. How many pounds of corn were produced if five plants produced the following amounts: $6\frac{1}{2}$, $4\frac{3}{4}$, $3\frac{7}{8}$, $4\frac{7}{32}$, and $5\frac{3}{16}$ pounds?

2–64. How much trim would you need for a window that is $6\frac{1}{2}$ by 8 feet?

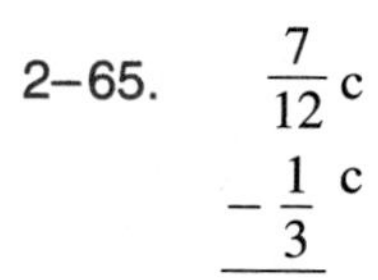

2–65. $\frac{7}{12}$ c, $-\frac{1}{3}$ c

2–66. $17\frac{1}{16}$ oz, $-14\frac{3}{8}$ oz

2–67. $16\frac{3}{8}$ in., $-\ 9\frac{15}{16}$ in.

2–68. What is the difference in thickness of $\frac{5}{8}$-inch and $\frac{1}{2}$-inch sheetrock?

2–69. How much of an 8-foot heater hose is left if you cut off a piece $3\frac{3}{4}$ feet long?

2–70. 32 rafters $\times\ 4\frac{1}{4}$ ft =

2–71. $\frac{1}{3}$ in. $\times \frac{3}{8}$ in. $\times \frac{64}{72}$ in. $\times$ 16 boxes =

2–72. $2\frac{1}{3}$ yd $\times\ 4\frac{3}{5}$ yd $\times\ 2\frac{5}{30}$ yd =

2–73. What is the total length of eight studs if each is $7\frac{3}{4}$ feet long?

2–74. If the travel time each way to work is $\frac{3}{4}$ hour, how many hours are spent in transit during a 5-day workweek?

2–75. 12-in. board $\div \frac{3}{16}$ in. strips =

2–76. $7\frac{1}{5}$ cm $\div$ 4 spaces =

2–77. $7\frac{1}{2}$ acres $\div\ 1\frac{1}{4}$ acres/lot =

2–78. If $3\frac{1}{2}$ pizzas are to be divided among 12 persons, how much will each person receive?

2–79. A room $15\frac{1}{2}$ feet long is to have three electrical outlets equally spaced. How far apart will they be?

2–80. $\dfrac{200 \text{ oz}}{\frac{3}{4}} =$

2–81. $\dfrac{3\frac{1}{4}\text{ in.} + 4\frac{1}{8}\text{ in.} + \frac{1}{2}\text{ in.}}{3}$

2–82. $\dfrac{7\frac{1}{2}\text{ in.} \times 15\frac{1}{2}\text{ in.} \times 576\text{ blocks}}{12\text{ in.} \times 12\text{ in.}} =$

2–83. What is the thickness of a pipe whose outside diameter is $3\frac{3}{8}$ inches and whose inside diameter is $2\frac{3}{4}$ inches?

2–84. A "2 by 4" is actually $1\frac{5}{8}$ inches by $3\frac{5}{8}$ inches. How much is planed off *each* side of the $3\frac{5}{8}$-inch sides?

2–85. A piece of pipe is $14\frac{1}{2}$ inches long. How much is left after cutting off pieces measuring $2\frac{1}{2}$, $1\frac{1}{16}$, and $3\frac{5}{32}$ inches, allowing $\frac{1}{16}$ inch for each cut?

2–86. A taper has a diameter of $2\frac{5}{16}$ inches on one end and $1\frac{5}{8}$ inches at the other. What is the difference between the diameters?

2–87. The volume of a rectangular solid (block) is found by multiplying the length times the height. What is the volume of a block of concrete $11\frac{3}{4}$ inches by 8 inches by $6\frac{1}{2}$ inches?

2–88. A cubic foot contains approximately $7\frac{1}{2}$ gallons. How many gallons of gasoline would go into a $2\frac{3}{4}$-cubic-foot gas tank?

2–89. Find length *A* given the blueprint shown in Figure 2–9.

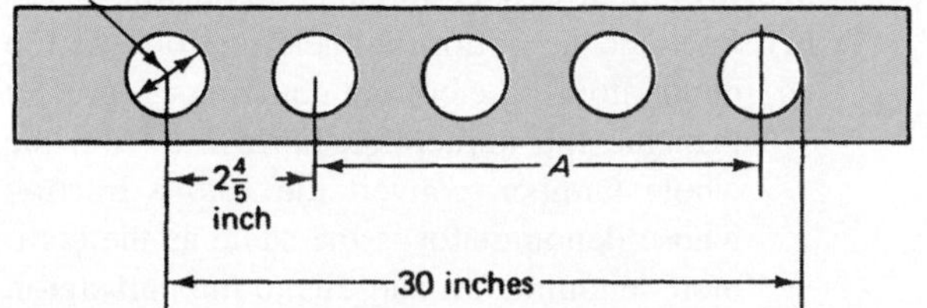

Figure 2–9

2–90. A 16-gallon fuel tank indicates it is $\frac{1}{4}$ full. What will it cost to fill the tank if fuel costs $\$1\frac{3}{4}$ per gallon?

THINK TIME

There are many different ways to determine the lowest common denominator for fractions. Some of the ways have been shown. Perhaps you know other methods or can create new ones. Here is a great way to find the lowest common denominator, or LCD, as it is known.

PROBLEM: Find the lowest common denominator for the fractions $\frac{1}{24} + \frac{3}{28} + \frac{7}{15}$.

SOLUTION: Place the denominators in an inverted division symbol as shown. Start by dividing the denominators by 2, a prime number. Place the results, 12 and 14, as well as 15, which could not be divided, in a new inverted division symbol. Repeat the process until no more divisions by 2 are possible, then divide by 3, another prime number, and continue this procedure until only prime numbers result.

$$
\begin{array}{r|rrr}
2 & 24 & 28 & 15 \\ \hline
2 & 12 & 14 & 15 \\ \hline
2 & 6 & 7 & 15 \\ \hline
3 & 3 & 7 & 15 \\ \hline
 & 1 & 7 & 5
\end{array}
$$

Then multiply the divisors and remaining prime numbers to obtain the lowest common denominator.

$$2 \times 2 \times 2 \times 3 \times 1 \times 7 \times 5 = 840$$

PROCEDURES TO REMEMBER

1. To reduce a fraction, divide the numerator and denominator by an equal quantity. *Example:*

$$\frac{12}{24} = \frac{12 \div 12}{24 \div 12} = \frac{1}{2}$$

2. To convert an improper fraction to a mixed number, divide the numerator by the denominator, write the quotient, and place the remainder over the denominator. *Example:*

$$\frac{5}{2} = \begin{array}{r} 2 \\ 2\overline{)5} \\ \underline{4} \\ 1 \end{array} = 1\frac{1}{2}$$

3. To convert a whole number or a mixed number to an improper fraction, multiply the denominator by the whole number and add the original number to create the new numerator. *Examples:*

$$2\frac{7}{8} = \frac{8 \times 2 + 7}{8} = \frac{23}{8} \text{ and } 7 = \frac{7}{1}\left(\frac{5}{5}\right) = \frac{35}{5}$$

4. To multiply fractions:
   (a) Write whole numbers over 1.
   (b) Convert mixed numbers to improper fractions.
   (c) Reduce when possible. Do this by dividing the same quantity into the numerator and denominator (top into bottom or bottom into top).
   (d) Multiply the numerators and denominators.
   (e) Reduce to common fraction or to a mixed number.
5. When dividing fractions:
   (a) Write whole numbers over 1.
   (b) Convert mixed numbers to improper fractions.
   (c) Invert the fraction to the right of the division symbol.
   (d) Reduce when possible.
   (e) Multiply the numerators and the denominators.
   (f) Convert the result to a common fraction or a mixed number.
6. To find the lowest common denominator:
   (a) Determine the prime factors of each denominator.
   (b) Using the prime factors of the first denominator and the prime factors from other denominators not found in each of the other denominators, multiply these together. The result will be the lowest common denominator (LCD).
   (c) Check to make certain each denominator divides evenly into the common denominator.
7. When adding fractions:
   (a) Write the fractions to be added.
   (b) If they do not have a common denominator, find the lowest common denominator.
   (c) Change all fractions to equivalent fractions with the lowest common denominator.
   (d) Add all numerators.
   (e) Place the sum of the numerators over the common denominator.
   (f) Reduce to a common fraction or mixed number.
8. When subtracting fractions:
   (a) Write the fractions to be subtracted.
   (b) If they do not have a common denominator, find the lowest common denominator.
   (c) Change all fractions to equivalent fractions with the lowest common denominator.
   (d) Subtract the bottom numerator from the top numerator. If the bottom numerator is greater than the top numerator, borrow 1 from the whole number, convert the 1 to a fraction whose denominator is the same as the common denominator, add this to the numerator,

then subtract the bottom numerator from the top numerator.
   (e) Place the difference over the lowest common denominator.
   (f) Reduce to a common fraction or mixed number.
9. Order of operation rules:
   (a) Do all operations inside the parentheses.
   (b) Evaluate the exponential expressions.
   (c) Do all multiplications and divisions from left to right.
   (d) Do all additions and subtractions from left to right.

CHAPTER SUMMARY

A *fraction* is a number that is used to compare a part to a whole. A fraction indicates one number divided by another number or a whole number over 1. Fractions are written to show the relationship of the part to the whole. An example of a fraction is $^5/_8$. The number below the line, 8, is called the *denominator* and tells the number of parts into which the whole has been divided. The number above the line, 5, is called the *numerator* and tells the number of parts of the whole.

The numerator and denominator are called the *terms* of the fraction. A fraction whose numerator is smaller than its denominator is called a *proper* fraction. A fraction whose numerator is larger than its denominator is called an *improper* fraction.

A whole number and a fraction written together is called a *mixed number.* A fraction that contains a fraction or fractions in the numerator or denominator is called a *complex* fraction. A *common denominator* is a number that is a multiple of the denominators of a set of fractions. The *lowest common denominator* is the smallest number that is a multiple of the denominators of a set of fractions. To *reduce* a fraction to its lowest terms, divide the numerator and the denominator by the same quantity. *Prime factors* are numbers that can be obtained by multiplying the quantities by 1.

A *prime number* is a whole number greater than 1 that cannot be evenly divided except by itself and 1. A *composite* number is a whole number greater than 1 that can be divided by a whole number other than itself.

CHAPTER TEST

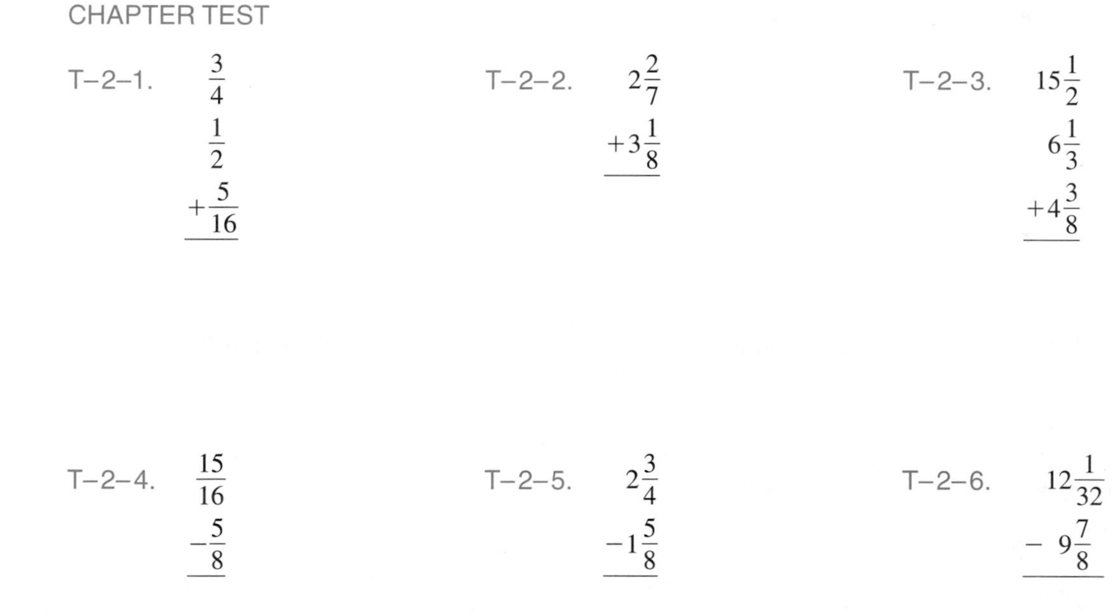

T-2-1. $\frac{3}{4} + \frac{1}{2} + \frac{5}{16}$

T-2-2. $2\frac{2}{7} + 3\frac{1}{8}$

T-2-3. $15\frac{1}{2} + 6\frac{1}{3} + 4\frac{3}{8}$

T-2-4. $\frac{15}{16} - \frac{5}{8}$

T-2-5. $2\frac{3}{4} - 1\frac{5}{8}$

T-2-6. $12\frac{1}{32} - 9\frac{7}{8}$

T–2–7. $\frac{5}{9} \times \frac{3}{10} =$

T–2–8. $7\frac{1}{3} \times 2\frac{1}{2} =$

T–2–9. $4\frac{5}{16} \times 5\frac{1}{3} =$

T–2–10. $\frac{1}{8} \div \frac{3}{4} =$

T–2–11. $\frac{25}{36} \div \frac{5}{9} =$

T–2–12. $5\frac{3}{8} \div 3\frac{1}{4} =$

T–2–13. $\dfrac{9}{3\frac{2}{3}} =$

T–2–14. $\dfrac{\frac{5}{8} \times \frac{16}{45}}{\frac{3}{4} \div \frac{4}{3}} =$

T–2–15. $\dfrac{\frac{2}{3} \times 7\frac{3}{4}}{\frac{1}{2} \times 37\frac{1}{16}} =$

T–2–16. $18\frac{2}{3} \times 8\frac{1}{4} \div 72 =$

T–2–17. $6\frac{1}{4} \times 8\frac{1}{2} \times 5\frac{1}{2} \div 36 =$

T–2–18. Find the dimension *A* for a set of eight steps in Figure 2–10.

T–2–19. Find the missing dimension *B* for a set of eight steps in Figure 2–10.

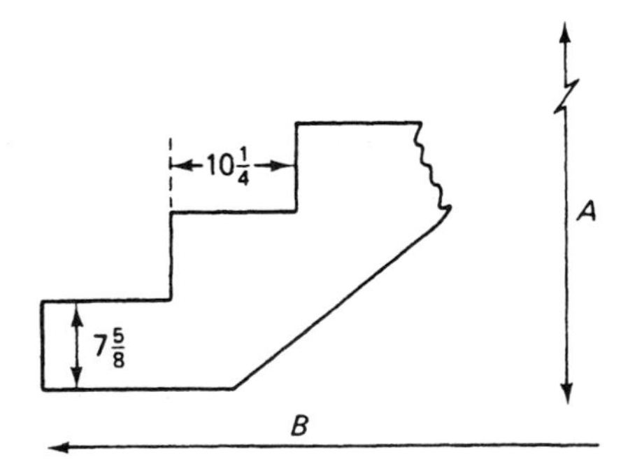

Figure 2–10

T–2–20. A truck is loaded with bales of scrap iron, each weighing $^1/_{20}$ ton. If the total weight of the bales is $5^1/_2$ tons, how many bales are on the truck?

T–2–21. Eight inches of plastic tubing weighs $^3/_4$ ounce. How much will 2 inches of the tubing weigh?

T–2–22. Twenty-four strips, each $1^7/_8$ inches wide, are to be ripped from a sheet of paneling. Twenty-four cuts are necessary and $^1/_{16}$ inch is lost on each cut. How wide must the paneling be?

T–2–23. A master stonemason spends the following hours on different jobs: $4^1/_2$, $8^1/_2$, $1^3/_4$, $2^1/_4$ and $5^3/_4$ At \$ $20^1/_2$ per hour, what will the charge for labor be?

T–2–24. Partners Marge, Paul, and Steve agreed to share gains or losses in proportion to their respective investments. Marge invested \$21,000; Paul, \$28,000; and Steve, \$35,000. If they are to share a gain of \$126,000, how much should each receive?

T–2–25. A freshly molded plastic boat bumper weighed $14^1/_2$ kilograms. After being shaped, it weighed $12^5/_8$ kilograms. The shaping cost is \$2 per kilogram, and the plastic removed in shaping can be sold for $\$^2/_3$ per kilogram. Find the net cost of shaping the boat bumper.

3

DECIMALS

OBJECTIVES

After studying this chapter, you will be able to:

3.1. Understand the definitions and perform the basic decimal operations.
3.2. Change fractions to decimals.
3.3. Change decimals to fractions.
3.4. Perform the basic arithmetic operations with decimals.
3.4. Solve applied problems using decimals.

SELF-TEST

This test will evaluate your ability to work with decimals. If you are able to solve these problems, move on to percentages. Show your work; this will indicate areas where you may need to improve.

Read the following numbers and write them as numerals.

S–3–1. Twelve and three tenths

S–3–2. Three and thirty-five thousandths

S–3–3. Four hundred five and sixty thousandths

S–3–4. Round 72.245 to the nearest hundredth.

S–3–5. Round 18.0015 to the nearest thousandth.

Convert the following fractions to decimals.

S–3–6. $\frac{3}{5}$

S–3–7. $\frac{3}{16}$

S–3–8. $\frac{1}{50}$

S–3–9. $1\frac{8}{9}$ (round to the nearest thousandth)

Convert the following decimals to common fractions.

S–3–10. 0.75

S–3–11. 0.375

S–3–12. Convert 0.9375 (to the nearest 16th)

S–3–13. $0.34\frac{2}{3}$

Perform the indicated operation.

S–3–14. $32.97 + 7.84 + 6.0042$

S–3–15. $0.007 + 7.0007 + 0.77$

S–3–16. $5.207 - 4.81$

S–3–17. 0.984×3.28

S–3–18. $123.57 \div 156.61597$

S–3–19. $0.25956 \div 0.412$

S–3–20. What is the difference between $^3/_4$ and $^5/_8$ expressed as a decimal?

S–3–21. There are seven steps in a flight of stairs. If each step is $8^1/_4$ inches high, what is the distance between the floors? Express your answer as a decimal?

S–3–22. The wall of a building is made of cement blocks $9^5/_8$ inches thick, with plaster on one side $^5/_{16}$ inch thick. What is the decimal thickness of the wall?

S–3–23. Northwest Airlines's stock is selling for $25^5/_8$ dollars. What will be the cost of 40 shares?

S–3–24. Number 12 wire is 0.081 inch in diameter. What is the diameter to the nearest 64th of an inch?

S–3–25. A piece of steel was $27^7/_8$ inches long. Four $4^5/_8$-inch pieces were cut off. How long was the remaining portion if the saw blade was 0.012 inch wide?

3.1 UNDERSTAND THE DEFINITIONS AND PERFORM THE BASIC OPERATIONS OF DECIMALS

Decimals or decimal fractions are used whenever you make a purchase in a store. When you check the deductions on your paycheck, decimals are used. The dollars and cents used in everyday living require basic knowledge of decimals. As the metric system becomes more widely used in business and industry, decimals will be used even more.

Since information is fed into computers to assist you in your occupation, you will need to know the mathematical language of the computer—decimals. Even though you may not work directly with computers, your occupation will rely heavily on the use of computers for time- and money-saving results. Decimals are the language for now and the future.

A number written in decimal notation is generally called a *decimal.* The position of a digit in a decimal determines its place value and name.

For example, the quantity 352.468 has two parts: the whole number part (352) and the decimal part (468). The number is read: three hundred fifty-two *and* four hundred sixty-eight thousandths (see Figure 3–1). Note the position of the last digit gives the decimal its name.

A small quantity like 0.0058 would be read as shown in Figure 3–2.

Note: A zero may be placed in front of the decimal point; however, this is not necessary.

A large quantity like 547,329.4632 would be read as shown in Figure 3–3.

A *decimal* or *decimal fraction* is a fraction whose denominator is 10 or some multiple of 10, such as 10, 100, 1000, and so on. Examples of decimal fractions are $^{7}/_{10}$, $^{35}/_{100}$, $^{783}/_{1000}$, and $^{5,281}/_{10,000}$ When writing decimal fractions, the denominator is not placed below the numerator as it is with fractions. We indicate the denominator with a *decimal point* (.) in the numerator. The decimal point is placed so there are as many digits to the right of the decimal point as there are zeros in the denominator.

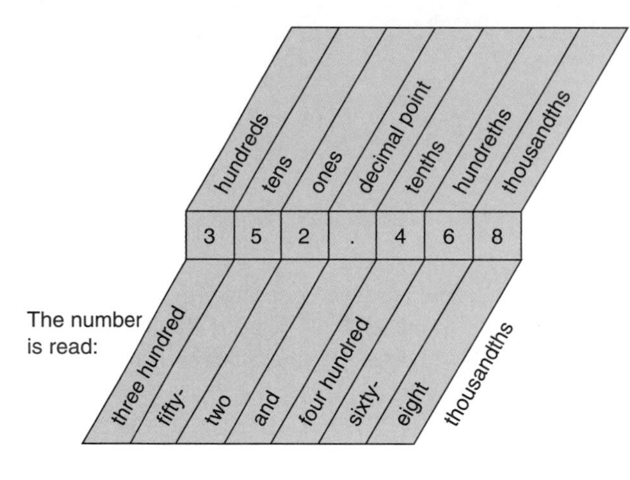

Figure 3–1

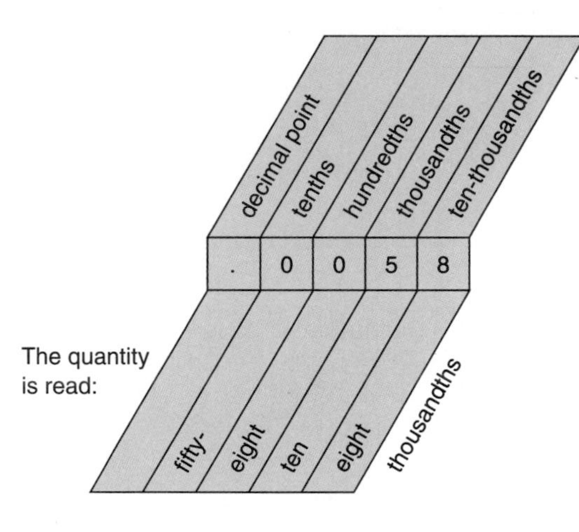

Figure 3–2

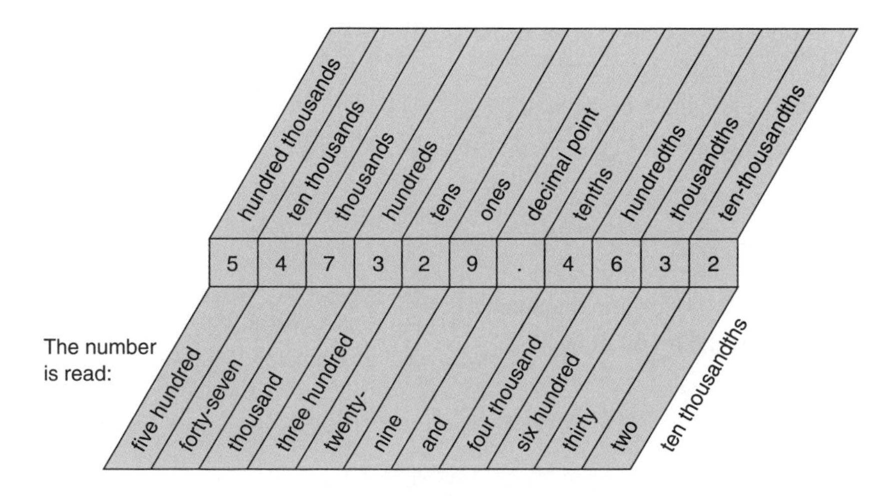

Figure 3–3

Thus $^{47}/_{100}$ is written 0.47, $^{783}/_{1000}$ is written 0.783, and $^{532}/_{100}$ is written 5.32. The proof of this simple procedure may be shown by dividing. Thus

$$\begin{array}{r} .47 \\ 100\overline{)47.00} \\ \underline{40\ 0} \\ 7\ 00 \\ \underline{7\ 00} \\ 0 \end{array}$$

Note: $^{47}/_{100}$ is called the fractional form and 0.47 is called the decimal form. A zero is placed to the left of the decimal point when there is no whole number. Thus 0.47 instead of .47, and 0.783 instead of .783. This is done to emphasize that the number is a decimal fraction.

When decimals are read, the decimal point may be called *point* or *and.* Thus 3.5 may be read "three point five" or "three and five tenths." The names of decimals are related to the denominator 10 or power of 10. Thus 0.3 is read "three tenths" and 0.74 is read "seventy-four hundredths." Names are assigned to each *place* value to the left and right of the decimal point. Study Figure 3–4. The two columns on the left name the number and its word form. The third column shows multiples of ten or the number of tens that produce the quantity that appears in the first column. The multiplication process shows the *power of 10.* The fourth column indicates the power of 10 when the number is written in a simpler form called the *exponential form.* Understanding the two columns on the right will help when you solve problems.

Thus the number 69,373.56354 is read "sixty-nine thousand, three hundred seventy-three and fifty-six thousand, three hundred fifty-four one hundred thousandths." If no decimal point is given, it is understood the decimal point is to the right of the last digit. Thus 12 = 12.0 and 528 = 528.0.

Rounding decimals is similar to rounding whole numbers. To round a decimal to a certain accuracy, determine

Number	*Name*	*Powers-of-10 Form*	*Exponential Form*
1,000,000	one million	$10 \times 10 \times 10 \times 10 \times 10 \times 10$	10^6
100,000	one hundred thousand	$10 \times 10 \times 10 \times 10 \times 10$	10^5
10,000	ten thousand	$10 \times 10 \times 10 \times 10$	10^4
1000	one thousand	$10 \times 10 \times 10$	10^3
100	one hundred	10×10	10^2
10	ten	10	10^1
1	one	1	10^0
0.1	one tenth	$\frac{1}{10}$	$\left(\frac{1}{10}\right)^1$ or 10^{-1}
0.01	one hundredth	$\frac{1}{10} \times \frac{1}{10}$	$\left(\frac{1}{10}\right)^2$ or 10^{-2}
0.001	one thousandth	$\frac{1}{10} \times \frac{1}{10} \times \frac{1}{10}$	$\left(\frac{1}{10}\right)^3$ or 10^{-3}
0.0001	one ten-thousandth	$\frac{1}{10} \times \frac{1}{10} \times \frac{1}{10} \times \frac{1}{10}$	$\left(\frac{1}{10}\right)^4$ or 10^{-4}
0.00001	one hundred-thousandth	$\frac{1}{10} \times \frac{1}{10} \times \frac{1}{10} \times \frac{1}{10} \times \frac{1}{10}$	$\left(\frac{1}{10}\right)^5$ or 10^{-5}
0.000001	one millionth	$\frac{1}{10} \times \frac{1}{10} \times \frac{1}{10} \times \frac{1}{10} \times \frac{1}{10} \times \frac{1}{10}$	$\left(\frac{1}{10}\right)^6$ or 10^{-6}

Figure 3–4 Place Values for Decimals

the place of accuracy desired, then look to the digit to the right of the place of accuracy. If the digit is less than 5, do not change the number and drop the remaining digits.

For example, round 338.43 to the nearest tenth. Since 3 is to the right of 4 (the tenths place) and less than 5, leave the 4 and drop the remaining digits.

Thus 338.43 $\overrightarrow{\text{(tenths)}}$ 338.4. (The arrow with the degree of accuracy indicates rounding.)

If the digit to the right of the place of accuracy is greater than 5, round the place of accuracy up by one digit. Thus round 579.378 to the nearest hundredth. Since the 8 is to the right of the hundredths place and greater than 5, round the 7 to an 8, and drop the remaining digits.

Thus 579.378 $\overrightarrow{\text{(hundredths)}}$ 579.38.

Other examples of rounding:

$$756.54 \overrightarrow{\text{(ones)}} 757$$

$$8984.555 \overrightarrow{\text{(hundredths)}} 8984.56$$

$$0.6574 \overrightarrow{\text{(thousandths)}} 0.657$$

EXERCISE 3–1 BASIC CONVERSION OF FRACTIONS AND DECIMALS PLUS ROUNDING

Write the following as fractions with a denominator of 10 or multiples of 10 (10,100,1000 . . .) and do not reduce.

3–1. 0.9

3–2. 0.85

3–3. 21.215

Write the following as decimals.

3–4. Forty hundredths

3–5. Eight tenths

3–6. Five and thirty-two thousandths

3–7. Thirty-six and twenty-four thousandths

Write the following decimals in words.

3–8. 0.909

3–9. 3.005

3–10. 5.0055

Write the following as fractions or mixed numbers with denominators of 10 or multiples of 10 and reduce.

3–11. Thirty-three and three tenths

3–12. Five hundred and four hundred-thousandths

3–13. Three and five hundred thirty-five ten-thousandths

3–14. Seventy-seven hundred-thousandths

3–15. Nine hundred thirty-four thousand and six ten-thousandths

3–16. Round 0.498 to the nearest tenth.

3–17. Round 5.3556 to the nearest hundredth.

3–18. Round 55.5555 to the nearest thousandth.

3–19. Round 1949.98 to the nearest tenth.

3–20. Round $36.545\overline{45}$ to the nearest thousandth.

3.2 CHANGING FRACTIONS TO DECIMALS

A common fraction may be converted to a decimal by dividing the denominator into the numerator.

Example 3–1: Convert a Proper Fraction to a Decimal

PROBLEM: Convert $^3/_4$ to a decimal.

SOLUTION: Divide the Denominator (4) into the numerator (3).

$$\begin{array}{r} .75 \\ 4\overline{)3.00} \\ \underline{2\,8} \\ 20 \\ \underline{20} \end{array}$$

Therefore, the fraction $^3/_4$ equals the decimal 0.75.

$$\frac{3}{4} = 0.75$$

Example 3–2: Convert a Proper Fraction to a Decimal

PROBLEM: Convert $^9/_{16}$ to a decimal.

SOLUTION: Divide the denominator (16) into the numerator (9).

$$\begin{array}{r} .5625 \\ 16\overline{)9.0000} \\ \underline{8\,0} \\ 1\,00 \\ \underline{96} \\ 40 \\ \underline{32} \\ 80 \\ \underline{80} \end{array}$$

Thus $$\frac{9}{16} = 0.5625$$

Example 3–3: Convert a Mixed Number to a Decimal

PROBLEM: Convert $15\frac{5}{6}$ to a decimal.

SOLUTION: Since 15 is a whole number, it may be placed to the left of the decimal point; then divide the numerator (5) by the denominator (6).

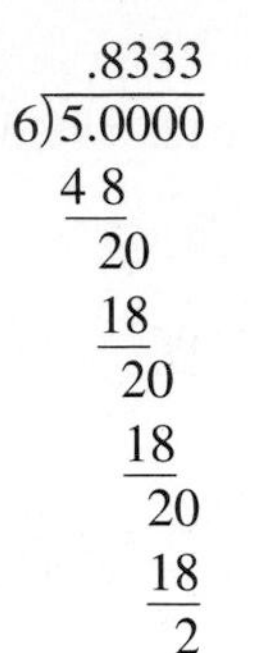

Note: Regardless of how long you continue to divide, you will always obtain a 3. This is called a *repeating decimal.* It may be written three ways: by placing a dot above the 3: $\dot{3}$; by placing a short line above the 3: $\overline{3}$; or by placing three dots immediately after the last 3: 3. . . .

Thus $15\frac{5}{6} = 15.8\dot{3} = 15.8\overline{3} = 15.83\ldots$

Example 3–4: Convert a Proper Fraction to a Decimal and Round to a Given Value

PROBLEM: Convert $\frac{2}{3}$ to a decimal and round to the nearest thousandth.

SOLUTION: Divide the denominator (3) into the numerator (2).

$$
\begin{array}{r}
.666\overline{6} \\
3\overline{)2.0000} \\
\underline{1\,8} \\
20 \\
\underline{18} \\
20 \\
\underline{18} \\
20
\end{array}
$$

Note: Regardless of how long this division continues, the result will be 6, thus a repeating decimal. Therefore $\frac{2}{3} = 0.666\overline{6} = 0.667$ when rounded to the nearest thousandth, since the digit to the right of the thousandth digit is greater than five.

EXERCISE 3–2 CONVERT FRACTION TO DECIMALS

Convert the following fractions to decimals and, indicate if they are repeating decimals.

3–21. $\frac{3}{10}$

3–22. $\frac{5}{8}$

3–23. $\frac{5}{16}$

3–24. $\frac{49}{100}$

3–25. $\frac{1}{9}$

3–26. $\frac{2}{32}$

3–27. $\frac{21}{32}$

3–28. $\frac{37}{8}$

3–29. $\frac{5}{9}$ (round to the nearest

3–30. $\frac{31}{64}$ (round to the nearest thousandth)

3.3 CHANGING DECIMALS TO FRACTIONS

In business and industry, it is often necessary to change decimals to fractions. One common need is to change decimals to fractions with denominators of fourths, eighths, sixteenths, thirty-seconds, or sixty-fourths. These changes need to be made rapidly, thus there are *tables of decimal equivalents.* The exact value is frequently not on the table. Therefore, the decimal must be rounded off to the nearest fraction, depending on the purpose of the conversion. These conversions will become more and more important as industry and business become more international and the metric system is used as the measurement system.

To convert any decimal to a fraction, simply write the decimal as a fraction. For example, 0.5 may be written as $^{5}/_{10}$, 0.79 as $^{79}/_{100}$, and 0.329 as $^{329}/_{1000}$. These fractions may be reduced by dividing the numerator and the denominator by the same quantity.

Example 3–5: Convert a Decimal to a Fraction and Reduce

PROBLEM: Convert 0.65 to a fraction and reduce to lowest terms.

SOLUTION: Convert the decimal to its decimal fraction form. Write as a fraction.

$$0.65 = \frac{65}{100}$$

Reduce the fraction by dividing the numerator and denominator by 5.

$$\frac{\overset{13}{\cancel{65}}}{\underset{20}{\cancel{100}}} = \frac{13}{20}$$

Thus $$0.65 = \frac{65}{100}$$

Example 3–6: Convert a Decimal to a Mixed Number and Reduce

PROBLEM: Convert 3.875 inches to a mixed number and reduce to lowest terms.

SOLUTION: Convert the decimal to its decimal fraction form.

$$3.875 = 3\frac{875}{1000}$$

Reduce the fraction by dividing the numerator and denominator by 25.

$$3\frac{\overset{35}{\cancel{875}}}{\underset{40}{\cancel{1000}}} = 3\frac{35}{40}$$

Reduce the fraction again by dividing the numerator and denominator by 5.

$$3\frac{\overset{7}{\cancel{35}}}{\underset{8}{\cancel{40}}} = 3\frac{7}{8}$$

Thus $$3.875 = 3\frac{7}{8}$$

Remember, several steps may be necessary to reduce a fraction to its lowest form.

Example 3–7: Convert a Decimal to a Fraction with a Given Denominator

PROBLEM: Convert 0.316 to the nearest sixteenth.

SOLUTION: Place 0.316 over 1 and multiply the numerator and denominator by 16.

$$\frac{0.316}{1} \times \frac{16}{16} = \frac{0.316 \times 16}{1 \times 16} = \frac{5.056}{16}$$

Round to the nearest sixteenth.

$$\frac{5.056}{16} = \frac{5}{16}$$

Thus, to the nearest sixteenth,

$$0.316 = \frac{5}{16}$$

SOLUTION: Arrange 7.98 (multiplicand) on the top and 2.3 (multiplier) on the bottom, multiply as with whole numbers. Then count the total number of places in the top and bottom numbers and place the decimal in the product the total places from left to right.

```
   7.98      2 places
×   2.3     +1 place
   2394
  1596
  18.354     3 places
```

Dividing decimals is similar to dividing whole numbers, except that placing the decimal point is very important. To divide decimals, move the decimal point in the divisor (the number in front of the division sign) to the right to make the divisor a whole number. Move the decimal point in the dividend the same number of places to the right, adding zeros if necessary. Then place the decimal point in the quotient directly above the decimal point in the dividend and divide like whole numbers.

Example 3–12: Divide Decimals

PROBLEM: $1776 \div 71.04 =$

SOLUTION: Place 1776 (dividend) inside the division sign, and 71.04 (divisor) in front of the division sign. Place a decimal point after 1776. Move the decimal point two places to the right in the divisor, and move the decimal point two places to the right in the dividend adding two zeros. Place the decimal in the quotient and divide.

```
              25.
71.04)1776.00.
       1420 8
        35520
        35520
```

Example 3–13: Divide Decimals and Round to a Given Value

PROBLEM: 286.5 mi ÷ 9.934 gal = (round to the nearest tenth)

SOLUTION: Place 286.5 miles inside the division sign, and 9.934 in front of the division sign. Move the 9.934 decimal point three places to the right, and move the 286.5 decimal point three places to the right. Add necessary zeros and divide. Since the quotient is to be rounded to the nearest tenth, the quotient needs to be carried to the hundredth, and then rounded to the nearest tenth.

```
               28.84
9.934)286.500.00
      198 68
       87820
       79472
        83480
        79472
         40080
         39736
```

Thus 28.84 rounds to 28.8, since the 4 in the hundredths place is less than 5.

EXERCISE 3–4 ARITHMETIC OPERATION OF DECIMALS

Perform the indicated operations.

3–51. 64.10 + 6.005 + 82.441

3–52. 6.284 + 3.41 + 80.521 + 32

3–53.
```
  10.35
  50.07
+155.46
```

3–54.
```
 962.44
  26.449
+  8
```

3–55.
```
   7.5294
  50.72
   9.0036
+143
```

3–56.
```
  30.04
− 20.99
```

3–57. 903.47 − 830.74

3–58. \$20.00 − \$14.79

3–59. 2.1 − 0.5328

3–60. 764.983 − 853.001

3–61. 0.73 × 16

3–62. 14.84 × 0.48

3–63. $14.79 × 18

3–64. 84.68 × 3.07

3–65. Find the product of 300.547 and 240.

3–66. 1.440 ÷ 24

3–67. $2.1\overline{)0.364}$

3–68. 74.6 ÷ 0.002

3–69. $958.80 ÷ 24

3–70. Find the quotient of 7038.66 and 2.19.

3–71. What is the total of your Target bill: $4.88, $0.97, $54.99, $19.92, $4.99, $2.99, $1.19, and $2.39?

3–72. How many miles were driven on a trip if the odometer read 69,843.4 at the start and 72,544.2 at the end?

3–73. What is your weekly pay if you worked 37.75 hours and earned $9.18 per hour?

3–74. How many gallons of gasoline, to the nearest tenth, can you purchase for $20 if gasoline costs $1.599 per gallon?

3–75. What is the cost per day, to the nearest cent, for a $4000 per month luxury condo in Naples, Florida? (Use 30 days per month.) What is the cost per hour?

3.5 SOLVE APPLIED PROBLEMS USING DECIMALS

As the world becomes more complex, those who will be making decisions will be expected to solve problems: many of these problems will involve decimals. Although calculators and computers are aides to solving problems, the reasoning process is vital to their solution.

Solving applied or word problems is the real joy of studying mathematics. It provides an opportunity for using all the various skills that have been learned.

Many people have difficulty, and some have a real fear, of solving applied or word problems. A good problem-solving strategy will be helpful in solving many problems. Following is a procedure that will be helpful:

- Define the problem.
- Determine what you are solving for.
- Estimate the answer.
- Set up and solve the problem.
- Test your answer.

Some examples of application problems will help show the reasoning and problem solving procedure.

Example 3–14: Applied Decimal Problem

PROBLEM: A beginning golfer purchased a set of irons for $99.95, a set of woods for $77.99, and 2 dozen balls at $14.59 per dozen. The sales tax on those items was $13.60.

If she pays with three $100 bills, how much change will she receive?

SOLUTION: First multiply 2 × $14.59 to find the cost of the golf balls, and then find the total bill by adding 2 × $14.59, $99.95, $79.99 and $13.60.

Then multiply 3 times $100 and subtract the total.

$$\begin{array}{rr} 2 \times \$14.59 = & \$29.18 \\ & 99.95 \\ & 79.99 \\ & +13.60 \\ \hline & \$222.72 \\ 3 \times \$100 = & \$300 \\ & -\ 222.72 \\ \hline & \$77.28 \end{array}$$

Example 3–15: Applied Decimal Problem

PROBLEM: Find the week's pay for a precision machinest who earns $18.94 per hour. The machinest works 48 hours during the week, and is paid time and a half for any hours more than 40.

SOLUTION: Multiply 40 times $18.94 to find the regular pay. Find the overtime hours by subtracting 40 from 48. Multiply this quantity by 1.5, which is equivalent to time and a half. This will determine the overtime pay. Then add the totals to determine the week's pay.

$$\begin{array}{rcr} 40 \times \$18.94 & = & \$757.60 \\ (48 - 40) \times 1.5 \times \$18.94 & = & +\ 227.28 \\ \hline & & \$1212.16 \end{array}$$

Example 3–16: Applied Decimal Problem

PROBLEM: Find the total tax bill on a $192,000 coach home, if the property tax is $10.55 per $1000 and the water/sewer/garbage tax is $2.58 per $500.

SOLUTION: Divide $192,000 by $1000 and multiply the result by $10.55 for the property tax. Similarly, divide $192,000 by $500 and multiply the result by $2.58 to find the water/sewer/garbage tax. Add the two to find the total tax bill.

$$\begin{array}{rcr} (\$192{,}000/\$1000) \times \$10.55 & = & \$2025.60 \\ (\$192{,}000/\$500) \times \$2.58 & = & +\ 990.72 \\ \hline & & \$3016.32 \end{array}$$

EXERCISE 3–5 APPLIED DECIMAL PROBLEMS

3–76. to 3–85. Given the check register on page 49, determine the missing values. Remember to add the deposits and subtract the checks.

PLEASE BE SURE TO DEDUCT ANY PER CHECK CHARGES OR SERVICE CHARGES THAT MAY APPLY TO YOUR ACCOUNT

CHECK NO.	DATE	CHECKS ISSUED TO OR DESCRIPTION OF DEPOSIT	[−] AMOUNT OF CHECK		√ T	[−] CHECK FEE (IF ANY)	[+] AMOUNT OF DEPOSIT		BALANCE	
									1321	85
2737	5/16	Normandale Swim & Tennis	111	92					3-76.	
3-77.	5/16	Target Dept. Store	3-78.						930	46
2739	5/16	Marshall Field's Dept. Store	69	85					3-79.	
2740	5/16	Hennepin Technical Centers	3-80.						850	86
2741	5/18	Deposit					350	00	3-81.	
2742	5/18	Avion Travel	290	00					3-82.	
2743	5/18	American Cancer Society	3-83.						905	86
2744	5/20	HTC (gas)	7	40					3-84.	
2745	5/20	St. Francis Shop	10	87					3-85.	

3–86. An oversized piston measures 4.165 inches. It is oversized by 0.025 inch. What is the regular size?

3–87. A weekly time card reads: 8.10, 7.94, 8.32, 8.14, and 8.68 hours. What is the total hours worked for the week?

3–88. A 22-ounce porterhouse steak costs $31.95 at McGarvey Brothers Steak House. What is the cost per ounce to the nearest cent?

3–89. Medora is paid $20 for cutting a lawn that takes 2.75 hours to cut. How much did she earn per hour to the nearest cent?

3–90. What is the cost of an 8-minute international phone call if the first three minutes costs $3.50 and each additional minute costs $0.35?

3–91. A 36-inch stainless steel rod has three 4.375-inch pieces cut from it. If each cut wastes 0.125 inch, how much of the rod is left?

3–92. Your entertainment center costs $699.50; however, you agreed to 30 monthly payments of $25.75. How much would have been saved if you would have paid the original price?

3–93. How many rivets are there in a 25-pound box if each rivet weighs 0.0625 pound?

3–94. A cash discount of $11.39 is allowed if a $569.74 bill is paid within 10 days. What is the bill after the discount?

3–95. A 14,520-square-foot home lot sells for $35,000. What is the cost per square foot to the nearest cent?

3–96. State business income tax is determined using the following procedure: $880 + 0.0255 × the profit. What is the state business income tax if a company had a profit of $1.73 million?

3–97. Each revolution of the wheel of a car travels 6.15 feet. How many revolutions are made in a mile, which is 5280 feet (round to the nearest revolution)?

3–98. An oil well flows at 1.85 barrels per hour. If the price of crude oil is $14.91 per barrel, how many dollars and cents will the well produce in 24 hours? In a week? In one year of 365 days?

3–99. The Northwest Athletic Club charges $94 per month for a husband/wife membership and $65 for a single membership. How much would be saved in a year if the husband converted to a single membership, since his wife never goes to the club?

3–100. Angella and Andy purchased a $160,000 home in St. Louis Park. Their monthly mortgage payment is $900.50 for 30 years. How much will they pay in 30 years? How much more will they pay to the bank than what they borrowed?

THINK TIME

It is important to know decimal and fractional equivalents. Figure 3–5 shows a mixture of decimals and fractions. Connect the pairs of equivalents with straight lines. When you have completed this, you will have a five-pointed star.

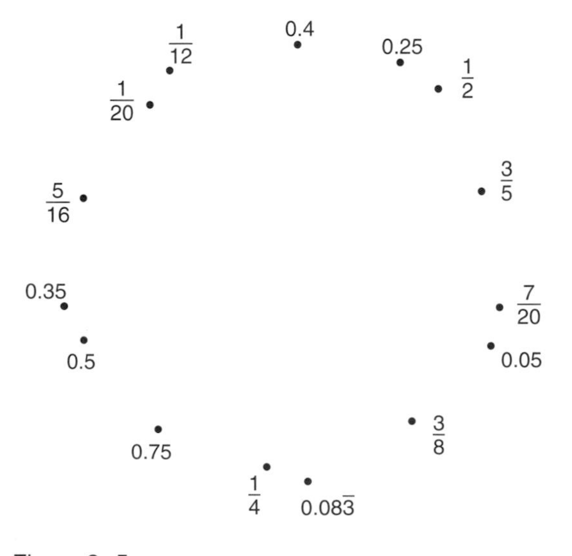

Figure 3–5

PROCEDURES TO REMEMBER

1. To change a fraction to a decimal:
   (a) Set up the division of the denominator into the numerator.
   (b) Place the decimal point.
   (c) Divide, carrying the division as many places as required.
2. To change a decimal to a fraction:
   (a) Write the decimal in its decimal fraction form in a power of 10.
   (b) Reduce the fraction to its lowest terms.

3. To change a decimal to a fraction with a specific denominator (eighth, sixteenth, etc.):
   (a) Write the decimal over 1.
   (b) Multiply the numerator and denominator by the specific denominator (eighth, sixteenth, etc.).
   (c) Round the numerator to the nearest whole number.

CHAPTER SUMMARY

A *decimal fraction* is a fraction whose denominator is 10 or a power of 10. The *decimal point* (.) separates the whole number from the decimal part of a number. The word *point* or *and* is used to indicate a decimal point when reading decimal quantities. A value that continues to repeat when one quantity is divided into another is called a *repeating decimal. Example:* $\frac{1}{3} = 0.\overline{3}$ or $\dot{3}$ or 0.333. . . . A repeating decimal is indicated by a bar or dot over the repeating digit or by three dots after the repeating digit. Decimal fractions may be written in two ways: in *fractional form* ($^{37}/_{100}$) or in *decimal form* (0.37).

A number that contains a decimal is said to be in *decimal notation* and is called a *decimal.* A *decimal* may have a whole number part, a decimal point, and a decimal part.

To write a decimal in words, write the whole number, then the *decimal point* or *and* for the decimal point; then write the decimal part, using the last digit as the name for the decimal. The same procedure is used when reading a decimal.

To *round* a decimal to a given accuracy, determine the place value desired. Then look to the digit to the right of the place value: if that digit is less than 5, write the number and drop the remaining digits; if the digit is greater than 5, round up the place of accuracy by one digit and drop the remaining digits.

To *add* and *subtract* decimals, write the numbers so the decimals are aligned vertically, place the decimal in the answer, and then perform the operations similar to those for whole numbers.

To *multiply* decimals write the decimal numbers as if they were whole numbers and multiply; then count the total number of decimal places to the right of the decimal in the multiplicand and the multiplier.

Dividing decimals is similar to dividing whole numbers, except for placing the decimal point. To divide decimals, if necessary, move the decimal point in the divisor to the right so that the divisor is a whole number. Then move the decimal point in the dividend the same number of places to the right, adding zeros if necessary. Place the decimal point above in the quotient, and divide like a whole number adding zeros as necessary.

To solve applied problems:

- Define the problem.
- If possible, draw a figure or chart indicating the given values on the figure or chart.
- Estimate the answer.
- Set up and solve.
- Test your answer.

CHAPTER TEST

Read the following numbers and write as numbers.

T–3–1. Seven and five hundredths

T–3–2. Seven and sixty-six thousandths

T–3–3. Three thousand and six thousandths

T–3–4. Round 37.4667 to the nearest hundredth.

T–3–5. Round 3345.45556 to the nearest thousandth.

Convert the following fractions to decimals.

T–3–6. $\frac{4}{5}$

T–3–7. $\frac{11}{32}$

T–3–8. $\frac{2}{25}$

T–3–9. $\frac{26}{27}$

T–3–10. $3\frac{3}{11}$ (round to the nearest thousandth)

Convert the following decimals to fractions.

T–3–11. 0.15

T–3–12. 0.1875

T–3–13. Convert 0.844 to the nearest thirty-second.

T–3–14. Convert $0.44\frac{4}{9}$ to a common fraction.

Perform the indicated operation.

T–3–15. $0.008 + 8.0088 + 7.89$

T–3–16. $3.274 - 2.99$

T–3–17. 0.843×2.86

T–3–18. $0.2056 \div 0.008$

T–3–19. Divide 200.64 by 6.66 (round to the nearest hundredth).

T–3–20. Find the decimal difference between $\frac{7}{8}$ and $\frac{5}{6}$ (to the nearest thousandth).

T–3–21. A machinist turned a $^{27}/_{64}$ -inch-diameter shaft from a piece of stock $^{1}/_{2}$ inch in diameter. Express the difference in diameters as a decimal.

T–3–22. A plumber made a run of pipe consisting of three pieces whose lengths were $15\,^{7}/_{16}$ inches, $10\,^{3}/_{8}$ inches, and $18\,^{7}/_{8}$ inches. Express the total length of pipe as a decimal.

T–3–23. An interactive laser disk stock is selling for $48 $^{7}/_{8}$. Find the cost of 55 shares?

T–3–24. A blueprint calls for a 0.84375-inch opening. To the nearest thirty-second, what drill should be used?

T–3–25. When plastic cures, it reduces in size by $^{1}/_{250}$. Find the length, in decimal form, that a 30-foot plastic pipe would be after curing.

4

PERCENTAGES

OBJECTIVES

After studying this chapter, you will be able to:

4.1. Understand the properties of percent and perform the basic operations with percent.
- Convert fractions to decimals to percent.
- Convert percents to decimals to fractions.

4.2. Solve the three types of percent problems.
- Find the amount when the percent and base are given.
- Find the percent when the base and amount are given.
- Find the base when the percent and amount are given.

4.3. Solve applied percent problems, including percent increase and decrease, discounts, taxes, commissions, and so on.

SELF-TEST

The skills to solve percent problems are similar to the skills used doing decimals. Your results from this test may indicate that you need to review this chapter. You can identify areas to review by checking your work.

S–4–1. Convert $^{11}/_{16}$ to a decimal.

S–4–2. Convert $5\,^{7}/_{8}$ to a decimal.

S–4–3. Convert 35% to a fraction.

S–4–4. Convert $37\,^{1}/_{2}$ % to a fraction.

S–4–5. Convert 0.37 to a percent.

S–4–6. Convert 0.08 to a percent.

S–4–7. Convert 0.005 to a percent.

S–4–8. Convert $^{1}/_{2}$ % to a fraction.

S-4-9. Convert $34\frac{2}{3}\%$ to a fraction.

S-4-10. Convert 0.115 to a percent.

S-4-11. $63\% = \frac{?}{100}$

S-4-12. Convert $\frac{7}{10}$ to a percent.

S-4-13. $\frac{6}{25} = \frac{?}{100}$

S-4-14. What is 30% of 50?

S-4-15. Find 39% of 1992.

S-4-16. If $13\frac{1}{2}\%$ of the transistors manufactured are defective, how many would be defective if 5000 were produced?

S-4-17. If fertilizer is 28% nitrogen, how much nitrogen is in a 22-pound bag?

S-4-18. What is $6\frac{3}{4}\%$ of \$2800?

S-4-19. 800 is 12% of what number?

S-4-20. What is the sales tax on a new automobile costing \$23,988.63 at a tax rate of 6%?

S–4–21. If 6.3% of your salary is deducted for the retirement fund, how much would be deducted from a gross weekly salary of $788?

S–4–22. What is the total interest on an automobile loan of $13,500 if the interest rate is $9\frac{1}{4}$ % per year for 3 years?

S–4–23. What is the total interest on a $400 life insurance loan at $5\frac{1}{4}$ % each year for 12 years?

S–4–24. If a transistor radio is marked down from $32 to $24, what is the percent of discount?

S–4–25. If the attendance at a church was 2300 at the beginning of the year, what would it be at the end of the year if it increased $12\frac{1}{2}$ %

4.1 UNDERSTAND THE PROPERTIES OF PERCENT AND PERFORM THE BASIC OPERATIONS WITH PERCENT

Percentages are used to indicate a relationship between numbers. Frequently, percentages are used to show the rate of change, the interest, and to determine taxes. Changes are indicated in percentages for increases or decreases in wages, crime, cost of living, production, natural resources, and so on. Thinking customers study the rates for interest on loans, carrying charges, and mortgages. Many dollars are saved by selecting lending agencies that offer the lowest rate of interest. Skills working with percentages will be helpful both on the job and for personal use.

Percent represents the number of parts out of 100 or the comparison of a number to another number. Generally, we compare a quantity or number by placing it as a numerator over the denominator, 100. The word *percent* is derived from the Latin word *percentum* and means "by the hundred." The symbol used for percent is %. A *percent* refers to a given number of parts of the whole, which is one hundred percent. Twenty-five percent may be written 25%; it means "25 out of 100."

The figure shows a whole square divided into 100 equal squares. The 100 squares are equal to 100%. The 25 shaded squares are equal to "25 out of 100" or 25%.

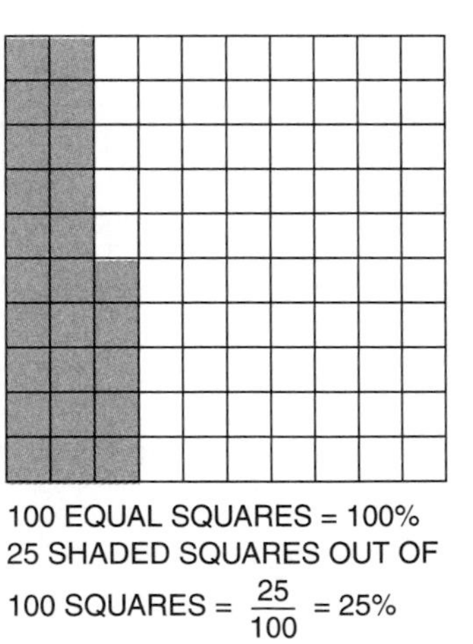

100 EQUAL SQUARES = 100%
25 SHADED SQUARES OUT OF
100 SQUARES = $\frac{25}{100}$ = 25%

To convert a percent to a decimal, remove the percent sign and move the decimal point two places to the left. This is equivalent to dividing the percent by 100.

To convert a decimal to a percent, move the decimal point two places to the right and write the percent sign (%) after the number. This is equivalent to multiplying the decimal by 100.

To change a fraction to a percent, convert the fraction to a decimal, and then move the decimal two places to the right and add the % symbol.

To change a percent to a fraction, remove the percent sign, place the number over 100, and then reduce to lowest terms. To understand these procedures better, work through the following examples.

Example 4–1: Change a Percent to a Decimal

PROBLEM: Change 15% to a decimal.

SOLUTION: Remove the percent sign and move the decimal point two places to the left.

$$15\% = 0.15$$

Example 4–2: Change a Decimal to a Percent

PROBLEM: Change 0.075 to a percent.

SOLUTION: Move the decimal point two places to the right and add the percent symbol.

$$0.075 = 7.5\% \text{ or } 7\frac{1}{2}\%$$

Example 4–3: Change a Fraction to a Percent

PROBLEM: Change 7 over 8 to a percent.

SOLUTION: Divide 8 into 7.

$$\frac{7}{8} = 0.875$$

Move the decimal point two places to the right and add the percent sign.

$$0.875 = 87.5\% \text{ or } 87\frac{1}{2}\%$$

Example 4–4: Change a Mixed Number to a Fraction

PROBLEM: Change $62\frac{1}{2}$ % to a fraction.

SOLUTION: Remove the percent sign and place $62\frac{1}{2}$ over 100. Since % means "out of 100," multiply by 10.

$$62\frac{1}{2}\% = \frac{62\frac{1}{2}}{100} = \frac{62.5}{100} = \frac{625}{1000}$$

Reduce to lowest terms.

$$\frac{\overset{25}{\cancel{625}}}{\underset{40}{\cancel{1000}}} = \frac{\overset{5}{\cancel{25}}}{\underset{8}{\cancel{40}}} = \frac{5}{8}$$

Example 4–5: Change a Complex Mixed Number to a Fraction

PROBLEM: Change $16\frac{2}{3}$% to a fraction.

SOLUTION: Remove the percent sign and place $16\frac{2}{3}$ over 100.

$$16\frac{2}{3}\% = \frac{16\frac{2}{3}}{100}$$

Change the numerator and denominator to improper fractions.

$$\frac{16\frac{2}{3}}{100} = \frac{\frac{50}{3}}{\frac{100}{1}}$$

Rewrite the complex fraction as a division-of-fractions problem.

$$\frac{\frac{50}{3}}{\frac{100}{1}} = \frac{50}{3} \div \frac{100}{1}$$

Invert, reduce, and multiply.

$$\frac{50}{3} \div \frac{100}{1} = \frac{\overset{1}{\cancel{50}}}{3} \times \frac{1}{\underset{2}{\cancel{100}}} = \frac{1}{6}$$

Therefore, $16\frac{2}{3}$ % is equal to the fraction $\frac{1}{6}$.

EXERCISE 4–1 CONVERTING: FRACTIONS, DECIMALS, AND PERCENTAGES.

Convert the following percents, decimals, and fractions as indicated.

4–1. Convert 37% to a decimal.

4–2. Convert 0.865 to a percent.

4–3. Convert 16.5 to a percent.

4–4. Convert 0.173 to a percent.

4–5. $^1/_8$ equals what percent?

4–6. Convert 0.3% to a fraction.

4–7. Convert 25.5% to a decimal.

4–8. Convert $33^1/_3$ % to a fraction.

4–9. Convert 37.5% to a fraction.

4–10. Convert 0.08 to a percent.

4–11. 0.0091 equals what percent?

4–12. Convert $^4/_7$ to a percent.

4–13. Convert $^3/_8$ to a percent.

4–14. Convert $28^4/_7$ % to a fraction.

4–15. Convert 0.036 to a percent.

4–16. Convert $^3/_{10}$ to a decimal.

4–17. $12^1/_2$ % equals what decimal value?

4–18. Convert $66^2/_3$ % to a fraction.

4–19. Convert $^5/_6$ to a percent.

4–20. Convert $^5/_{12}$ to a percent.

4.2 SOLVE THE THREE TYPES OF PERCENT PROBLEMS

There are three types of percentage problems. Examples are as follows: (a) What is 20% of 540? (b) Thirty-seven is what percent of 97? (c) Sixty-three is 25% of what number? All problems concerning percentages may be traced back to these examples.

Percent problems involve three things: the *base,* the *rate of percent,* and the *percentage.* These concepts can best be explained by an example. A man earns $200 a week and saves 30% of what he earns. Thus he saves $60 a week. In this problem, $200 is the base, 30% is the rate of percent, and $60 is the percentage. To find the percentage or amount saved, take 30% of $200, which is $60; 30% of $200 is $60 because 30% of $200 means $0.30 \times \$200 = \$60$.

The following formulas are to be used to find percentage, base, and rate.

1. To find the *percentage,* use the formula

   percentage = base times rate $\quad p = b \times r$

2. To find the *base,* use the formula

   base = percentage divided by rate $\quad b = \dfrac{p}{r}$

3. To find the *rate,* use the formula

   rate = percentage divided by base $\quad r = \dfrac{p}{b}$

Sometimes it is difficult to determine if a number represents the rate, base, or percentage. Some helpful hints concerning percentages are as follows:

1. The rate always has a percent sign or the word *percent* with it.
2. The base is the number that follows the word *of* in a word problem and refers to the entire quantity.
3. The percentage is the result of finding the rate when the base is given.

Example 4–6: Percentage Equals Base Times Rate

PROBLEM: Find 30% of 360.

SOLUTION: Convert 30% to a decimal.

$$30\% = 0.30$$

Since *of* means multiply, multiply as shown.

$$0.30 \times 360 = p$$
$$0.30 \times 360 = 108$$

Thus 30% of 360 is 108.

Example 4–7: Percentage Equals Base Times Rate ($p = b \times r$)

PROBLEM: Find 121% of $1500.

SOLUTION: Convert 121% to a decimal.

$$121\% = 1.21$$

Since *of* means multiply, multiple as shown.

$$1.21 \times \$1500 = p$$
$$1.21 \times \$1500 - \$1815$$

Thus 121% of $1500 is $1815.

Example 4–8: Base Equals Percentage Divided by Rate $\left(b = \dfrac{p}{r}\right)$

PROBLEM: 50 is what percent of 75?

SOLUTION: Write $50 = p \times 75$. Since *is* means equal and *of* means multiply, write $50 = p \times 75$ and $\dfrac{50}{75} = p$; then divide $0.66^2/_3\% = p$.

$$p = 66^2/_3\%$$

Thus 50 is $66^2/_3\%$ of 75.

Example 4–9: Rate Equals Percentage Divided by Base $\left(r = \dfrac{p}{b}\right)$

PROBLEM: 18% of what is 360?

SOLUTION: Convert 18% to a decimal. Write 18% = 0.18.

Since *of* means multiply and *is* means equal, write $0.18 \times r = 360$; then write $r = \dfrac{360}{0.18}$ and divide.

$$r = 2000$$

Thus 18% of 2000 = 360.

EXERCISE 4–2 SOLVING THREE TYPES OF PERCENTAGE PROBLEMS

4–21. What is 42% of 536?

4–22. What is 9% of 15?

4–23. What is 275% of 828?

4–24. Find 76% of $584.

4–25. What is 0.4% of 5280?

4–26. What percent of 38 is 19?

4–27. 76 is what percent of 216?

4–28. What percent of 45 is 144?

4–29. What percent of 17.3 is 4.325?

4–30. 32.8 is what percent of 8.2?

4–31. 15% of what is 285?

4–32. 48 is 320% of what?

4–33. 30 is 2.5% of what?

4–34. 20.8% of what is 104?

4–35. $16\frac{2}{3}$ % of what is 12?

4.3 SOLVE APPLIED PERCENT PROBLEMS, INCLUDING PROFIT, DISCOUNT, INCREASE/DECREASE, TAXES, COMMISSIONS, AND OTHER APPLICATIONS

There are numerous applications of percent in business, industry, and everyday life. Applied problems involving percent are not always stated as basic percent problems, but they can frequently be translated to one of the three basic percent problems. It is important to be able to apply your skills with percents to everyday problems. The next group of examples will give you an opportunity to develop your skills working with percents, starting with business applications of percentage.

Example 4–10: Commission

PROBLEM: Find the commission for selling a $5895.99 leather sofa if the rate of commission is 6.5%.

SOLUTION: *Commission,* the amount of payment a sales person receives, is determined in the same way as sales tax, by multiplying the rate of commission by the amount of the sale. Thus convert the commission rate (6.5%) to a decimal, and multiply by the amount of the sale.

$$6.5\% = 0.065$$
$$0.065 \times \$5895.99 = 383.23935$$

Round to the nearest cent; thus $383.24 is the commission.

Example 4–11: Selling Price

PROBLEM: Determine the selling price of a small motorcycle if the cost to the dealer is $1224 and the profit is 30% of the cost.

SOLUTION: Change 30% to a decimal.

$$30\% = 0.30$$

Multiply $1224 by 0.30 to find the profit.

$$\$1224 \times 0.30 = \$367.20$$

Then add the profit to the dealer's cost to find the selling price.

$$\begin{array}{rl} \$1224.00 & \text{cost} \\ +\ \ 367.20 & \text{profit} \\ \hline \$1591.20 & \end{array}$$

ALTERNATIVE SOLUTION: Because the selling price is 100% of the dealer's cost plus 30% profit, we could add these percents and then multiply the dealer's cost by this percent. Follow these procedures to solve the problem. Add 100% and 30%.

$$\begin{array}{r} 100\% \\ +\ 30\% \\ \hline 130\% \end{array}$$

Convert 130% to a decimal.

$$130\% = 1.30$$

Multiply the dealer's cost ($1224) by 1.30 to find the selling price:

$$\text{selling price} = \$1224 \times 1.30 = \$1591.20$$

Example 4–12: Discount

PROBLEM: Find the cost of a pair of snow tires that were priced at $96 but were reduced 20%.

SOLUTION: Discounts are given as a percent off the original price. To find the discount, multiply the original price by the rate of discount (20%), then subtract the discount ($19.20) from the original selling price ($96) to find the discounted price ($78.80). Change 20% to a decimal.

$$20\% = 0.20$$

Multiply the original selling price ($96) by the decimal (0.20):

$$\text{discount} = \$96 \times 0.20 = \$19.20$$

Subtract the discount ($19.20) from the original price of $96.

$96.00	original price
− 19.20	reduction
$76.80	

ALTERNATIVE SOLUTION: The reduced price is actually 80% of the original price. Subtract 20% from 100% to get this.

$$100\% - 20\% = 80\%$$

Change 80% to a decimal.

$$80\% = 0.80$$

Multiply 0.80 by the original price ($96) to determine the reduced price:

$$\text{reduced price} = \$96 \times 0.80 = \$76.80$$

Interest is money paid for the use of money. It is determined by a percent of the amount of money being loaned. Generally, interest is determined on the basis of one year. Banks and other lending agencies pay interest on the amount placed in the bank, the rate of interest being a percent per year. The base is called the *principal,* or the amount of money loaned. The *percentage* is the interest paid. Interest is found by multiplying the rate *times* the principal *times* the amount of time, which is generally by the year but could be a fractional part of a year. For example, 5 years 9 months would be written $5\,^9/_{12}$ or 5.75 years. The formula to find interest is $i = prt$.

Example 4–13: Interest

PROBLEM: Find the interest on $2500 for 1 year at $12\,^1/_2\,\%$ per year.

SOLUTION: Interest equals principal times the rate times the time, or $i = prt$. Identify the given quantities.

$$i = \$2500 \times 12\frac{1}{2}\% \times 1\ \text{(year)}$$

Change $12\,^1/_2\,\%$ to a decimal.

$$12\frac{1}{2}\% = 0.125$$

Multiply the principal by the rate by the time.

$$\begin{aligned} i &= prt \\ &= \$2500 \times 0.125 \times 1 \\ &= \$312.50 \end{aligned}$$

The interest on $2500 at $12\,^1/_2\,\%$ for 1 year is $312.50.

Example 4–14: Principal Equals the Interest Divided by the Rate and Time

PROBLEM: Interest of $120 is paid at a rate of 5% per year. How much money (principal) is in the account?

SOLUTION:

$$\text{Principal} = \frac{\text{interest (\$120)}}{\text{rate (5\%)} \times \text{time (1 year)}} \quad \text{or}$$

$$p = \frac{i}{rt}$$

Identify the given quantities.

$$\begin{aligned} p &= \frac{i}{rt} \\ &= \$\frac{120}{5\% \times 1} \end{aligned}$$

Change 5% to a decimal.

$$5\% = 0.05$$

Divide the percentage ($120) by the rate times the time to find the principal.

$$\begin{aligned} p &= \$\frac{120}{(0.05)(1)} \\ &= \$2400 \end{aligned}$$

Thus $2400 is the principal in the savings account.

Example 4–15: Rate Equals the Interest Divided by the Principal and Time

PROBLEM: Find the rate of interest when $7955 is paid for 1 year on a mortgage of $86,000.

SOLUTION:

$$\text{Rate of interest} = \frac{\text{interest (\$7955)}}{\text{principal (\$86,000)} \times \text{time (1 year)}} \quad \text{or}$$

$$r = \frac{i}{pt}$$

Identify the given quantities.

$$i = \$7955 \text{ and } p = \$86,000$$

$$r = \frac{\$7955}{(\$86,000)(1)}$$

Divide the interest ($7955) by the principal ($86,000) times the time (1 year).

$$r = \frac{\$7955}{(\$86,000)(1)} = 0.0925$$

Change the decimal (0.0925) to a percent.

$$0.0925 = 9.25\%$$

The rate of interest on the mortgage is 9.25% or $9\frac{1}{4}\%$.

Example 4–16: Time Equals the Interest Divided by the Principal and Rate

PROBLEM: On a $10,000 municipal bond, the total interest received is $5300. If the rate of interest is $6\frac{5}{8}\%$, how many years has the interest been paid to total $5300?

SOLUTION:

$$\text{Time} = \frac{\text{interest (\$5300)}}{\text{principal (\$10,000)} \times \text{rate}\left(6\frac{5}{8}\%\right)} \quad \text{or}$$

$$t = \frac{i}{pr}$$

Identify the given quantities.

$$t = \frac{\$5300}{(\$10,000)\left(6\frac{5}{8}\%\right)}$$

Change $6\frac{5}{8}\%$ to a decimal.

$$6\frac{5}{8} = 0.06625$$

Evaluate

$$t = \frac{\$5300}{(\$10,000)(0.06625)} = 8 \text{ years}$$

Thus interest of $6\frac{5}{8}\%$ has been paid for 8 years on the $10,000 bond.

Companies generally give discounts for volume buying. An additional discount is given for cash payment or payment within a short period of time. Finding the percent increase or decrease, profit or loss, and the percent of change is necessary when solving applied problems.

Example 4–17: Discounts

PROBLEM: Find the price paid for $240 of wool fabrics if discounts of 20% and 2% are given.

SOLUTION: Find the 20% discount by multiplying the original price ($240) by the rate of discount (0.20).

$$d = \$240 \times 0.20 = \$48$$

Subtract the first discount from the original price.

$240.00	original price
− 48.00	first discount
$192.00	discount price

Find the 2% cash discount by multiplying the discount price ($192) by the rate of discount (0.02).

$$d = \$192 \times 0.02 = \$3.84$$

Then subtract the cash discount from the discount price.

$192.00	discount price
− 3.84	cash discount
$188.16	price paid

ALTERNATIVE SOLUTION: Since the discounted price is 80% of the original price, subtract 20% from 100% and multiply 0.80 by the original price. Multiply the decimal 0.80 by the original price ($240) to determine the discounted price:

$$\text{discounted price} = \$240 \times 0.80 = \$192$$

Then subtract 2% from 100% and change to a decimal. Multiply the decimal 0.098 by the discounted price $192 to determine the price paid:

$$\text{price paid} = \$192 \times 0.98 = \$188.16$$

Example 4–18: Percent Increase

PROBLEM: Find the percent increase in yield of wheat by applying fertilizer, if the yield increases from 48 bushels to 64 bushels.

SOLUTION: Find the amount of change.

$$\begin{array}{rl} 64 & \text{bushels yield with fertilizer} \\ -48 & \text{bushel yield without fertilizer} \\ \hline 16 & \text{bushel increase} \end{array}$$

Divide the amount of change (16) by the yield without fertilizer (48) to find the percent change.

$$\text{percent change (increase)} = \frac{\text{change}}{\text{original}} = \frac{16}{48}$$

Note: $^{16}/_{48}$ should be reduced to $^{1}/_{3}$ or $33^{1}/_{3}\%$. Then change the decimal to a percent.

$$0.333 = 33\frac{1}{3}\%$$

Example 4–19: Shooting Percentage

PROBLEM: A basketball player makes 10 out of 14 shots. What percent does he make?

SOLUTION: The percentage is 10 and the base is 14. The rate or percent must be found. To find the rate, use the formula

$$\text{rate} = \frac{\text{percentage}}{\text{base}} \quad \text{or} \quad r = \frac{p}{b}$$

Identify the given quantities of the formula.

$$r = \frac{p}{b} = \frac{10}{14}$$

Second, divide the percentage 10 by the base 14 to determine the rate.

$$\begin{aligned} r &= \frac{10}{14} \\ &= 0.7142 \end{aligned}$$

Change the decimal to the nearest tenth of a percent.

$$0.7142 = 71.4\%$$

Thus the basketball player made 71.4% of his shots. That is, his shooting average in basketball is 71.4%.

The percent of increase or decrease is used in many applications to show an example of the amount of change. Whenever the percent change (increase or decrease) is used to indicate the percent of increase or decrease, find the amount of change and divide by the original amount. Thus an increase of a stock from \$100 to \$120 is an increase of \$20; then divide \$20 by \$100; that equals 0.20 or an increase of 20%.

Example 4–20: Increase or Decrease

PROBLEM: During the summer of 2000, the price of gasoline rose from \$1.399 to \$1.899 at your local service station. What is the percent increase?

SOLUTION: Subtract the new price (\$1.899) from the old price (\$1.399) to determine the amount of increase. Then divide the amount of increase (\$0.50) by the original amount and round to the nearest tenth of a percent.

$$\begin{array}{r} \$1.899 \\ -\ 1.399 \\ \hline \$\ .50 \end{array}$$

Then $\frac{\$.50}{\$1.399} = 0.3573 = 35.7\%$ increase.

EXERCISE 4–3

Solve the following applied percent problems to improve your skills with practical problems.

4–36. The profit on an electric calculator is 55%. What is the selling price if the cost is \$18?

4–37. Find the state sales tax on a new car priced at \$14,995 if the tax rate is 4%.

4–38. A dress costing \$27 is reduced 15%. What is the reduced price?

4–39. A football team has an 87.5% winning percentage. How many games of a 16-game schedule did the team win?

4–40. A grain buyer bought wheat at \$5.40 a bushel and sold it at an 8% loss. What was the selling price of the wheat?

4–41. The 6% sales tax on a pontoon boat amounted to \$473.70. What was the cost of the boat?

4–42. The city sales tax on a color television set was \$28 for an \$800 set. Determine the sales tax rate.

4–43. A mixture for concrete consists of 3 parts sand, 2 parts gravel, and 1 part cement. What is the percent of cement in the mixture?

4–44. A small business building is insured for 90% of its value at an insurance rate of $^1/_4$ % per year. The building is valued at \$140,000. Determine the annual premium.

4–45. Find the interest on \$580 at 8% for 1 year.

4–46. Find the interest on \$1250 at $9^1/_2$ % for 1 year.

4–47. Find the interest on \$17,500 at $7^1/_4$ % for 20 years.

4–48. Find the interest on \$4050 at 12.5% for $3^1/_2$ years.

4–49. Find the monthly payments on a \$13,500 automobile at $10^1/_2$ % for 3 years.

4–50. Find the rate of discount for an AM–FM radio priced at $49.95 that was reduced to $39.95 for a special sale.

4–51. Find the interest on a $250 savings account for 5 years 9 months at $7\frac{1}{2}$ %.

4–52. If an electric bill is paid within 15 days, a discount of $1\frac{1}{2}$% is given. What would a customer pay on a $33 electric bill if it was paid within 15 days?

4–53. A dealer allowed $2875 for a used car. He then sold the car for $3000. What was his percent of gain?

4–54. Find the rate if the principal is $8000 and the interest for 1 year is $400.

4–55. What is the rate of interest if the principal is $3300 and the interest received after 8 years is $924?

4–56. Kay Cobb's Furniture offered $\frac{1}{3}$ off a mattress–box spring combination. Find the sale price if the original price was $639.

4–57. Angella bought two dresses on sale at 30% off and saved $41 from the marked price. What was the original price of the dresses?

4–58. A grain dealer bought peanut hearts for $8.70 per bag and sold them for $11.60. Find his percent of profit.

4–59. How long must $30,000 be invested at $7\frac{1}{2}$ % to yield $6750?

4–60. Find the sale price of an outboard motor that was marked 22% off a list price of $589.

4–61. How much would you save if you received a $3\frac{1}{2}$ % discount for paying cash for a $595 TV set?

4–62. A lumber company allows an 8% contractor discount. If a contractor bought lumber worth $8256, how much would the contractor pay for the lumber after the discount?

4–63. A hardware store sold a radial arm saw for $392.50. The store bought the saw for 65% of the selling price. What did the store pay for the saw?

4–64. A real estate agent, Kay Cobb, bought two lots for $22,000 and $18,000. She sold the first lot at a profit of 8% but lost 10% on the second. How much did she gain or lose on these deals?

4–65. A soap is advertised as $99^{44}/_{100}$ % pure. How many grams of impurities would there be in 5000 grams?

4–66. A service charge of $1^{1}/_{2}$ % is charged on the unpaid balance of charge accounts. Determine the interest on a bill for $59.95.

4–67. A real estate salesperson receives a 3.4% commission on the sale of a house. How much commission would she receive for selling a $244,500 house?

4–68. A trucking firm charges 3% commission for hauling and collects another $1^{1}/_{3}$ % for insurance. What are the fees for a $12,000 cargo?

4–69. A certain crude oil is 14.875% sulfur. How many gallons of sulfur would there be in 8000 barrels of crude oil if each barrel contained 31.5 gallons?

4–70. If 37.42% is a "factor" for hot-dog buyers at baseball games, determine the profit on hot dogs for the season. Eighty-one games are played and the average attendance is 24,276. Hot dogs sell for $3.25 each and the profit on the hot-dog sales is 62%.

4–71. At a baseball game, 37.42% of the people bought hot dogs. How much money was collected if 27,753 people attended the game and the hot dogs sold for $3.25?

4–72. A tolerance of ±0.8% is allowable for a car axle. How much variation is allowable on $^{7}/_{8}$ -inch axle?

4–73. What are the upper and lower limits of tolerance of a 910-ohm resistor if the tolerance factor is ± 0.5%?

4–74. Find the efficiency of an electric motor that draws 54 watts from an electrical supply and delivers 46.6 watts to its load.

4–75. Minnesota iron ore is 18.1% pure. How much iron can be refined from 150 tons of iron ore?

4–76. The list price of an office chair is \$229.89. The interior decorator's discount is 40%. What does the decorator pay for the chair if the state sales tax is 4%?

4–77. A 2-meter length of nylon rope was given a tensile test and stretched to 2.23 meters before breaking. Determine the percent of the original length it was stretched.

4–78. A drill-press operator was supposed to drill a hole 0.3125 inch in diameter. The hole drilled measured 0.34375. Find the percent of error.

4–79. Albert bought a lakeshore cabin for \$23,600, and later sold it for 25% more than he paid for it. The realtor charged a 6% fee on the selling price. What profit did Albert make?

4–80. A lawn fertilizer is 22% nitrogen. If a 50-pound bag costs \$2.98, find the price per pound for nitrogen.

4–81. If $1\frac{1}{4}$ pounds of salt are dissolved in 5.5 gallons of water that weigh 45.8 pounds, what is the percent of salt in the solution to the nearest hundredth?

4–82. How much would you need to invest at 7.5% if you wanted an income of \$1200 per month?

4–83. The cost of living rose 6.3%, 8.1%, and 5.8% each year over a 3-year period. The raise in pay amounted to 25% over 3 years. How much is a person ahead or behind at the end of 3 years if his initial yearly salary is \$20,500?

4–84. The depreciation of a new car is approximately 18% of the original price the first year, 12.5% of the original price the second year, and 8% of the original price the third year. The \$22,900 car loan is at 13.5% per year for 3 years. Determine the depreciation and interest for 3 years.

4–85. A salesperson drove 410.22 freeway miles on 12.9 gallons of gasoline and 327.25 city miles on 11.9 gallons. What is the percent (to the nearest tenth) decrease in miles per gallon from freeway to city driving?

T–4–9. Convert 0.32% to a common fraction.

T–4–10. Convert $7\frac{3}{4}$ % to a decimal.

T–4–11. $78\% = \frac{?}{100}$

T–4–12. Convert $\frac{7.5}{10}$ to a percent.

T–4–13. $\frac{3}{33\frac{1}{3}} = \frac{?}{100}$

T–4–14. Find 25% of 78.9.

T–4–15. Find $7\frac{1}{4}$ % of 5764.

T–4–16. A family sold a lot for $20,000 and gained 25% on the transaction. What did the lot cost the family?

T–4–17. What is the percent increase from the application of fertilizer on a wheat field whose yield was 24 bushels before and 38 bushels after?

T–4–18. A man bought a house for $182,000, and 5 years later sold it for $102,500. Find the percent of gain.

T–4–19. A new automobile depreciated 24% and was worth $18,800 at the end of the year. What was it worth at the beginning of the year?

T–4–20. A lathe is set to operate at 4200 revolutions per minute. The belt slippage is 18%. What is the actual speed of the machine?

T–4–21. Brucie works at $17.32 per hour for 22 hours per week. Find her take-home pay if 27.2% is deducted for taxes, social security, and other benefits.

T–4–22. A set of golf clubs originally priced at $475 was discounted at 40%, 30%, and 10%. What was the final sale price?

T–4–23. The retail price of a radial tire is $155. If the tire store is allowed a 25% profit and the manufacturer earns a 20% profit, what is the cost of making the tire?

T–4–24. How much money would you need to invest at $7\frac{1}{2}$% to have an income of $750 per month?

T–4–25. A plastic material loses 4.25% of its weight in drying. What was its original weight if its weight after drying is 134.05 grams?

5

POWERS, ROOTS, AND ORDER OF OPERATIONS

OBJECTIVES

After studying this chapter, you will be able to:

5.1. Understand powers of 10, scientific notation, and conversions.
5.2. Understand definitions and properties of powers or roots and evaluation.
5.3. Understand square-root calculations and application.
5.4. Understand order of operations agreement.

SELF-TEST

This test will determine your need to review powers of 10 and square roots. Show your work; this will indicate areas where you may need to improve.

S–5–1. Write 5000 as a power of 10.

S–5–2. Write 589,000 as a power of 10.

S–5–3. Convert 7×10^5 to a number.

S–5–4. Convert 2.4×10^6 to a number.

S–5–5. Convert 0.0000052×10^8 to a number.

S–5–6. Write 0.0002 as a power of 10.

S–5–7. Convert 4.54×10^{-1} to a number.

S–5–8. Convert 779×10^{-4} to a number.

S–5–9. Evaluate 21^2.

S–5–10. Evaluate 6^3.

S–5–11. Evaluate $\left(\frac{3}{8}\right)^3$.

S–5–12. Evaluate 3^6.

S–5–13. Find the square root of 141,000 to the nearest tenth.

S–5–14. Determine the square root of 0.120409.

S–5–15. Find the square root of 24,409 to the nearest thousandth.

S–5–16. Find the diagonal of a square with 15-foot sides to the nearest foot.

S–5–17. Find the hypotenuse of a right triangle if one leg is 12 units and the other leg is 5 units.

S–5–18. Find the cube root of 343.

S–5–19. Find the length, to the nearest tenth, of the side of a square with a 16.97-meter diagonal.

S–5–20. Find the side of a square, to the nearest tenth, if the diagonal is 18.5.

S–5–21. Evaluate $24 \div (8 - 4)$.

S–5–22. Evaluate $600 \div 50 - 20 + 18$.

S–5–23. Evaluate $3^3 + 5 - 8 \times 4$.

S–5–24. Evaluate $4^3 - 3^3 - 2^2$.

S–5–25. Find the number of miles are traveled in a 48-minute flight if the jet is traveling at 600 miles per hour.

5.1 POWERS OF TEN, SCIENTIFIC NOTATION, AND CONVERSIONS

The power-of-10 concept was discussed in Chapter 3. In this chapter we develop power-of-10 concepts further, and learn scientific notation and how to determine roots. Knowledge of powers of 10 helps estimate answers when working with very large or very small numbers. The hand-held calculator is the most efficient way to work with powers of 10. However, it is necessary to have an understanding of how to convert numbers to powers and to convert powers-of-10 quantities to numbers.

When working with very large or very small numbers it is helpful to convert numbers to a *power of 10.* Converting to *scientific notation* is the process of rewriting a number as a number between 1 and 10 times the appropriate power of 10. A number is said to be in scientific notation if it is expressed as the product of a single digit and some power of 10. For example, 4,000,000 could be written $4 \times 1{,}000{,}000 = 4 \times 10^6$. The decimal point of any number can be shifted to the left or right depending on the power of 10. For example, in $5.25 \times 10^3 = 5250$ the 3 adjacent to the 10 is read "10 to the third power" or "10 to the third," which means to move the decimal point three places to the right, since $10^3 = 1000$. Also, $7.46 \times 10^{-2} = 0.0746$; the -2, read "10 to the negative 2 power," means to move the decimal point two places to the left, since $10^{-2} = 0.01$.

Writing quantities in engineering notation is rewriting them using powers of ten, just like scientific notation. Occasionally an appropriate System International (SI) prefix is attached (see Chapter 17). For example, 7.5 kW means 7.5×10^3 W or 7500 watts (since $k = 1000 = 10^3$), and 8 microseconds = 0.000008 second = 8×10^{-6} seconds, which is written as 8 μs in engineering notation. This notation is used in technical fields, particularly in electronics and computers. Figure 5–1 shows powers of 10. Note the power of 10 is the same digit as the number of zeros in the original numbers.

Greater than 1	*Less than 1*
$1 = 10^0$	
$10 = 10^1$	$10^{-1} = \frac{1}{10^1} = \frac{1}{10} = 0.1$
$100 = 10^2$	$10^{-2} = \frac{1}{10^2} = \frac{1}{100} = 0.01$
$1000 = 10^3$	$10^{-3} = \frac{1}{10^3} = \frac{1}{1000} = 0.001$
$10{,}000 = 10^4$	$10^{-4} = \frac{1}{10^4} = \frac{1}{10{,}000} = 0.0001$
$100{,}000 = 10^5$	$10^{-5} = \frac{1}{10^5} = \frac{1}{100{,}000} = 0.00001$
$1{,}000{,}000 = 10^6$	$10^{-6} = \frac{1}{10^6} = \frac{1}{1{,}000{,}000} = 0.000001$

Figure 5–1

Note the negative power of 10 is the number of places the decimal would need to be moved to the right so that the first digit would be more than 1.

A number is written in scientific notation when the first factor is a single digit and the second is a power of 10. Thus

$$5000 = 5 \times 1000 = 5 \times 10^3$$
$$63{,}700 = 6.37 \times 10{,}000 = 6.37 \times 10^4$$
$$0.007 = 7 \times 0.001 = 7 \times 10^{-3}$$

The following examples show how to convert numbers to powers of 10 and how to convert power-of-10 quantities to numbers.

Example 5–1: Convert a Number to a Power of Ten

PROBLEM: Write 9000 as a power of 10.

SOLUTION: $9000 = 9 \times 1000 = 9 \times 10^3$

The power of 10 is equal to the number of zeros in the original number.

Example 5–2: Convert from a Power of Ten to a Number

PROBLEM: Write the value of 5.23×10^5.

SOLUTION: $5.23 \times 10^5 = 5.23 \times 100{,}000 = 523{,}000$

The number of zeros is equal to the number of places moved to the right or the power of 10.

Example 5–3: Convert from a Small Number to a Power of Ten

PROBLEM: Write 0.000008 as a power of 10.

SOLUTION: $0.000008 = 8 \times 0.000001 = 8 \times 10^{-6}$

The negative 6 indicates the number of places the decimal needs to be moved to the left.

Example 5–4: Convert from a Power of Ten to a Number

PROBLEM: Write 5.69×10^{-4} as a number.

SOLUTION: $5.69 \times 10^{-4} = 5.69 \times 0.0001 = 0.000569$

The negative 4 indicates how many places to the left the decimal point should be moved.

EXERCISE 5–1 POWERS OF TEN AND SCIENTIFIC NOTATION

Write each of the following as a power of 10.

5–1. 3000

5–2. 150

5–3. 46,000

5–4. 567,000

5–5. 8,760,000,000

Convert the following to numbers.

5–6. 4×10^3

5–7. 3.9×10^2

5–8. 6.66×10^5

5–9. 0.000036×10^6

5–10. 1.234×10^4

Write each of the following as a power of 10.

5–11. 0.007

5–12. 0.00086

5–13. 0.29

5–31. Evaluate 0.5^2.

5–32. Evaluate 0.22^2.

5–33. Evaluate 0.8^3.

5–34. Evaluate 0.102^2.

5–35. Evaluate 0.12^3.

5–36. Evaluate 3.15^2.

5–37. Evaluate 10.1^3.

5–38. Evaluate 0.4^4.

5–39. Evaluate 2.5^5.

5–40. Evaluate $8^2 \times 8^2$.

5.3 SQUARE-ROOT CALCULATIONS AND APPLICATIONS

A calculator is used to find square root. This section demonstrates the longhand procedure to determine square root. The example illustrates the cumbersome and difficult procedure, and makes you appreciate the benefits of a calculator.

As you observe the longhand method of calculating square root as shown in the example, keep in mind this is one of the methods that may be used. You should also note that most square-root calculations determine only an approximate value. The extent of accuracy will depend on the individual problems. This example is included only to illustrate the difficulty of calculating square root. As the numbers become more complex, the calculation becomes more difficult.

Example 5–9: Longhand Square-Root Calculation

PROBLEM: Calculate the square root of 576.

SOLUTION: Set up the square-root process. Place 576 under the radical symbol. Divide the digits into groups of two, starting from the decimal point and moving to the left.

$$\sqrt{5_{\wedge}76.}$$

From the first group, 5, find the largest perfect square smaller than 5. This is 2, since $2^2 = 4$. Place the 2 above the number 5 and the number 4 below the number 5. Subtract the number 4 from the number 5 and bring down the next group, 76.

$$\begin{array}{r} 2\ \ \ . \\ \sqrt{5_{\wedge}76.} \\ \underline{4} \\ 1\ 76 \end{array}$$

Multiply 20 by 2, the number above the radical sign. Then divide the result of 20×2, or 40, into 176. This will give you the approximate quotient, which is 4. Place 4 above the number 6 on top of the radical sign.

$$\begin{array}{r} 4 \\ 40\overline{)176.00} \\ \underline{160} \end{array} \qquad \begin{array}{r} 2\ \ 4 \\ \sqrt{5_{\wedge}76} \\ \underline{4} \\ 1\ 76 \end{array}$$

Then add the approximate quotient 4 to 40, $4 + 40 = 44$. Then multiply 44 by 4 and place the result below the 176. Subtract the resulting number from 176.

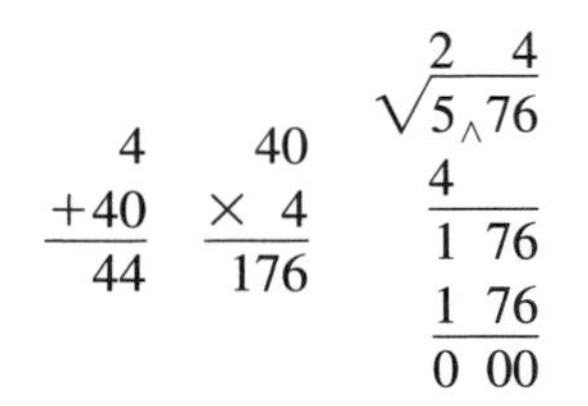

Since the resulting subtraction of 176 from 176 equals zero, we have a perfect square of 24. Check your answer by multiplying 24 by 24.

$$\begin{array}{r} 24 \\ \times 24 \\ \hline 96 \\ 48 \\ \hline 576 \end{array}$$

Once you understand the procedure, the process can be condensed as follows:

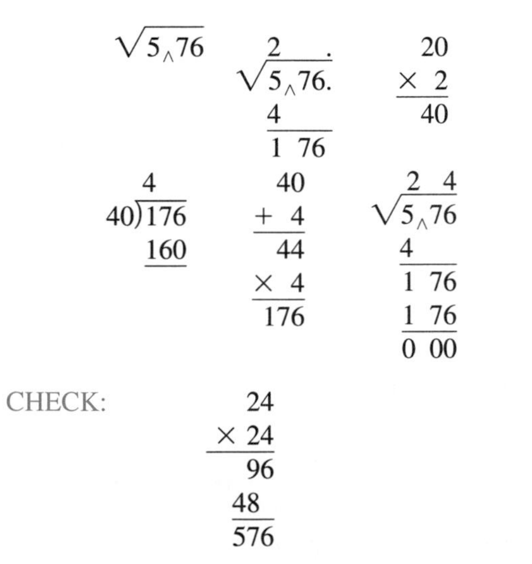

The same procedure is followed for quantities that are not perfect squares; however, the process becomes more complex. Unless the quantity under the radical is a perfect square, the result will never come out even.

Square-root skills are necessary to determine unknown distances. For example, carpenters use square root to find the length of rafters. The formula for finding the length of a rafter is based on the following formula: The square of the hypotenuse of a right triangle is equal to the square of one side plus the square of the other side. Study Figure 5–2. The hypotenuse is equal to the square root of a side (base) squared plus the other side (altitude) squared. Study the following formulas.

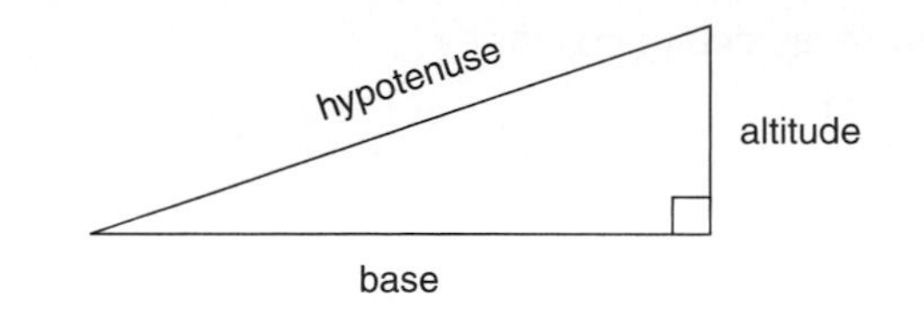

Figure 5–2

$$\text{hypotenuse} = \sqrt{\text{base}^2 + \text{altitude}^2}$$
$$\text{base} = \sqrt{\text{hypotenuse}^2 - \text{altitude}^2}$$
$$\text{altitude} = \sqrt{\text{hypotenuse}^2 - \text{base}^2}$$

These formulas are valid only for *right* triangles.

Example 5–10: Square-Root Application

PROBLEM: Find the length of a rafter to the nearest $\frac{1}{8}$ inch if the rise is 6 feet and the run is 12 feet. See Figure 5–3.

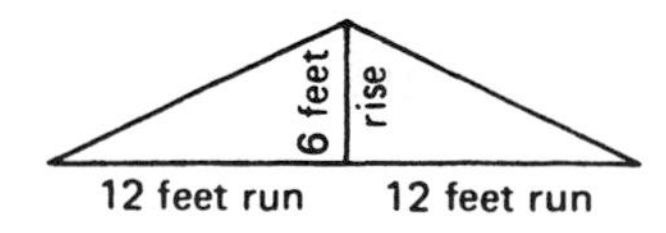

Figure 5–3

SOLUTION: To find the length of a rafter, take the square root of the square of the rise (6 feet) plus the square of the run (12 feet). Use the formulas shown.

$$\begin{aligned} \text{Rafter length} &= \sqrt{(\text{rise})^2 + (\text{run})^2} \\ &= \sqrt{6^2 + 12^2} \\ &= \sqrt{36 + 144} \\ &= \sqrt{180} \end{aligned}$$

Using a calculator, key 180 and press [√]. The display will then show 13.416408, which means 13.416408 feet.

To convert to the nearest $\frac{1}{8}$ inch, from 13.416408 feet, subtract 13 feet.

$$13.416408 - 13 = 0.416408 \text{ ft}$$

Then multiply 0.416408 by 12 inches (12 inches = 1 foot).

$$0.416408 \text{ ft} \times 12 \text{ in./ft} = 4.9968943 \text{ in.}$$

From 4.9968943 inches, subtract 4 inches.

$$4.9968943 \text{ in.} - 4 \text{ in.} = 0.9968943 \text{ in.}$$

Multiply 0.9967943 inch by $^8/_8$ to convert the decimal part of an inch to eighths.

$$0.9968943 \times \frac{8}{8} = \frac{7.9751546}{8}$$

This rounds to $^8/_8$ inch or 1 inch; thus the 4.9968943 becomes 5 inches to the nearest $^1/_8$ inch. The rafter would be 13 feet 5 inches.

EXERCISE 5–3 SQUARE ROOTS AND SQUARE ROOT APPLICATIONS

Use a calculator to solve the following. Where appropriate, calculate the roots to the same number of digits as in the original number.

5–41. $\sqrt{121}$

5–42. $\sqrt{256}$

5–43. $\sqrt{1024}$

5–44. $\sqrt{3136}$

5–45. $\sqrt{5476}$

5–46. $\sqrt{708{,}964}$

5–47. $\sqrt{0.0576}$

5–48. $\sqrt{38.4400}$

Calculate the square root to the nearest hundredth.

5–49. $\sqrt{0.6561}$

5–50. $\sqrt{0.002116}$

5–51. $\sqrt{38}$

5–52. $\sqrt{149}$

5–53. $\sqrt{85}$

5–54. $\sqrt{5280}$

5–55. $\sqrt{0.1234}$

5–56. $\sqrt{10{,}000}$

5–57. $\sqrt{30.603024}$

5–58. $\sqrt{0.60516}$

5–59. $\sqrt{1989}$

5–60. $\sqrt{91.550451}$

5–61. Find the length of the diagonal (hypotenuse) of a 24-meter square to the nearest tenth.

5–62. Find the length of the diagonal (hypotenuse) of a 14.5-centimeter square to the nearest hundredth.

5–63. Find the length of the diagonal of the rectangle in Figure 5–4 to the nearest tenth.

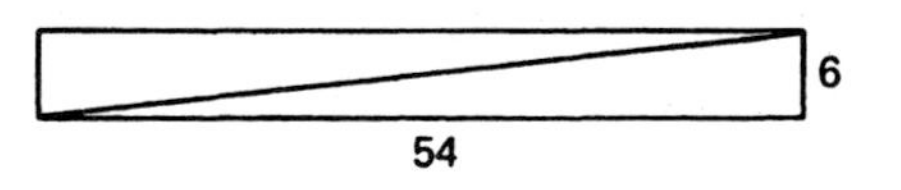

Figure 5–4

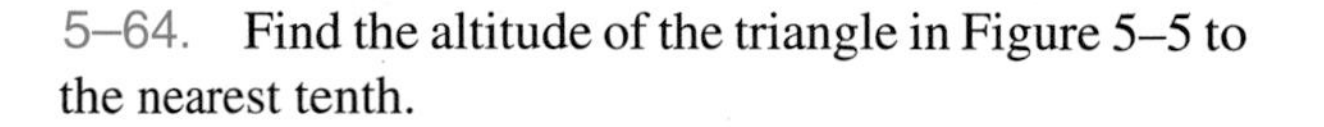

5–64. Find the altitude of the triangle in Figure 5–5 to the nearest tenth.

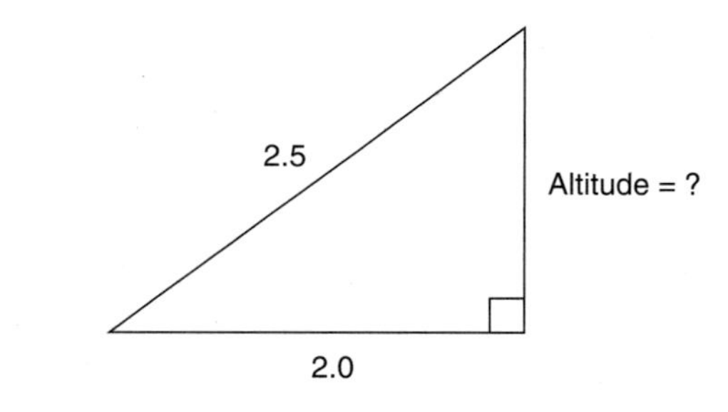

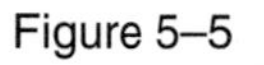

Figure 5–5

5–65. Find the hypotenuse of the triangle in Figure 5–6 to the nearest thousandth.

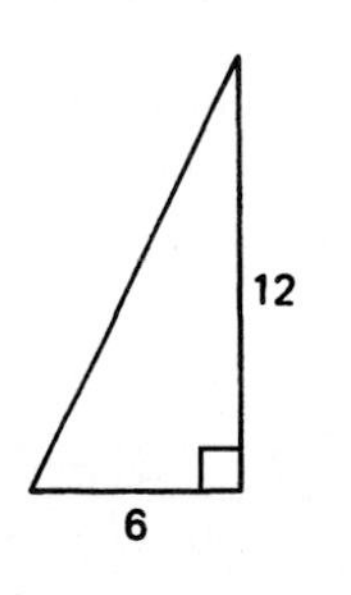

Figure 5–6

5–66. Find the altitude of a right triangle whose hypotenuse is 20 and whose base is 16.

5–67. Find the hypotenuse of the triangle in Figure 5–7 to the nearest tenth.

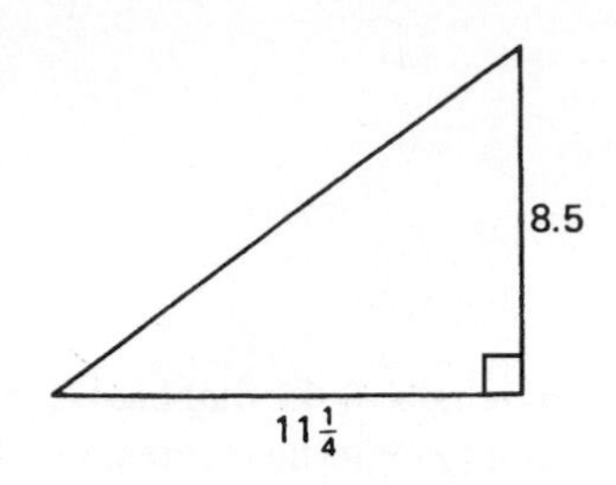

Figure 5–7

5–68. Find the rise if the rafter is 37.1 feet and the run is 36 feet to the nearest tenth of a foot.

5–69. Find the length of the rafter for the roof in Figure 5–8 to the nearest hundredth.

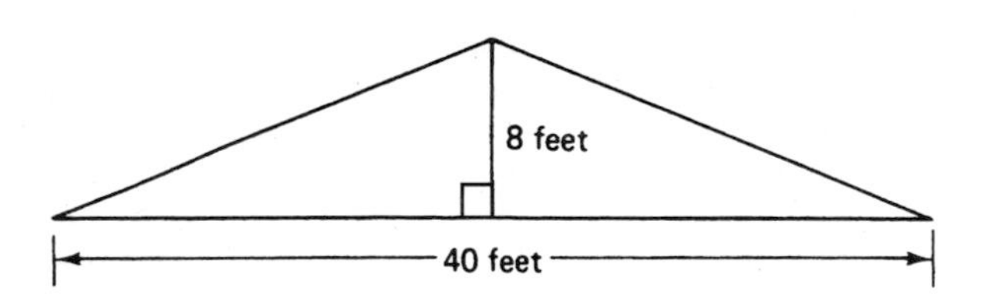

Figure 5–8

5–70. Good proportioning in printing requires the diagonal be two times the width of the page. To the nearest hundredth, find the length of the page if the width is 24 centimeters.

5.4 ORDER OF OPERATIONS AGREEMENT

Occasionally, several arithmetic operations are indicated in one numerical expression. The way a mechanic replaces tools in a tool box does not matter, but the way the mechanic replaces the parts of an engine does matter.

To avoid any confussion in the evaluation of a mathematical expression, mathematicians throughout the world have adopted a standard order of operation for arithmetic calculations.

The rules are as follows:

1. Do all operations within the signs of grouping first: parentheses ( ), brackets [ ], or braces { } from left to right.
2. Evaluate all quantities with exponents.
3. Do all multiplications and divisions from left to right.
4. Do all additions and subtractions from left to right.

Rules for the Order of Operations

A way to remember the order of operations is to remember: "Please excuse my dear Aunt Sally." This may sound strange, but the connections are as follows: *P* for parentheses from "please"; *E* for exponents from "excuse"; *M* for multiply from "my"; *D* for divide from "dear"; *A* for add from "Aunt"; and *S* for subtract from "Sally."

Thus

P—parentheses

E—exponents

M and *D*—multiply and divide from left to right

A and *S*—add and subtract from left to right

Example 5–11: Order of Operations ($\times$, $\div$, $+$)

PROBLEM: Evaluate $24 \times 3 \div 12 + 8$.

SOLUTION: Remember: *P, E, M, D, A, S.* Since there are no signs of grouping or quantities with exponents, multiply, divide, and add from left to right. Evaluate

$$\underbrace{24 \times 3}\div 12 + 8 =$$
$$\underbrace{72 \quad \div 12} + 8 =$$
$$6 \quad + 8 = 14$$

Example 5–12: Order of Operations (exponents, $\times$, $+$, $-$)

PROBLEM: Simplify $5^2 + 6 - 3 \times 5$.

SOLUTION: Since there are no signs of grouping, evaluate the quantity with the exponent, then multiply, and finally add and subtract from left to right.

$$\begin{aligned} &\underbrace{5^2} + 6 - 3 \times 5 = \\ &25 + 6 - \underbrace{3 \times 5} = \\ &\underbrace{25 + 6} - 15 \quad = \\ &31 - \quad 15 \quad = 16 \end{aligned}$$

Order of Operation (parentheses, exponents, ×, ÷, +, −)

PROBLEM: Simplify $2(40 \div 4) \div 5 + 8^2 - 30[11 - 9]$.

SOLUTION: Do the operation inside the signs of grouping first from left to right. Then evaluate the exponents, multiply, divide, add, and subtract.

$$\begin{array}{lllll}
P & 2(\underline{40 \div 4}) \div 5 + 8^2 & - 30(\underline{11 - 9}) & = & \\
 & 2(10) \div 5 + \underline{8^2} & - 30(2) & = & \\
E & 2(10) \div 5 + 64 & - \underline{30(2)} & = & \\
M & \underline{20 \div 5} \quad + 64 & - 60 & = & \\
D & \underline{4 + 64} & - 60 & = & \\
A & \underline{68} & \underline{- 60} & = & \\
S & \quad 8 & & = & 8
\end{array}$$

An applied example of order of operations is finding the perimeter of a rectangle when the dimensions are given in feet and the answer is needed in inches.

Example 5–14: Applied Order of Operations (× and +)

PROBLEM: Find the number of inches of weather stripping needed for a $2\frac{1}{2}$-foot by $6\frac{2}{3}$-foot door.

SOLUTION: Note the weather stripping would be on the top and two sides, and there are 12 inches in a foot. Thus

$$2\left(6\frac{2}{3} \times 12 \text{ in.}\right) + 2\frac{1}{2}(12 \text{ in.}) =$$

$$2\left(\frac{20}{\cancel{3}} \times \frac{\overset{4}{\cancel{12}} \text{ in.}}{1}\right) + \frac{5}{\underset{1}{\cancel{2}}}\left(\frac{\overset{6}{\cancel{12}} \text{ in.}}{1}\right) =$$

$$2(80 \text{ in.}) + 30 \text{ in.} =$$

$$160 \text{ in.} + 30 \text{ in.} = 190 \text{ in.}$$

Example 5–15: Applied Order of Operations

PROBLEM: Find the square feet of sheet rock needed for a 16 by $8\frac{1}{2}$ foot wall if there are three windows and each window is 48-by-24 inches.

SOLUTION: Find the total square feet of the wall, and subtract the area of the windows. The area is found by multiplying the length times the width.

Thus

$$\left(16 \text{ ft} \times 8\frac{1}{2} \text{ ft}\right) - 3\left(\frac{48 \text{ in.}}{12} \times \frac{24 \text{ in.}}{12}\right) =$$

$$(16 \text{ ft} \times 8.5 \text{ ft}) - 3(4 \text{ ft} \times 2 \text{ ft}) =$$

$$136 \text{ sq ft} - 3(8 \text{ sq ft}) =$$

$$136 \text{ sq ft} - 24 \text{ sq ft} = 112 \text{ sq ft}$$

EXERCISE 5–4 ORDER OF OPERATIONS

Solve the following order of operation problems. Remember: "Please excuse my dear Aunt Sally" (*PEMDAS*).

5–71. $5 - 3 + 6$

5–72. $2 + 8 \times 6$

5–73. $48 \div 8 - 2$

5–74. $\dfrac{36 - 27}{9 - 6}$

5–75. $32 - (65 - 33)$

5–76. $6^2 - 30 + 2$

5–77. $\frac{80}{16} - \frac{90}{18}$

5–78. $3^3 - 3^2 - 3$

5–79. $5^2 - 18 - (3^2 - 2)$

5–80. $(100 \div 10) \div 5$

5–81. $4^2 \times 4 + 2 - 18(5^2 - 2^4)$

5–82. $5^3 + 36 \div 3^2 - 2^4 - 3^2$

5–83. $3(1000 \div 10 \div 10) - 2[500 \div 50]$

5–84. $200 \div (15 \div 3) + [144 \div 72]$

5–85. $4^2 + 39 \div 13 - [5]^2 + 5^3$

5–86. $6(20 - 2^4) \div (2^5 + 2^4)$

5–87. $\frac{24 + 12}{6 + 12} + \frac{48}{12} - 12^2 \div 32$

5–88. Find the square yards of carpeting required for a room 24 feet by 33 feet. (Area is found by multiplying the length times the width.)

5–89. Compare and place the correct symbol (=, >, <).

$(32 - 16) - 11 \stackrel{?}{\_} 32 - (16 - 11)$

5–90. How many cubic yards of concrete are needed for a driveway that is 4 inches high, 15 feet wide, and 30 yards long?

Note: Convert all measurements to yards, and then multiply the length, width, and height.

THINK TIME

Many different methods can be used to approximate the square root of a number. Study the approximation procedure shown. A basic function calculator will be very helpful in finding the approximation.

PROBLEM: Find the approximate square root of 58.

SOLUTION: Recall the perfect squares smaller and larger than 58.

$$7 \times 7 = 49 \text{ and } 8 \times 8 = 64$$

Make 7 the first approximation, $A_1 = 7$. Divide 58 by 7, and add 7; then find the average as shown. This result will give the second approximation, A_2.

$$A_2 = \frac{\frac{58}{7} + 7}{2} = \frac{8.2857143 + 7}{2}$$

$$= \frac{15.285714}{2} = 7.6428571$$

Now divide 58 by A_2, the second approximation, add A_2, and find their average. This will give the third approximation, A_3.

$$A_3 = \frac{\frac{58}{7.6428571} + 7.6428571}{2}$$

$$= \frac{15.2311642}{2} = 7.6158244$$

Note: $(7.6158211)^2 = 58.000731$; thus the approximation is accurate to the nearest thousandth. If your first approximation is near the correct value, the second approximation will be reasonably accurate. This procedure may be used to find square roots.

PROCEDURES TO REMEMBER

1. To determine the scientific notation:
   (a) To convert a number to scientific notation, write the first digit of the number, place a decimal point, add the digits that follow, and then write the power of 10 according to the number of places the decimal point needs to be moved to achieve the original number. *Example:* $52{,}000 = 5.2 \times 10^4$.
   (b) To convert a power-of-10 quantity to a number, write the number, and move the decimal point according to the place of the power—to the right for positive powers and to the left for negative powers. *Examples:* $6.53 \times 10^4 = 65{,}300$; $7.84 \times 10^{-3} = 0.00784$.
2. To evaluate quantities with exponents
   (a) Longhand—use the base quantity as a factor as many times as the exponent indicates and multiply

   $$4^3 = 4 \times 4 \times 4 = 64$$

   (b) With a calculator—use the [x^y] key for 4^3. Press [4], then press [x^y], then [3] and [=]; the display should read 64.
3. To calculate the square root of a number:
   (a) Place the number under the radical symbol.
   (b) Divide the digits of the number into groups of two, starting from the decimal point and moving in both directions as necessary.
   (c) Begin with the group on the far left; find the largest perfect square that is equal to or less than the number in the group.
   (d) Place this root number above the group and above the radical symbol.
   (e) Multiply this root number by itself and place the result below the first group on the left.
   (f) Subtract, and bring down the next group to the right of the first group.
   (g) Multiply 20 by the number above the radical symbol.
   (h) Divide this approximate number into the dividend created in step (f). This will give you an approximate quotient.
   (i) Place this number above the second group above the radical symbol.
   (j) Multiply this group by the last digit of the number, and enter it under the number created in step (f).
   (k) Repeat steps (f) through (j) until the square root has been found or satisfactorily approximated by adding zeros to form groups of two as necessary to the right of the numbers to the right of the decimal point.
4. To evaluate expressions with several arithmetic operations:

Rules for the Order of Operations

1. Do all operations within the signs of grouping: parentheses ( ), brackets [ ], and braces { } from left to right.
2. Evaluate all quantities with exponents from left to right.
3. Do all multiplications and divisions from left to right.
4. Do all additions and subtractions from left to right.

CHAPTER SUMMARY

1. *Scientific notation* is the procedure of rewriting a number as a single-digit number, placing the decimal point, adding the remaining digits (generally limited to three digits), and indicating the appropriate power of 10—right for the positive powers and left for negative powers.
2. To calculate the *square root* of a number means to determine what number, multiplied by itself, will equal the original number.

3. *Factors* are numbers that are multiplied together to produce the product.
4. An *exponent* is a number written above and to the right of another number. It indicates how many times that number is used as a factor.
5. To *square* a number means to multiply the number by itself.
6. To *cube* a number means that the number is used as a factor three times.
7. When a factor is used three or more times, it is called a *power.*
8. The *radical* symbol, $\sqrt[3]{78}$, indicates that the cube root is to be determined.
9. The *index of the root,* $\sqrt[3]{\ }$, is located inside the radical sign.
10. When no index number is given, square root is understood.
11. The number under the radical, $\sqrt{78}$, is called the *radicand.* Thus 78 is the radicand.
12. Numbers that have even square roots are called *perfect squares.*
13. The diagonal of a square is equal to $\sqrt{2}$ times the length of a side.
14. The hypotenuse of a right triangle equals the square root of the square of the base plus the square of the altitude.
15. The base of the right triangle is equal to the square root of the square of the hypotenuse minus the square of the altitude.
16. The altitude of a right triangle is equal to the square root of the hypotenuse minus the square of the base.
17. Order of Operations procedure:

    1. Perform operations inside signs of grouping.
    2. Simplify any expressions with exponents.
    3. Multiply and divide from left to right.
    4. Add or subtract from left to right.

18. Order of Operation (all from left to right):

    "Please—parentheses
    Excuse—exponents
    My—multiply
    Dear—divide
    Aunt—add
    Sally"—subtract

CHAPTER TEST

T–5–1. Write 9000 as a power of 10.

T–5–2. Write 789,000,000 as a power of 10.

T–5–3. Convert 8×10^6 to a number.

T–5–4. Convert 4.2×10^5 to a number.

T–5–5. Convert 0.000000056×10^9 to a number.

T–5–6. Write 0.005 as a power of 10.

T–5–7. Convert 5.63×10^{-2} to a number.

T–5–8. Convert 357×10^{-3} to a number.

T–5–9. Evaluate $(23.2)^2$.

T–5–10. Evaluate $(7)^3$.

T–5–11. Find the square of 1.4142 to the nearest hundredth.

T–5–12. Evaluate $(4)^4$.

T–5–13. Calculate the square root of 141,376.

T–5–14. Calculate the square root of 0.22373 to the nearest thousandth.

T–5–15. Calculate the square root of 1609 to the nearest hundredth.

T–5–16. Find the diagonal of an 8-by-10-inch rectangle to the nearest tenth.

T–5–17. Find the hypotenuse of a right triangle if one side is 12.5 units and the other side is 16.4 units (to the nearest tenth).

T–5–18. Find the cube root of 216.

T–5–19. Find the length of the side of a square if the diagonal is 10.6066 (to the nearest tenth).

T–5–20. Calculate the longest line that could be drawn on a basketball court that is 88 feet by 50 feet to the nearest hundredth foot.

T–5–21. Evaluate $2 \times (6 + 4 \times 9)$.

T–5–22. Evaluate $800 \div 32 - 16 + 3^4$.

T–5–23. Evaluate $7 + 5^3 \times (5 - 3)$.

T–5–24. Evaluate $5^3 - 4^3 - 3^3$.

T–5–25. Find the cost of 2 bottles of \$1.99 soda and 3 bags of \$1.89 chips if you have \$.80 coupons for the soda and \$.55 coupons for the chips.

6

BASIC MEASUREMENT: U.S./METRIC

OBJECTIVES

After studying this chapter, you will be able to:

6.1. Understand ruler measurement: U.S./metric
6.2. Understand basic units of measurement and conversion: U.S./metric
6.3. Understand arithmetic operations with units of measurement: U.S./metric
6.4. Understand conversions and applications: U.S./metric and metric/U.S.

SELF-TEST

This test will determine your need to develop your skills with basic measurement. If you are able to understand the definitions, perform the measurements, and do the problems, move on to Chapter 7, Data, Graphs, and Basic Statistics.

Identify the following lettered points on the given rulers in Figures 6–1 and 6–2.

S–6–1. W = ________

S–6–2. X = ________

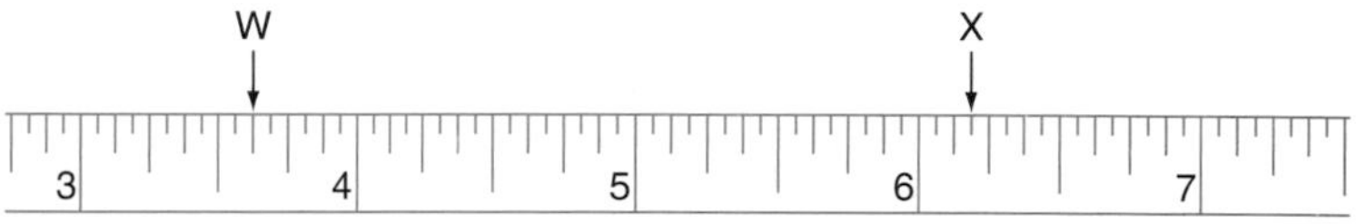

W X
3 4 5 6 7

Figure 6–1

S–6–3. Y = ________

S–6–4. Z = ________

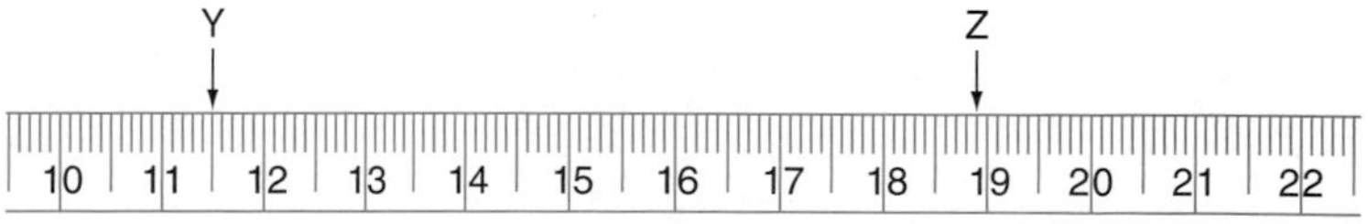

Y Z
10 11 12 13 14 15 16 17 18 19 20 21 22

Figure 6–2

S–6–5. Convert 72 inches to feet.

S–6–6. Convert 2.75 tons to pounds.

S–6–7. Convert 2 gallons to fluid ounces.

S–6–8. Convert 5.4 kilometers to meters.

S–6–9. Convert 240 milliliters to liters.

S–6–10. Convert 5280 meters to kilometers.

S–6–11. Convert 3.98 grams to milligrams.

S–6–12. Find 4 feet minus 28 inches.

S–6–13. Find 5 gallons 2 quarts divided by 4.

S–6–14. Add

	18 meters 34 centimeters
+	5 meters 89 centimeters

S–6–15. Multiply 54 inches by 5, and convert to feet and inches.

S–6–16. Multiply $4\frac{1}{2}$ pints by 20, and convert to quarts.

S–6–17. Divide 94 kilometers 1400 meters by 18.

S–6–18. Multiply 355 milliliters by 24, and convert to liters.

S–6–19. A swimming pool uses 450 milliliters of chlorine per day. How many liters will be used in 30 days?

S–6–20. Find the cost of a 4-pound 12-ounce sirloin steak that costs $6.95 per pound.

S-6-21. A 1-pound 6-ounce box of Cap'n Crunch® costs $3.97. Find the cost per ounce.

S-6-22. How many gallons of Diet Pepsi are in a 24-can pack if each can contains 12 fluid ounces?

S-6-23. A person walks 1 kilometer 800 meters in $\frac{1}{2}$ hour. How many miles (to the nearest tenth) could she walk in 2 hours?

S-6-24. A plane flies from New York to San Francisco, a distance of 2560 air miles, in 5 hours. How many kilometers per hour is this?

S-6-25. How many gallons of oil are needed per year for a car if the oil is changed every three months and 5 quarts are used for each change?

6.1 RULER MEASUREMENT: U.S. AND METRIC

Measuring anything is comparing the object or quantity being measured to a previously established standard unit. *Linear measurement* means finding the distance between two points and attaching a unit of length.

When we count the number of objects, we get an *exact number.* However, when we measure, we get an *approximate number.* Measurement is approximate because there could always be a more accurate measuring device.

Nearly every occupation requires some form of measurement. For example: the amount of flour, the length of a field goal, the size of a window, the length of a windshield wiper, the stopping distance of a car, gallons of gasoline, and so on. Whenever measurement is done, common sense should prevail.

The length of something is given, to the nearest division, on the scale of the measuring instrument. The accuracy of the measurement will always depend on the scale of the measurement device.

Thus an appropriate measurement instrument should always be selected. For example, one would not measure the size of a car's gas tank in fluid ounces, the distance home with a 12-inch ruler, or the amount of cooking oil for a cake with a gallon container.

The following procedures are suggested for measuring with common sense and reason, which are the most important factors.

Procedures of Measurement

1. Select an appropriate measuring device.
2. Study the scale of the measuring tool by counting the divisions between each of the spaces shown and determine the value of each division.
3. Find the value of all the subdivisions until you know the value of the smallest subdivision.
4. Read the value of the quantity being measured, "rounding" to the nearest subdivision.

The *precision* of a measurement depends on the instrument used to do the measuring. The precision of the instrument is determined by the smallest subdivision on the scale.

The longest lines on the 6-inch ruler in Figure 6–3 indicate 1-foot divisions. In order of decreasing length, the other lines indicate $\frac{1}{2}$-inch, $\frac{1}{4}$-inch, $\frac{1}{8}$-inch and $\frac{1}{16}$-inch divisions.

The longest lines on the metric ruler in Figure 6–4 indicate centimeters (abbreviated as *cm*), and the shortest lines indicate millimeters (abbreviated as *mm*). The medium-length lines indicate 5-millimeter divisions. The ruler shows that 10 millimeters equals 1 centimeter, and

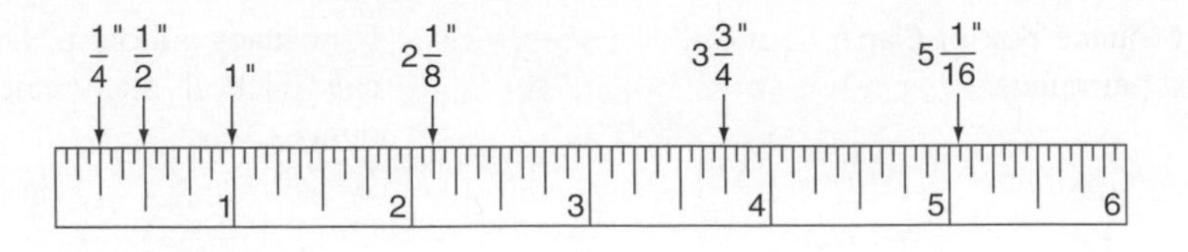

Figure 6–3

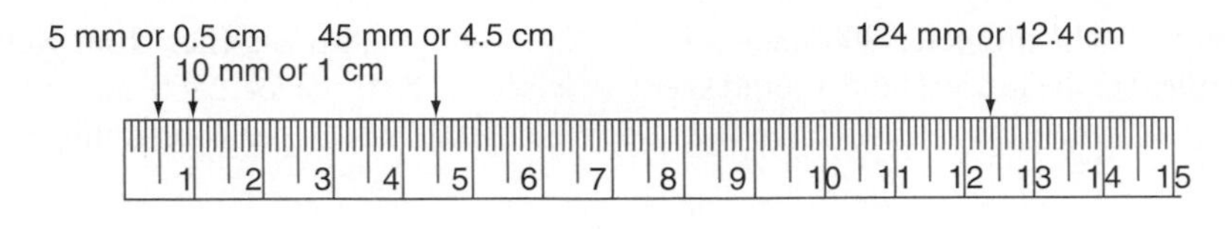

Figure 6–4

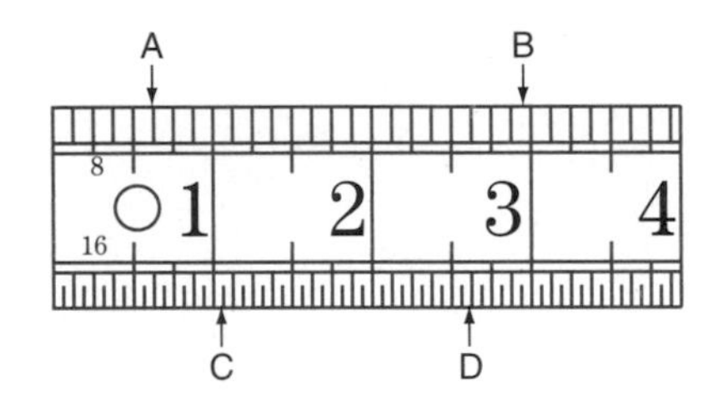

Figure 6–5

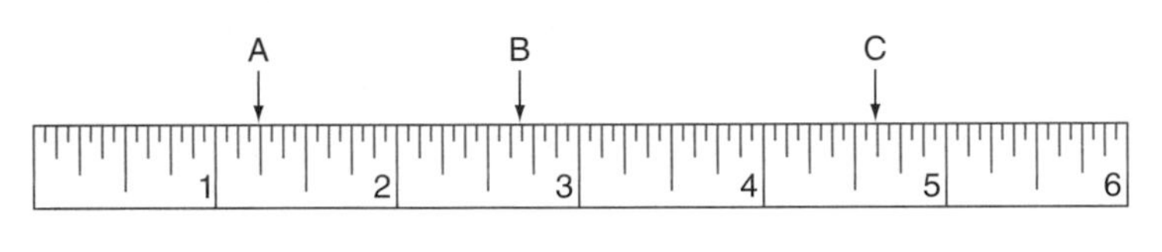

Figure 6–6

5 millimeters equals 0.5 centimeter. Converting from one unit of measure to another is easy in the metric system.

The measurement process is shown in the following example. Given the inch-scale ruler (see Figure 6–5), read the values of the letters from A to D, to the nearest 8th of an inch.

When reading a ruler, observe the scale of the ruler. Note the "8" in the upper left-hand corner, which means the top scale is divided into 8ths.

In the lower left-hand corner is "16," which indicates that the bottom scale is divided into 16ths. This is a unique ruler with the given scales.

If the scale is not shown on the ruler, the scale is determined by number of subdivisions between each major division. Thus to determine the value of "A," note the upper scale is divided into 8ths, count 5 spaces from the left to right, and read "A" as $^5/_8$ inch.

For the letter "B," since the upper portion of the ruler is divided into 8ths and letter "B" is more than half the distance between $2\,^7/_8$ and 3, read letter "B" as 3 inches.

Concerning the letter "C," since the lower portion of the ruler is divided into 16ths, and letter "C" is one subdivision beyond 1 inch, the letter "C" is read as $1\,^1/_{16}$ inch.

The letter "D" is between 2 and 3 inches and is 10 subdivisions beyond 2; thus it is read as $2\,^{10}/_{16}$ inches, which is the same as $2\,^5/_8$ inches.

The following examples show how to read U.S. and metric measure.

Example 6–1: Read a U.S. Measurement Ruler

PROBLEM: Read the measure of "A," "B," and "C" on the given ruler (see Figure 6–6).

SOLUTION: Observe that "A" is at a place that divides the inch into 4 equal parts; it is the first such division after the 1-inch mark, and is read as $1\frac{1}{4}$ inches.

Letter "B" is at a place that divides the inch into 16 equal parts; it is the 11th such division after the 2-inch mark, and is read as $2\frac{11}{16}$ inches.

Letter "C" is at a place that divides the inch into 8 equal parts; it is the 5th such division after the 4-inch mark, and is read as $4\frac{5}{8}$ inches.

Example 6–2: Read a Metric Ruler

PROBLEM: Read the measure of "E," "F," and "G" on the 20-centimeter ruler (see Figure 6–7).

SOLUTION: Observe that "E" is at the 2-centimeter or 20-millimeter place, and is read as 2 centimeters or 20 millimeters.

The letter "F" is at a place that divides the centimeter into 10 equal parts; it is the 5th such division after the 6-centimeter mark, and is read as 6.5 centimeters or 65 millimeters.

Letter "G" is at a place that divides the centimeter into 10 equal parts; it is the third such division after the 9-centimeter mark, and is read as 9.3 centimeters or 93 millimeters.

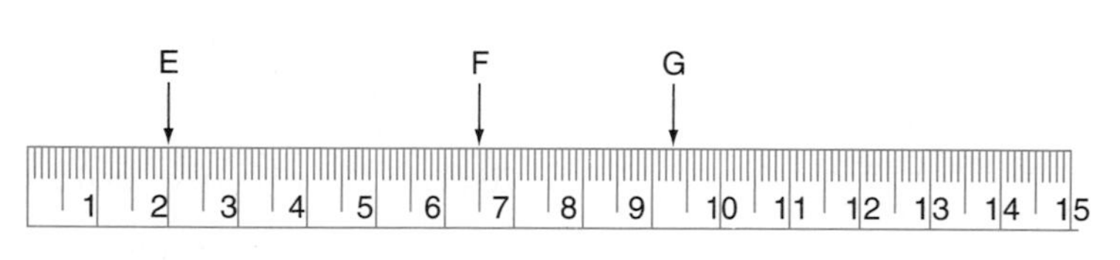

Figure 6–7

EXERCISE 6–1 U.S. AND METRIC MEASURE

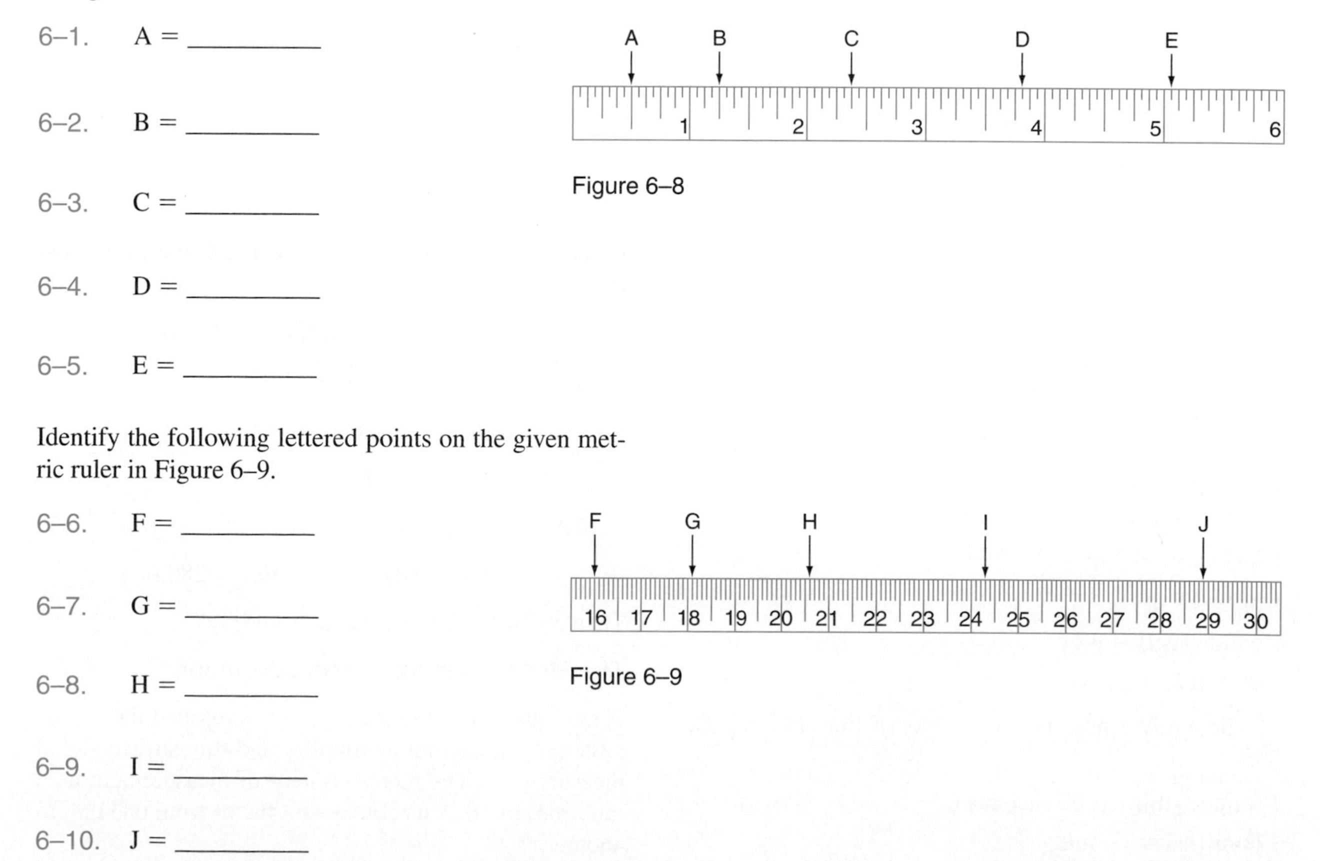

Identify the following lettered points on the 6-inch ruler in Figure 6–8.

6–1. A = ________

6–2. B = ________

6–3. C = ________

6–4. D = ________

6–5. E = ________

Figure 6–8

Identify the following lettered points on the given metric ruler in Figure 6–9.

6–6. F = ________

6–7. G = ________

6–8. H = ________

6–9. I = ________

6–10. J = ________

Figure 6–9

U.S. System of Measurement

Anything that cannot be counted must be measured. Therefore, standards have been established for various types of measures according to their size and required accuracy.

The "scales of measurement" used in the United States originated in England. Previously they were known as the English System, but they are now known as the U.S. System, since England converted to the metric system in the 1960s.

The common units of measurement of length in the U.S. System are the inch, foot, yard, and mile. There are other units, which are used for special measurement. For example, in land measurement the "rod" is the unit of measurement. A rod is $16\frac{1}{2}$ feet long. In horse racing the "furlong" is used. A furlong is 660 feet long. At sea, the "knot" is used as a measure. A knot is 6080 feet long. However, the most common measurements in the U.S. System are inches, feet, yards, and miles.

The equivalences between units of length in the U.S. System are:

1 foot (ft) = 12 inches (in.)

1 in. = $0.08\overline{3}$ ft

1 yard (yd) = 3 ft = 36 in.

1 ft = $0.\overline{3}$ yd

1 mile (mi) = 5280 ft

1 ft = 0.000189 mi

The equivalences between units of weight in the U.S. System are:

1 pound (lb) = 16 ounces (oz)

1 oz = 0.0625 lb

1 ton (t) = 2000 lb

1 lb = 0.0005 t

The equivalences between units of volume in the U.S. System are:

1 cup (c) = 8 fl oz

1 fl oz = 0.125 c

1 pint (pt) = 16 fl oz

1 fl oz = 0.0625 pt

1 quart (qt) = 2 pt

1 pt = 0.5 qt

1 gallon (gal) = 4 qt

1 qt = 0.25 gal

The equivalences between units of time in the U.S. System are:

1 minute (min) = 60 seconds (s)

1 hour (hr) = 60 min

1 day (d) = 24 hr

1 week (wk) = 7 d

These equivalences can be used to convert from one unit to another. To change from one unit to another, replace the unit with the appropriate equivalent measure and multiply. The following examples will show the procedure for converting from one unit to another.

Example 6–3: Converting Units Within the U.S. System

Convert feet to inches.

$$6 \text{ ft} = 6(\text{ft}) = 6(12 \text{ in.}) = 72 \text{ in.}$$

Replace 1 ft with 12 in. and multiply.

Convert pounds to ounces.

$$3.5 \text{ lb} = 3.5(\text{lb}) = 3.5(16 \text{ oz}) = 56 \text{ oz}$$

Replace 1 lb with 16 oz and multiply.

Convert gallons to quarts.

$$8.75 \text{ gal} = 8.75(\text{gal}) = 8.75(4 \text{ qt}) = 35 \text{ qt}$$

Replace 1 gal = 4 qt and multiply.

Convert minutes to hours.

$$480 \text{ min} = 480(\text{min}) = 480(0.01\overline{6} \text{ hr}) = 8 \text{ hr}$$

Replace 1 min = $0.01\overline{6}$ hr and multiply.

For some conversions, it may be necessary to do more than one conversion.

Convert quarts to fluid ounces.

$$4 \text{ qt} = 4(\text{qt}) = 4(2 \text{ pt}) = 8 \text{ pt}$$

Replace 1 qt = 2 pt and multiply. Then convert pints to fluid ounces.

$$8 \text{ pt} = 8(\text{pt}) = 8(16 \text{ fl oz}) = 128 \text{ fl oz}$$

Replace 1 pt = 16 fl oz and multiply.

Convert 8 yards to inches.

$$8 \text{ yd} = 8(\text{yd}) = 8(3 \text{ ft}) = 24 \text{ ft}$$

Replace 1 yd = 3 ft and multiply.

$$24 \text{ ft} = 24(\text{ft}) = 24(12 \text{ in.}) = 288 \text{ in.}$$

Then replace 1 ft = 12 in. and multiply.

The Metric System of Measurement

A commission in France in 1789 developed the metric system in an attempt to simplify and standardize global measurement. The metric system of measurement uses multiples of 10 as its conversion factor from one unit to another.

Prefixes from Greek and Latin are used to denote basic units. The same prefixes are used for length, weight, and volume. The prefixes are:

kilo = 1000	1 kilometer (km) = 1000 meters (m)
hecto = 100	1 hectometer (hm) = 100 meters (m)
deca = 10	1 decameter (dam) = 10 meters (m)
	1 meter (m) = 1 m
deci = 0.1	1 decimeter (dm) = 0.1 m
centi = 0.01	1 centimeter (cm) = 0.01 m
milli = 0.001	1 millimeter (mm) = 0.001 m

The most commonly used prefixes are *kilo, centi,* and *milli.*

The basic unit of length in the metric system is the *meter.* One meter is one ten-millionth of the distance from the North Pole to the equator. The equivalences between units of length in the metric system are:

- A meter is a little more than a yard or the distance from a door knob to the floor.
- The thickness of a dime is about a millimeter.
- The basic unit of weight (mass) in the metric system is the gram.

The equivalences between units of weight in the metric system are:

1 kilogram (kg) = 1000 grams (g)
1 hectogram (hg) = 100 g
1 decagram (dag) = 10 g
1 gram = 1 g
1 decigram (dg) = 0.1 g
1 centigram (cg) = 0.01 g
1 milligram (mg) = 0.001 g

The basic unit of volume (capacity) in the metric system is the liter. The equivalences between units of volume in the metric system are:

1 kiloliter (kl) = 1000 liters (L)
1 hectoliter (hl) = 100 L
1 decaliter (dal) = 10 L
1 liter = 1 L
1 deciliter (dl) = 0.1 L
1 centiliter (cl) = 0.01 L
1 milliliter (ml) = 0.001 L

Note that when only the liter symbol is used, a cap "L" is used. However, this may vary with different authors.

These equivalences can be used to convert from one unit to another. To change from one unit to another, replace the unit with the appropriate equivalent measure and multiply. In Chapter 17, the "decimal shift" method of conversion will be demonstrated. The following examples will show the procedure for converting from one unit to another.

Example 6–4: Converting Units within the Metric System

Convert kilometers to meters.

78 km = 78(km) = 78(1000 m) = 78,000 m

Replace 1 km = 1000 m and multiply.

Convert grams to kilograms.

5280 g = 5280(g) = 5280(0.001 kg) = 5.28 kg

Replace 1 g = 0.001 kg and multiply.

Convert centimeters to meters.

3200 cm = 3200(cm) = 3200(0.01 m) = 32 m

Replace 100 cm = 1 m and multiply.

Convert milliliters to liters.

8870 ml = 8870(ml) = 8870(0.001 L) = 8.87 L

Replace 1 ml = 0.001 L and multiply.

Convert centigrams to grams.

656 cg = 656(cg) = 656 (0.01 g) = 6.56 g

Replace 1 cg = 0.01 g and multiply.

EXERCISE 6–2: CONVERSIONS—U.S. AND METRIC

6–11. Convert 36 feet to yards.

6–12. Convert 60 inches to feet.

6–13. Convert 5.5 tons to pounds.

6–14. Convert 62 feet to yards.

6–15. Convert 12.5 gallons to quarts.

6–16. Convert 18480 feet to miles.

6–17. Convert 5.75 pounds to ounces.

6–18. Convert 38 quarts to gallons.

6–19. Convert 2.5 gallons to fluid ounces.

6–20. Convert 3960 feet to miles.

6–21. Convert 850 centimeters to meters.

6–22. Convert 7850 meters to kilometers.

6–23. Convert 32 kilometers 528 meters to kilometers.

6–24. Convert 1.52 kilograms to grams.

6–25. Convert 588 milliliters to liters.

6–26. 7.5 L = __________ ml

6–27. 10,000 mm = __________ m

6–28. 50,000 cm = __________ km

6–29. 0.85 metric ton = __________ kg

6–30. 0.25 m = __________ cm

6.3 ARITHMETIC OPERATIONS WITH UNITS OF MEASUREMENT

A measurement includes a number and a unit. A number with a unit attached is called a *denominate number.* Some measurements include more than one unit of measure; for example, 5 feet 7 inches, 8 pounds 10 ounces, and 8 hours 30 minutes.

It is important to represent numbers in their appropriate and accepted form. For example, we would not say a person is 80 inches tall, but we would convert this measurement to a denominate number which is a number with one or more units. Since we speak of a person's height in feet and inches, convert 80 inches into feet and inches. Thus to convert 80 inches into feet and inches, divide 80 inches by 12, since 12 in. = 1 ft.

$$\begin{array}{r@{\,}l} & \text{6 ft} \quad \text{8 in.} \\ 12 \big) & \overline{\text{80 in.}} \\ & \underline{\text{72 in.}} \\ & \text{8 in.} \end{array}$$

Note: 12 in. = 1 ft and 12 × 7 = 72, 80 in. − 72 in. = 8 in. Thus the person is 6 feet 8 inches.

Also, we would not say that we drive 44,880 feet to work, but we would convert to miles by dividing 44,880 feet by 5280 feet, since there are 5280 feet in a mile. Thus 44,880 ft = 8.5 mi.

It is important to select the appropriate measurement according to what is measured, and, if necessary, convert it to the proper form. When performing arithmetic operations, the results may need to be converted to the accepted form of measurement.

In the metric system, when working with two or more denominate numbers, convert to the larger dimension. For example, 6 kilometers 450 meters would be written 6.45 kilometers, since 450 meters converts to 0.45 meter.

$$450 \text{ m} = 450(\text{m}) = 450(0.001 \text{ km}) = 0.45 \text{ m}$$

replacing 1 m = 0.001 km

$$5850 \text{ ml} = 5850(\text{ml}) = 5850(0.001 \text{ L}) = 5.85 \text{ L}$$

since 1 ml = 0.001 L

Example 6–5: Convert Inches to Feet

PROBLEM: Convert 27 inches to feet and inches.

SOLUTION: Divide 27 inches by 12, since 12 in. = 1 ft, and carry the remainder in inches.

$$\begin{array}{r@{\,}l} & \text{2 ft} \quad \text{3 in.} \\ 12 \big) & \overline{\text{27 in.}} \\ & \underline{-\text{24 in.}} \\ & \text{3 in.} \end{array}$$

or $\dfrac{\text{27 in.}}{12} = \text{2 ft 3 in.}$

Example 6–6: Convert Ounces to Pounds

PROBLEM: Convert 80 ounces to pounds.

SOLUTION: Divide 80 ounces by 16, since 16 oz = 1 lb.

$$\frac{80 \text{ oz}}{16 \text{ oz/lb}} = 5 \text{ lb}$$

or multiply 80 ounces by $\dfrac{1 \text{ lb}}{16 \text{ oz}}$

$$\frac{80 \text{ oz}}{1} \times \frac{1 \text{ lb}}{16 \text{ oz}} = 5 \text{ lb}$$

or multiply 80 ounces by 0.0625 since 1 oz = 0.0625 lb.

Example 6–7: Convert Quarts to Gallons

PROBLEM: Simplify 30 quarts to gallons.

SOLUTION: Divide 30 quarts by 4, since there are 4 quarts in a gallon.

$$\begin{array}{r@{\,}l} & \text{7 gal} \quad \text{2 qt} \\ 4 \big) & \overline{\text{30 qt}} \\ & \underline{\text{28 qt}} \\ & \text{2 qt} \end{array}$$

or multiply 30 quarts by $\dfrac{1 \text{ gal}}{\text{qt}}$

$$\frac{30 \text{ qt}}{1} \times \frac{1 \text{ gal}}{4 \text{ qt}} = 7.5 \text{ gal}$$

or multiply 30 quarts by 0.25 since 1 qt = 0.25 gal.

Example 6–8: Adding Units of Length—U.S. System

PROBLEM: Add the following: 16 feet 8 inches, 10 feet 2 inches, and 8 feet 9 inches.

SOLUTION: Arrange the quantities in vertical order, add the quantities, and convert to simplest form.

$$\begin{array}{rr} & \text{16 ft 8 in.} \\ & \text{10 ft 2 in.} \\ + & \text{8 ft 9 in.} \\ \hline & \text{34 ft 19 in.} \\ + & \text{1 ft 7 in.} \\ \hline \text{Thus} & \text{35 ft 7 in.} \end{array}$$

Note: 19 in. = 1 ft 7 in.

Example 6–9: Subtract Units of Time—U.S. System

PROBLEM: Find the difference between 24 hours and 9 hours 45 minutes.

SOLUTION: Place the 24 hours on top and 9 hours 45 minutes on the bottom. Borrow 1 hour or 60 minutes from 24 hours, place 60 minutes in the minute column and subtract.

$$\begin{array}{rr} 23\text{ hr} & \\ \not{2}\not{4}\text{ hr} & 60\text{ min} \\ -\ 9\text{ hr} & 45\text{ min} \\ \hline 14\text{ hr} & 15\text{ min} \end{array}$$

Example 6–10: Subtract Units of Volume—U.S. System

PROBLEM: Subtract 2 gallons 3 quarts from 8 gallons 1 quart.

SOLUTION: Arrange the quantities in appropriate columns and subtract. Since 3 quarts is more than 1 quart, borrow 4 quarts (1 gal = 4 qt) from 8 gallons, add to 1 quart and subtract.

$$\begin{array}{rr} 7\text{ gal} & 5\text{ qt} \\ \not{8}\text{ gal} & \not{1}\text{ qt} \\ -2\text{ gal} & 3\text{ qt} \\ \hline 5\text{ gal} & 2\text{ qt} \end{array}$$

Example 6–11: Divide Units of Weight—U.S. System

PROBLEM: Divide 8 pounds 13 ounces of gourmet coffee among three people.

SOLUTION: Divide 8 pounds 13 ounces by 3, making the appropriate simplifications.

$$\begin{array}{rrr} & 2\text{ lb} & 15\text{ oz} \\ 3) & 8\text{ lb} & 13\text{ oz} \\ & -6\text{ lb} & \\ & 2\text{ lb} = & +32\text{ oz} \\ & & 45\text{ oz} \end{array}$$

Note: 2 lb = 32 oz since 1 lb = 16 oz.

Example 6–12: Applied Multiplication Length—U.S. System

PROBLEM: Thirty-two 28-inch spokes are needed for a deck railing. How many feet of cedar two-by-two's are needed?

SOLUTION: Multiply 32 × 28 in. to find the total inches, and then convert to feet by dividing by 12, since 12 in. = 1 ft, 32 × 28 in. = 896 in.

$$\frac{896\text{ in.}}{12\text{ in./ft}} = 74.\overline{6} = 76\text{ ft}$$

Round up to 76 feet, since lumber is sold in multiples of 2 feet.

Example 6–13: Applied Conversion—Weight—U.S. System

PROBLEM: Find the cost of a 6-pound 12-ounce prime rib roast that is priced at $6.99 per pound.

SOLUTION: Convert the 9 ounces to pounds, add to the 6 pounds, and multiply by $6.99 per pound. Multiply $12 \times \frac{1\text{ lb}}{16\text{ oz}}$, since 1 lb = 16 oz.

$$\frac{12\text{ oz}}{1} \times \frac{1\text{ lb}}{16\text{ oz}} = 0.75\text{ lb}$$

Thus 6 lb 12 oz = 6.75 lb

Multiply $\frac{6.75\text{ lb}}{1} \times \frac{\$6.99}{1\text{ lb}} = \$47.18$ and round to the nearest cent.

Example 6–14: Convert to Simplest Form—Metric System

PROBLEM: Convert 5 liters 380 milliliters to liters.

SOLUTION: Convert 380 milliliters to liters.

$$380\text{ ml} = 380\text{ (ml)} = 380(0.001\text{ L}) = 0.38\text{ L}$$

replace 1 ml = 0.001 L

Thus 5 L 380 ml = 5.38 L

Example 6–15: Applied Multiplication/Cost—Metric System

PROBLEM: Find the shipping cost of 80 small electric motors that each weigh 148.5 grams if the shipping cost is $4.75 per kilogram.

SOLUTION: Multiply 80 × 148.5 g, convert to kilograms, and multiply by $4.75/kg.

$$80 \times 148.5\text{ g} = 11{,}880\text{ g} = 11.88\text{ kg}$$

Round up the 11.88 to 12 kilograms, since the shipping rate of shipping is $4.75 per kilogram. Thus 12 kg × $4.75/kg = $57.00.

EXERCISE 6–3 U.S. AND METRIC ARITHMETIC OPERATIONS

6–31. Convert 81 inches to feet and inches.

6–32. Convert 8500 feet to miles and feet.

6–33. Add 8 feet 8 inches, 2 feet 5 inches, and 4 feet 10 inches.

6–34. Multiply 4 pounds 8 ounces by 4.

6–35. Subtract 8 feet 9 inches from 18 feet.

6–36. Divide 5 gallons by 16.

6–37. Multiply $8\frac{1}{2}$ hours by 6; give the answer in minutes.

6–38. Convert $5\frac{1}{2}$ pints to fluid ounces.

6–39. Convert 15 cups to quarts.

6–40. Convert 24,800 pounds to tons; answer to the nearest tenth.

6–41. Convert 86 centimeters to millimeters.

6–42. 14,046 meters equals how many kilometers?

6–43. How many grams are there in 16,560 milligrams?

6–44. 8565 milliliters equals how many liters?

6–45. 8 kilograms 344 grams equals how many kilograms?

6–46. Multiply 2.5 milliliters by 1500; give the answer in liters.

6–47. Subtract 250 milliliters from 5 liters.

6–48. Divide 35 kilograms by 70; give the answer in grams.

6–49. Add 1.3 kilograms, 600 grams, and 1,000,000 grams.

6–50. Divide 350 liters 3500 milliliters by 7; give the answer in milliliters.

6–51. How many feet of weather stripping are needed for a door that is 84 inches by 36 inches, if the weather stripping is placed on the sides and top only, not on the bottom?

6–52. If four 2-foot 10-inch pieces are cut from a 12-foot board, how many feet and inches are left?

6–53. How many quarts of orange juice are in a container that has twenty-four 20-fluid-ounce bottles?

6–54. Medora carried two 3-gallon containers of water to a campsite. If water weighs $8\frac{1}{3}$ pounds per gallon, how many pounds did she carry?

6–55. A manuscript weighs 3 pounds 7 ounces; find the cost of mailing at $0.38 per ounce.

6–56. A patient should receive 2 grams of calcium per day. If each calcium tablet contains 500 milligrams, how many tablets a day should the patient take?

6–57. An aluminum can weighs 0.75 ounce. How many cans would be needed to have 3 pounds of aluminum?

6–58. The circumference of the earth is approximately 40,000 kilometers. If a plane is flying at 850 kilometers per hour, how long would it take to go around the world (round to the nearest tenth of an hour)?

6–59. Each patient should receive a 2-milliliter flu shot. How many liters are needed for 2500 patients?

6–60. At a reception for 60 people, each person drank two 400-milliliter cups of coffee. At $3.95 per liter, what is the cost of the coffee?

6.4 CONVERSIONS AND APPLICATIONS—U.S./METRIC AND METRIC/U.S.

More than 90 percent of the world's population uses the metric system. Thus it is necessary to convert from one system to another. Today nearly everything is "dual dimensioned," meaning both U.S. and metric measures are listed. However, being able to make conversions is a good skill to better understand the measurement systems. For example, a can of soda is labeled 12 fluid ounces and 355 milliliters; spring water is labeled 0.5 liter or 1 pint 0.9 fluid ounce.

Since conversions from one system to another are not exact, they should be stated as approximate. To make the conversion easier to learn, we will use the following basic conversions:

Units of Length

1 inch = 25.4 millimeters

1 millimeter = 0.0394 inch

1 inch = 2.54 centimeters

1 centimeter = 0.394 inch

1 meter = 3.28 feet

1 foot = 0.305 meter

1 meter = 39.37 inches

1 inch = 0.0254 meter

1 mile = 1.61 kilometers

1 kilometer = 0.62 mile

Units of Weight

1 ounce = 28.35 grams

1 gram = 0.0353 ounce

1 pound = 454 grams

1 gram = 0.0022 pound

1 kilogram = 2.2 pounds

1 pound = 0.455 kilogram

1 metric ton = 2200 pounds

1 pound = 0.00045 metric ton

Units of Volume

1 milliliter = 0.034 fluid ounce

1 fluid ounce = 29.38 milliliter

1 liter = 1.06 quarts

1 quart = 0.94 liter

1 gallon = 3.79 liters

1 liter = 0.264 gallon

These equivalences can be used to convert from one system to the other. Note there are no equivalences for time, since both systems are the same.

To change from one unit of measure to the other, replace the unit with the approximate equivalent measure and multiply. Generally, after the conversions, the results are rounded to the nearest tenth or hundredth.

The following will illustrate the "equivalence method" of conversion.

Example 6–16: Convert 200 Miles to Kilometers

PROBLEM: Convert 200 miles to kilometers.

SOLUTION: Rewrite 200 miles, replace 1 mi = 1.61 km and multiply.

$$200 \text{ mi} = 200 \text{ (mi)} = 200 \text{ (1.61 km)} = 322 \text{ km}$$
$$\text{Thus } 200 \text{ mi} = 322 \text{ km}$$

Example 6–17: Convert Pounds to Kilograms

PROBLEM: Convert 125 pounds to kilograms.

SOLUTION: Rewrite 125 pounds, replace 1 lb = 0.455 kg and multiply.

$$125 \text{ pounds} = 125(\text{lb}) = 125(0.455 \text{ kg}) = 56.8 \text{ kg}$$

Thus $125 \text{ lb} = 56.8 \text{ kg}$

Example 6–18: Convert Gallons to Liters

PROBLEM: Convert 55 gallons to liters.

SOLUTION: Rewrite 55 gallons, replace 1 gal = 3.79 L and multiply.

$$55 \text{ gallons} = 55(\text{gal}) = 55(3.79 \text{ L}) = 208.5 \text{ L}$$

Thus $55 \text{ gal} = 208.5 \text{ L}$

Example 6–19: Convert Centimeters to Inches

PROBLEM: Convert 210 cm to inches.

SOLUTION: Rewrite 210 cm, replace 1 cm = 0.394 in. and multiply.

$$210 \text{ cm} = 210(\text{cm}) = 210(0.394 \text{ in.}) = 82.7 \text{ in.}$$

Thus $210 \text{ cm} = 82.7 \text{ in.}$

Example 6–20: Convert Kilometers to Miles

PROBLEM: Convert 100 kilometers to miles.

SOLUTION: Rewrite 100 kilometers, replace 1 km = 0.62 mi and multiply.

$$100 \text{ km} = 100(\text{km}) = 100(0.62 \text{ mi}) = 62 \text{ mi}$$

Thus $100 \text{ km} = 62 \text{ mi}$

Example 6–21: Convert Liters to Quarts

PROBLEM: Convert 2 liters to quarts.

SOLUTION: Rewrite 2 liters, replace 1 L = 0.94 qt and multiply.

$$2 \text{ L} = 2(\text{L}) = 2(0.94 \text{ qt}) = 1.88 \text{ qt}$$

Thus $2 \text{ L} = 1.88 \text{ qt}$

Example 6–22: Application—Volume—Metric/U.S.

PROBLEM: Some doctors recommend athletes drink about 600 milliliters of fluid each hour during workouts and games. How many fluid ounces should a player consume during a $2\frac{1}{2}$-hour workout?

SOLUTION: Convert milliliters to fluid ounces by replacing 1 ml = 0.034 fluid ounce and multiply; then multiply by 2.5 for the $2\frac{1}{2}$-hour workout

$$600 \text{ ml} = 600\ (0.034 \text{ fl oz}) = 20.4 \text{ fl oz}$$
$$20.4 \text{ fl oz } (2.5) = 51 \text{ fl oz or about } 1\tfrac{1}{2} \text{ qt.}$$

Example 6–23: Application—Length and Weight—Metric/U.S.

PROBLEM: Ryan was born on May 28, 2001, in Paris. He was 50 millimeters tall and weighed 3 kilograms 600 grams. How tall was he in inches and what was his weight in pounds?

SOLUTION: Convert centimeters to inches by replacing 1 cm = 0.394 in. and multiply. Then convert kilograms to pounds by replacing 1 kg = 2.2 lb and multiply.

$$50 \text{ mm} = 50(\text{mm}) = 50(0.394 \text{ in.}) = 19.7 \text{ in.}$$
$$\text{replace } 1 \text{ mm} = 0.394 \text{ in.}$$
$$3 \text{ kg } 600 \text{ g} = 3.6 \text{ kg}$$
$$3.6 \text{ kg} = 3.6(\text{kg}) = 3.6(2.2) = 7.92 \text{ lb}$$
$$\text{replace } 1 \text{ kg} = 2.2 \text{ lb}$$

Thus Ryan weighed 7.92 lb and was 19.7 in. tall when he was born.

EXERCISE 6–4 U.S. AND METRIC APPLICATIONS

6–61. Convert 220 yards to meters.

6–62. Convert 128 pounds to kilograms.

6–63. Convert 2 liters of soda to fluid ounces.

6–64. Convert 2500 kilograms of bananas to pounds?

6–65. Two-hundred-ten-centimeter cross-country skis are how many inches long?

6–66. The Olympic 10,000-meter run equals how many yards?

6–67. Convert 5500 pounds to metric tons.

6–68. How many liters will a 18.5 gallon fuel tank hold?

6–69. One cup equals how many milliliters?

6–70. How many fluid ounces of chlorine are needed to treat a swimming pool for 30 days if 1500 milliliters are used per day?

6–71. Determine the width of 35-millimeter film in inches to the nearest hundredth.

6–72. The *Exxon Valdez* spilled 10,000,000 gallons of crude oil in Prudhoe Bay in Alaska in 1989. How many kiloliters of crude oil were spilled?

6–73. A gallon of gasoline costs $1.799. Find the cost per liter to the nearest thousandth.

6–74. A 175-millimeter Howitzer is equivalent to how many inches to the nearest hundredth?

6–75. Vitamin C generally comes in 500-milligram tablets. How many tablets should one take per day if 1 gram is the recommended amount?

6–76. An average human being has about 6 quarts of blood in their body. How many liters is this?

6–77. The winning height of the MIAC men's high jump was 6 feet 9 inches. Convert this to meters.

6–78. A piece of thirty-by-thirty-centimeter tile weighs 550 grams. How many pounds would a box of 144 of these tiles weigh?

6–79. A teaspoon of table salt contains about 2000 milligrams of sodium. How many grams of sodium is this?

6–80. The approximate speed of light is 300,000 kilometers per second. Convert this to miles per second.

6–81. A turkey costs $1.09 per pound. Determine the cost per kilogram.

6–82. The highway speed limit in Canada is 100 kilometers per hour. How many miles per hour is this to the nearest tenth?

6–83. How many pounds of fertilizer are needed for 300 apple trees, if each tree receives 450 grams of fertilizer?

6–84. A lumber delivery truck from Canada traveled 556 kilometers on 88.5 liters of fuel. Determine the miles per gallon.

6–85. The Sofitel, a French hotel, sells coffee only by the liter for receptions. How many 8-ounce cups are there in a 5-liter container?

THINK TIME

Our world is becoming more metric. We are moving to a global economy, which will require an understanding for standard measurement. In the meantime, many objects are dual-dimensioned, labeled with U.S. and metric measure.

Make a list of items—including food, clothing, sporting goods, tools, and so on—that are dual-dimensioned.

Item	*U.S. Measure*	*Metric Measure*
Can of soda	12 fluid ounces	350 milliliters

PROCEDURES TO REMEMBER

The procedures of measurement are:

1. Select an appropriate measuring tool.
2. Study the scale of the measuring tool.
3. Find the value of all the subdivisions until the smallest subdivision is reached; this will determine the accuracy of measurement.
4. Read the value of the quantity being measured and "round" to the nearest subdivision.

5. To change from one unit of measurement to another unit, replace the unit with the appropriate equivalent measure and multiply. For example: 1 pound = 16 ounces.

   5 pounds = __________ ounces
   5 (pounds) = __________ ounces
   5 (16 ounces) = 80 ounces

6. In the metric system, when performing arithmetic operation, convert to the larger dimension.

CHAPTER SUMMARY

1. *Linear measurement* means finding the distance between two points and attaching an appropriate unit of length.
2. An *exact number* results when objects are counted.
3. An *approximate number* results when an object is measured.
4. The *precision* of a measurement depends on the instrument used to do the measuring and is determined by the smallest subdivision on the instrument's scale.
5. A number with a unit attached is called a *denominate number.*
6. To convert from one system of measurement to another, use the appropriate conversion factor and round accordingly.

CHAPTER TEST

Identify the following lettered points on the given rulers in Figures 6–10 and 6–11.

T–6–1. P = __________

T–6–2. Q = __________

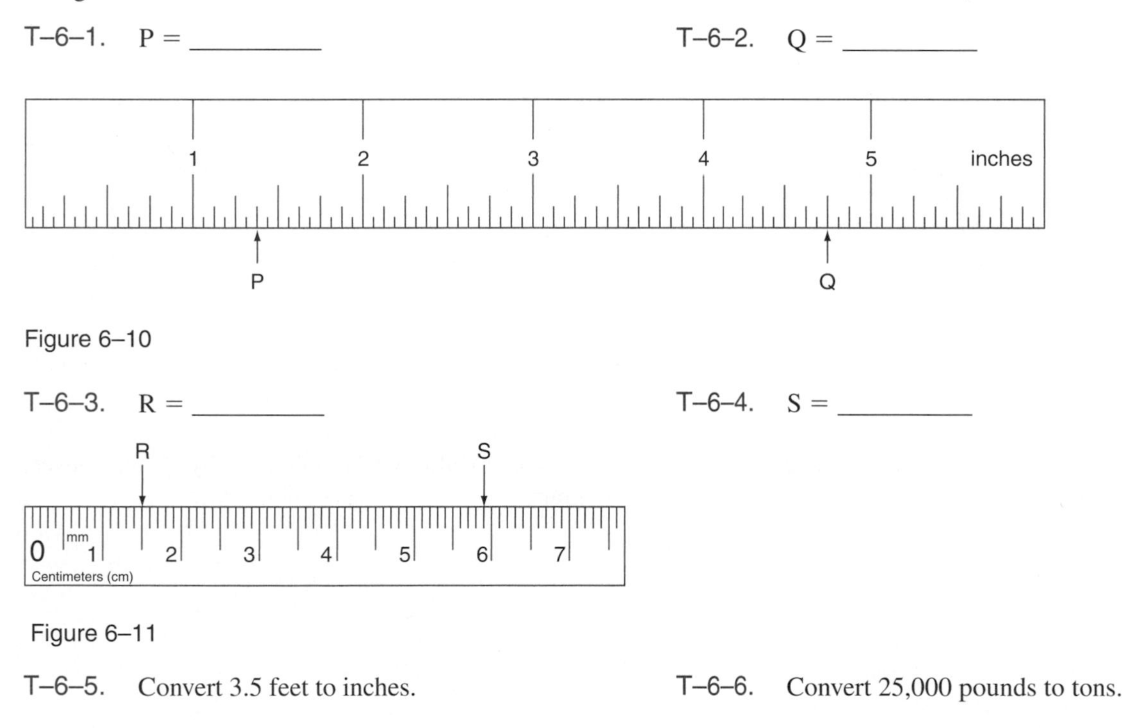

Figure 6–10

T–6–3. R = __________

T–6–4. S = __________

Figure 6–11

T–6–5. Convert 3.5 feet to inches.

T–6–6. Convert 25,000 pounds to tons.

T–6–7. Convert 384 fluid ounces to gallons.

T–6–8. Convert 5850 meters to kilometers.

T–6–9. Convert 7.8 liters to milliliters.

T–6–10. Convert 7500 grams to kilograms.

T–6–11. Convert 22,750 milligrams to grams.

T–6–12. Subtract 3 feet 7 inches from 8 feet.

T–6–13. Divide 6 gallons 2 quarts by 4.

T–6–14. Add 168 kilometers 569 meters to 345 kilometers 231 meters.

T–6–15. Multiply 12 fluid ounces by 12, and convert to quarts and pints.

T–6–16. Convert 8800 milliliters to liters.

T–6–17. Divide 45 kilograms by 12 and convert to grams and centigrams.

T–6–18. Multiply 875 milligrams by 30 and convert to grams.

T–6–19. How many 200-milliliter servings of soda can be served from a 2-liter container?

T–6–20. Find the total cost of a 3-kilogram-520-gram roast that is priced at $11.95 per kilogram.

T–6–21. Swimming burns 550 calories per hour. How long would a person need to swim to burn the 825 calories from a piece of pecan pie?

T–6–22. The distance around the world is 24,887 miles. How many kilometers is this to the nearest tenth?

T–6–23. How many minutes should a 5-pound 8-ounce ham be roasted if 1 hour is required per 2 pounds?

T–6–24. The record ski jump for women at Bear Mountain is 115 meters. This is equivalent to how many feet and inches, rounded to the nearest inch?

T–6–25. Find the cost per ounce of a 510-gram container of peanut butter that costs $1.79.

7

DATA, GRAPHS, AND BASIC STATISTICS

OBJECTIVES

After studying this chapter, you will be able to:

7.1. Understand the definitions and basic operations of tables and charts.
7.2. Understand and read graphs, including pictographs, bar graphs, line graphs, circle graphs, frequency distributions, and histograms.
7.3. Understand the definitions and basic operations of statistics, including range, mean, median, mode, quartiles, and percentiles.

SELF-TEST

The skills needed to solve data, graph, and basic statistical problems are different from those learned in earlier chapters, but will require knowledge of many of those skills. Your results from this test may indicate that you need to review this chapter. You can identify areas to review by completing the self-test and checking your work.

S–7–1. Given the December 25, 2002 temperatures in South Dakota, what was the range of temperature?

South Dakota

Aberdeen	22/–7
Huron	31/4
Pierre	36/5
Rapid City	38/10
Sioux Falls	30/6
Watertown	17/4

S–7–2. From problem S–7–1, find the average high temperature of the six cities.

S–7–3. From problem S–7–1, find the average low temperature of the six cities.

Given the bar graph **1996 Summer Olympic Medals**, complete problems S–7–4 through S–7–6.

S–7–4. How many total medals did the United States win?

S–7–5. How many silver medals did Russia win?

S–7–6. How many total medals were won by the United States and Russia?

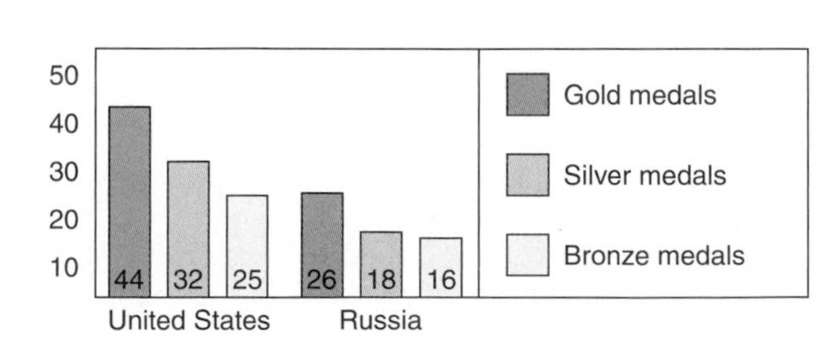

1996 Summer Olympic Medals

Given the circle graph **Advertising Budget of the New Shampoo,** complete problems S–7–7 through S–7–9.

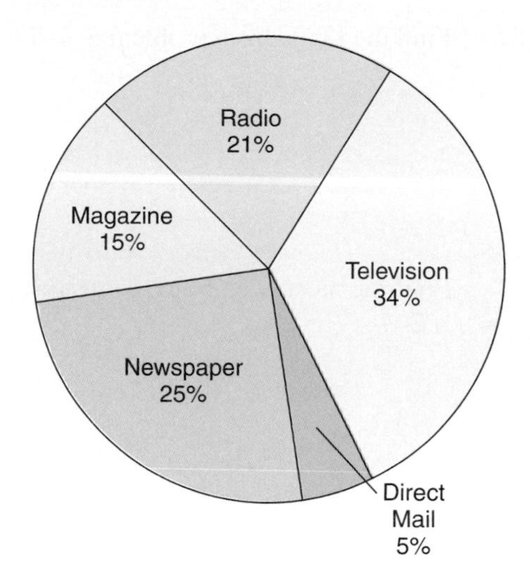

Advertising Budget of the New Shampoo

S–7–7. If the total budget is $150,000, how much is spent on radio advertising?

S–7–8. What percent of the budget is used for print material?

S–7–9. How much is spent on electronic advertising if the total budget is $250,000?

S–7–10. Find the average of the following: 55, 59, 62, 72, 66, 76.

S–7–11. Find the range of the following final math scores: 93, 86, 59, 76, 87, 92, 97, 77, 81, 85, 91, 54, 88, 78, 85, 89, 91, 77, 86, 84, 73, 90.

S–7–12. Find the mean of the following temperatures: 0, 7, 2, −8, −7, 4, −5.

S–7–13. Find the range given the following prices of $1\frac{1}{2}$-pound loaf of bread at various markets: $2.09, $1.59, $1.89, $1.39, $1.99, $1.79, $1.89, $1.49, $1.69, $2.19, $1.75, $1.85.

S–7–14. Find the mode from problem S–7–13.

S–7–15. Determine the median from problem S–7–13.

S–7–16. Find the average price per loaf to the nearest cent from problem S–7–13.

S–7–17. Find the mode given the weights of the offensive line-up of a National Football League team: 260, 210, 211, 274, 308, 302, 314, 359, 308, 199, 204.

S–7–18. Find the range from problem S–7–17.

S–7–19. Find the median from problem S–7–17.

S–7–20. Find the mean weight to the nearest pound from problem S–7–17.

S–7–21. To earn a B in a math class a student must average 85 on 4 tests. Her scores on the first 3 tests were 88, 84, and 75. What is the lowest score she can receive on the last test to earn a B?

S–7–22. Find the median given the following scores on a math placement:

83	69	68	75	49
81	65	69	71	72
79	83	79	46	68
85	55	71	66	74
54	60	89	78	95

S–7–23. Find the first quartile from problem S–7–22.

S–7–24. Find the third quartile from problem S–7–22.

S–7–25. Find the 90th percentile from problem S–7–22.

7.1 DATA: TABLES AND CHARTS

Statistics is the branch of mathematics that collects data and facts for decision making. As the world continues to become more complex and much more information becomes available, data and facts need to be used to make decisions. The purpose of using statistics is to study and understand information in order to make meaningful decisions in a systematic way.

In this chapter, you will learn various ways to use data and information to help find "the right answer." Much information can be provided in very simple ways. For example, a bus, train, or airline schedule provides information regarding the place of departure and arrival plus the times of operation. Another basic use of a chart is an "eye chart" which is used by a doctor to determine the appropriate prescription.

Reading a Schedule

Information is obtained merely by reading a schedule; we do not need to study the data. For example, from the portion of the bus schedule given in Table 7–1, Bus 52B Inbound leaves Southdale Park and Ride at 6:53, 7:08, 8:13, and 9:18 A.M.

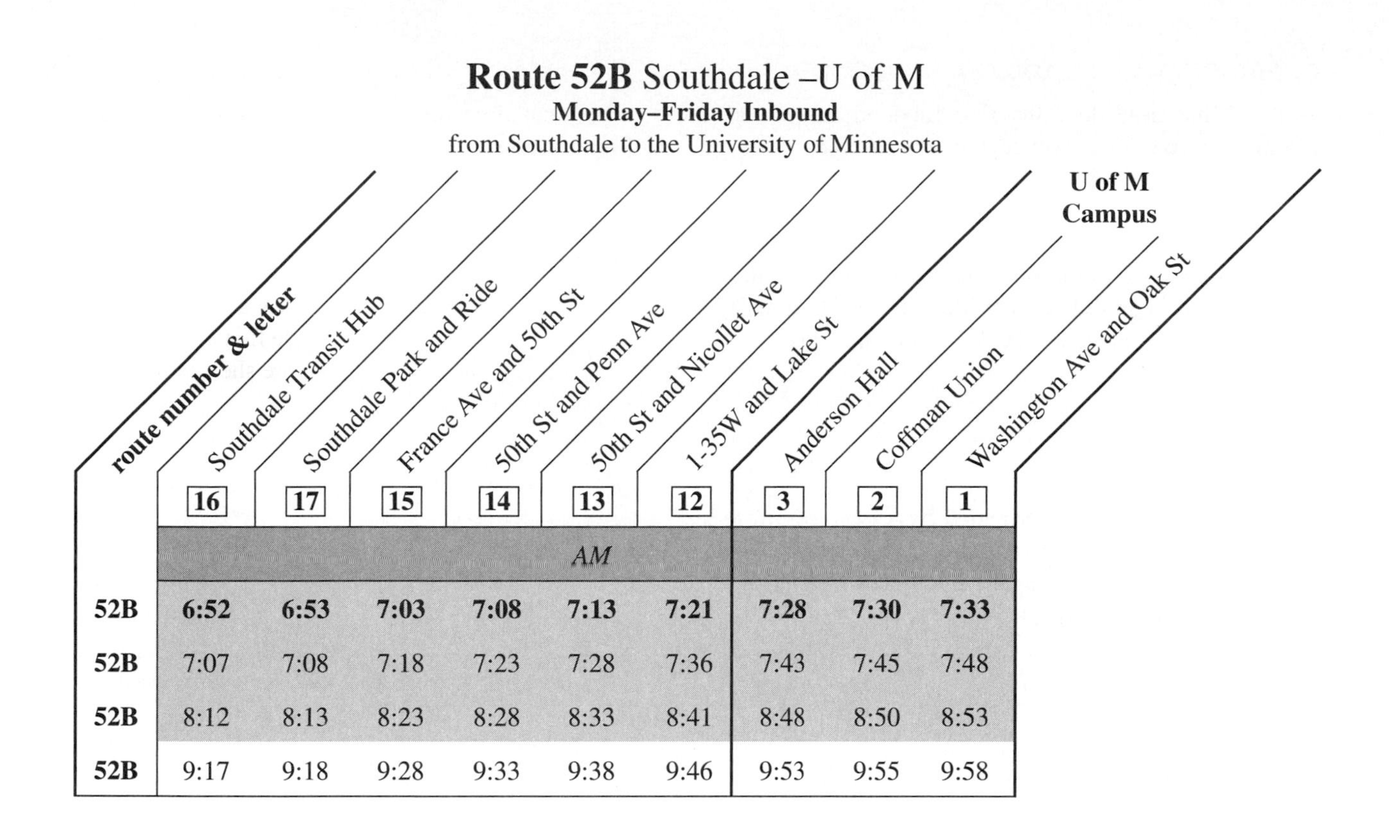

Route 52B Southdale –U of M

Monday–Friday Inbound

from Southdale to the University of Minnesota

route number & letter	Southdale Transit Hub 16	Southdale Park and Ride 17	France Ave and 50th St 15	50th St and Penn Ave 14	50th St and Nicollet Ave 13	I-35W and Lake St 12	U of M Campus: Anderson Hall 3	U of M Campus: Coffman Union 2	U of M Campus: Washington Ave and Oak St 1
	AM								
52B	**6:52**	**6:53**	**7:03**	**7:08**	**7:13**	**7:21**	**7:28**	**7:30**	**7:33**
52B	7:07	7:08	7:18	7:23	7:28	7:36	7:43	7:45	7:48
52B	8:12	8:13	8:23	8:28	8:33	8:41	8:48	8:50	8:53
52B	9:17	9:18	9:28	9:33	9:38	9:46	9:53	9:55	9:58

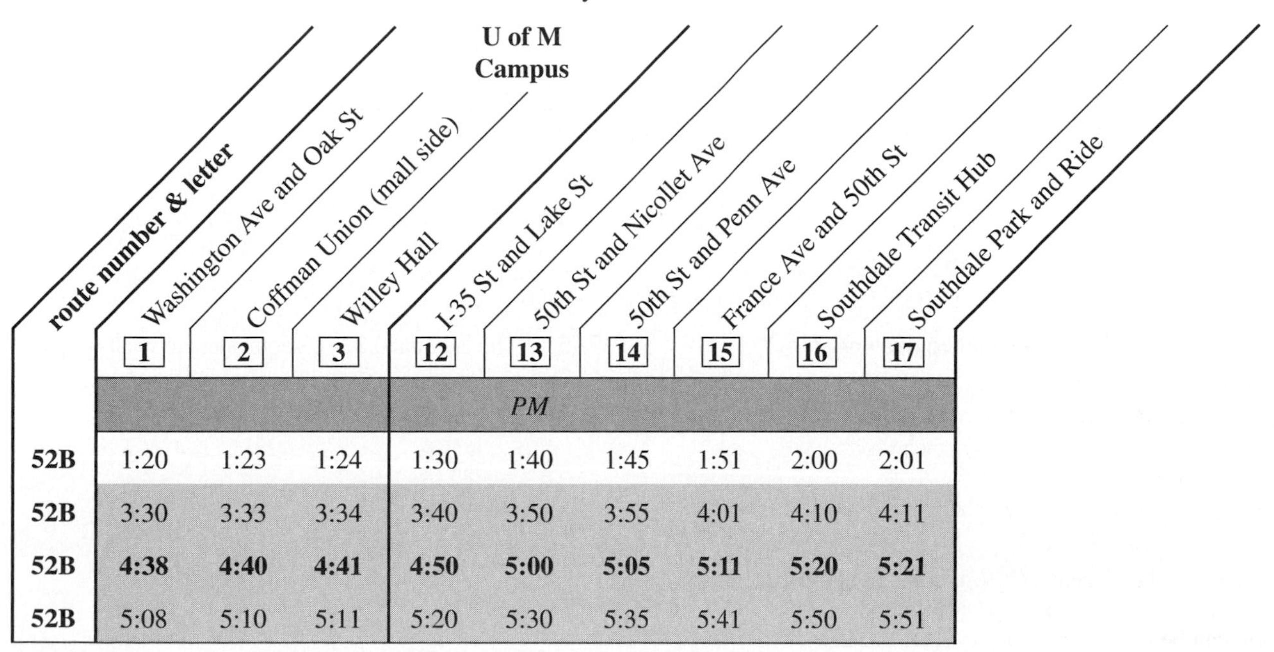

Monday–Friday Outbound

from the University of Minnesota to Southdale

route number & letter	U of M Campus: Washington Ave and Oak St 1	U of M Campus: Coffman Union (mall side) 2	U of M Campus: Willey Hall 3	I-35 St and Lake St 12	50th St and Nicollet Ave 13	50th St and Penn Ave 14	France Ave and 50th St 15	Southdale Transit Hub 16	Southdale Park and Ride 17
					PM				
52B	1:20	1:23	1:24	1:30	1:40	1:45	1:51	2:00	2:01
52B	3:30	3:33	3:34	3:40	3:50	3:55	4:01	4:10	4:11
52B	**4:38**	**4:40**	**4:41**	**4:50**	**5:00**	**5:05**	**5:11**	**5:20**	**5:21**
52B	5:08	5:10	5:11	5:20	5:30	5:35	5:41	5:50	5:51

Table 7–1

EXERCISE 7–1: READING A SCHEDULE

7–1. What time does the first Inbound bus of the morning leave France Avenue and 50th Street?

7–2. What time does the last Outbound bus leave Coffman Union?

7–3. According to the schedule, how many minutes travel time is required from Willey Hall to Southdale Park and Ride if one takes the 3:34 Outbound bus?

Reading a Table

Information may also be provided by a table of information; for example, a table of fraction-decimal-percentage equivalents. Note the fractions are converted to their decimal equivalents.

Fraction		*Decimal*		*Percent*	*Fraction*		*Decimal*		*Percent*
$^1/_{20}$	=	.05	=	5%	$^3/_5$	=	.60	=	60%
$^1/_8$	=	$.12^1/_2$	=	$12^1/_2\%$	$^5/_8$	=	$.62^1/_2$	=	$62^1/_2\%$
$^1/_6$	=	$.16^2/_3$	=	$16^2/_3\%$	$^3/_4$	=	.75	=	75%
$^1/_5$	=	.20	=	20%	$^4/_5$	=	.80	=	80%
$^1/_3$	=	$.33^1/_3$	=	$33^1/_3\%$	$^5/_6$	=	$.83^1/_3$	=	$83^1/_3\%$
$^3/_8$	=	$.37^1/_2$	=	$37^1/_2\%$	$^7/_8$	=	$.87^1/_2$	=	$87^1/_2\%$
$^2/_5$	=	.40	=	40%	$^9/_{10}$	=	.90	=	90%
$^1/_2$	=	.50	=	50%					

Table 7–2 Given the Fraction-Decimal-Percent Conversion Chart
Note: $^3/_8 = 0.37^1/_2 = 37^1/_2\%$.

EXERCISE 7–2: READING A FRACTION-DECIMAL-PERCENT CHART

7–4. What is the decimal equivalent of $^5/_{16}$?

7–5. The decimal $0.83^1/_3$ is equivalent to what fraction?

7–6. The percent $62^1/_2\%$ is equivalent to what fraction?

Reading a Chart/Table

Another example of a chart or table is the U.S./metric conversion table shown in Table 7–3.

Units of Length	*Units of Weight*	*Units of Volume*
1 in. = 2.54 cm or 1 cm = 0.394 in.	1 oz = 28.35 g or 1 g = 0.0353 oz	1 L = 1.06 qt or 1 qt = 0.94 L
1 m = 3.28 ft or 1 ft = 0.305 m	1 lb = 454 g or 1 g = 0.00220 lb.	1 gal = 3.79 L or 1 L = 0.264 gal
1 m = 39.37 in. or 1 in. = 0.0254 m	1 kg = 2.2 lb or 1 lb = 0.455 kg	
1 mi = 1.61 km or 1 km = 0.62 mi		

Table 7–3 U.S./Metric Conversion Chart
Note: The table gives the U.S./metric equivalents for length, weight, and volume. For example, 1 m = 39.37 in. and 1 gal = 3.79 L.

EXERCISE 7–3: READING A U.S./ METRIC CHART

7–7. How many kilometers are there in a mile?

7–8. How many grams are in a pound?

7–9. One foot equals how many meters?

7–10. What percent greater is a liter than a quart?

7.2 GRAPHS

Frequently, data from a table or chart are converted to a graph. The reason is that the graph gives "a picture of the information," and is easier and quicker to understand. Graphs may be used to "tell a story." There are several types of graphs that may be used. The person making the graph should select the type of graph that will deliver the information most effectively.

Pictographs

The most basic graph is called a pictograph. A *pictograph* uses a symbol to represent and present information. Generally, a pictograph will give a quick picture of data. Each symbol represents a number of units. Partial symbols represent a fractional part of the given unit. The pictograph will include a scale that indicates the value of the picture. Reading a pictograph is a simple procedure. The selected symbol (picto) represents a certain value that is arranged on horizontal lines to indicate the amount. The title of the pictograph is important, because it should "tell what the pictograph is showing."

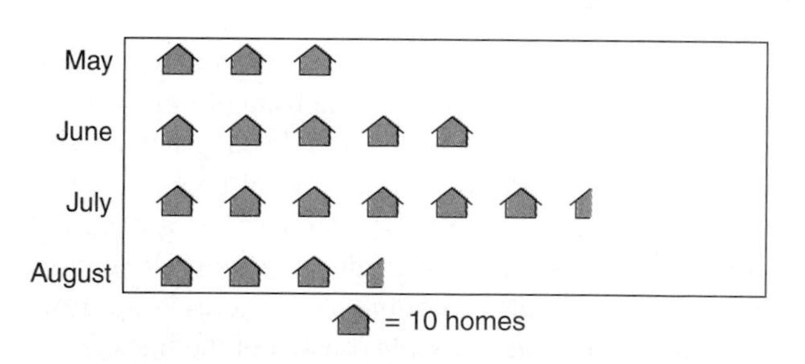

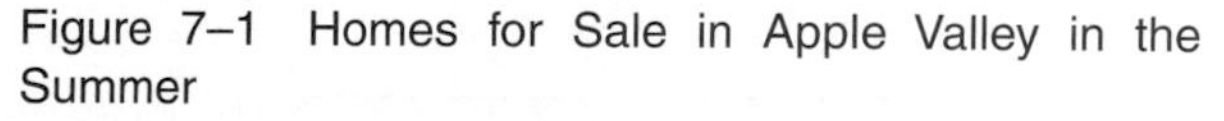

Figure 7–1 Homes for Sale in Apple Valley in the Summer

See the pictograph of "Homes for Sale in Apple Valley in the Summer" in Figure 7–1.

Note the legend indicates [house symbol] equals 10 homes. Therefore, there are 65 homes for sale in July, since there are 6 "whole houses" and 1 "half a house." There are 35 houses for sale in August.

To prepare a pictograph:

- Label the pictograph with a proper title that indicates what the pictograph is showing.
- Draw and label the vertical and horizontal lines.
- Decide what each symbol is, and define a value for each symbol.
- Draw the pictograph indicating the correct information.

EXERCISE 7–4: READING A PICTOGRAPH

Given the pictograph in Figure 7–2, complete exercises 7–11 through 7–13.

7–11. How many women played basketball?

7–12. What is the range of participation from soccer to football?

7–13. How many women participated in all sports at Mandan High?

Basketball
Football
Soccer
Softball
Swimming
Track
= 20 women

Figure 7–2 Mandan High School Women in Sports

Bar Graphs

Another commonly used graph is the bar graph. A *bar graph* is used to compare quantities using bars to represent data (vertical line) and time (horizontal line). Generally, the horizontal scale is some form of time, and the vertical scale is the amount. A double bar graph may be used to display data for the purpose of comparing one year to another or one brand to another. The title of a bar graph should tell what the graph is showing. Reading a bar graph is a simple procedure (see Figures 7–3). However, be sure to note the scale of each of the lines.

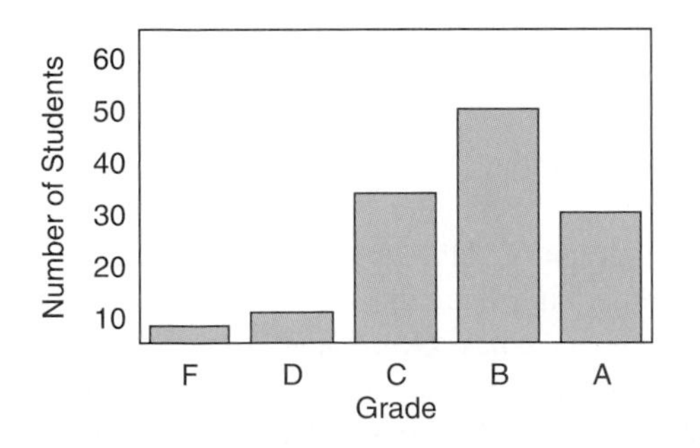

Figure 7–3 Grades for an Easy Math Test

According to the bar graph, 30 students earned A's. This is easy to read, because the bar aligns with the vertical number 30. However, it is difficult to read the number of students earning C's. Thus, you must estimate. Since the bar ends between 30 and 40, a reasonable estimate would be 32. You could estimate that eight students earned F's.

To construct a bar graph:

- Give the graph a proper title that indicates what the graph shows.
- Draw and label the vertical and horizontal lines.
- Show the scale on the lines—generally the vertical represents amount and horizontal represents time.
- Draw the bars and shade or color accordingly.

Note: The horizontal line or scale is known as the reference scale. The vertical line or scale is called the frequency scale.

EXERCISE 7–5: READING A BAR GRAPH

Given the bar graph in Figure 7–4, complete exercises 7–14 through 7–16.

7–14. From Figure 7–4, what is the average income of a person in her 20s?

7–15. From Figure 7–4, what is the range of income from the prime earning years from the 20s to the 60s?

7–16. From Figure 7–4, what is the average earnings from the 20s to the 60s.

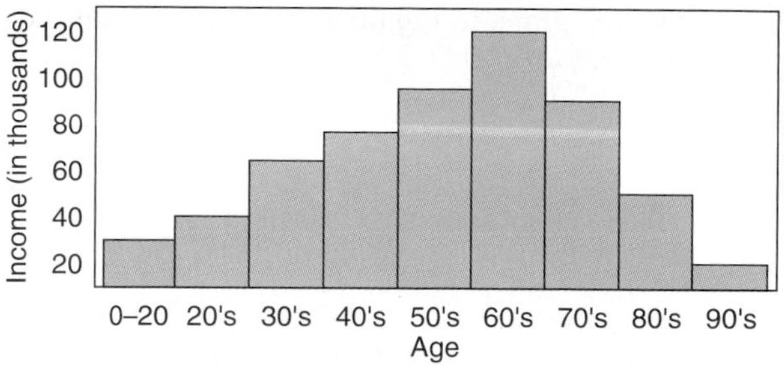

Figure 7–4 Average Individual Income Compared to Age

Line Graphs

If the midpoints of the top of the bars in a bar graph were connected, a line graph would result. Line graphs are frequently used to show movement of a quantity.

A *line graph* (or broken line graph) is similar to a bar graph, except the data are represented by points instead of bars. A line graph is used to show trends or changes over time. A double line graph may be used to compare or display two different quantities.

The title of a line graph should tell what the graph is showing. Reading a line graph is easy (see Figure 7–5). However, if it is a double line graph, you must be careful to note the different lines, which may be solid, dotted, or dashed; and may be in different colors.

Note the line graph is a double line graph, with "——" indicating 2000 and "------" indicating 2002. Thus the average 2002 temperature in May was 30°F, whereas the average December 2000 temperature was −10°F. The average temperature for the year can be found by adding the temperature of each month and dividing by 12. Thus the average temperature for 2002 is

$$(-10 + (-10) + 5 + 15 + 30 + 40 + 45 + 60 + 55 + 25 + 15 + 0)/12 = 270/12 = 22.5$$

Therefore, the average temperature for 2002 is 22.5°F.

To construct a line (broken line) graph:

- Label the graph with an appropriate title indicating what the graph is attempting to show.
- Draw and label the vertical and horizontal lines.
- Show the proper type of line for the given lines (solid, dotted, dashed, etc.).
- Place the *dots or X's* at the proper points, and draw the lines.

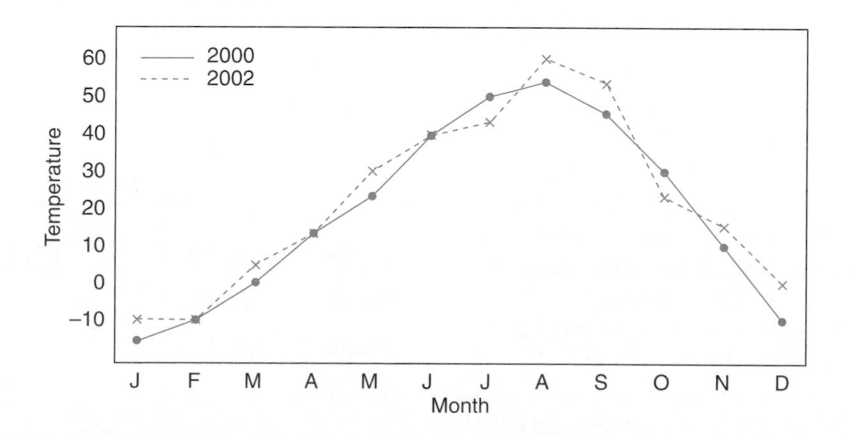

Figure 7–5 Average Monthly Temperature in an Alaskan Eskimo Village

EXERCISE 7–6: READING A LINE GRAPH

Given the line graph in Figure 7–6, complete exercises 7–17 through 7–19.

7–17. How much did ZIP-IT earn in 2002?

7–18. How much and in what year did TRO have the greatest gain?

7–19. How many times and in what year did both companies earn the same amount?

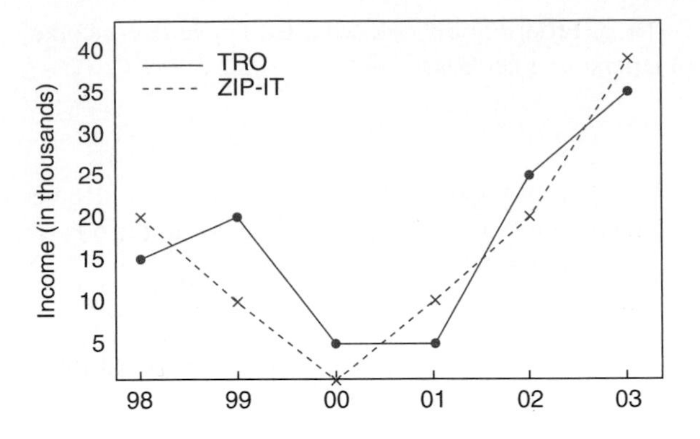

Figure 7–6 Comparison of Income of Two Start-up Software Companies

Histograms

A special type of bar graph is called a histogram. A histogram is used when there are several values that can be arranged into groups; for example, finding the gasoline mileage for several cars, the total points scored in a football game, or the salaries of a company.

A *histogram* is a special type of bar graph. The width of the bars corresponds to the range of values which are called *class intervals*. The height of the bars reflects the number of frequencies of data in each group or class interval. Generally, a histogram is developed from a set of data with more than 20 values. The data are arranged into class intervals, which include more than one value. The number of class intervals is frequently between 5 and 8. The size of the class interval is found by finding the range of values and then adjusting the size of the class interval so there are between 5 and 8 class intervals.

For example, if the range of mileage for 65 cars was from 52 to 12, the range would be 40 because $52 - 12 = 40$. Thus the class intervals for the mileage data could be: 10–14, 15–19, 20–24, 25–29, 30–34, 35–39, and 40–44.

Note the midpoints of the class intervals would be 12, 17, 22, 27, 32, 37, and 42.

Once the class intervals are established, a tally of each value is recorded and this tally is summarized into the class frequency (see Table 7–4).

Note the cumulative frequency is determined by adding the frequencies of the class intervals; thus 5 is the first; $5 + 8 = 13$, so 13 is the second; $13 + 14 = 27$, so 27 is the third; and so on.

Class Interval	*Tally*	*Class Frequency*	*Cumulative Frequency*
10–14	卌	5	5
15–19	卌 III	8	13
20–24	卌 卌 IIII	14	27
25–29	卌 卌 卌 II	17	44
30–34	卌 卌 I	11	55
35–39	卌 III	8	63
40–44	II	2	65

Table 7–4 Class Interval—Cumulative Frequency Form

EXERCISE 7–7: READING A HISTOGRAM

Given the histogram in Figure 7–7, complete exercises 7–20 through 7–22.

7–20. How many cars averaged between 10 and 15 miles per gallon?

7–21. How many cars averaged more than 35 miles per gallon?

7–22. How many cars averaged between 20 and 40 miles per gallon?

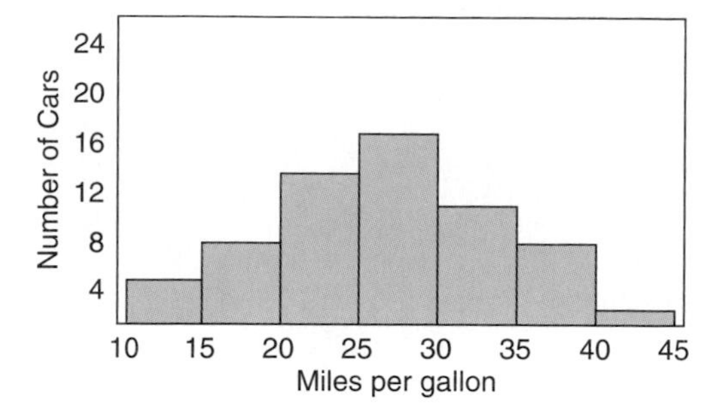

Figure 7–7 Average Gas Mileage of 65 Cars

If lines are drawn from the midpoints of the class intervals, a *frequency polygon* is formed.

Circle Graphs

Another graph frequently used is a circle graph. A circle graph may be used to show how a dollar is spent, how 24 hours are used, or how a budget is distributed.

A *circle graph* (or *pie chart*) represents data by the size of the sectors or the size of the pieces of pie. The size of each sector is proportional to the size of the data it represents. Circle graphs are used to compare parts to the whole circle.

Each sector size is determined by a percent of the whole circle. That percent is multiplied by 360° to determine the size of the sector in degrees.

A circle graph is easy to read because the sector sizes give an instant picture as parts of the whole circle (see Figure 7–8).

Thus the student studies 6 hours per day, since 25% of 24 hours is 6 hours ($0.25 \times 24 = 6$); the student sleeps 8 hours per day, since $33\frac{1}{3}\% = \frac{1}{3}$ and $\frac{1}{3} \times 24$ hours $= 8$.

To construct a circle graph:

- Determine the percent each sector (pie slice) is of the whole.
- Since the circle contains 360°, multiply each percent of the sector by 360°; this will determine the number of degrees for each sector.
- Use a compass to draw the circle, and a protractor to measure the size in degrees of each sector of the circle.
- Remember to give the circle graph a title, which should tell what the circle graph is showing.

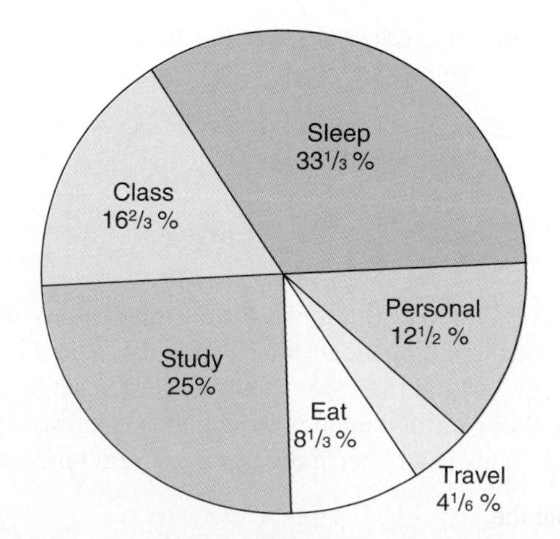

Figure 7–8 How a Student Spends a 24-hour Day

EXERCISE 7–8: READING A CIRCLE GRAPH

Use the information from Figure 7–9 to complete exercises 7–23 through 7–25.

7–23. How many dollars are contributed to Retirement/Savings?

7–24. How many dollars are paid for taxes?

7–25. If you earned $5000, given the information in the circle graph, find your yearly take-home pay.

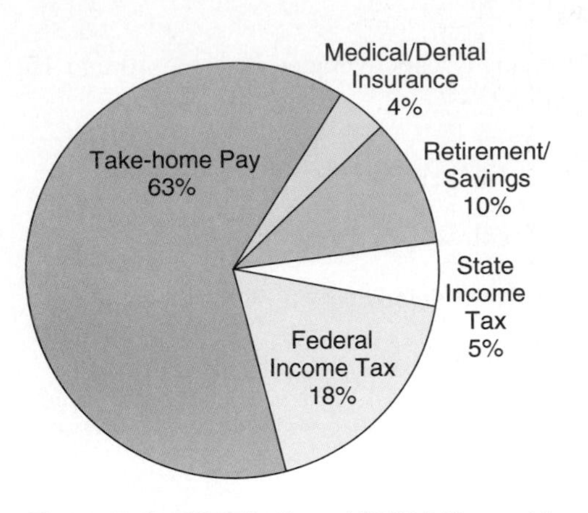

Figure 7–9 Distribution of $5000 Earned Income

7.3 MEASURES OF CENTRAL TENDENCY

When studying a set of data, the most basic statistic is the range. *Range* is defined as the spread from the largest to the smallest in a set of data.

Given that $65,000 is the highest teachers' salary and $34,000 is the lowest, the range would be $65,000 − $34,000 = $29,000. Thus $29,000 would be the range in teachers' salaries. This is a general number, but it gives the span from largest to smallest. Another basic statistic is called the *mean* or *average.* The *mean* or *average* is found by dividing the sum of a set of data by the number of items.

The mean or average of seven teachers' salaries would be found as follows:

$$34{,}000 + 38{,}000 + 46{,}000 + 56{,}000 + 56{,}000 + 65{,}000 + 48{,}000 =$$

$$\frac{343{,}000}{7} = \$49{,}000$$

Thus $49,000 is the mean, or average, salary of the seven teachers. The mean can be shown as $\bar{x}$, and is read as "*x* bar," which indicates the average of a set of data.

Another central tendency in statistics is known as the median. This is a better indicator of central tendency in cases where there is an extremely large or extremely small data-value point. Such an extreme value could skew the mean. *Skew* means to unbalance the facts or move the average or mean to a position that does not reflect the data. The *median* of a set of data is determined by arranging the data in numerical order (smallest to largest), and finding the middle value.

The median is useful when one value in a group or set of data is much larger, or much smaller, than the rest of the data.

If the data has an odd number of items, the median is the middle number. If the data has an even number of items, the median is the average of the two middle numbers, and is found by adding the two middle numbers and dividing by two. Regarding the teachers' salaries previously shown, arrange the salaries in ascending order (smallest to largest): $34,000 + $38,000 + $46,000 + $48,000 + $56,000 + $56,000 + $65,000. Since there are seven items, the fourth item would be the median. Thus $48,000 would be the median of the seven teachers' salaries.

Another measure of central tendency is the mode. The *mode* of a set of data is the quantity (item) that occurs most frequently. Thus, given the teachers' salaries $34,000, $38,000, $46,000, $48,000, ($56,000), ($56,000) $65,000, $56,000 is the mode of this set of data since $56,000 occurs twice. If no quantity in the group is repeated, the set of data has no mode. However, with a large set of data, more than one mode may result.

The following examples will illustrate the various ways to find central tendency statistics.

Example 7–1: Find the Range

PROBLEM: Find the range in temperatures in Flin Flon, Manitoba, for a week in February.

S	M	T	W	T	F	S
0°	−8°	11	14°	5°	−12°	2°

SOLUTION: Find the difference between the highest and lowest temperatures. Since the highest was +14° and the lowest was −12°, the range is 26°.

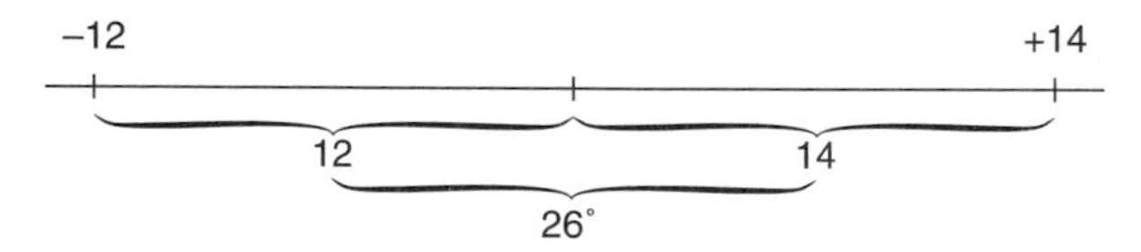

Example 7–2: Find the Mean

PROBLEM: Find the mean or average temperature for a week in December in Naples, Florida, given the following data: 73°, 80°, 82°, 84°, 76°, 88°, 83°.

SOLUTION: Add the seven temperatures and divide by 7.

$$73° + 80° + 82° + 84° + 76° + 88° + 83° = \frac{566°}{7}$$
$$= 80.857°$$
$$= 80.9°$$

Since accuracy cannot be obtained by calculation, round to the nearest tenth. Therefore, the average temperature for the week would be 80.9°.

Note: When doing any calculation, common sense should prevail. Thus it would make sense that the temperature would be rounded to the nearest tenth or one more digit than the original measurements.

Example 7–3: Find the Median

PROBLEM: Find the median calls per day given the number of daily service calls for an air conditioning service company: 24, 30, 28, 24, 36, 39, 31, 33, 38, 37.

SOLUTION: Arrange the quantities in ascending (smallest to largest) order; then count the number of calls by writing the number below the actual number to determine the center value.

24	24	28	30	31	33	36	37	38	39
1st	2nd	3rd	4th	5th	6th	7th	8th	9th	10th

Since there are ten quantities, to find the median, add the 5th and 6th quantities (31 and 32) and divide by 2 to find the median.

$$\frac{31 + 33}{2} = \frac{64}{2} = 32$$

Thus 32 is the median of the given numbers.

Example 7–4: Find the Mode

PROBLEM: Given the points scored by Kevin Garnett of the Minnesota Timberwolves in an eight-game road trip 24, 29, 31, 23, 26, 32, 33, 26, find the mode.

SOLUTION: Arrange the numbers in ascending order, and circle the number that occurs most frequently.

23, 24, (26), (26) 29, 31, 33

Since 26 occurs two times, 26 is the mode of this data.

EXERCISE 7–9: FINDING RANGE, MEAN, MEDIAN, AND MODE

Given the following prices of a gallon of 2% milk at various retailers complete exercises 7–26 through 7–29.

Grocery Stores	*Convenience Stores*	*Gas Stations*
$3.39	$3.19	$2.99
$2.89	$3.09	$3.19
$2.79	$2.99	$3.29
$3.09	$3.39	$3.49

7–26. Find the range of the milk prices.

7–27. Find the mean of the milk prices.

7–28. Find the median of the milk prices.

7–29. Find the mode of the milk prices.

Given the following SAT scores from a select senior class complete exercise 7–30 throgh 7–33.

1200	1205	1261
1151	1076	1321
1212	994	1199
1304	1154	1291
1329	1044	1126

7–30. Find the range of the SAT scores.

7–31. Find the mean of the SAT scores.

7–32. Find the median of the SAT scores.

7–33. Find the mode of the SAT scores.

7–34. The prices for an ink cartridge for a printer at 5 stores were: $23.95, $31.99, $27.95, $29.95 and $30.95. Find the range in price.

7–35. From problem 7–34, find the mean or average of the ink cartridge prices.

7–36. From problem 7–34, find the median cost of the ink cartridges.

7–37. From problem 7–34, find the mode of the given data?

7–38. Find the median number of students in the Edina Elementary Schools if the enrollment in each grade is 690, 780, 990, 1040, 703, 960, 820, and 1005.

7–39. The reaction time of drivers to certain stimuli was measured at 0.53, 0.46, 0.50, 0.49, 0.52, 0.53, 0.44, and 0.55 seconds. Determine the mean reaction time.

7–40. Three math teachers reported average grades of 79, 74, and 82 in their classes, which consisted of 32, 25, and 17 students respectively. Determine the mean grade for all three classes.

Quartiles and Percentiles

Other types of statistics used to define and understand data are quartiles and percentiles. Basically, *quartiles* separate a set of data into four groups: the bottom fourth, the second fourth, the third fourth, and the top fourth.

Percentiles separate data into groups of one hundredths. Thus when a statement is made that "the student's score was in the 95th percentile," it means that 95% of the student's scores are below this score and 5% are above this score. Percentiles are used for extremely large populations, generally those of over 1000. They are used not only to understand data but also to project information and predict results. To better comprehend quartiles and percentiles, it is essential to understand how they are determined.

To determine quartiles for a small set (50 or less) of data:

- Arrange the set of data in order from smallest to largest (similar to finding the median).
- Count off the first one-fourth of the data; if there are 24 scores, it would be the 6th score; this score is known as the first quartile or Q_1.
- Count off the next one-fourth of the scores or the 12th score (with 24 items), which would be the second quartile (median) or Q_2.
- Count off the next one-fourth of the scores or the 18th score (with 24 items), which would produce the third quartile or Q_3.

Example 7–5: Finding Quartiles with a Small Sample

PROBLEM: Given the following scores: 11, 13, 15, 3, 4, 18, 16, 18, 14, 11, 4, 9, 9, 6, 7, 10. Find the quartiles.

SOLUTION: Arrange the scores in order from smallest to largest. Then count off by groups of 4, since there are 16 scores.

1	*2*	*3*	*4*	*1*	*2*	*3*	*4*	*1*	*2*	*3*	*4*	*1*	*2*	*3*	*4*
3	4	4	(6)	7	9	9	(10)	11	11	13	(14)	15	16	18	18
			Q_1				Q_2				Q_3				

Thus the quartile scores are: $Q_1 = 6$, $Q_2 = 10$, $Q_3 = 14$.

To determine percentiles and quartiles for extremely large populations (more than 1000).

- Arrange the set of data in order from smallest to largest by establishing class intervals and record.
- Determine the total number of scores by completing the cumulative frequency procedure. Then multiply the percent (percentile) by the total number of scores to determine the actual score that satisfies the percentile.

  Remember that the scores are not always exact numbers. (To obtain the exact values of specific percentiles, a procedure known as *interpolation* is used.) For example, if there are 350 scores, the 75th percentile (75%) score is to be found.

$$75\% \times 350 \text{ or } 0.75 \times 350 = 245$$

Thus the 245 score value would be the 75th percentile.

- Count off the first one hundred scores of the data or multiply the cumulative total by 1%; if there were 1000 data items, it would be the 10 lowest scores ($1000 \times 0.01 = 10$). The 75th percentile of 1000 scores would be the 750th score ($1000 \times 0.75 = 750$). *Note:* The 75th percentile is the same as the 3rd quartile or Q_3.

Example 7–6: Reading a Percentile Table

PROBLEM: Given Table 7–5 find the 33rd and 50th percentiles and estimate the 95th percentile.

SOLUTION: According to Table 7–5 the 33rd percentile falls in the 40–49 class interval, since 100 of the 300 cumulative scores are below or included in this interval, as is the $33^1/_3\%$ of the cumulative percent.

Half ($^{150}/_{300}$) of the scores are below or are included in the 50–59 class interval. Therefore, the 50th percentile is in the 50–59 class interval. Note this would also be the 2nd quartile (Q_2) and the median.

Since the 80–89 class interval includes 97% ($96^2/_3\%$) of the scores, the 95th percentile will be in the 70–79 class interval. Remember that the procedure for obtaining the exact value of the 95th percentile is done by interpolation, which is covered in a more advanced text.

Class Interval	*Frequency*	*Cumulative Frequency*	*Cumulative*	*Percent*
0–9	5	5	$^{5}/_{300} = 1^2/_3\%$	2%
10–19	15	5+ 15 = 20	$^{20}/_{300} = 6^2/_3\%$	7%
20–29	20	20+ 20 = 40	$^{40}/_{300} = 13^1/_3\%$	13%
30–39	25	65	$^{65}/_{300} = 21^2/_3\%$	22%
40–49	35	100	$^{100}/_{300} = 33^1/_3\%$	33%
50–59	50	150	$^{150}/_{300} = 50\%$	50%
60–69	60	210	$^{210}/_{300} = 70\%$	70%
70–79	55	265	$^{265}/_{300} = 88^1/_3\%$	88%
80–89	25	290	$^{290}/_{300} = 96^2/_3\%$	97%
90–100	10	300	$^{300}/_{300} = 100\%$	100%

Table 7–5 Reading Placement Test Scores

EXERCISE 7–10: READING AND UNDERSTANDING QUARTILES AND PERCENTILES

Use Table 7–5 to solve problems 7–41 to 7–50.

7–41. How many scores are in the 30–39 class interval?

7–42. How many total scores are less than the 33rd percentile?

7–43. What is the class interval of the 88th percentile?

7–44. Estimate the class interval of the 75th percentile.

7–45. What is the class interval of the 2nd Quartile?

7–46. If 70 and above is a passing score, how many passed the exam?

7–47. If the passing score was lowered to 50, what percent of the students would pass?

7–48. Estimate how many total scores were over the 90th percentile.

7–49. Given Table 7–5, if 70 is the minimum score for a student to move to the next grade, project the number of students in a 750-student elementary school who would move to the next grade.

7–50. Given Table 7–5, an accelerated advance placement physics class accepts only those students in the 90–100 class interval on the Reading Placement Test. How many students in the 576 Edina High junior class would be accepted?

Use Table 7–6 on the following page to solve problems 7–51 to 7–65.

7–51. What is the class interval of the 30th percentile?

7–52. What is the class interval of the 75th percentile?

7–53. Estimate the 90th percentile.

7–54. Estimate the median speed.

7–55. What is the class interval of the 3rd Quartile?

7–56. Estimate the 95% percentile to the nearest mile per hour.

7–57. How many vehicles were traveling between 55 and 74 miles per hour?

7–58. Given the speed limit of 65 miles per hour, how many total vehicles were exceeding the speed limit?

Class Interval Miles per hour	*Frequency*	*Cumulative Frequency*	*Calculation of Cumulative Frequency*	*Cumulative %*
40–44	32	32	$^{32}/_{1000} = 0.032$	3.2%
45–49	54	86	$^{86}/_{1000} = 0.086$	8.6%
50–54	60	146	$^{146}/_{1000} = 0.146$	14.6%
55–59	69	215	$^{215}/_{1000} = 0.215$	21.5%
60–64	85	300	$^{300}/_{1000} = 0.300$	30.0%
65–69	104	404	$^{404}/_{1000} = 0.404$	40.4%
70–74	186	590	$^{590}/_{1000} = 0.590$	59.0%
75–79	160	750	$^{750}/_{1000} = 0.750$	75.0%
80–84	156	906	$^{906}/_{1000} = 0.906$	90.6%
85–89	94	1000	$^{1000}/_{1000} = 1.00$	100%

Table 7–6 Freeway Speeds Measured by Radar

7–59. If the freeway speed limit is the 59th percentile, how many vehicles were traveling below the speed limit?

7–60. If the freeway speed is reduced to 54 miles per hour, what percentile of vehicles were exceeding the speed limit?

7–61. Suppose the minimum freeway speed limit is more than 50 miles per hour, what percentile of drivers are below the posted minimum speed?

7–62. Estimate the 60th percentile to the nearest mile per hour.

7–63. If the fine for speeding is $10 per mile over the 69 mile per hour posted limit, find the total fine for all the speeders in the 80–84 class interval.

7–64. If the fine for driving 80 mph or over is $50 per mile over the posted limit, find the total fine for 80–84 mph and 85–89 mph law breakers.

7–65. Make three recommendations as to what could be done to reduce the number of drivers exceeding the speed limit.

THINK TIME

The selling price of homes in an exclusive suburb are: \$525,000; \$465,000; \$529,999; \$485,000; \$450,990; \$535,779; \$1,750,995; \$579,995; \$495,875; \$499,999; \$505,000; and \$505,995. Determine the measures of central tendency (range, median, mode, and average). Which measure of central tendency best describes the value of the homes in this suburb? Why is your measure the most realistic answer?

PROCEDURES TO REMEMBER

1. The *range* of a set of data is determined by finding the difference between the largest and smallest value.
2. To find the *average* or *mean,* add all the values in a set of data and divide by the number of values.
3. The *median* of a set of data is determined by arranging the data in numerical order (smallest to largest) and finding the middle value.
4. The *mode* of a set of data is determined by studying the data and selecting the values that appear most frequently. A set of data may have more than one mode.
5. To determine *quartiles* arrange the data; in ascending order (smallest to largest) then count off the first one-fourth of the data, this score is the first quartile (Q_1) or lower quartile. Then count off the second fourth of the data; this score is the second quartile (Q_2), also known as the median. Finally, count off the third fourth of the data; this score is the third quartile (Q_3). The score above the Q_3 is called the upper quartile.
6. Determining *percentiles* is a similar procedure to finding quartiles; however, percentiles are used for extremely large populations. To find percentiles, arrange the data in ascending order, then count off the first one-hundredth of the data values; this score is the first percentile (P_1). Then count off the next one-hundredth of the data values; this score is the second percentile (P_2). The remaining percentiles follow this same pattern.

CHAPTER SUMMARY

1. *Statistics* is the branch of mathematics that collects and analyzes data for decision making.
2. A *graph* is a "picture" used to show a numerical relationship.
3. A *pictograph* uses symbols to represent and present information.
4. A *bar graph* is used to compare quantities. It uses bars to represent data.
5. A *line graph* or broken line graph is similar to a bar graph, except that the data are represented by points instead of bars.
6. A *histogram* is a special type of bar graph in which the bar includes a numerical interval. The width of the bars of a histogram correspond to a range of values, called *class intervals.*
7. *Class frequency* is the number of items in a class interval.
8. Lines drawn from the midpoints of the class intervals form a *frequency polygon.*
9. A *circle graph* or pie chart shows how a whole is broken into parts. It represents data by the size of sectors or the pieces of pie.
10. *Range* is defined as the spread from the largest to the smallest in a set of data.
11. The *mean* or *average* is found by dividing the sum of the set of data by the number of items, a single number that describes a set of data.
12. *Skew* means to unbalance the facts or move the average or mean to a position that does not reflect the data.
13. The *median* is the middle number in a set of data. Median is determined by arranging the data in numerical order (smallest to largest) and finding the middle value.
14. The *mode* is the quantity that occurs most frequently in a set of data.
15. *Quartiles* separate a set of data into four groups: the bottom fourth (or bottom percentile), the second fourth, the third fourth, and the top fourth (or top percentile).
16. *Percentiles* separate data into groups of one hundredths.

CHAPTER TEST

Solve the following problems using the skills learned while completing this chapter.

T–7–1. Given the temperatures in eight cities in Canada on a day in 2003, determine the range of temperatures.

Calgary	21/−4	Regina	23/ 18
Edmonton	11/−14	Toronto	33/ −9
Montreal	31/10	Vancouver	44/40
Quebec	24/ −4	Winnipeg	35/−13

T–7–2. From problem T–7–1, find the average low temperature for the eight Canadian cities.

T–7–3. From problem T–7–1, find the average high temperature for the eight Canadian cities.

Given the line graph **The State Speed Limits of Commercial Trucks,** complete problems T–7–4 through T–7–5.

T–7–4. How many states have a speed limit of 65 miles per hour or more?

T–7–5. How many states have a speed limit between 60 miles per hour and 70 miles per hour?

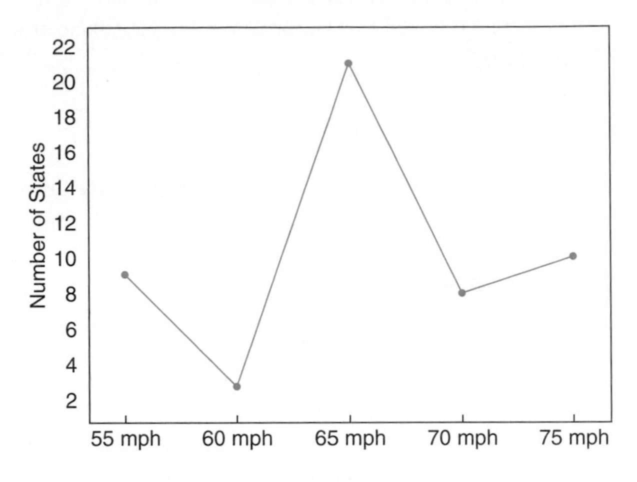

The State Speed Limits of Commercial Trucks

T–7–6. From the graph in problem T–7–4, how many states have a speed limit?

Given the circle graph **Retiring Investors Portfolio,** complete problems T–7–7 through T–7–9.

T–7–7. If the total portfolio is $225,000, how much is invested in financial stocks?

T–7–8. If the total portfolio is reduced to $195,000, how much is invested in technology and telecomm?

T–7–9. How many degrees of the circle graph are in the health portion?

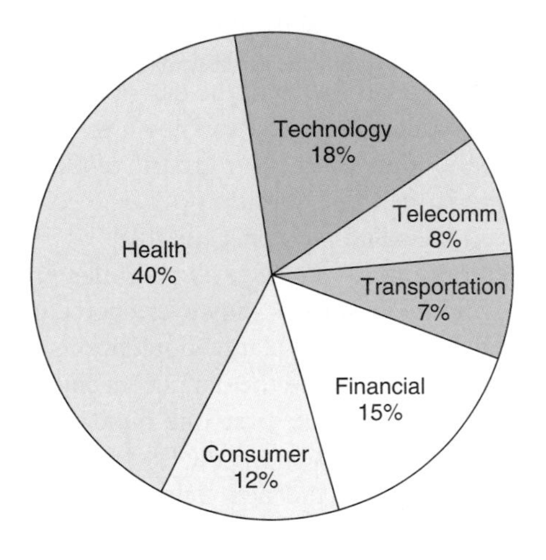

Retiring Investors Portfolio

T–7–10. Find the average of the following spelling scores: 82, 83, 61, 90, 82, 63, 69, 64, 95, 97, 93.

T–7–11. Find the range of the given real estate sales persons: $74,160; $88,563; $46,246; $90,763; $85,321; $55,463.

T–7–12. Find the average of the following yearly stock prices: +5, −16, +27, −8, −3, +31, −3, +16, −23, +7.

T–7–13. Find the range of the following hourly rates for various factory jobs: $7.65, $8.44, $5.57, $9.24, $10.56, $11.04, $9.21, $9.24, $6.67, $6.79, $8.09, $9.12.

T–7–14. Find the mode from the data in problem T–7–13.

T–7–15. Find the median from the data in problem T–7–13.

T–7–16. Find the mean from the data in problem T–7–13.

T–7–17. Find the range of the following basketball scores: 85, 53, 61, 84, 25, 36, 30, 74, 53, 66, 78, 34.

T–7–18. Find the mean from the data in problem T–7–17.

T–7–19. Find the mode from the data in problem T–7–17.

T–7–20. Find the median from the data in problem T–7–17.

T–7–21. The cross-country truck driver drove 248 miles during the first four hours, what will she need to average for the next 3 hours if the total trip is 452 miles?

T–7–22. Find the median given the following IQ scores of a group of students:

100	104	93	86	67
121	116	99	105	110
84	98	117	111	127
101	127	93	110	116
79	85	106	115	105

T–7–23. Find the first quartile from the data in problem T–7–22.

T–7–24. Find the third quartile from the data in problem T–7–22.

T–7–25. Find the 80th percentile from the data in problem T–7–22.

DEFINITIONS AND BASIC OPERATIONS OF ALGEBRA

OBJECTIVES

After studying this chapter, you will be able to:

8.1. Understand the definitions, properties, and basic operations of algebra.
8.2. Simplify algebraic expressions.
8.3. Understand exponents and their operation.
8.4. Convert written problems to equations.

SELF-TEST

The test will help you evaluate your basic algebraic skills. Show your work so that any difficulties can be noted and help obtained.

S–8–1. Combine: $+42 + (-26)$

S–8–2. Combine: $-17 + (+6)$

S–8–3. Combine: $-11xy + 7xy$

S–8–4. Combine: $+764 + (-231)$

S–8–5. Divide: $\frac{-144}{-3}$

S–8–6. Divide: $\frac{126}{-21}$

S–8–7. Multiply: $(-3)(-4)(-5)(-6)$

S–8–8. Multiply: $(-29)(-11)(+5)(-13)$

S–8–9. Simplify: $189 - (69 \div 23) \times 2 + 44$

S–8–10. Simplify: $8 \div 8 \times 8 \div 8 \times 8$

S–8–11. Simplify: $-8 - 4(x - 2y) + 6y$

S–8–12. Simplify: $-x + 3y(y - 3) + 7x$

S–8–13. Simplify: $11 - [6 \div 10(9 - 6 - 3) + 2]$

S–8–14. Simplify: $35 - [9z - (7z + 2) + 3(5z + 3)]$

S–8–15. Multiply: $y^2y^3y^4$

S–8–16. Multiply: $(3x^2)(2x^3)$

S–8–17. Divide: $\frac{25x^3y^4}{5x^2y^3}$

S–8–18. Divide: $\frac{3xy^5z^2}{18x^2yz^3}$

S–8–19. Divide: $\frac{-36a^3b^2c^1}{-6ab^2c^3}$

S–8–20. Multiply: $(-5a^2)^3$

Express the following as algebraic expressions.

S–8–21. The sum of 3, a, and b^3.

S–8–22. The quotient of seven more than h and 7.

S–8–23. The area of a triangle equals one-half the product of the base (b) times the altitude (a).

S–8–24. The side (a) opposite of the 30° angle of 30°–60°–right triangle equals one-half the hypotenuse (c).

S–8–25. The cost to rent money per day is interest (i), equals the principal (p) times the rate (r) times the time in days (d) divided by 360.

8.1 DEFINITIONS, PROPERTIES, AND BASIC OPERATIONS OF ALGEBRA

Algebra is a continuation of arithmetic, and a branch of mathematics that uses symbols (often letters) to represent numbers. The rules and procedures of arithmetic apply to algebra.

Many problems that would be difficult to solve with arithmetic can be solved using algebra. Knowledge of algebra is essential to solving many problems in business and industry.

Algebra may be considered a language of symbols, numbers, and operations with numbers.

A *numeral* is a symbol that names a number according to a system of numeration. Arithmetic and algebraic numbers are: 0, 1, 2, 3, 4, 5, 6, 7, 8, 9.

A *definite number* has an exact meaning or value; for example, the number of people in a room.

A *measured number* is a quantity obtained by measuring and is approximate; for example, the weight of a person.

Integers are numerals, and include both positive and negative numbers, as shown on the number line.

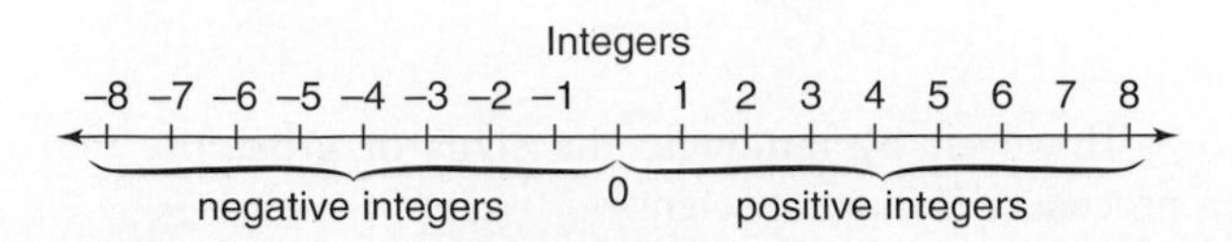

A *rational number* can be expressed in the form $^a/_b$, where b is not zero. Rational numbers are generally called fractions.

An *irrational* number cannot be expressed in the form of $^a/_b$. The square roots of numbers other than perfect squares are examples of irrational numbers—they never come out even. An *algebraic expression* uses signs and symbols to represent numbers and quantities. A *numerical algebraic expression* contains only numbers. The quantities separated by a positive or negative sign are called *terms*. Thus $5xy + 3c - z$ is an algebraic expression, and the terms are $+5xy$, $+ 3c$, and $-z$. In the term $6xy$, the 6, x, and y are *factors* of the term. The 6 is the *numerical coefficient;* the x and y are *literal coefficients.* In the example x^5, x is the *base* and 5 is the *exponent* or *power.*

A *monomial* is an algebraic expression that has one term, for example, $5x$. A *binomial,* $4k - j$, has two terms; a *trinomial,* $l + k - j$, has three terms; and a *polynomial,* $k + m + n + s$, has two or more terms. *Like terms* are quantities that have the same literal factors, $4k + 5k - 2k$. *Unlike terms* are terms that have unlike literal factors or powers. Algebra has the same operations as arithmetic (addition, subtraction, multiplication, and division). Multiplication may be expressed in several different ways. The following symbols all indicate multiplication: $a \times b = a \cdot b = (a)\ (b) = a(b) = (a)b$. The *signs of quantity* indicate whether a number is *positive* (+) or *negative* (−). When someone gives you \$5, the number is positive; when you give someone \$5, the number is considered negative. A plus sign in front of a number indicates that it is positive; a negative or minus sign in front of a number indicates a negative. When neither a plus nor a minus sign appears in front of a number or quantity, it is considered positive.

The *absolute value* of a number is the distance between zero and that number on the number line. Absolute value is always positive. The symbol for absolute value is 11, two parallel lines one on each side of the number. Therefore, the absolute value of $|-3| = 3$ and $|+3| = 3$.

−3 +3

← 3 units →← 3 units →

Thus $|-7 + 4| = |-3| = +3$ and $|-52| = 52$

Often terms are grouped together. Special signs are used to indicate groups. Quantities within a group are to be treated as one quantity. Grouping symbols are as follows:

Parentheses	( )
Brackets	[ ]
Braces	{ }
Vinculum or bar	——

Parentheses, brackets, and braces are placed around groups, and a vinculum, or bar, is placed over a group. A vinculum or bar is generally used in writing complex fractions, the division sign, or a radical sign.

Addition of Signed Values

An algebraic sum may have a positive or negative value. The algebraic sum of two quantities with like signs is the sum of their values, with the same sign in front of that value. The algebraic sum of two or more quantities with unlike signs is the difference of their values, with the sign of the larger quantity. These concepts can be seen in the following examples.

Example 8–1: Addition of Positive Numbers

PROBLEM: Add +3, +2, and +4.

SOLUTION: Plot the given numbers on the number line starting at zero and moving to the right for each number. The number line in Figure 8–1 shows the result.

Example 8–2: Addition of Positive and Negative Numbers

PROBLEM: Find the value of +5, −4, and −3.

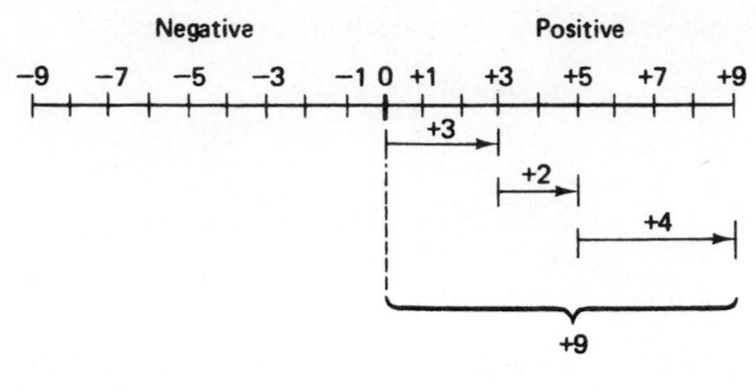

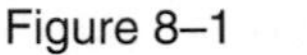
Figure 8–1

SOLUTION: Plot the given numbers on the number line in the order they are given, moving the direction the sign indicates. For example, move to the right from zero for +5 and to the left for −4 and −3. When all the numbers have been plotted, note the distance from zero to the right or left. In this problem the distance to the left of zero is negative 2. Thus the result is −2. See Figure 8–2.

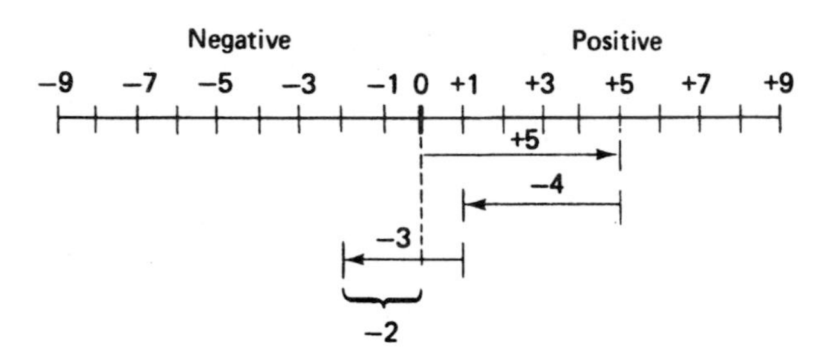

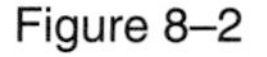
Figure 8–2

In algebra, subtraction is generally not performed as it is in arithmetic. Most algebraic expressions are written linearly and we remove the signs of grouping and combine like quantities—this is equivalent to subtraction. With algebra, when expressions are written in linear form, remove the signs of grouping and combine the quantities. This is the way subtraction is performed in algebra. When removing signs of grouping, note that like signs are positive and unlike signs are negative.

By definition, subtraction means finding the difference between two quantities. If the numbers are signed, it is still necessary to find the difference. This can be shown on the number line. For example, $+3 - (-5) =$

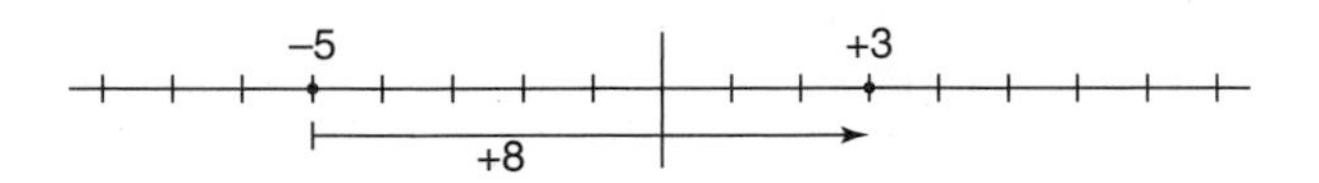

However, by removing the signs of grouping, the procedure is more efficient.

$$+3 \underbrace{-(-5)} =$$
$$+3 \quad +5 \quad = 8$$

A practical application of these concepts is temperature. If the temperature on January 15 drops from 30° during the day to −10° at night, the change is −40°. This is shown on the following thermometer:

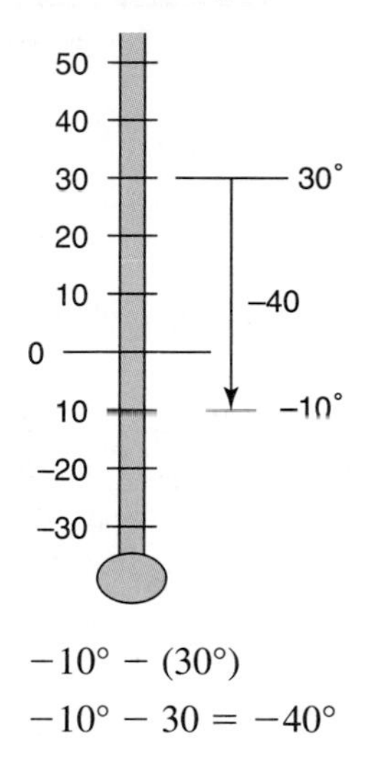

$$-10° - (30°)$$
$$-10° - 30 = -40°$$

Example 8–3: Addition and Subtraction—Remove Signs of Grouping and Evaluate

PROBLEM: Evaluate 8.4 − (−3.5).

SOLUTION: Remove the signs of grouping and combine like quantities. When removing signs of grouping, remember like signs are positive and unlike signs are negative.

$$8.4 \underbrace{-(-3.5)} =$$
$$8.4 + 3.5 \quad = 11.9$$

Example 8–4: Addition and Subtraction—Remove Several Signs of Grouping and Evaluate

PROBLEM: Evaluate −24 + (8) − (−32) − (20).

SOLUTION: Remove the signs of grouping and combine like quantities. Remember like signs are positive and unlike signs are negative.

$$-24 + (8)\underbrace{-(-32)} - (+20) =$$
$$\underbrace{-24 + 8} + 32 - 20 =$$
$$\underbrace{-16 + 32} - 20 =$$
$$+16 - 20 = -4$$

Multiplication and Division of Signed Values

When quantities with unlike signs are multiplied or divided, the results are negative. When terms with like signs are multiplied or divided, the results are positive.

Example 8–5: Multiplication of Positive Numbers

PROBLEM: If you earned $14 each day for 5 days at a part-time job, how much money did you earn?

SOLUTION: Multiply 5 days × $14.

$$5 \times 14 = 70$$

Thus you have earned $70. See Figure 8–3.

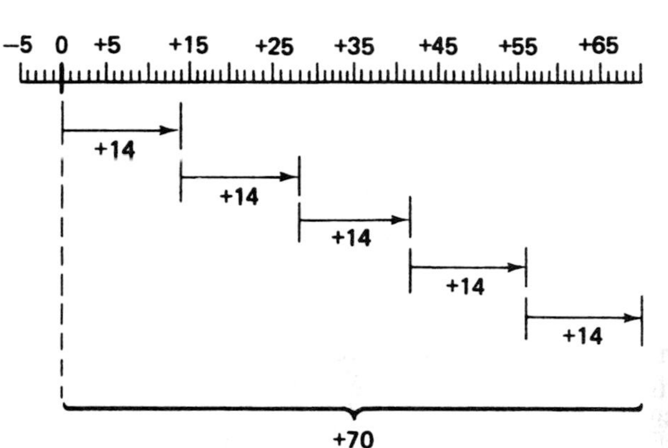

Figure 8–3

Example 8–6: Multiplication of Positive and Negative Numbers

PROBLEM: If you spend $3 a day for 7 days, what change will there be in your account?

SOLUTION: Because you are spending, you are subtracting from your account; therefore, you multiply −3 times 7.

$$-\$3 \times 7 = -\$21$$

So you would have −$21 in your account. Study the number line in Figure 8–4.

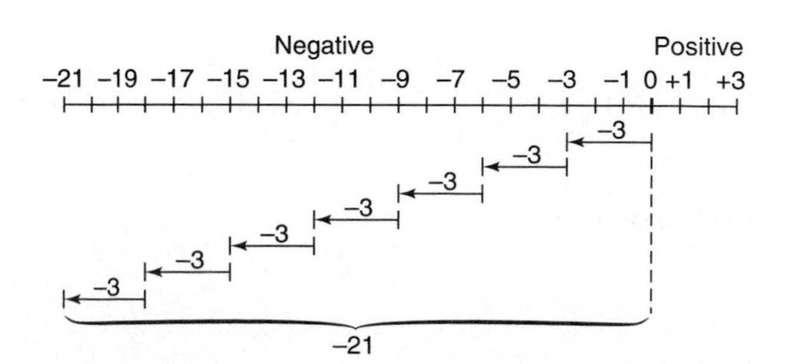

Figure 8–4

Example 8–7: Multiplication of Two Negative Numbers

PROBLEM: Multiply -2×-3.

SOLUTION: Subtraction is the reverse of addition; think of this multiplication problem as -3 subtracted from zero 2 times. Remember that like signs result in a positive sign.

$$\begin{aligned}(-2) \times (-3) &= -(-3) - (-3) \\ &= +3 + 3 \\ &= 6\end{aligned}$$

Study the number line shown in Figure 8–5.

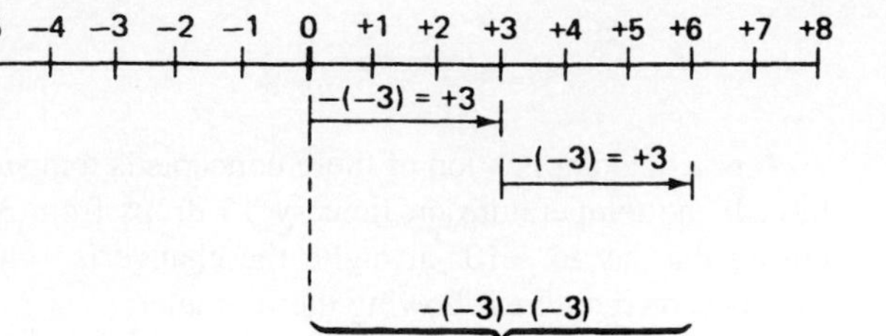

Figure 8–5

Example 8–8: Division of Positive and Negative Numbers

PROBLEM: Divide -42 by 7.

SOLUTION: Write the problem in a fractional form and simplify.

$$\frac{-42}{7} = \frac{-6}{1} = -6$$

EXERCISE 8–1 PERFORM OPERATIONS WITH SIGN NUMBERS

Solve the following problems and indicate the sign of the result.

8–1. $+5 + 3 =$

8–2. $+8 - 16 =$

8–3. $-32 + 39 =$

8–4. $+15 + 3 =$

8–5. $-9 - (+15)$

8–6. $-4 - (-16) =$

8–7. $-5 \times +4 =$

8–8. $-6 \times -3 =$

8–9. $+8 \times -3 =$

8–10. $+16 \times +4 =$

8–11. $-4 \times -4 \times -4 =$

8–12. $-5 + 3 - 2 + 4 =$

8–13. $9 - 5 + 6 - 1 =$

8–14. $-11 + 10 - 11 + 10 =$

8–15. $-34 + 72 =$

8–16. $5 \times -3 \times -2 \times -1 =$

8–17. $\frac{28}{-7} =$

8–18. $\frac{-39}{13} =$

8–19. $\frac{-36}{-6} + \frac{-6}{+6} =$

8–20. $\frac{-20}{+5} + \frac{-25}{-5} =$

8.2 SIMPLIFYING ALGEBRAIC EXPRESSIONS

The removal of grouping symbols is a necessary operation in simplifying an algebraic expression. The sign in front of the grouping determines what operation is to be performed. When there is a negative sign in front of the grouping symbol, change all the signs within the group to the opposite sign and remove the grouping symbol. When a positive sign is in front of the grouping symbol, do not change any of the signs within the grouping symbols as you simplify the expression.

When removing several sets of grouping signs, start with the innermost set of groupings and work to the outermost, making sign changes as necessary. After the grouping symbols have been removed, complete the arithmetic operations. First, perform all multiplication and division operations from left to right. Then perform all addition and subtraction operations, from left to right. Finally, combine all like quantities.

The *signs of relation* show the value of one term with respect to another. The equal sign (=) means "is equal to." For example, $143 = 143$. The equal sign with a slash through it (≠) means "is not equal to." For example, $12 \neq 11$. An arrowhead is used to indicate that a difference exists between two quantities. For example, $5 > 4$ indicates that 5 "is greater than" 4 and $6 < 32$ indicates that 6 "is less than" 32. The arrowhead always points to the smaller quantity.

Example 8–9: Evaluating an Algebraic Expression

PROBLEM: Evaluate $3(16 \div 2) - 6$.

SOLUTION: First perform the division inside the parentheses. $3(16 \div 2) - 6 =$

$$3\left(\frac{\overset{8}{\cancel{16}}}{\underset{1}{\cancel{2}}}\right) - 6 =$$

$$3(8) - 6 =$$

Complete the multiplication, then combine 24 and −6.

$$3(8) - 6 = 18$$
$$24 - 6 = 18$$

Example 8–10: Simplifying an Algebraic Expression with Sets of Grouping

PROBLEM: Simplify $5 + \{3x - [2 - (5 + 5x)]\}$.

SOLUTION: First, remove the innermost parentheses, changing every sign within the parentheses because it is preceded by a negative sign.

$$5 + \{3x - [2 - (5 + 5x)]\} =$$
$$5 + \{3x - [2 - 5 - 5x]\} =$$

Then remove the brackets, changing every sign within the brackets as indicated.

$$5 + \{3x - [2 - 5 - 5x]\} =$$
$$5 + \{3x - 2 + 5 + 5x\} =$$

Remove the braces, but do not change any signs because the braces are preceded by a positive sign. Then combine like quantities.

$$5 + \{3x - 2 + 5 + 5x\} =$$
$$5 + 3x - 2 + 5 + 5x =$$

$$5 + 3x - 2 + 5 + 5x =$$
$$+ 8 + 8x =$$

Example 8–11: Evaluating an Algebraic Expression with Several Sets of Grouping

PROBLEM: Simplify

$$\{24 \div 6 \div 2 - (3x - 6[3 - 2x]) + 4\}$$

SOLUTION: First, remove the innermost brackets by multiplying the enclosed terms by −6.

$$\{24 \div 6 \div 2 - (3x - 6[3 - 2x]) + 4\} =$$
$$\{24 \div 6 \div 2 - (3x - 18 + 12x) + 4\} =$$

Second, remove the parentheses, changing every sign enclosed by the parentheses, because of the negative sign in front of the parentheses.

$$\{24 \div 6 \div 2 - (3x - 18 + 12x) + 4\} =$$
$$\{24 \div 6 \div 2 - 3x + 18 - 12x + 4\} =$$

Then perform the division operations, moving from left to right as shown.

$$\{24 \div 6 \div 2 - 3x + 18 - 12x + 4\} =$$
$$\left\{\frac{24}{6} \div 2 - 3x + 18 - 12x + 4\right\} =$$
$$\{4 \div 2 - 3x + 18 - 12x + 4\} =$$
$$\left\{\frac{4}{2} - 3x + 18 - 12x + 4\right\} =$$
$$\{2 - 3x + 18 - 12x + 4\} =$$

Remove the braces, but do not change any signs. Since there is no sign in front of the braces, it is understood to be positive.

$$2 - 3x + 18 - 12x + 4 =$$

Then combine like quantities.

$$2 - 3x + 18 - 12x + 4 =$$
$$24 - 15x = 24 - 15x$$

EXERCISE 8–2 SIMPLIFY ALGEBRAIC EXPRESSIONS

Simplify the following by performing the operations indicated. Show all necessary work.

8–21. $3 \times 4 + 5 \times 6 =$

8–22. $8 \times 7 - 6 \times 6 =$

8–23. $(16 \div 2) - 8 =$

8–24. $60 \div (5 \times 3) =$

8–25. $60 \div 5 \times 3 =$

8–26. $8 \div 8 \times 8 \div 8 =$

8–27. $(8 \div 8) \times (8 \div 8) =$

8–28. $24 - (90 \div 18) + 5 =$

8–29. $390 \div 13 - 45 + 10 =$

8–30. $x + y - z + z - x + y =$

8–31. $a + b - (3a + 5b) - \{a - b\} =$

8–32. $(6a + 2b) - (4b - 3b \quad 2b) =$

8–33. $3k + 2 - [2j + 5 - 6j + 3k - 3] =$

8–34. $4r - 8y - (3r - 4y) + 4r - 32y =$

8–35. $\{(r + 3t - 8) - (r - 3t + 8)\} =$

8–36. $\{-3a - 2b - (c - a)\} =$

8–37. $128 \div 2 \div 2 \div 2 \div 2 =$

8–38. $4a - [- 5b - (- 12b + 3a)] - \{a + 3b\} =$

8–39. $7b - 4\{2b + 6k - [- 2b + 6k] - 4b\} =$

8–40. $8x - [4y - (-3x + 3y) - 2x] =$

8.3 UNDERSTANDING EXPONENTS AND THEIR OPERATION

An *exponent* is a number written above and to the right of a quantity; it indicates how many times the *base number* is used as a factor. Exponents are used to abbreviate algebraic expressions: for example, $k \cdot k \cdot k \cdot k = k^4$. The letter k is the base number or base quantity, and 4 is the exponent. Study the following examples, which show the Laws of exponents.

1. When multiplying quantities with like bases containing exponents, add the exponents.

$$(j^3)(j^2) = j^{3+2} = j^5$$

or $(j \cdot j \cdot j)(j \cdot j) = j \cdot j \cdot j \cdot j \cdot j = j^5$

Thus $\boxed{(a^m)(a^n) = a^{m+n}}$

or $(3)^3(3)^2 = (3)^5 = (3)(3)(3)(3)(3) = 243$

2. When dividing quantities with like bases containing exponents, subtract the exponent of the denominator from the exponent of the numerator.

$$\frac{y^5}{y^2} = \frac{\overset{1}{\cancel{y}} \cdot \overset{1}{\cancel{y}} \cdot y \cdot y \cdot y}{\underset{1}{\cancel{y}} \cdot \underset{1}{\cancel{y}}}$$

or $\dfrac{y^5}{y^2} = y^{5-2} = y^3$

Thus $\boxed{\dfrac{a^m}{a^n} = a^{m-n}}$

or $\dfrac{(3)^5}{(3)^2} = (3)^{5-2} = (3)^3 = 27$

3. A quantity with an exponent of zero is equal to 1.

$$1 = \frac{1}{1} = \frac{\overset{1}{\cancel{a}} \cdot \overset{1}{\cancel{a}} \cdot \overset{1}{\cancel{a}} \cdot \overset{1}{\cancel{a}}}{\underset{1}{\cancel{a}} \cdot \underset{1}{\cancel{a}} \cdot \underset{1}{\cancel{a}} \cdot \underset{1}{\cancel{a}}} = \frac{a^4}{a^4} = a^{4-4} = a^0 = 1$$

Thus $\boxed{\dfrac{a^m}{a^m} = a^{m-m} = a^0}$

The following examples illustrate how exponents should be solved.

Example 8–12: Multiplying Quantities with Exponents

PROBLEM: Multiply $x^3 \cdot x^6$.

SOLUTION: $(x^3)(x^6) = x^{3+6} = x^9$ because

$$\underbrace{x \cdot x \cdot x}_{(x^3)} \cdot \underbrace{x \cdot x \cdot x \cdot x \cdot x \cdot x}_{(x^6)} = x^9$$

When you multiply quantities with the same bases, you add the exponents of like terms.

Example 8–13: Dividing Quantities with Exponents

PROBLEM: Divide $y^8 \div y^5$.

SOLUTION: $y^8 \div y^5 = y^{8-5} = y^3$ because

$$\frac{y^8}{y^5} = \frac{\overset{1}{\cancel{y}} \cdot \overset{1}{\cancel{y}} \cdot \overset{1}{\cancel{y}} \cdot \overset{1}{\cancel{y}} \cdot \overset{1}{\cancel{y}} \cdot y \cdot y \cdot y}{\underset{1}{\cancel{y}} \cdot \underset{1}{\cancel{y}} \cdot \underset{1}{\cancel{y}} \cdot \underset{1}{\cancel{y}} \cdot \underset{1}{\cancel{y}}} = y^3$$

When you divide the bases, subtract the exponents (denominator from numerator) of the like terms.

Example 8–14: Raising a Quantity to a Power

PROBLEM: Simplify $(6x^2y^3)^2$.

SOLUTION: Write the quantity in the parentheses two times and multiply.

$$\begin{aligned}(6x^2y^3)^2 &= (6x^2y^3)(6x^2y^3)\\ &= (6)(6)x^{2+2}y^{3+3}\\ &= 36x^4y^6\end{aligned}$$

When you raise to a power, multiply the numerical factors and add the exponents of the like bases.

EXERCISE 8–3 PERFORM OPERATIONS WITH EXPONENTS

Perform the operations indicated and evaluate when necessary.

8–41. $m^3 \cdot m^4 =$

8–42. $x^3 \cdot x^4 \cdot x^2 =$

8–43. $5^4 \cdot 5^1 =$

8–44. $3 \cdot 3 \cdot 3^2 =$

8–45. $a^5 \cdot a^{10} \cdot a^{15} =$

8–46. $xy \cdot xy \cdot xy =$

8–47. $x^3y^3 \cdot x^4y^4 \cdot x^1y^2 =$

8–48. $(xyz)(xyz) =$

8–49. $(xy)^3 =$

8–50. $(-4)^2(-4)^1(-4)^3 =$

8–51. $\dfrac{x^5}{x^4} =$

8–52. $\dfrac{x^3y^2}{x^2y^2} =$

8–53. $(x^3)^4 =$

8–54. $(x^3y^2)^3 =$

8–55. $-(x^1y^2z^3)^4 =$

8–56. $\left(\frac{y^3}{y^2}\right)^3 =$

8–57. $\left(\frac{2x^2}{4y^3}\right)^2 =$

8–58. $\left(\frac{3^2x^2y^3}{x^1y^2}\right)^2 =$

8–59. $\frac{12r^1s^2t^3}{6rst} =$

8–60. $\left(\frac{16x^2r^4s^5}{2^4x^2s^5r^4}\right)^2 =$

8.4 CONVERT WRITTEN PROBLEMS TO EQUATIONS

One of the more difficult tasks in mathematics is to change a written problem to a formula or an equation. Some of the reasons for this difficulty are (1) a lack of understanding of the problem and what is to be answered; (2) a lack of the skill to translate words and ideas into symbols; or (3) a lack of everyday application and use of the formulas or equations. Mathematical operations are often expressed by words and phrases that mean the same thing. Some of these are as follows:

Addition: add; sum; combine; total.

Subtraction: find the difference; subtracted from; is how much greater; decreased by; the difference between; is less than.

Multiplication: find the product; square the quantity; cube the quantity; double the quantities.

Division: divided by; find the quotient.

As you work the following word problems, keep in mind what the words tell you to do, how the words are translated into symbols, and how the problem is solved.

Example 8–15:

PROBLEM: Translate to an equation: The sum of three numbers doubled is 38.

SOLUTION: Define the numbers:

x = the first number
y = the second number
z = the third number

Sum means to add and *double* means to multiply all the numbers by 2. Therefore,

$$2(x + y + z) = 38$$

Example 8–16:

PROBLEM: Translate to an equation: The distance traveled depends on the time and the speed of travel.

SOLUTION: Define the quantities involved:

d = the distance
r = the rate of travel
t = the time traveled

Since the distance traveled depends on the hours traveled times the rate of travel, we have

$$d = rt$$

Example 8–17:

PROBLEM: Translate to an equation: The volume of a cone is $^1/_3$ times the radius times the radius times the height times π.

SOLUTION: Define the given quantities: V = volume, r = radius, h = height, $\pi = 3.142$. Thus

$$V = \frac{1}{3}(r)(r)(h)(3.142)$$

$$\text{or } V = \frac{3.142r^2h}{3}$$

$$V = 1.047r^2h$$

EXERCISE 8–4 CONVERT WRITTEN PROBLEMS TO EQUATIONS

Translate the following word problems into formulas or equations. Review the previous examples if necessary.

8–61. Find the sum of two numbers minus a third number.

8–62. Subtract y from the sum of x^3 and z^4.

8–63. The product of two numbers divided by a third number equals 6.

8–64. To convert liters to quarts, multiply by 1.05.

8–65. To convert feet to inches, multiply by 12.

8–66. The perimeter of a square equals the sum of the lengths of the sides.

8–67. The area of a square equals the length of one side times the length of another side.

8–68. The perimeter of a rectangle equals the sum of the lengths of the sides.

8–69. The circumference of a circle equals the diameter times π.

8–70. Miles per gallon equals the miles driven divided by the number of gallons used.

8–71. Four times k plus 2 times p equals f.

8–72. The surface area of a cube equals 6 times the area of each face.

8–73. The sum of three consecutive numbers is 37.

8–74. Square the sum of a plus b minus c cubed.

8–75. The quotient of j and k times l equals t minus s.

8–76. The number of gallons of water in a container is equal to the product of length, width, height, and 7.5.

8–77. The hypotenuse squared of a right triangle equals the sum of the lengths of the other two sides squared.

8–78. The strength of a board depends on the product of the width and the depth squared, divided by the length.

8–79. The weight in pounds of water in a tank equals the product of the length, width, height, and 62.5.

8–80. The total cost of carpeting a room depends on the area in square yards times the cost per square yard. Determine the cost if the room's dimensions are x yards by y yards. Let C be the total cost and let c be the cost per yard.

THINK TIME

One of the symbols frequently used in mathematics is zero, 0. Several facts about zero are important when doing algebraic operations.

1. Zero is the point between positive and negative numbers. It is the only quantity that is neither positive nor negative.
2. When determining the powers of 10, each additional zero to the right of the number 10 indicates an additional power of 10. For example:

   $10 = 10^1$, $100 = 10^2$, $1000 = 10^3$

3. The addition or subtraction of zero to or from a number results in the given number.
4. When any number is multiplied by 0, the product is 0.
5. When 0 is divided by any other number, the quotient is 0.
6. Division of any number by 0 is *impossible* or *undefined.*
7. Any quantity to the zero power equals 1.

PROCEDURES TO REMEMBER

1. The algebraic sum of quantities with *like* signs is the sum of their values, with the same sign.
2. The algebraic sum of quantities with *unlike* signs is the difference of their values, with the sign of the greater value.
3. When terms with like signs are multiplied or divided, the results are positive.
4. When terms with unlike signs are multiplied or divided, the results are negative.
5. When removing grouping symbols with a positive sign in front, do not change any signs within the group.
6. When removing grouping symbols with a negative sign in front, change each of the signs within the group to the opposite sign indicated.
7. When removing several sets of grouping symbols, start with the innermost set of groupings and work to the outermost, making sign changes as necessary.
8. After grouping signs have been removed and all necessary sign changes have been made, perform all multiplication and division operations from left to right and then perform all addition and subtraction operations, moving from left to right.
9. When multiplying quantities with like bases with exponents, add the exponents.
10. When dividing quantities with like bases with exponents, subtract the exponent of the denominator from the exponent of the numerator. When a quantity is moved from the numerator to the denominator or vice versa, change the sign of the exponent.

CHAPTER SUMMARY

1. A *numeral* is a symbol that names a number according to a system of numeration.
2. A *definite number* has an exact meaning or value.
3. A *measured number* is an approximate quantity obtained by measuring.
4. *Integers* are numerals, and include both positive and negative numbers.
5. A *rational* number can be expressed in the form $^a/_b$, where b is not zero.
6. An *irrational number* cannot be represented as a ratio of integers.
7. An *algebraic expression* is an expression that uses signs and symbols to represent numbers and quantities.
8. A *numerical algebraic expression* is made up entirely of numbers.
9. A *term* is part of an expression and is separated from other terms by a positive or negative sign.
10. One or more *factors* may be combined to form a term.
11. A *numerical coefficient* refers to the numerical part of a term.
12. A *literal coefficient* refers to literal parts of a term.
13. The *base* is the quantity that is used as a factor when working with exponents.
14. An *exponent* or *power* indicates the number of times the base is multiplied by itself.
15. A *monomial* is an algebraic expression that has one term.
16. A *binomial* is an algebraic expression that has two terms.
17. A *trinomial* is an algebraic expression that has three terms.
18. A *polynomial* is an algebraic expression that has two or more terms.
19. *Like* terms are quantities that have the same literal factors.
20. *Unlike* terms are quantities that have unlike literal factors.
21. *Signs of quantity* indicate whether a number is positive or negative.
22. *Parentheses, brackets, braces,* and *vincula* or *bars* are used to group terms.
23. *Signs of operation* indicate what operation should be performed.
24. The equal sign (=) means "is equal to" and an equal sign with a slash through it (≠) means "is not equal to."
25. An arrowhead is used to indicate that a difference exists between two quantities. For example, $8 > 4$ indicates that 8 "is greater than" 4, and $8 < 9$ indicates that 8 "is less than" 9. The arrowhead always points to the smaller quantity.
26. A quantity with a zero exponent is equal to 1.

27. Addition means to "combine," "total," "add," and "sum."
28. Subtraction means to "find the difference," "subtracted from," "is how much greater than," "decreased by," and "the difference between."
29. Multiplication means to "find the product," "square the quantity," "cube the quantity," and "double the quantities."
30. Division means to "divide by" and "find the quotient."
31. The *absolute value* of a number is the distance between zero and the number on the number line and is always positive.

CHAPTER TEST

T–8–1. Combine: $-35 + 19$

T–8–2. Combine: $-140 + 93 + 120 - 62$

T–8–3. Combine: $-23 + (18) =$

T–8–4. Combine: $-98 - (-46) =$

T–8–5. Multiply: $(-4)(+5)(-6) =$

T–8–6. Multiply: $(-30)(-2)(-2) =$

T–8–7. Divide: $\dfrac{-111}{+3}$

T–8–8. Divide: $\dfrac{+27}{-3}$

T–8–9. Combine: $-66ab + (76ab)$

T–8–10. Simplify: $84 + 36 \times 8 \div 4 \times 2 - 12 =$

T–8–11. Simplify: $8 - 0(6 \div 3) - 16 \div 2 =$

T–8–12. Simplify: $3x - 7 - 2(x - 5) + 15 =$

T–8–13. Simplify:
$4 + \{10 - [-6 - (-7 + 4) - 2]\} =$

T–8–14. Simplify:
$(5x - 2) - 2\{3x - [5x + 7(3x - 4) + 3]\} =$

T–8–15. Multiply: $-b^3 \cdot b^5 =$

T–8–16. Multiply: $(12x^3)(12x^2y^3) =$

T–8–17. Divide: $\dfrac{-64x^3y^5z^2}{-16x^3y^4z} =$

T–8–18. Divide: $\dfrac{45j^3k^3l^3}{-5j^3k^2l^1} =$

T–8–19. Solve: $(-4j^3)^4 =$

Express the following as algebraic expressions.

T–8–20. The difference between x and y divided by r^3.

T–8–21. The sum of a and b minus c multiplied by x^2 divided by y^4 minus 7.

T–8–22. The product of 36 minus z times j.

T–8–23. The area of a trapezoid equals one-half the sum of the bases (b_1 and b_2) times the altitude (a).

T–8–24. The perimeter of a $60° - 60° - 60°$ triangle equals three times one side (s).

T–8–25. The cost (C) to operate a car equals the cost per gallon (c) times the gallons (g) plus insurance (i) plus oil (o) plus tires (t) plus maintenance $c(m)$ plus other expenses.

9

BASIC EQUATIONS

OBJECTIVES

After studying this chapter, you will be able to:

9.1. Understand the definitions and basic operations of equations.
9.2. Solve basic equations using multiplication and division.
9.3. Solve literal equations.

SELF-TEST

This basic equations test will determine what you need to study in this chapter. Solve for the unknown, showing all necessary work. Check your work.

S–9–1. $x - 5 = 3$

S–9–2. $4x + 5 = 17$

S–9–3. $x + 3 = 10$

S–9–4. $2x + 3 = x + 4$

S–9–5. $4b = 36$

S–9–6. $17x - 7x = x + 18$

S–9–7. $4x + 5 - 7 = 2x + 6$

S–9–8. $9y - 19 = y - 2y + 11$

S–9–9. $4x - 10 = 2x + 2$

S–9–10. $9x + 9 = -3x + 15$

S–9–11. $300x - 250 = 50x + 750$

S–9–12. $2.5x + 0.5x = 1.5x + 1.5$

S–9–13. $x + 2x + 3 - 4x = 5x - 9$

S–9–14. $2y + 3y - 4 = 5y + 6y - 16$

S–9–15. $75z - 150 = 80z - 300$

S–9–16. $(4x + 6) - 2x = (x - 6) + 24$

S–9–17. $15y - [3 - (4y + 4) - 57] = -(2 + y)$

S–9–18. $4t - (12t - 24) + 38t - 39 = 0$

S–9–19. $47r - 17 = 235 - 37r$

S–9–20. $ax + b = 7$. Solve for x.

S–9–21. $3(b - y) = 5(b - 2y)$. Solve for y.

S–9–22. $5(x - 2b) = -3x + 2(x + 2b)$. Solve for x.

S–9–23. $\frac{3}{4}(3a - 2x) = \frac{1}{4}x + \frac{1}{4}a$. Solve for x.

S–9–24. $5.2(x + 3) + 3.7(2 - x) = 3$. Solve for x.

S–9–25. $(x + 3)^2 = (x - 4)^2$. Solve for x.

9.1 DEFINITIONS AND BASIC OPERATIONS OF EQUATIONS

We use many formulas or literal equations to solve problems. As technology continues to advance, more formulas will be developed to solve new problems. It is essential that technical workers have knowledge of formulas and methods used to solve problems. In this chapter we give explanations, illustrations, and an opportunity to work with formulas to become skilled in their use. Many people fail to understand the relationship between formulas and equations. What a technical worker calls a formula, the mathematician calls a literal equation.

As you learn the fundamental operations of equations, be aware that these are the same procedures used to solve formulas used in business and industry.

An *equation* is a statement with two algebraic expressions that are equal. For example, $3x = 12$ has two algebraic expressions, $3x$ and 12, that are equal. A *formula* is a rule or procedure used in mathematics, science, or industry to solve problems. Formulas are generally composed of letters, symbols, and constant terms. *Literal quantities* are letters that represent numbers or unknown quantities.

Historically, the symbol for justice and equality has been the balance scale, often called the "scales of justice." The scales of justice can serve as an aid to understanding equations. See Figure 9–1. The balance point A is equivalent to the equal sign (=) of an equation. To keep the scales balanced, if you add, subtract, multiply, or divide on one side of the scale, you must do the same operation on the other side of the scale. Failure to do the same operation on each side will result in an imbalance or error.

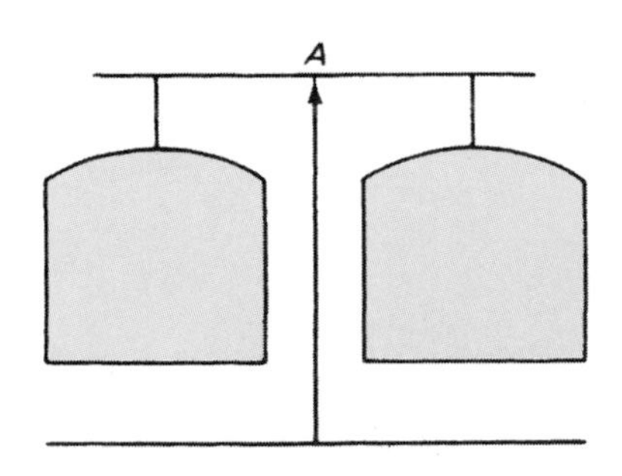

Figure 9–1

Study Figure 9–2. Note in each equation, what is given on the left side is equal to what is given on the right side.

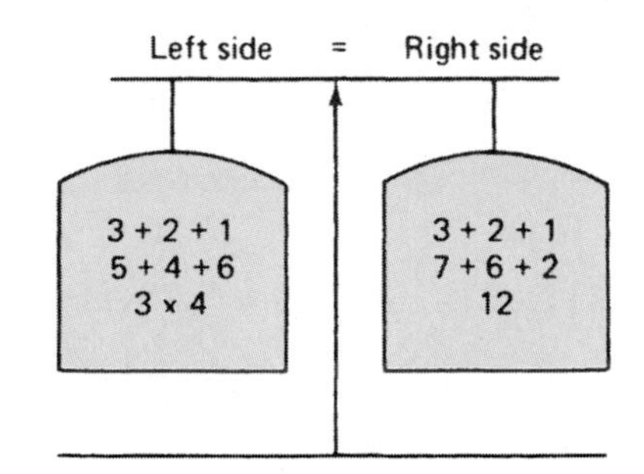

Figure 9–2

The quantities on the left side of the equation are known as the *left member.* The quantities on the right side are known as the *right member.* The equal sign in the center is the *sign of equality* or the balance point of the equation. Remember, whatever mathematical operation you do on one side of the equation must be done on the other side. Failure to practice this principle will result in an "injustice." On the job, incorrect use of a formula may result in an injustice to yourself or to your customer.

The procedures used in solving equations are the same as those used in solving formulas. Using the skills you will learn solving equations will help you in solving formulas.

The four basic operations used in solving equations are the same as the fundamental operations of arithmetic: addition, subtraction, multiplication, and division. However, when used to solve equations, they are called the *axioms* or *rules* of algebra.

Study Figures 9–3 to 9–6. Note that the balance scale remains balanced as weights are added, subtracted, multiplied, or divided in equal amounts.

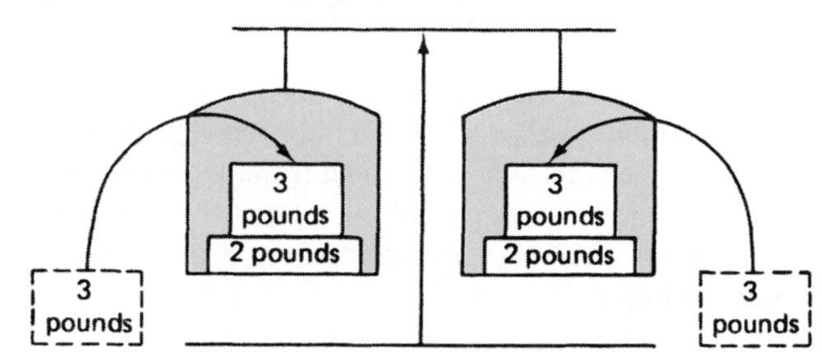

Figure 9–3 Add equal weight to each side.

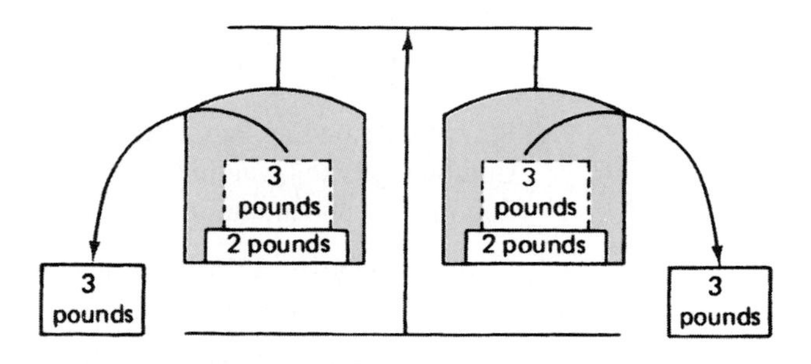

Figure 9–4 Subtract equal weight from each side.

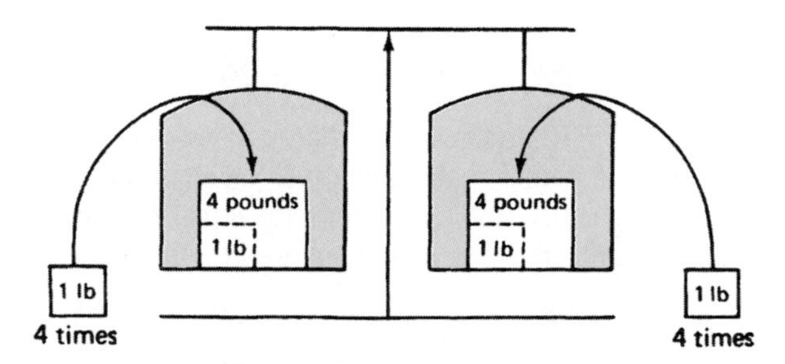

Figure 9–5 Multiply the weight on each side by an equal quantity.

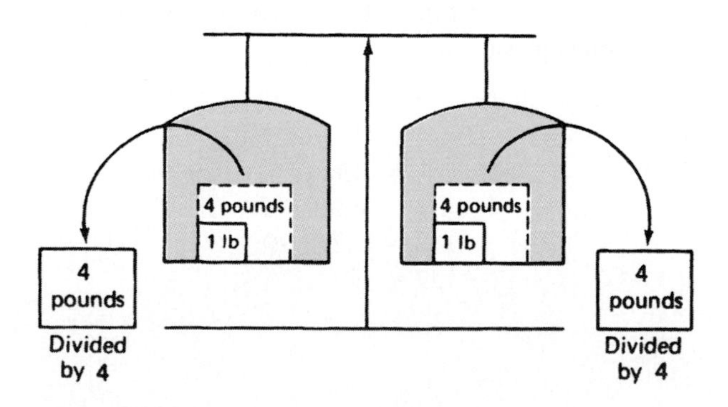

Figure 9–6 Divide the weight on each side by an equal quantity.

The solution or value of the unknown quantity is called the *root* of the equation. When the root is found, the problem has been solved or the *solution set* has been determined. Always check your solutions to prevent costly mistakes or errors. Always check your solutions to be certain you have the correct balance or that both sides of the equation are equal.

The following examples illustrate the basic axioms of algebra: addition, subtraction, multiplication, and division. Note there is often more than one way to solve an equation.

Solving Equations Using Addition or Subtraction

Addition Axiom

The quantity added to the left side of the equation must also be added to the right side.

Example 9–1: Solve the Equation Using Addition

PROBLEM: Find the value of X in the equation $X - 3 = 2$.

SOLUTION A: Solve this equation by *adding* 3 to each side of the equation.

$$\begin{array}{rcr} X - 3 & = & 2 \\ +\ 3 & = & +3 \\ \hline X \quad & = & 5 \end{array}$$

CHECK: Substitute 5 for X in the original equation and evaluate.

$$X - 3 = 2$$
$$5 - 3 = 2$$
$$2 = 2$$

SOLUTION B: This problem could be solved by transposition, which is the process of moving a quantity from one side of the equal sign to the other side. When a quantity is moved to the opposite side, its sign is changed. This procedure is equivalent to adding the same quantity to each side of the equation. You could solve this equation by moving the -3 to the other side of the equal sign and changing it to $+3$. Thus

$$X - 3 = 2$$
$$X = 2 + 3$$
$$X = 5$$

Subtraction Axiom

The quantity that is subtracted from the left side of the equation must also be subtracted from the right side.

Example 9–2: Solve the Equation Using Subtraction

PROBLEM: Find the value of X in the equation $X + 4 = 6$.

SOLUTION A: Solve this equation by *subtracting* 4 from each side of the equation.

$$\begin{array}{rcr} X + 4 & = & 6 \\ -4 & = & -4 \\ \hline X & = & 2 \end{array}$$

CHECK: Substitute 2 for X in the original equation and evaluate.

$$X + 4 = 6$$
$$2 + 4 = 6$$

SOLUTION B: This problem may be solved by transposition, by moving the $+4$ to the opposite side of the equation sign and changing its sign from plus to minus. This is equivalent to subtracting 4 from each side of the equation.

Thus

$$X + 4 = 6$$
$$X = 6 - 4$$
$$X = 2$$

Example 9–3: Solve the Equation Using Subtraction and Addition

PROBLEM: Find the value of X in the equation $8X - 5 = 7X + 19$.

SOLUTION A: Solve this equation by *subtracting* $7X$ from each side of the equation and then *adding* 5 to each side of the equation.

$$\begin{array}{rcr} 8X - 5 & = & 7X + 19 \\ -7X \quad & = & -7X \quad \\ \hline X - 5 & = & +19 \end{array}$$

$$\begin{array}{rcr} X - 5 & = & 19 \\ +5 & = & +5 \\ \hline X & = & 24 \end{array}$$

CHECK: Substitute 24 for X in the original equation and evaluate.

$$8X - 5 = 7X + 19$$
$$8(24) - 5 = 7(24) + 19$$
$$192 - 5 = 168 + 19$$
$$187 = 187$$

SOLUTION B: This problem may be solved by transposition: $+7X$ is moved to the opposite side of the equal sign and its sign is changed from positive to negative. Then the -5 is moved to the opposite side of the equal sign and its sign is changed from negative to positive. Observe:

$$8X - 5 = 7X + 19$$
$$8X - 7X - 5 = +19$$
$$8X - 7X = 19 + 5$$
$$X = 24$$

Example 9–4: Solve by Removing Parentheses, Combining Like Quantities, Subtraction, and Addition

PROBLEM: Find the value of X in the equation $3(3X - 4) = 4(2X + 6) - 20$.

SOLUTION: Remove the parentheses by completing the multiplication indicated.

$$3(3X - 4) = 4(2X + 6) - 20$$
$$9X - 12 = 8X + 24 - 20$$

Then move -12 and $+ 8X$ to the opposite side of the equation and combine the quantities. Remember when moving a quantity from one side of the equal sign to the other to change the sign on the quantity.

$$9X - 12 = 8X + 24 - 20$$
$$9X - 8X = 12 + 24 - 20$$
$$X = 16$$

CHECK: Substitute 16 for X in the original equation and evaluate.

$$3(3X - 4) = 4(2X + 6) - 20$$
$$3[3(16) - 4] = 4[2(16) + 6] - 20$$
$$3[48 - 4] = 4[32 + 6] - 20$$
$$3[44] = 4[38] - 20$$
$$132 = 152 - 20$$
$$132 = 132$$

EXERCISE 9–1 SOLVE BASIC EQUATIONS USING ADDITION AND SUBTRACTION

Solve the following equations using either method. Check your solutions by substituting your answer for the unknown in the original equation. Study the examples if you need help. Remember that the same operation must be done on both sides of the equation.

9–1. $X + 3 = 5$

9–2. $X - 6 = 10$

9–3. $3X - 4X = -2X + 10$

9–4. $8y - 12 = 7y - 11$

9–5. $X - 7 = 13$

9–6. $3 + a = 10$

9–7. $k - 12 = 3$

9–8. $r - 4 = -18$

9–9. $X + 14 = 8$

9–10. $2(X + 10) = X + 30$

9–11. $3(X - 1) + 2X = 2(3X + 4)$

9–12. $4(3 - 2X) = 7(5 - X)$

9–13. $3(3X + 3) = 4(2X - 4)$

9–14. $3(X - 8) = 2(X - 14)$

9–15. $6(2x) - 16 = 4x + 32$

9–16. $2(18 - 4x) = 4(2x + 4)$

9–17. $5(2x - 5) = 4x + 5$

9–18. $8(x - 4) = 2x + 10$

9–19. $4(x + 8) = 6(x + 6) + 4$

9–20. $7(x - 1) - 2 = 5(x + 1)$

Solve Basic Equations Using Multiplication and Division

Not all equations can be solved by addition and subtraction procedures. For example, in the equation $5X = 15$, the unknown is greater than 1 unit; thus the unknown should be divided by 5 so that it becomes 1 unit. However, remember to divide both sides of the equation by 5.

> **Division Axiom**
>
> Each side of the equation must be divided by the same quantity.

Example 9–5: Solve Using Division

PROBLEM: Find the value of X in the equation $5X = 15$.

SOLUTION: Divide both sides of the equation by 5.

$$5X = 15$$

$$\frac{\overset{1}{\cancel{5}}X}{\underset{1}{\cancel{5}}} = \frac{\cancel{15}}{\underset{1}{\cancel{5}}}$$

$$X = 3$$

CHECK: Substitute $X = 3$ into the original equation and evaluate.

$$5X = 15$$
$$5(3) = 15$$
$$15 = 15$$

To solve $\frac{2}{3}X = 12$, an equation where the unknown is a fraction, multiply both sides by $\frac{3}{2}$, the reciprocal of $\frac{2}{3}$. If the unknown is a fraction, multiply *both* sides of the equation by the reciprocal of the fraction. Check by substituting the root into the original equation.

> **Multiplication Axiom**
>
> Each side of the equation must be multiplied by the same quantity.

Example 9–6: Solve Using Multiplication

PROBLEM: Find the value of X in the equation $\frac{3}{2}X = 12$.

SOLUTION: Multiply both sides of the equation by $\frac{3}{2}$, the reciprocal of $\frac{3}{2}$.

$$\frac{2}{3}X = 12$$

$$\left(\frac{\cancel{3}^{1}}{\cancel{2}_{1}}\right)\frac{\cancel{2}^{1}}{\cancel{3}_{1}}X = \frac{\cancel{12}^{6}}{1}\left(\frac{3}{\cancel{2}_{1}}\right)$$

$$X = 18$$

CHECK: Substitute $X = 18$ into the original equation and evaluate.

$$\frac{2}{3}X = 12$$

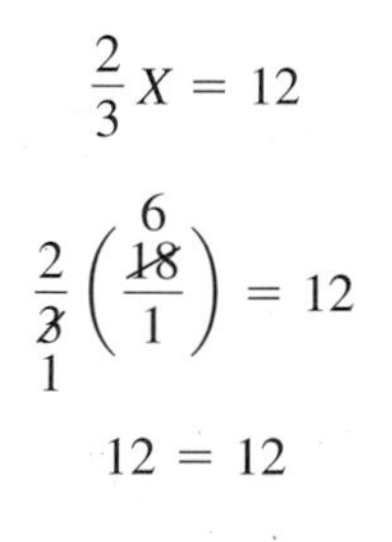

$$\frac{2}{\cancel{3}_{1}}\left(\frac{\cancel{18}^{6}}{1}\right) = 12$$

$$12 = 12$$

Solving Equations Using Combined Methods

Study the following examples, which use more than one of the algebra axioms.

Example 9–7: Solve by Removing Parentheses, Combining Terms, Adding, and Dividing

PROBLEM: Find the value of X in the equation

$$8X - 5(4X + 3) = (-3) - 4(2X - 7)$$

SOLUTION: Remove the parentheses by multiplying.

$$8X - 20X - 15 = (-3) - 8X + 28$$

Combine the like quantities as indicated.

$$8X - 20X - 15 = (-3) - 8X + 28$$
$$-12X - 15 = -8X + 25$$

Transpose the quantities as indicated.

$$-12X - 15 = -8X + 25$$
$$-25 - 15 = -8X + 12X$$

Combine like quantities as indicated.

$$\underbrace{-25 - 15} = \underbrace{-8X + 12X}$$
$$-40 = 4X$$

Then divide both sides by 4.

$$-40 = 4X$$

$$\frac{\cancel{-40}^{-10}}{\cancel{4}_{1}} = \frac{\cancel{4}^{1}X}{\cancel{4}_{1}}$$

$$-10 = X$$

CHECK: Substitute $X = -10$ into the original equation and evaluate.

$$8X - 5(4X + 3) = (-3) - 4(2X - 7)$$
$$8(-10) - 5[4(-10) + 3] = (-3) - 4[2(-10) - 7]$$
$$-80 - 5[-40 + 3] = (-3) - 4[-20 - 7]$$
$$-80 - 5[-37] = (-3) - 4[-27]$$
$$-80 + 185 = -3 + 108$$
$$105 = 105$$

Example 9–8: Solve by Removing Parentheses, Combining Terms, Adding and Dividing

PROBLEM: Find the value of k in the equation

$$4k + 3[2k - 4(k - 2)] = 72 - 6k$$

SOLUTION: Remove the inside parentheses.

$$4k - 6k + 24 = 72 - 6k$$

Combine like quantities.

$$4k + 3[-2k + 8] = 72 - 6k$$

Then remove the outside parentheses.

$$4k + 3[2k - 4k + 8] = 72 - 6k$$

Combine like quantities.

$$-2k + 24 = 72 - 6k$$

Transpose quantities as necessary so that the positive unknown is on the left side of the equation.

$$-2k + 6k = 72 - 24$$

Combine like quantities.

$$4k = 48$$

Divide both sides by +4.

$$4k = 48$$

$$\frac{\cancel{4}^{1}k}{\cancel{4}_{1}} = \frac{\cancel{48}^{12}}{\cancel{4}_{1}}$$

$$k = 12$$

CHECK: Substitute $k = 12$ in the original equation and evaluate.

$$4k + 3[2k - 4(k - 2)] = 72 - 6k$$
$$4(12) + 3[2(12) - 4(12 - 2)] = 72 - 6(12)$$
$$48 + 3[24 - 4(10)] = 72 - 72$$
$$48 + 3[24 - 40] = 0$$
$$48 + 3[-16] = 0$$
$$48 - 48 = 0$$
$$0 = 0$$

Example 9–9: Solve by Removing Parentheses, Combining Terms, Subtracting and Dividing

PROBLEM: Find the value of the y in the equation

$$16.5 - 1.5(2y - 0.5) - 15.6 + 2.1(y + 0.3) = 0.03$$

SOLUTION: Remove the parentheses.

$$16.5 - 3.0y + 0.75 - 15.6 + 2.1y + 0.63 = 0.03$$

Combine like quantities.

$$-0.9y + 2.28 = 0.003$$

Move 2.28 to the right side of the equation; remember to change the sign.

$$-0.9y = 0.03 - 2.28$$

Combine like quantities.

$$-0.9y = -2.25$$

Divide both sides by -0.9.

$$\frac{\overset{1}{\cancel{-0.9}}y}{\underset{1}{\cancel{-0.9}}} = \frac{\overset{2.5}{\cancel{-2.25}}}{\underset{1}{\cancel{-0.9}}}$$

$$y = 2.5$$

CHECK: Substitute $y = 2.5$ into the original equation and evaluate.

$$16.5 - 1.5(2y - 0.5) - 15.6 + 2.1(y + 0.3) = 0.03$$
$$16.5 - 1.5(2(2.5) - 0.5) - 15.6 + 2.1(2.5 + 0.3) = 0.03$$
$$16.5 - 1.5(5 - 0.5) - 15.6 + 2.1(2.8) = 0.03$$
$$16.5 - 1.5(4.5) - 15.6 + 5.88 = 0.03$$
$$16.5 - 6.75 - 15.6 + 5.88 = 0.03$$
$$0.03 = 0.03$$

Example 9–10: Solve by Removing Parentheses, Combining Terms, Subtracting and Dividing

PROBLEM: Find the value x in the equation

$$(x + 3)(x + 3) - 8 = (7 + x)(4 - x) + 2x^2$$

SOLUTION: Remove the parentheses by multiplying as shown and combine like quantities.

$$(x + 3)(x + 3) - 8 = (7 + x)(4 - x) + 2x^2$$

$$x^2 + 3x + 3x + 9 - 8 = 28 - 7x + 4x - x^2 + 2x^2$$
$$x^2 + 6x + 1 = 28 - 3x + x^2$$

Move the unknowns to the right side of the equal sign and combine the quantities.

$$x^2 + 6x + 1 = 28 - 3x + x^2$$
$$x^2 - x^2 + 6x + 3x = 28 - 1$$
$$0x^2 + 9x = 27$$

Solve for x by dividing both sides by 9.

$$\frac{\overset{1}{\cancel{9}}x}{\underset{1}{\cancel{9}}} = \frac{\overset{3}{\cancel{27}}}{\underset{1}{\cancel{9}}}$$

$$x = 3$$

CHECK: Substitute $x = 3$ into the original equation and evaluate.

$$(x + 3)(x + 3) - 8 = (7 + x)(4 - x) + 2x^2$$
$$(3 + 3)(3 + 3) - 8 - 8 = (7 + 3)(4 - 3) + 2(3)^2$$
$$(6)(6) - 8 = (10)(1) + 2(9)$$
$$36 - 8 = 10 + 18$$
$$28 = 28$$

EXERCISE 9–2 SOLVE BASIC EQUATIONS

Some of the steps in previous problems may be combined when you become more comfortable at solving equations. However, while learning, it is a good idea to show each operation you perform.

9–21. $3a - 9 = 9$

9–22. $5L + 3 = 2 + 4L$

9–23. $\frac{2}{3}X - 9 = \frac{1}{3}X$

9–24. $k - 10 = 5 + 4k$

9–25. $2t + 8 = t + 13$

9–26. $4d + 15 = 6d + 5$

9–27. $12R - 18 + 42 = 10(R + 1)$

9–28. $4X + 8 = 2(4X + 2)$

9–29. $11X + 3 - 4X = 16 - 2X + 2$

9–30. $18 - 5(3 - 2z) = z - 4(z + 9)$

9–31. $4 + 3(J - 7) = 16 + 2(J + 1) + 3J$

9–32. $4(a - 5) - 3(a - 2) = 2(a - 1)$

9–33. $(2x)^2 + 2(x) + 1 = 4x^2 + 9$

9–34. $5X - (3X - 2) = 10$

9–35. $(a + 1)(a - 2) = a^2 + 5$

9–36. $\frac{1}{5}n + 7 = 13$

9–37. $\frac{1}{3}X + \frac{1}{4}X = \frac{7}{2}$

9–38. $10(y - 2) - 10(2 - y) = 4y - 40$

9–39. $(b + 4)(5 - 2b) - b(10 - 2b) = -6$

9–40. $(a + 5)(a - 4) + 4a^2 = (5a + 3)(a - 4) + 2(a - 4) + 64$

9–41. $\frac{3}{4}x + \frac{1}{3}x = 4x - 35$

9–42. $9(k + 2) - (2k + 19) = 4k - 10$

9–43. $(h + 6)(h - 6) = (9 - h)(4 + 3h) + 4h^2 - 3$

9–44. $(2c + 3)(c - 2) - c^2 = (c + 8)(c + 2)$

9–45. $(y + 1)(y^2 - y + 1) = y^3 - 8y - 31$
(*Hint:* Perform the multiplication as shown in Example 9–10.)

Be sure that you checked all your solutions. If not—
CHECK THEM!

9.3 SOLVING LITERAL EQUATIONS

A *literal equation* is an equation with some or all of the quantities represented by letters. Letters from the beginning of the alphabet usually represent known quantities, and letters toward the end of the alphabet represent unknown quantities. When solving an equation for x, y, or z where the rest of the equation contains literal factors, you are solving a literal equation.

The same procedures are used to solve literal equations that are used to solve other equations. Furthermore, the same procedures are used to solve formulas that are used to solve the basic equations and literal equations—the algebra axioms of addition, subtraction, multiplication, and division. Letters in a literal equation are handled as though they were numbers in a regular equation. Perform the same operations with literal equations that you performed with the basic equations.

When you have learned how to work with literal equations, you will be able to work with the formulas. This section will help you gain the knowledge and skill you need to work with the formulas.

Remember that the early letters in the alphabet represent known quantities and should be treated as numbers. Study the following examples to learn how to solve literal equations.

Example 9–11: Solve the Literal Equation Using Division

PROBLEM: Find the value of x in the equation $ax = b$.

SOLUTION: Divide both sides of the equation by a.

$$ax = b$$

$$\frac{\overset{1}{\cancel{a}}x}{\underset{1}{\cancel{a}}} = \frac{b}{a}$$

$$x = \frac{b}{a}$$

CHECK: Substitute $x = {}^{b}/_{a}$ into the original equation and evaluate.

$$ax = b$$

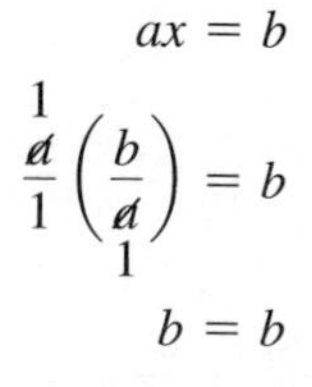

$$\frac{\overset{1}{\cancel{a}}}{1}\left(\frac{b}{\underset{1}{\cancel{a}}}\right) = b$$

$$b = b$$

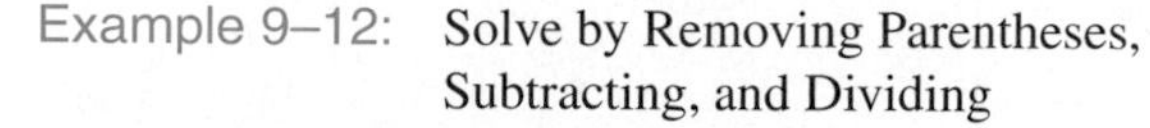

Example 9–12: Solve by Removing Parentheses, Subtracting, and Dividing

PROBLEM: Find the value of x in the equation $b(x + 1) = c$.

SOLUTION: Remove the parentheses and transpose $+b$; x will remain on the left side of the equal sign.

$$bx + b = c$$

$$bx = c - b$$

Then divide both sides by b.

$$bx = c - b$$

$$\frac{\overset{1}{\cancel{b}x}}{\underset{1}{\cancel{b}}} = \frac{c - b}{b}$$

$$x = \frac{c - b}{b}$$

CHECK: Substitute $x = \frac{c - b}{b}$ into the original equation and evaluate.

$$b(x + 1) = c$$

$$b\left(\frac{c - b}{b} + 1\right) = c$$

Perform the operations inside the parentheses first. Note that 1 is changed to ${}^{b}/_{b}$, which also equals 1.

$$b\left(\frac{c - b}{b} + \frac{b}{b}\right) = c$$

$$b\left(\frac{c - b + b}{b}\right) = c$$

$$\frac{\overset{1}{\cancel{b}}}{1}\left(\frac{c}{\underset{1}{\cancel{b}}}\right) = c$$

$$c = c$$

Example 9–13: Solve by Combining Terms, Subtracting, and Dividing

PROBLEM: Find the value of y in the equation

$$9y + 12k = 3y + 9k - y$$

SOLUTION: Combine like quantities.

$$9y - 12k = 2y + 9k$$

Transpose the unknown quantities so that all y quantities are on the left side of the equation.

$$9y - 2y = 9k + 12k$$

Combine like quantities.

$$7y = 21k$$

Solve for y by dividing both sides by 7.

$$7y = 21k$$

$$\frac{\overset{1}{\cancel{7}y}}{\underset{1}{\cancel{7}}} = \frac{\overset{3}{\cancel{21}k}}{\underset{1}{\cancel{7}}}$$

$$y = 3k$$

CHECK: Substitute $y = 3k$ into the original equation and evaluate.

$$9y - 12k = 3y + 9k - y$$
$$9(3k) - 12k = 3(3k) + 9k - 3k$$
$$27k - 12k = 9k + 9k - 3k$$
$$15k = 15k$$

Example 9–14: Solve by Removing Parentheses, Subtracting, Adding, Rewriting, and Dividing

PROBLEM: Find the value of y in the following equation:

$$x(y - 5) + z(y + 6) = x + y$$

SOLUTION: Remove the parentheses.

$$xy - 5x + zy + 6z = x + y$$

Transpose the quantities so that all the y quantities are on the left side and all the other quantities are on the right side and combine.

$$xy - 5x + zy + 6z = x + y$$
$$xy + zy - y = x + 5x - 6z$$
$$xy + zy - y = + 6x - 6z$$

Rewrite as follows:

$$y(x + y - 1) = 6(x - z)$$

(This is known as factoring the common quantity from each side of the equation.) Then divide both sides by $(x + z - 1)$.

$$\frac{y\overset{1}{\cancel{(x + z - 1)}}}{\underset{1}{\cancel{(x + z - 1)}}} = \frac{6(x - z)}{(x + z - 1)}$$

$$y = \frac{6(x - z)}{(x + z - 1)}$$

CHECK: The checking process is more difficult than working the problem, so check by reworking the complete problem.

EXERCISE 9–3 SOLVE LITERAL EQUATIONS

Solve and check the following exercises. Remember, the beginning letters of the alphabet are considered as known numbers and the later letters of the alphabet as unknowns.

9–46. $6abx = 9abc$

9–47. $k - nx = j$

9–48. $3az = 5az + 2$

9–49. $7x + 8a + 22a = 0$

9–50. $5a^3 x = 30$

9–51. $4w = (4a)(8b)$

9–52. $5(p - x) = 3(p - x)$

9–53. $20a = 14a - (x + 2a)$

9–54. $\frac{2w}{3} - 4 = 20$

9–55. $7x + 8a - 3x = 3a + 2x$

9–56. $\frac{N}{7} + 2 = 8$

9–57. $ax + b - c = 0$

9–58. $8y + (d - 2y) = 4(d - y)$

9–59. $\frac{5}{6}(3a - 2x) + \frac{1}{6}x = a$

9–60. $(2x - b)(2x + b) + b(b - 8) = 4x(x - 5) - 48b$

THINK TIME

Develop some formulas that could be used in your job. Practice using these formulas to solve current or potential problems. Try to think of areas where formulas may be used if there are no formulas currently available. If you have the opportunity, discuss the formulas with a supervisor. You may find that there are formulas that could be used to solve problems and make work easier and more efficient.

PROCEDURES TO REMEMBER

1. To solve equations by addition and subtraction:
   (a) Remove parentheses, starting with the innermost set of parentheses.
   (b) Combine like quantities on each side of the equation.
   (c) Using either the addition and subtraction axioms or the transposition method, move the unknown quantities to one side of the equation and the known quantities to the other side of the equation.
   (d) Combine like quantities.
   (e) Solve for the unknown.
   (f) Check your solution by substituting the value of the unknown into the original equation and evaluating.
2. To solve equations by multiplication and division:
   (a) Remove parentheses, starting with the innermost set of parentheses.
   (b) Combine like quantities; starting from left to right, perform multiplication and division, then addition and subtraction, from left to right.

(c) Transpose quantities as necessary so that the positive unknown is on one side of the equal sign and the known quantities are on the other.
(d) Combine like quantities.
(e) Solve for the unknown by either the multiplication or division axiom.
(f) Check your solution by substituting the value of the unknown into the original equation and evaluating.

3. To solve literal equations:
   (a) Handle all literal factors as numbers.
   (b) Remove parentheses, starting from the innermost set of parentheses.
   (c) Combine like quantities; starting from left to right, perform multiplication and division, then addition and subtraction, from left to right.
   (d) Move the unknown quantity or quantities to one side of the equal sign so that the unknown quantities are positive.
   (e) Combine like quantities.
   (f) Solve for the unknown quantity using either the multiplication or division axiom.
   (g) Make certain that the unknown is a positive value. This may be done by multiplying each side of the equation by -1 or making an interchange of the left and right members.
   (h) Check your solution by substituting the value of the unknown into the original equation and evaluating.

CHAPTER SUMMARY

1. An *equation* is a statement that two algebraic expressions are equal.
2. A *formula* is a rule or procedure referring to a relationship expressed as an equation by letters, symbols, and constant terms.
3. *Literal quantity* means letters or symbols representing numbers.
4. The *left side* of an equation is the quantity to the left of the equal sign, and the *right side* is the quantity to the right of the equal sign.
5. *Addition axiom:* A quantity added to the left side of the equation must also be added to the right side.
6. *Subtraction axiom:* A quantity subtracted from the left side of the equation must also be subtracted from the right side.
7. *Multiplication axiom:* Each side of the equation must be multiplied by the same quantity.
8. *Division axiom:* Each side of the equation must be divided by the same quantity.
9. The solution or the value of the unknown that satisfies an equation is called the *root* of the equation or the *solution set.*
10. A *literal equation* is an equation in which some or all of the known quantities are represented by letters instead of numbers; frequently, both letters and numbers are included.
11. *Transposition* is the process of moving a quantity from one side of the equal sign to the other. When a quantity is moved from one side of the equation to the other side, the sign of the quantity is changed. This is because equivalent quantities are being added or subtracted from both sides.
12. A *reciprocal* is a quantity that, when multiplied by the quantity, is equal to 1.

CHAPTER TEST

Solve for the unknowns and check.

T–9–1. $x - 7 = 17$

T–9–2. $2x - 3 = 13$

T–9–3. $x + 1 = -8$

T–9–4. $3k + 15 = 5k + 11$

T–9–5. $\frac{J}{5} = -20$

T–9–6. $8x = -72 + 168$

T–9–7. $4x - 6 = 5x - 10$

T–9–8. $8y = 35 + y$

T–9–9. $4z - 2(3 + z) = z - 4$

T–9–10. $6n - 8 = 2(2n - 1)$

T–9–11. $y + 3(y - 2) = 2(y + 4)$

T–9–12. $3(x - 4) = 2(x - 2)$

T–9–13. $3x - 2 = 5x + 7$

T–9–14. $\frac{1}{2}x + 8 = 15$

T–9–15. $9 + \frac{x}{8} = 11$

T–9–16. $\frac{2}{3}y + \frac{1}{6} = \frac{1}{2}(y - 8 + 5)$

T–9–17. $4.3y + 2.5 = -0.7y + 22.5$

T–9–18. $y - 2a = 3y + 4a$

T–9–19. $2[(y - 3) - 3(5 - 3y)] = 4(3 + y)$

T–9–20. $3(m - 10) - 2(45 - m) = 0$

T–9–21. $\frac{1}{4x} = 16$

T–9–22. $5(a - x) = 3(a - 3x)$

T–9–23. $8(x - 3) - 5(3x + 1) - 6 = 0$

T–9–24. $3(2y - 4) + 2(y - 3) = 5(y + 3)$

T–9–25. $y(y + 4) + 11 = 8(y + 2) + (y - 4)(y - 5)$

10

FORMULA EVALUATION

OBJECTIVE

10.1. After studying this chapter, you will be able to: Understand the definitions and evaluate formulas using addition and subtraction.

10.2. Define formulas and evaluate using multiplication.

10.3. Evaluate formulas using division and other operations.

SELF-TEST

This formula evaluation test will determine what you need to study in this chapter. This evaluation covers simple and complex formulas. Evaluate the following formulas using the given values. Be sure to show your work.

S–10–1. If $C = fgh$ and $f = 8$, $g = 9$, $h = 10$, find C.

S–10–2. If $y = f^2 + 3g - h$ and $f = 8$, $g = 9$, $h = 10$, find y.

S–10–3. If $m = \frac{h}{3} + \sqrt{g}$ and $h = 10$, $g = 9$, solve for m.

S–10–4. If $k = f^2 + g^2 + h^2 - fg - gh$ and $f = 8$, $g = 9$, $h = 10$, find h.

S–10–5. If $p = \frac{(fg)^2}{h}$ and $f = 8$, $g = 9$, $h = 10$, find h.

S–10–6. When $v = \frac{abh}{2}$ and $a = 2$, $b = 7.5$, $h = 15\frac{1}{4}$, find v.

S–10–7. When $V = \pi r^2 h$ and $\pi = 3.14$, $r = 16$, $h = 5$, find V.

S–10–8. When $A = \frac{(B + b)a}{2}$ and $B = 10$, $b = 32$, $a = 28$, find A.

S–10–9. Solve $c = \sqrt{a^2 + b^2}$ when $a = 10$ and $b = 24$.

S–10–10. Solve $a = \sqrt{c^2 - b^2}$ when $c = 52$ and $a = 48$.

S–10–11. Solve $H = \dfrac{F + 1}{1.5}$ when $F = 14$.

S–10–12. Solve $D = \sqrt{\dfrac{HP}{6}}$ when $HP = 54$.

S–10–13. Solve $B = \sqrt{(h + a)(h - a)}$ when $h = 30$ and $a = 18$.

S–10–14. Solve $C = \dfrac{5(F - 32)}{9}$ when $F = 95$.

S–10–15. Solve $P = 0.44d^2Kn$ when $d = \dfrac{1}{2}$, $K = 100$, and $n = 3$.

S–10–16. Solve $L = \dfrac{EFD}{2(D - d)}$ when $E = 0.60$, $F = 130$, $D = 6$, and $d = 4$.

S–10–17. Find $i = prt$ when $p = \$8000$, $r = 8\%$, and $t = 5\frac{1}{2}$ years.

S–10–18. Find $A = 0.7854d^2$ when $d = 0.875$.

S–10–19. Find $V = \frac{4}{3}\pi r^3$ when $\pi = \frac{22}{7}$ and $r = 7$.

S–10–20. Find $A = \sqrt{s(s-a)(s-b)(s-c)}$ if $s = \frac{1}{2}(a+b+c)$, $a = 13$, $b = 11$, and $c = 16$.

10.1 UNDERSTAND DEFINITIONS AND EVALUATE FORMULAS USING ADDITION AND SUBTRACTION

Many students wonder, "How will I ever use algebra once I finish school?" Strange as it may seem, algebra is a tool of business and industry. In fact, formulas can be traced back to application of early arithmetic.

Consider the problem posed to an employee of the electric company. She was given the task of building a fence around an electric power substation. To find the amount of fencing she would need, she used a formula learned as a child. The formula: $P = 2l + 2w$. In other words, the distance around the substation, or the perimeter (P), is equal to 2 times the length ($2l$) plus 2 times the width ($2w$). Knowing this formula saved time—she did not need to make all of the measurements. By measuring one side and the width, she could use the formulas to determine the perimeter.

Formulas save time and money. They are essential to the world of work. Formulas are algebra.

In Chapter 9 a definition of a formula was given. Here is another way to consider how a formula may be defined. A *formula* uses mathematical language to express the relationship between two or more variables. For example, the formula for finding the number of miles a car will travel on 1 gallon of gasoline is

$$\text{Miles per gallon} = \frac{\text{miles traveled}}{\text{gallons of gasoline used}}$$

$$\text{mpg} = \frac{m}{g}$$

Evaluating a formula is not difficult. It is important to have the formula in the proper form and follow given procedures.

Most students who have difficulty with mathematics in general, and formula evaluation in particular, fail to read the problem correctly. Thus they are unable to perform the designated operations. This is shown by skipping essential steps, failing to show the work, and having a lack of understanding of what answer is to be found.

As you complete the following pages, carefully study the formulas given in the examples, noting the steps involved in reaching a solution. As you are given an opportunity to work with formulas, follow the same steps as in the examples. Be certain to show your work, and check the examples when necessary.

Evaluating Formulas Using Addition or Subtraction

The following are examples of formulas that use the operations of addition and subtraction.

Example 10–1: Find the Perimeter

Perimeter of a Triangle

$$P = a + b + c$$

The formula for the perimeter (distance around the outside) of a triangle is $P = a + b + c$. This means that the perimeter of a triangle is equal to side a plus side b plus side c. See Figure 10–1.

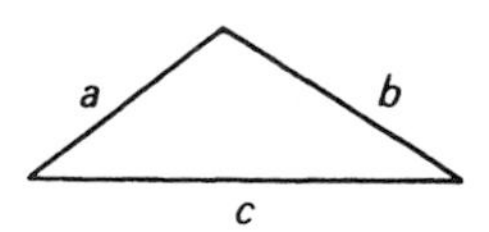

Figure 10–1

PROBLEM: What is the perimeter of a triangle when $a = 5$, $b = 6$, and $c = 9$?

SOLUTION:

Write the formula:	$P = a + b + c$
Substitute the givens:	$P = 5 + 6 + 9$
Evaluate:	$P = 20$

We call this process the *evaluation* of a formula.

Example 10–2: Evaluate the Formula

PROBLEM: The formula for a type of friction is $K = x + y - z$. Evaluate the formula if $x = 3$, $y = 7$, and $z = 4$.

SOLUTION:

Write the formula:	$K = x + y - z$
Substitute the givens:	$K = 3 + 7 - 4$
Evaluate:	$K = 6$

EXERCISE 10–1 SOLVE USING FORMULAS

Evaluate the following formulas using addition and subtraction.

10–1. Find the perimeter of the triangle shown in Figure 10–2.

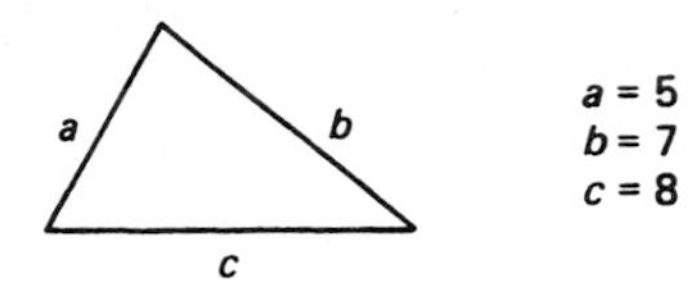

Figure 10–2

10–2. Find the perimeter of the trapezoid shown in Figure 10–3.

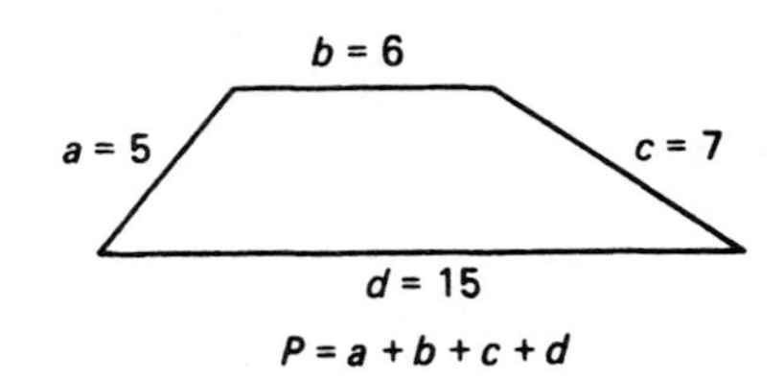

Figure 10–3

10–3. Find the perimeter of a 6-inch square. See Figure 10–4.

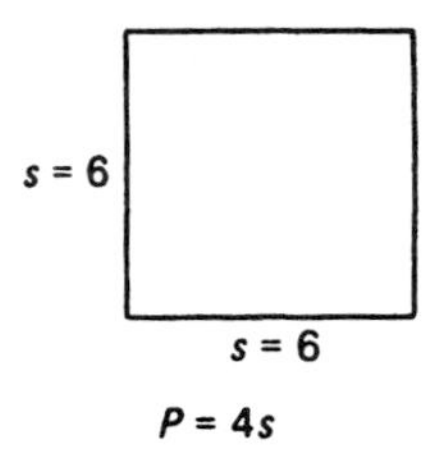

Figure 10–4

10–4. Find the missing dimension x in Figure 10–5.

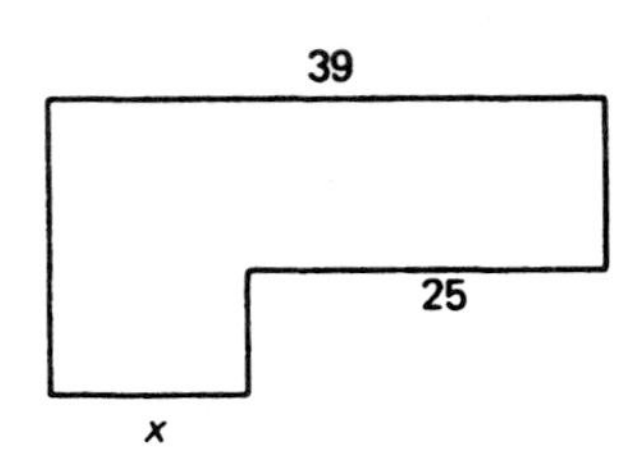

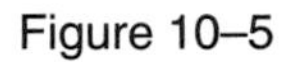

Figure 10–5

10–5. Find the missing dimension y in Figure 10–6.

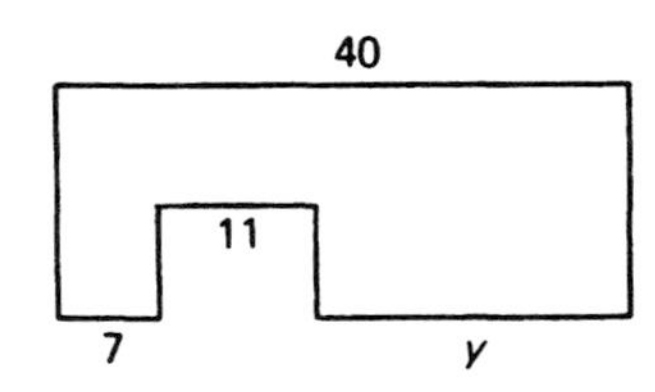

Figure 10–6

10–6. Solve the formula $K = x + y - z$ for y when $K = 20$, $x = 5$, and $z = 4$.

10–7. Find the perimeter of the property in Figure 10–7.

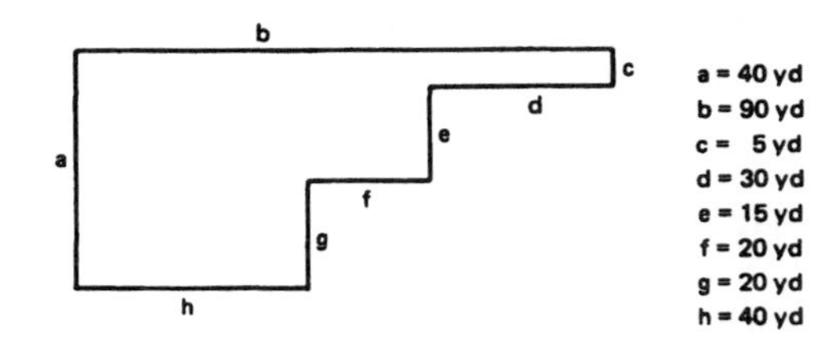

Figure 10–7

10–8. Using the formula for determining profit, $p = g - c - l - j$, find the cost of manufacturing the products when $p = \$20,000$, $g = \$57,000$, $l = \$10,000$, and $j = \$1500$.

10–9. Using a formula for balancing a checkbook, $B = A - a - b - c$, find the amount of check a when $B = \$99$, $A = \$257$, $b = \$77$, and $c = \$19$.

10–10. Using a formula for finding caloric intake, $C = a + b + c + d + e$, find the value of e when $C = 401$, $a = 75$, $b = 110$, $c = 130$, and $d = 22$.

10.2 DEFINE FORMULAS AND EVALUATE USING MULTIPLICATION

The formulas in the preceding section used addition and subtraction. The formulas in this section use multiplication.

Example 10–3: Find the Area of a Rectangle

> **Area of a Rectangle**
>
> The formula for finding the area of a rectangle is $A = lw$. This means that the area (A) of a rectangle is equal to the length (l) times the width (w).

PROBLEM: Find the area of a rectangle when the length (l) is 10 feet and the width (w) is 8 feet.

SOLUTION:

Write the formula:	$A = lw$
Substitute the givens:	$A = (10 \text{ ft})(8 \text{ ft})$
Evaluate:	$A = 80 \text{ sq ft}$

Example 10–4: Find the Area of a Square

> **Area of a Square**
>
> The formula for finding the area of a square is $A = s^2$. This means that the area (A) of a square is equal to the length of one side (s) times the length of another side (s), or $A = (s)(s)$ or s^2.

PROBLEM: Find the area of a square when $s = 4$ feet.

SOLUTION:

Write the formula:	$A = s^2$
Substitute the givens:	$A = (4 \text{ ft})(4 \text{ ft})$
Evaluate:	$A = 16 \text{ sq ft}$

Example 10–5: Find the Circumference of a Circle

> **Circumference of a Circle**
>
> The formula for finding the circumference (distance around) a circle is $C = d\pi$. This means that the circumference (C) is equal to the diameter (d) times π, or 3.142.

PROBLEM: Find the circumference of a circle when $d = 7$ feet.

SOLUTION:

Write the formula:	$C = d\pi$
Substitute the givens:	$C = (7 \text{ ft})(3.142)$
Evaluate:	$C = 22 \text{ ft}$

Example 10–6: Find the Distance

> **Distance**
>
> The formula for finding distance is $d = rt$. This means that distance (d) equals the rate (r) times the time (t).

PROBLEM: Find the distance traveled when $r = 55$ miles per hour and $t = 240$ minutes.

SOLUTION:

Write the formula: $d = rt$

Substitute the givens:

$$d = \left(\frac{55 \text{ mi}}{\cancel{\text{hr}}}\right)\left(\frac{1 \cancel{\text{hr}}}{\underset{1}{\cancel{60}} \cancel{\text{min}}}\right)\left(\frac{\overset{4}{\cancel{240}} \cancel{\text{min}}}{1}\right)$$

Evaluate: $d = 220 \text{ miles}$

Note: Only like units may be multiplied. Convert to like units when necessary.

Example 10–7: Find the Interest

> **Interest**
>
> The formula for finding interest is $i = prt$. This means that the amount of interest (i) is equal to the principal (p) times the rate of interest (r) times the time in years (t).

PROBLEM: Find the interest when $p = \$5000$, $r = 0.08$, and $t = 3$ years.

SOLUTION:

Write the formula:	$i = prt$
Substitute the givens:	$i = (\$5000)(0.08)(3)$
Evaluate:	$i = \$1200$

EXERCISE 10–2 EVALUATE THE FORMULAS

Solve the following problems, performing the indicated operations.

10–11. Find the area of a rectangle that is 15 inches by 11 inches.

$A = lw$ $l = 15$ in. $w = 11$ in.

10–12. Find the distance traveled if the rate is 40 miles per hour and the time is 20 hours.

$d = rt$ $r = 40$ mph $t = 20$ hr

10–13. Find the interest if the principal is \$1000, the rate is 7%, and the time is 25 years.

$i = prt$ $r = 0.07$ $t = 25$ years

10–14. Find the volume of a rectangular solid that is 30 feet by 12 feet by 8 inches.

$V = lwh$ $l = 30$ ft $w = 12$ ft $h = 8$ in.

10–15. Find the circumference of a circle 8 yards in diameter.

$C = d\pi$ $d = 8$ yd $\pi = 3.1416$

10–16. Convert 27 feet to inches (1 foot = 12 inches).

10–17. Find the circumference of a circle if the radius is 34 inches.

$C = 2\pi r$ $r = 34$ in. $\pi = 3.14$

10–18. Find the volume of a rectangular solid with a 36-square-foot base and a height of 24 inches. *Remember to convert inches to feet* (12 inches = 1 foot).

$V = Bh$ $B = 36$ sq ft $h = 24$ in.

10–19. Find the perimeter of a rectangle 7 feet long and 3 feet wide.

$p = 2l + 2w$ $l = 7$ ft $w = 3$ ft

10–20. Find the area of a 41-foot square.

$A = s^2$ $s = 41$ ft

10–21. Find the circumference in *feet* of a circle when the radius is 28.5 inches.

$C = 2\pi r$ $r = 28.5$ in. $\pi = 3.142$

10–22. Find the surface area of a 3-foot cube.

$S = 6e^2$ $e = 3$ ft

10–23. Find the surface area of a rectangular solid 6 feet long by 5 feet wide and 3 feet high.

$S = 2lw + 2hl + 2hw$ $l = 6$ ft $w = 5$ ft $h = 3$ ft

10–24. Find the simple interest for $298 at 5.5% for 3 years.

$i = prt$ $p = \$298$ $r = 5.5\%$ $t = 3$ years

10–25. Find the diameter, to the nearest tenth of a foot, of a tree with a circumference of 282.75 inches. *Remember to convert inches to feet* (12 inches = 1 foot).

$d = \dfrac{C}{\pi}$ $C = 282.75$ in.

10.3 EVALUATING FORMULAS USING DIVISION AND OTHER OPERATIONS

The formulas included thus far required using addition, subtraction, and multiplication. More difficult formulas may include operations of division and roots. Study the following examples, making special note of the procedures involved.

Example 10–8: Find the Area of a Triangle

> **Area of a Triangle**
>
> The formula for determining the area of a triangle is $A = \frac{bh}{2}$ or $\frac{1}{2}bh$ or $0.5bh$. This means that the area (A) of the triangle is equal to the base (b) times the height (h) divided by 2.

PROBLEM: Find the area of a triangle when the base is $b = 4$ feet, and the height is $h = 12$ feet. See Figure 10–8.

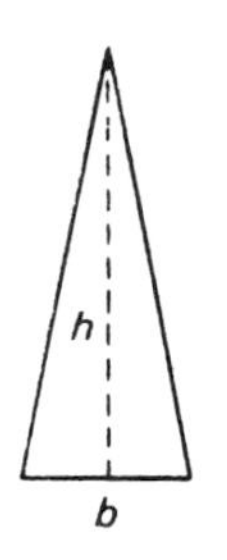

Figure 10–8

SOLUTION:

Write the formula: $A = \frac{bh}{2}$

Substitute the givens: $A = \frac{(4\text{ ft})(12\text{ ft})}{2}$

Evaluate: $A = 24$ sq ft

Example 10–9: Find the Area of a Trapezoid

> **Area of a Trapezoid**
>
> The formula for finding the area of a trapezoid is $A = \frac{1}{2}(b_1 + b_2)h$. This means that the area (A) of the trapezoid is equal to one half the sum of base 1 (b_1 is read "b sub 1") and base 2 (b_2) times the height (h).

PROBLEM: Find the area of the trapezoid: $b_1 = 10$ inches, $b_2 = 20$ inches, and $h = 5$ inches. See Figure 10–9

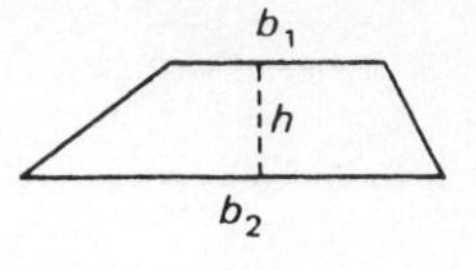

Figure 10–9

SOLUTION:

Write the formula: $A = \frac{1}{2}(b_1 + b_2)h$

Substitute the givens: $A = \frac{1}{2}(10\text{ in.} + 20\text{ in.})(5\text{ in.})$

$A = 0.5(30\text{ in.})(5\text{ in.})$

Evaluate: $A = 75$ sq in.

Example 10–10:

> **Depth of Thread**
>
> The formula for finding the depth of a certain type of thread is
>
> $$D = d - \frac{1.732}{N}$$
>
> This means that the depth of the thread (D) is equal to the depth of one thread (d) minus 1.732 divided by the number of threads (N).

PROBLEM: Find the depth of the thread when $d = 0.75$ and $N = 5$.

SOLUTION:

Write the formula: $D = d - \frac{1.732}{N}$

Substitute the givens: $D = 0.075 - \frac{1.732}{5}$

$D = 0.75 - 0.3464$

Evaluate: $D = 0.4036$

Example 10–11: Find the Height of an Isoceles Triangle

> **Height of Isosceles Triangle**
>
> The formula for finding the height of an isosceles triangle given the equal sides and base is
>
> $$h = \sqrt{a^2 - \frac{b^2}{4}}$$
>
> This means that the height (h) of the isoceles triangle is equal to the square root ($\sqrt{}$) of one of the equal sides squared (a^2) minus the base squared divided by 4, written as $\frac{b^2}{4}$.

PROBLEM: Find the height of an isoceles triangle when $a = 26$ feet and $b = 20$ feet. Use a calculator to find the square root.

SOLUTION:

Write the formula: $h = \sqrt{a^2 - \frac{b^2}{4}}$

Substitute the givens: $h = \sqrt{(26)^2 - \frac{(20)^2}{4}}$

$h = \sqrt{676 - \frac{400}{4}}$

$h = \sqrt{676 - 100}$

$h = \sqrt{576}$

Evaluate: $h = 24$ ft

EXERCISE 10-3 SOLVE THE COMPLEX FORMULAS

Solve the following problems, making certain to perform the indicated operations. Remember to convert measurements to the same unit.

10–26. $A = \frac{1}{2}(b_1 + b_2)h.$

Find A when $b_1 = 2$ feet, $b_2 = 36$ inches, and $h = 24$ inches.

10–27. $E = \frac{360}{n}.$

Find E when $n = 4$.

10–28. $A = \frac{0.7854nd^2}{360°}.$

Find A when $n = 60°$ and $d = 12$ inches.

10–29. $K = \frac{L(D - d)}{l}.$

Find K when $L = 3.5$, $D = 0.5$, $d = 0.25$, and $l = 0.25$.

10–30. $M = (D - 1.732P) + 3W.$

Find M when $D = 7.5$, $P = 10$, and $W = 0.06$.

10–31. $A = 0.7854d^2.$

Find A when $d = 11$ feet.

10–32. $S = 2\pi r(r + h).$

Find S when $\pi = \frac{22}{7}$, $r = 2$ feet, and $h = 1\frac{1}{2}$ feet.

10–33. $C = \frac{5}{9}(F - 32°).$

Find C when $F = 131°$.

10–34. $V = \frac{4\pi r^3}{3}$.

Find V when $r = 6$ yards and $\pi = 3.14$.

10–35. $F = \frac{9}{5}C + 32°$.

Find F when $C = 115°$.

10–36. $W = \frac{Kbd^2}{L}$.

Find W when $K = 6000$, $b = 6$, $d = 8$, and $L = 24$.

10–37. $S = \frac{gt^2}{2}$.

Find S when $g = 32$ and $t = 4$.

10–38. $S = \frac{n}{2}(a + l)$.

Find S when $a = 10$, $l = 6$, and $n = 50$.

10–39. $H = \frac{DV}{375}$.

Find H when $D = 187.5$ and $V = 200$.

10–40. $K = 2a - 5(n - 1)$.

Find K when $a = 8$ and $n = 3$.

10–41. $R = \frac{1}{\frac{1}{r_1} + \frac{1}{r_2}}$.

Find R when $r_1 = 3$ and $r_2 = 4$.

10–42. $A = \frac{D}{0.5(D - d)}$.

Find A when $D = 10$ and $d = 4$.

10–43. $A = 3.1416(R + r)(R - r)$.

Find A when $R = \frac{3}{2}$ and $r = \frac{3}{8}$.

10–44. $E = \frac{(N)P}{NP + R}$.
Find E when $N = 2$, $P = 4$, and $R = 5.3$.

10–45. $L = \sqrt{b^2 + h^2}$.
Find L when $b = 4.75$ and $h = 3.625$.

10–46. $b = \sqrt{h^2 - a^2}$.
Find b when $h = 10\frac{1}{2}$ and $a = 8\frac{3}{4}$.
(*Hint:* Convert $10\frac{1}{2}$ to 10.5 and $8\frac{3}{4}$ to 8.75.)

10–47. $d = \sqrt{\frac{4A}{\pi}}$.
Find d when $A = 490.875$ and $\pi = 3.14159$.

10–48. $r = \sqrt{\frac{A}{\pi}}$.
Find r when $A = 28.26$ and $\pi = 3.14$.

10–49. $l = 2\sqrt{h(2r - h)}$.
Find l when $h = 6$ and $r = 15$.

10–50. $A = \sqrt{s(s - a)(s - b)(s - c)}$.
Find A when $s = \frac{1}{2}(a + b + c)$ and $a = 8$, $b = 6$, and $c = 4$.

The exercises you have completed have involved formula evaluation. In the next chapter we discuss how to manipulate or transform formulas into the desired forms.

THINK TIME

Write three formulas used in your regular daily living. Using these formulas, with the knowledge and skills learned in this chapter, solve for the unknowns. Think of additional situations in which you could use these formulas. Practice solving problems with the stated formulas.

PROCEDURES TO REMEMBER

1. To evaluate formulas:
   (a) Read and study the formula to determine its purpose.
   (b) Reread the formula, making certain of the given quantities and the desired quantity.
   (c) Substitute the given quantities into the formula, making certain that you have used the correct dimensions.
   (d) Estimate your answers so that you will have a reasonable idea of the possible solution.
   (e) Evaluate; show all necessary mathematical operations to prevent errors.
   (f) Be sure that your solution is written in the correct dimensions.

CHAPTER SUMMARY

A formula uses mathematical language to express the relationship between two or more variables. Thus a formula is a mathematical abbreviation for an accepted operational procedure. Formula evaluation is the process of obtaining the value of a formula from the given variables.

CHAPTER TEST

Evaluate the following using the given quantities.

T–10–1. When $x = 5$, $y = 3$, and $z = 2$, $\dfrac{x + y}{z} = ?$

T–10–2. When $a = 15$, $b = 14$, and $c = 12$, $a^2 + b^2 - c^2 = ?$

T–10–3. When $x = 1$, $y = 3$, and $z = 2$, $x^2 + y^2 + z^3 = ?$

T–10–4. When $w = -2$ and $k = 3$, $\dfrac{2w + 5k}{7} = ?$

T–10–5. When $x = 1$, $y = 3$, $z = 2$, and $w = -2$, $xy + yz + zw = ?$

T–10–6. Evaluate $\dfrac{1}{x} + \dfrac{3}{y}$ when $x = 1$ and $y = -1$.

T–10–7. Evaluate $(x + z)(w + k)$ when $x = 1$, $z = 2$, $w = -2$, and $k = 3$.

T–10–8. Evaluate $(ab)^2(c - b)^2$ when $a = 4$, $b = 0.5$, and $c = 2.5$.

T–10–9. Evaluate $(a + b)^2(a - b)^2$ when $a = 6$ and $b = 4$.

T–10–10. Evaluate $(x + y - z)^3$ when $x = 5$, $y = 12$, and $z = 7$.

T–10–11. **The formula for the area of an ellipse is $A = \pi ab$. Find A when $\pi = 3.14$, $a = 20$, and $b = 12$.**

T–10–12. The formula for the perimeter of an ellipse is $P = \pi(a + b)$. Find P when $\pi = 3.14$, $a = 24$, and $b = 21$.

T–10–13. **If $l = 2\sqrt{h(2r - h)}$, find l when $h = 4$ and $r = 10$. This is the formula for finding the length of a segment when you know the radius of the circle and the height of the segment.**

T–10–14. The formula for the area of a ring is $A = \pi(R + r)(R - r)$. Find A when $R = \frac{7}{8}$, $r = \frac{3}{4}$, and $\pi = 3.14$. (*Hint:* Convert the fractions to decimals.)

T–10–15. The formula for the radius of a circle, given the area, is $r = \sqrt{\frac{A}{\pi}}$. Find r when $\pi = 3.14$ and $A = 502.4$.

T–10–16. The formula for the volume of a sphere is $V = \frac{4\pi r^3}{3}$. Find the volume of a sphere (in cubic yards) when the radius is 12 feet.

T–10–17. **The formula for the volume of a cylinder is $V = \pi r^2 h$. Find the volume of a cylinder when $\pi = 3.14$, $r = 5$, and $h = 12$.**

T–10–18. The formula for the area of a sector of a circle is $A = \frac{\theta}{360}\pi r^2$. Find A when $\theta = 75$, $\pi = 3.14$, and $r = 8$.

T–10–19. The formula for the diameter of a circle given the area is $d = \sqrt{A \div \frac{\pi}{4}}$. Find d when $A = 12.56$ and $\pi = 3.14$.

T–10–20. The formula for the area of a triangle, given all the sides, is

$$A = \sqrt{s(s - a)(s - b)(s - c)}$$

$$S = \frac{1}{2}(a + b + c)$$

Find A when $a = 28$, $b = 25$, and $c = 17$.

11

FORMULA TRANSPOSITION

OBJECTIVES

After studying this chapter, you will be able to:

11.1. Solve formulas using equation solving procedures.
11.2. Solve formulas involving roots.
11.3. Solve complex formulas using several operations.

SELF-TEST

This test on formula transposition will determine what you need to study in this chapter. If you are able to solve the formulas, you are ready to move to the next chapter. Be sure to show your work so that you can determine where you made errors and what you need to do to correct your mistakes.

S–11–1. Solve $P = a + b + c$ for c.

S–11–2. Solve $V = \frac{1}{3}Bh$ for B.

S–11–3. Solve $L = 2\pi rh$ for h.

S–11–4. Solve $P = \frac{w}{t}$ for t.

S–11–5. Solve $HP = \frac{fl}{t}$ for t.

S–11–6. Solve $H = \frac{D^2N}{2.5}$ for D.

S–11–7. Solve $V = \frac{1}{3}\pi r^2h$ for h.

S–11–8. Solve $A = \frac{h}{2}(b_1 + b_2)$ for b_2.

S–11–9. Solve $S = 2\,\pi rh + 2\,\pi r^2$ for h.

S–11–10. Solve $V = \pi r^2 h$ for r.

S–11–11. Solve $V = \frac{4}{3}\pi r^3$ for r.

S–11–12. Solve $DS = ds$ for d.

S–11–13. Solve $S = \frac{CD}{12d}$ for D.

S–11–14. Solve $W = \frac{5bd^2K}{2L}$ for K.

S–11–15. Find a when $T = 384$ and $T = 6a^2$.

S–11–16. Find r when $A = \frac{5\pi}{16}$, $R = \frac{3}{4}$, and $A = \pi R^2 - \pi r^2$.

S–11–17. Find b when $P = 51\pi$, $a = 32$, and $P = \pi(a + b)$.

S–11–18. Find r when $V = \pi R^2 h - \pi r^2 h$, $R = 8$, $h = 10$, $V = 77.5\pi$, and $\pi = 3.14$.

S–11–19. Solve $S = 4\pi r^2$ for r; then find r when $S = 484\pi$.

S–11–20. Find d if $V = \frac{32\pi}{3}$ and $V = \frac{\pi d^3}{6}$.

S–11–21. Solve $A = \dfrac{\theta \pi r^2}{360}$ for r.

S–11–22. Solve $r = \dfrac{\left(\dfrac{w}{2}\right)^2 + h^2}{2h}$ for w.

S–11–23. Solve $V = \dfrac{\pi h}{3}(a^2 + ab + b^2)$ for h; then find h when $a = 24$, $b = 18$, and $V = 3552$.

S–11–24. Solve $P = \pi\sqrt{2(a^2 + b^2)}$ for b.

S–11–25. Solve $\dfrac{1}{R} = \dfrac{1}{r_1} + \dfrac{1}{r_2} + \dfrac{1}{r_3}$ for r_3 and find r_3 when $R = \dfrac{12}{13}$, $r_1 = 2$, and $r_2 = 3$.

11.1 SOLVING FORMULAS USING EQUATION-SOLVING PROCEDURES

Given formulas and formulas found in handbooks are not always in the desired form. Therefore, it is important to know how to solve formulas so they will work for you. This is not a difficult task if you work through the process step by step. Many students find solving formulas or transposition of formulas difficult because they fail to read the formulas correctly. For example, the formula $A = {}^1\!/_2(b_1 + b_2)h$ could be misread as $A = {}^1\!/_2 b_1 + 2 \times h$. Failure to complete work inside the parentheses first will result in an incorrect answer. Correct reading of formulas can help to prevent confusion and error.

Reading the original formula correctly becomes very important when solving formulas. By reading the formula correctly, you will be able to solve the formula correctly. The solutions are often made using equation-solving procedures. This means that division is used to solve an equation containing multiplication. Study the following examples, and note how inverse operations are used to solve the equations.

Example 11–1: Solving a Formula Using Division

Area of a Rectangle

The formula for the area of a rectangle equals the length times the width, or $A = lw$.

PROBLEM: Solve the formula for w.

SOLUTION: Remember, division is the inverse operation of multiplication. Thus divide both sides by l to find w.

$$A = lw$$

$$\frac{A}{l} = \frac{\cancel{l}w}{\cancel{l}}$$

$$\frac{A}{l} = w$$

The width (w) equals the area (A) divided by the length (l).

$$w = \frac{A}{l}$$

Example 11–2: Applying the Solved Formula

PROBLEM: Find w from Example 11–1 when $A = 44$ square feet and $l = 11$ feet.

SOLUTION:

Write the formula: $w = \frac{A}{l}$

Substitute the givens: $w = \frac{\overset{4}{\cancel{44}} \text{ sq ft}}{\underset{1}{\cancel{11}} \text{ ft}}$

Evaluate: $w = 4$ ft

Example 11–3: Solving a Formula Using Addition and Multiplication

This example involves the operations of addition (subtraction) and multiplication (division).

Perimeter of a Rectangle

The perimeter of a triangle equals 2 times the width plus 2 times the length, or

$$P = 2l + 2w$$

PROBLEM: Solve the formula for w.

SOLUTION: Isolate the quantity w on one side of the equation by subtracting $2l$ from each side of the equation and combining like quantities.

$$P = 2l + 2w$$
$$P - 2l = 2l + 2w - 2l$$
$$P - 2l = 2w$$

Solve the equation for w by dividing each side of the equation by 2 or multiplying by $^1/_2$.

$$\frac{\overset{1}{\cancel{2}}w}{\underset{1}{\cancel{2}}} = \frac{P - 2l}{2}$$

$$w = \frac{P - 2l}{2}$$

Example 11–4: Applying the Solved Formula

PROBLEM: Find w from Example 11–3 when $P = 38$ and $l = 12$.

SOLUTION:

Write the formula: $w = \frac{P - 2l}{2}$

Substitute the givens: $w = \frac{38 - 2(12)}{2}$

$w = \frac{38 - 24}{2}$

Evaluate: $w = \frac{\overset{7}{\cancel{14}}}{\underset{1}{\cancel{2}}} = 7$

Example 11–5: Solving a Formula Using Multiplication

This example involves the operations of multiplication and division.

Area of a Triangle

The area of a triangle equals one half the base times the height, or $A = \frac{1}{2}bh$.

PROBLEM: Solve the formula for h.

SOLUTION: When a formula involves a fraction, change the fraction to a whole number by multiplication. First find the lowest common denominator, which is 2. Then multiply each term on both sides of the equation by 2.

$$A = \frac{1}{2}bh \text{ or } A = \frac{bh}{2}$$

$$2A = (\overset{1}{\cancel{2}})\frac{1}{\underset{1}{\cancel{2}}}bh$$

Then to solve for h, divide each side of the equation by b or multiply by $^1/_b$.

$$2A = bh$$

$$\frac{2A}{b} = \frac{\overset{1}{\cancel{b}}h}{\underset{1}{\cancel{b}}}$$

$$h = \frac{2A}{b}$$

Example 11–6: Applying the Solved Formula

PROBLEM: Find h for Example 11–5 when $A = 360$ square feet and $b = 40$ feet.

SOLUTION:

Write the formula: $h = \frac{2A}{b}$

Substitute the givens: $h = \dfrac{2(360 \text{ sq ft})}{40 \text{ ft}}$

Evaluate: $h = \dfrac{\overset{18}{\cancel{720 \text{ sq ft}}}}{\underset{1}{\cancel{40 \text{ ft}}}} = 18 \text{ ft}$

EXERCISE 11–1 SOLVE THE SIMPLE FORMULAS AND EVALUATE

Solve the following problems. Make sure to read carefully and solve for the correct letter. Show all necessary work.

11–1. Solve $Q = \dfrac{WL}{T}$ for T. Find T when $Q = 250$, $W = 125$, and $L = 20$.

11–2. Solve $i = prt$ for t. Find t when $p = \$500$, $r = 0.08$, and $i = \$120$.

11–3. Solve $C = 2\pi r$ for r. Find r when $C = 154$ and $\pi = \dfrac{22}{7}$.

11–4. Solve $p = a + b + c$ for c. Find c when $p = 100$, $a = 21$, and $b = 17$.

11–5. Solve $V = Bh$ for B. Find B when $V = 1000$ cubic feet and $h = 8$ feet.

11–6. Solve $Q = 0.000477\ EIT$ for T. Find T when $Q = 954$, $E = 1000$, and $I = 20$.

11–7. Solve $C = \dfrac{5}{9}(F - 32°)$ for F. Find F if $C = 100°$.

11–8. Solve $x - y = xy$. Find y when $x = 8$.

11–9. Solve $S = \pi rs$ for s. Find s when $S = 4352$, $\pi = 3.14$, and $r = 462$.

11–10. Solve $P = 2a + b$ for a. Find a when $P = 28$ and $b = 12$.

11–11. Solve $i = prt$ for r. Find r when $i = \$80$, $p = \$500$, and $t = 2$ years.

11–12. Solve $C = \pi d$ for d. Find d when $C = 120$ feet and $\pi = 3.14$, to the nearest hundredth.

11–13. Solve $L = \dfrac{\pi r A}{180}$ for A. Find A when $L = 314$, $r = 60$, and $\pi = 3.14$.

11–14. Solve $T = \frac{1}{2}ps + B$ for B. Find B when $T = 580$, $p = 58$, and $s = 12$.

11–15. Solve $T = 2B + pH$ for H. Find H when $T = 860$, $B = 45$, and $p = 11$.

11.2 SOLVING FORMULAS INVOLVING ROOTS

The procedures of transposition are more difficult when finding roots. Review what you have learned about the extracting roots in previous chapters and what you have learned about transposition in this chapter. Carefully study the following examples.

Example 11–7: Solving a Formula Using Division and Square Roots

> **Surface Area of a Cube**
>
> The surface area of a cube is 6 times the area of one side of the cube, or $A = 6e^2$.

FORMULA: The surface area of a cube equals 6 times the length of one edge squared: $A = 6e^2$.

PROBLEM: Solve the formula for e.

SOLUTION: Divide both sides by 6 (or multiply by $^1/_6$) to isolate e^2 on one side of the equation.

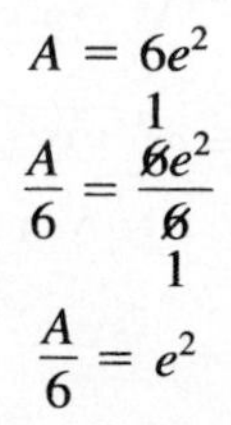

To solve for e, take the square root of each side of the equation.

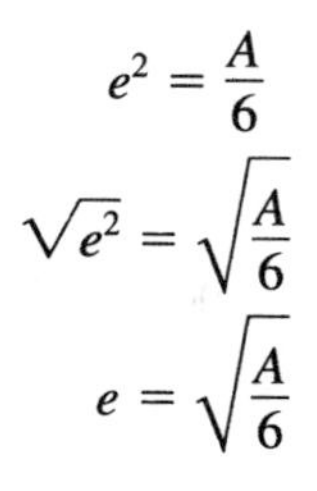

Example 11–8: Applying the Solved Formula

PROBLEM: Find e for Example 11–7 if $A = 216$ square feet.

SOLUTION:

Write the formula: $e = \sqrt{\dfrac{A}{6}}$

Substitute the givens: $e = \sqrt{\dfrac{\overset{36}{\cancel{216}}\text{ sq ft}}{\underset{1}{\cancel{6}}}}$

$e = \sqrt{36\text{ sq ft}}$

Evaluate: $e = 6$ ft

Example 11–9: Solving a Formula Using Division and Square Root

Area of a Circle

The area of a circle equals π (approximately 3.14) times the radius of the circle squared, or $A = \pi r^2$.

PROBLEM: Solve the formula for r.

SOLUTION: Divide both sides by π to isolate r^2 on one side of the equation.

$$A = \pi r^2$$

$$\frac{A}{\pi} = \frac{\overset{1}{\cancel{\pi}} r^2}{\underset{1}{\cancel{\pi}}}$$

$$\frac{A}{\pi} = r^2$$

Solve the formula for r by taking the square root of each side of the equation.

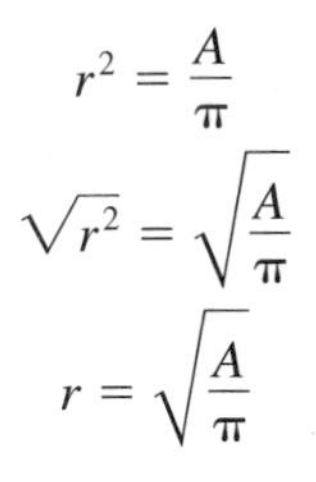

$$r^2 = \frac{A}{\pi}$$

$$\sqrt{r^2} = \sqrt{\frac{A}{\pi}}$$

$$r = \sqrt{\frac{A}{\pi}}$$

Example 11–10: Applying the Solved Formula

PROBLEM: Find r from Example 11–9 if the area of a circle is 50.24 and $\pi = 3.14$.

SOLUTION:

Write the formula:	$r = \sqrt{\frac{A}{\pi}}$
Substitute the givens:	$r = \sqrt{\frac{50.24 \text{ sq yd}}{3.14}}$
Evaluate:	$r = \sqrt{16 \text{ sq yd}}$
	$r = 4 \text{ yd}$

Example 11–11: Solving a Formula Using Multiplication and Cube Root

Volume of a Sphere

The volume of a sphere equals 4 times π times the radius of the sphere cubed divided by 3, or

$$V = \frac{4\pi r^3}{3}$$

PROBLEM: Solve the formula for r.

SOLUTION: Note that r is cubed; thus we will need to find the cube root of r^3. First, clear the fraction, $\frac{4\pi r^3}{3}$, by multiplying both sides of the equation by 3.

$$V = \frac{4\pi r^3}{3}$$

$$(3)V = \frac{\overset{1}{\cancel{(3)}}\, 4\pi r^3}{\underset{1}{\cancel{3}}}$$

$$3V = 4\pi r^3$$

Isolate r^3 by dividing both sides of the equation by 4π.

$$3V = 4\pi r^3$$

$$\frac{3V}{4\pi} = \frac{\overset{1}{\cancel{4\pi}} r^3}{\underset{1}{\cancel{4\pi}}}$$

$$\frac{3V}{4\pi} = r^3$$

Solve the formula for r by taking the cube root of each side of the equation.

Note: Scientific calculators will give the cube root by using the x^y key.

$$r^3 = \frac{3V}{4\pi}$$

$$\sqrt[3]{r^3} = \sqrt[3]{\frac{3V}{4\pi}}$$

$$r = \frac{\sqrt[3]{3V}}{4\pi}$$

Example 11–12: Applying the Solved Formula

PROBLEM: Find r from Example 11–11 if $V = 523.333$ cubic units and $\pi = 3.14$.

SOLUTION: $r = \sqrt[3]{\frac{3V}{4\pi}}$

$$r = \sqrt[3]{\frac{(3)\,(523.333)\text{ m}^3}{4(3.14)}}$$

$$r = \sqrt[3]{\frac{1570 \text{ m}^3}{12.56}}$$

$$r = \sqrt[3]{125 \text{ m}^3}$$

$$r = 5 \text{ m}$$

Note: To obtain the cube root with a calculator, key 125, and press the x^y key; then key $0.\overline{3}$ (which is the $\frac{1}{3}$ power) and press the = key. (Some calculators may use a different procedure.)

Example 11–13: Solving a Formula Using Squaring, Addition, and Square Root

Height of Isosceles Triangle

The height of an isosceles triangle is equal to the square root of one of the equal sides squared minus the base squared divided by 4, or

$$h = \sqrt{a^2 - \frac{b^2}{4}}$$

PROBLEM: Solve the formula for *a*.

SOLUTION: To find the value of *a*, square both sides of the equation.

$$(h)^2 = \left(\sqrt{a^2 - \frac{b^2}{4}}\right)^2$$

$$h^2 = a^2 - \frac{b^2}{4}$$

Then add $\frac{b^2}{4}$ to both sides of the equation.

$$\frac{b^2}{4} + h^2 = a^2 - \frac{b^2}{4} + \frac{b^2}{4}$$

$$\frac{b^2}{4} + h^2 = a^2$$

Take the square root of each side.

$$\sqrt{a^2} = \sqrt{\frac{b^2}{4} + h^2}$$

Thus

$$a = \sqrt{\frac{b^2}{4} + h^2}$$

Example 11–14: Applying the Solved Formula

PROBLEM: Find the value of *a* in Example 11–13 when $h = 24$ and $b = 20$.

SOLUTION:

Write the formula: $a = \sqrt{\frac{b^2}{4} + h^2}$

Substitute the givens: $a = \sqrt{\frac{(20)^2}{4} + (24)^2}$

Evaluate: $a = \sqrt{\frac{400}{4} + 576}$

$$a = \sqrt{676}$$

$$a = 26 \text{ units}$$

EXERCISE 11–2 SOLVE THE FORMULAS AND EVALUATE

Solve the following problems. Study each problem carefully to make sure that you understand for what quantity the problem is to be solved. If you have difficulty, study the examples.

11–16. Solve $A = \frac{\pi r^2}{2}$ for *r*. Find *r* when $A = 157$ square units and $\pi = 3.14$.

11–17. Solve $x = ky^2$ for *y*. Find *y* when $x = 128$ and $k = 8$.

11–18. Solve $F = \frac{kM}{d^2}$ for *d*. Find *d* when $k = 28$, $M = 8$, and $F = 14$.

11–19. Solve $S = \frac{at^2}{2}$ for *t*. Find *t* when $a = 4$ and $S = 128$.

11–20. Solve $F = \frac{kmM}{d^2}$ for M. Find M when $d = 3$, $F = 42$, $k = 14$, and $m = \frac{1}{3}$.

11–21. Solve the mass-energy equation $e = mc^2$ for c. Find c if $e = 6348$ and $m = 12$.

11–22. Solve $V = e^3$ for e. This is the formula for the volume of a cube. Find e when $V = 729$ cubic yards.

11–23. Solve $S = 4\pi r^2$, which is the formula for the surface area of a sphere, for r. Find r when $S = 784\pi$ square feet.

11–24. Solve the volume of a sphere formula, $V = \frac{\pi d^3}{6}$, for d. Find d when $V = \frac{171.5\pi}{3}$.

11–25. Solve $r = \sqrt{\frac{A}{\pi}}$ for A. This is the formula for the radius of a circle when you know the area. Find A when $r = 7$ and $\pi = \frac{22}{7}$.

11–26. Solve the algebra equation $(y - a)^2 - b = 4ab$ for $(y - a)$. Find y when $a = 6$ and $b = 4$.

11–27. Solve $A = \frac{\theta \pi r^2}{360}$ for r. This is the formula for the area of a sector of a circle. Find r when $A = 20\pi$ and $\theta = 72$.

11.3 SOLVING COMPLEX FORMULAS

Many formulas are used in business and industry. However, research and technology bring new problems with new formulas to solve these problems. Thus, no matter how many formulas you may learn, either in school or on the job, there will always be new formulas. It is important to learn the procedures to solve formulas in order to apply the procedures to solve future problems. Formulas become extremely complex, but once the techniques of solving formulas are learned, your confidence and ability will grow.

This group of formulas is more complex. It is important to use the skills you have learned to solve these formulas and literal questions.

Example 11–15: Applied Formula Solved Using Multiplication and Addition

Current Flowing through Armature of a Generator

The amount of current flowing through the armature of a generator can be determined by the formula

$$I = \frac{E - e}{R}$$

PROBLEM: Solve this formula for E.

SOLUTION: Clear the fraction by multiplying each side of the equation by R. The numerator of the fraction represents more than one quantity. Remember to place parentheses around the quantity $(E - e)$.

$$I = \frac{(E - e)}{R}$$

$$(R)\,I = \frac{(E - e)}{\cancel{R}}\cancel{(R)}$$

Add e to each side of the equation.

$$RI = E - e$$
$$e + RI = E - e + e$$
$$E = e + RI$$

Example 11–16: Applied Formula Solved Using Multiplication, Addition, Factoring, and Division

> **Drop in Voltage**
>
> The drop in voltage can be found by using the formula
>
> $$\frac{E}{e} = \frac{R + r}{r}$$

PROBLEM: Solve for r.

SOLUTION: Clear the fraction by multiplying each side of the equation by the lowest common denominator, er. Remember to place parentheses around $(R + r)$.

$$\frac{E}{e} = \frac{R + r}{r}$$

$$(\cancel{e}r)\frac{E}{\cancel{e}} = \frac{(R + r)}{\cancel{r}}(e\cancel{r})$$

$$(r)\,E = (R + r)\,(e)$$

$$rE = Re + re$$

Subtract re from each side of the equation.

$$rE = Re + re$$
$$rE - re = Re + re - re$$
$$rE - re = Re$$

The quantities that contain r can be written in another way.

$$rE - re = Re$$
$$r(E - e) = Re$$

Now divide both sides by the quantity $(E - e)$ to find r.

$$\frac{r\cancel{(E - e)}}{\cancel{(E - e)}} = \frac{Re}{(E - e)}$$

$$r = \frac{Re}{E - e}$$

Example 11–17: Applied Formula Solved Using Multiplication, Addition, Factoring, and Division

> **Sum of a Geometric Progression**
>
> The sum of a geometric progression can be determined by the formula
>
> $$S = \frac{r - a}{r - 1}$$

PROBLEM: Solve for r.

SOLUTION: Clear the fraction from the equation by multiplying each side by the lowest common denominator, $r - 1$.

$$S = \frac{r - a}{r - 1}$$

$$(r - 1)S = \frac{(r - a)}{\cancel{(r - 1)}}\cancel{(r - 1)}$$

$$(r - 1)S = r - a$$
$$rS - S = r - a$$

Thus add $+S$ and $-r$ to each side of the equation and solve.

$$rS - S + S - r = r - a + S - r$$
$$rS - \cancel{S} + \cancel{S} - r = \cancel{r} - a + S - \cancel{r}$$
$$rS - r = S - a$$

Rewrite $rS - r$ by factoring r from each term on the left side of the equation.

$$r(S - 1) = S - a$$

Now divide by the quantity $(S - 1)$ to find r.

$$r(S - 1) = S - a$$

$$\frac{r\cancel{(S - 1)}}{\cancel{(S - 1)}} = \frac{S - a}{(S - 1)}$$

$$r = \frac{S - a}{S - 1}$$

Example 11–18: Applied Formula Solved Using Multiplication, Subtraction, Factoring, and Division

Parallel Resistance

The parallel resistance can be found by the formula

$$\frac{1}{R} = \frac{1}{r_1} + \frac{1}{r_2}$$

PROBLEM: Solve for r_1.

SOLUTION: Remove the fractions by multiplying each side of the equation by the lowest common denominator, Rr_1r_2.

$$\frac{1}{R} = \frac{1}{r_1} + \frac{1}{r_2}$$

$$\overset{1}{\cancel{R}}r_1r_2\left(\frac{1}{\underset{1}{\cancel{R}}}\right) = R\overset{1}{\cancel{r_1}}r_2\left(\frac{1}{\underset{1}{\cancel{r_1}}}\right) + Rr_1\overset{1}{\cancel{r_2}}\left(\frac{1}{\underset{1}{\cancel{r_2}}}\right)$$

$$r_1r_2 = Rr_2 + Rr_1$$

Move all of the terms that include r_1 to one side of the equation, by subtracting Rr_1 from each side of the equation.

$$r_1r_2 = Rr_2 + Rr_1$$
$$r_1r_2 - Rr_1 = Rr_2 + Rr_1 - Rr_1$$
$$r_1r_2 - Rr_1 = Rr_2$$

Then factor r_1 from each term on the left side of the equation or rewrite as shown.

$$r_1r_2 - Rr_1 = Rr_2$$
$$r_1(r_2 - R) = Rr_2$$

Divide each side of the equation by $(r_2 - R)$.

$$r_1(r_2 - R) = Rr_2$$

$$r_1\frac{\overset{1}{\cancel{(r_2 - R)}}}{\cancel{(r_2 - R)}} = \frac{Rr_2}{(r_2 - R)}$$

$$r_1 = \frac{Rr_2}{r_2 - R}$$

Study this example and carefully note each step. These procedures will help you solve other formulas.

EXERCISE 11–3 SOLVE THE COMPLEX FORMULAS AND EVALUATE

Solve the following problems. Review the procedures in the examples if you have any difficulty. Show all necessary work.

11–28. The formula for determining latent heat vaporization is $Q = \frac{WL}{T}$. Solve this formula for T when $Q = 100$, $W = 5$, and $L = 800$.

11–29. The formula for finding the average speed of a uniformly accelerating body is $V = \frac{V_t + V_o}{2}$. Solve the formula for V_o and find V_o when $V = 312$ and $V_t = 112$.

11–30. The formula for determining temperature conversion is $T = \frac{1}{a} + t$. Solve this formula for a and find a when $T = 96$ and $t = 46$.

11–31. Using the electrical equivalent heat formula, $Q = 0.000477\ EIT$, solve for T and find T when $E = 1000$, $I = 200$, $Q = 95.4$.

11–32. Solve the thickness-of-pipe formula, $A = \frac{m}{t}(p + t)$, for t and find t when $A = 1$, $m = \frac{1}{4}$, and $p = 12$.

11–33. The formula for finding the area of an ellipse is $A = \pi ab$. Solve the formula for b and find b when $A = 113.04$ square units and $a = 18$.

11–34. Using the formula for finding the theoretical amount of air required to burn solid fuel, $M = 10.5C + 35.2\left(w - \frac{C}{8}\right)$, solve for C and find C when $M = 249.2$ and $w = 5$.

11–35. Solve the photographic enlargement formula, $\frac{1}{x} + \frac{1}{nx} = \frac{1}{f}$, for x and find x when $f = 12$ and $n = 3$.

11–36. The formula for finding the tap-size drill for a U.S. Standard thread is $S = T - \frac{1.299}{N}$. Solve the formula for N and find N when $T = 5.433$ and $S = 5$.

11–37. Solve the differential pulley formula, $W = \frac{2PR}{R - r}$, for R and find R when $P = 12$, $r = 75$, and $W = 48$.

11–38. Solve the algebraic equation, $x - y = xy$, for y and find y when $x = \frac{1}{8}$.

11–39. Using the prismoidal formula $V = \frac{h}{6}(B + 4M + b)$, solve for M and find M if $B = 25$, $b = 5$, $h = 36$, and $V = 420$.

11–40. Solve the formula for expansion of gases, $V_1 = V_0(1 + 0.5t)$, for t and find t when $V_1 = 112.5$ and $V_0 = 3$.

11–41. A formula for finding magnetic intensity is $H = \frac{0.4\pi NI}{L}$. Solve the formula for I and find I when $N = 4$, $L = 1.256$, $H = 2$, and $\pi = 3.14$.

11–42. Using the formula for finding the perimeter of an ellipse, $P = \pi\sqrt{2(a^2 + b^2)}$, solve for b and find b when $P = 16\pi$ and $a = 9.59$.

11–43. Solve the formula for finding the width of a segment of a circle, $W = 2\sqrt{h(2r - h)}$, for r and find r when $h = 8$ and $W = 8\sqrt{8}$.

11–44. The formula for finding the volume of a torus is $V = 2\pi^2 R r^2$. Solve the formula for r and find r when $R = 49$, $\pi = \frac{22}{7}$, and $V = 968$ square units.

11–45. The formula for finding the total surface area of a right circular cylinder is $T = 2\pi r(r + H)$. Solve for H and find H when $T = 320\pi$ and $r = 8$.

11–46. Solve the formula for finding the volume of a sphere, $V = \frac{\pi d^3}{6}$, for d and find d when $V = 288\pi$.

11–47. The formula for finding the distance an object travels when acted upon by gravity is $S = 16t^2 + vt$. Solve this formula for v and find v when $S = 1000$ feet and $t = 3$ seconds.

11–48. Using the formula for finding the distance an object travels when acted upon by gravity and an initial velocity, $S = 16t^2 + Vt$, solve for V and find V when $S = 1100$ and $t = 4$.

11–49. The formula for finding the exposed surface area of a cylinder is

$$S = \left(\frac{\pi d^2}{2} + \frac{\pi dl}{r}\right)\left(\frac{4rc}{d^2 l}\right)$$

Find S when $d = 4$, $l = 12$, $c = 2$, and $r = 2$. Leave the result in terms of π.

11–50. The formula for finding the volume of a cylinder is $V = \pi h(R + r)(R - r)$. Solve for V when $R = 8$, $h = 20$, and $r = 4$. Leave the answer in terms of π.

11–51. Solve $F = \frac{Nmv^2}{3}$ for v.

11–52. Solve $T = ph + 2A$ for h.

11–53. Solve $S = \frac{E - IR}{0.220}$ for I.

11–54. Solve $\frac{1}{f} = \frac{1}{p} + \frac{1}{q}$ for q.

11–55. Solve $I = \frac{En}{R + nr}$ for r.

11–56. Solve $E = RI + \frac{rI}{n}$ for I.

11–57. Solve $A = \frac{2}{3}hw$ for h.

11–58. Solve $V = 2\pi^2 Rr^2$ for R.

11–59. Solve $V = \pi R^2 h - \pi r^2 h$ for r^2.

11–60. Solve $T = \frac{(P + p)\,s}{2}$ for p.

11–61. Solve $V = \frac{4\pi r^3}{3}$ for r.

11–62. From problem 11–61, find r when $V = 1333.333\pi$.

11–63. Given $F = \frac{16mx}{T^2}$, find x.

11–64. From problem 11-63, find x when $m = 6$, $T = 4$, and $F = 90$.

11–65. Solve $\frac{1}{R} = \frac{1}{r_1} + \frac{1}{r_2} + \frac{1}{r_3}$ for r_3.

11–66. Solve $w = \frac{2PR}{R - r}$ for R.

11–67. Solve $P = \pi\sqrt{2(a^2 + b^2)}$ for a.

11–68. From problem 11–67, find b when $P = 12\pi$ and $a = 4.796$.

11–69. Solve $S = 16t^2 + Vt$ for V.

11–70. Solve $V = \frac{\pi d^3}{6}$ for d.

THINK TIME

Make a list of three formulas used in your occupation or at home. Study the formulas carefully to ensure your ability to work with them efficiently. Think of problems you may have on the job and how you may use the formulas. Use the formulas to solve these problems. Ask your instructor or a friend to check your work.

Formula transposition skills are vital to many occupations; the following practice problems are for your further development.

PROCEDURES TO REMEMBER

1. To solve formulas:
   (a) Read the formula to understand the purpose of the formula.
   (b) Reread the formula, noting the unknown quantity to be found.
   (c) Determine the procedures needed to solve the formula for the unknown quantity and estimate the answer.
   (d) Remember the "*scales of justice*" as the mathematical procedures are selected to solve the formula. When you add, subtract, multiply, or divide or take a root on one side of the equation, you must perform the same operation on the other side of the equation.
   (e) Show all necessary steps and work as you perform the mathematical operations. This will help you to prevent careless errors.
   (f) Check your work to make certain that you have reached a reasonable solution and that your answer is stated in the correct form (i.e., square feet, cubic inches, etc.).

CHAPTER SUMMARY

Formula transposition is the process of solving a formula for the desired quantity by changing its original form mathematically.

CHAPTER TEST

T–11–1. Solve $P = 2l + 2w$ for w.

T–11–2. Solve $A = \frac{bh}{2}$ for b.

T–11–3. Solve $C = 2\pi r$ for r.

T–11–4. Solve $Q = 0.7\ ELT$ for L.

T–11–5. Solve $a - b = ab$ for b.

T–11–6. Solve $P = 2a + b$ for a.

T–11–7. Find s, to the nearest hundredth, when $S = 4000$ and $r = 462$ in the formula $S = \pi rs$ ($\pi = 3.14$).

T–11–8. Find r when $i = \$800$, $p = \$5000$, and $t = 2$ in the formula $i = prt$.

T–11–9. Solve $T = 2B + pH$ for H.

T–11–10. Solve $S = \frac{\pi rs}{2}$ for r.

T–11–11. Solve $C = \frac{5(F - 32)}{9}$ for F.

T–11–12. Given the formula in problem T–11–11, find F if $C = 100$.

T–11–13. Solve $L = \frac{\pi r A}{180}$ for A.

T–11–14. Given the formula in problem T–11–13, find A when $L = 314$ and $r = 60$.

T–11–15. Solve $A = 6e^2$ for e.

T–11–16. Solve $A = \pi r^2$ for r.

T–11–17. Solve $S = 4\pi r^2$ for r.

T–11–18. Given the formula in problem T–11–17, find r when $S = 900\pi$.

T–11–19. Find y when $a = 6$ and $b = 4$ in the formula $(y - a) - b^2 = 2ab$. (First, solve the formula for y.)

T–11–20. Find e when $V = 343$ cubic inches given the formula $V = e^3$.

T–11–21. Solve $V = \frac{\pi r^2 h}{3}$ for h.

T–11–22. Solve $M = \frac{YI}{r}$ for r.

T–11–23. Solve $\frac{p_1V_1}{T_1} = \frac{p_2V_2}{T_2}$ for T_1.

T–11–24. Solve $A = \frac{h(B + b)}{2}$ for b.

T–11–25. Solve $F = \frac{4^2n^2Wx}{g}$ for n.

12

RATIO AND PROPORTION

OBJECTIVES

After studying this chapter, you will be able to:

12.1. Understand the definitions and basic operations with ratios.
12.2. Understand proportions and their solution.
12.3. Understand and perform the applications of ratios and proportions.

SELF-TEST

The completion of this self-test will indicate whether you need to further develop the skills taught in this chapter. Read each problem carefully and solve. Be sure to show your work so that errors may be identified. Write the ratios in fractional form and simplify.

S–12–1. 39 to 3

S–12–2. $18y$ to $3y$

S–12–3. $1\frac{3}{4}$ to $1\frac{1}{2}$

S–12–4. 5 hours to 50 minutes

Solve the following proportions problems for the unknown.

S–12–5. $\frac{4}{10} = \frac{N}{40}$

S–12–6. $\frac{4}{12} = \frac{Y}{9}$

S–12–7. $\frac{2.5}{K} = \frac{10}{6}$

S–12–8. $4:X = X:16$

S–12–9. $27:x = 18:12$

S–12–10. $y:15 = 75:15$

S–12–11. What number is to 14 as 2.5 is to 6?

S–12–12. Nylon tricot sells at the rate of $8.37 for 3 yards. How much will $\frac{1}{2}$ yard cost?

S–12–13. How much will 3 pounds of coffee cost if 2 pounds of coffee cost $4.92?

S–12–14. A company needs five workers to produce 31 air conditioners per day. How many workers would be needed to produce 217 per day?

S–12–15. A motorist averages 450 miles per day at a speed of 45 miles per hour. If she increases her speed to 55 miles per hour, how many miles would she average per day?

S–12–16. A stock pays a dividend of $5.50 per share every 182.5 days. What would the dividends amount to in 3 years?

S–12–17. A restaurant serves 175 people on the average each evening. Of those served, 85 ordered steak. The restaurant owner remodeled so that he now serves 245 people on the average each evening. If the ratio remains the same, how many steaks will he serve?

S–12–18. If you travel 594 miles in 11 hours, how long will it take to travel 486 miles?

S–12–19. A fast-food restaurant has an average daily gross sales of $8000 and a daily labor cost of $900. If the labor cost increases to $972, what would the gross sales need to be to keep the same profit ratio?

S–12–20. How far is it across the pond shown in Figure 12–1 from point *A* to point *B*?

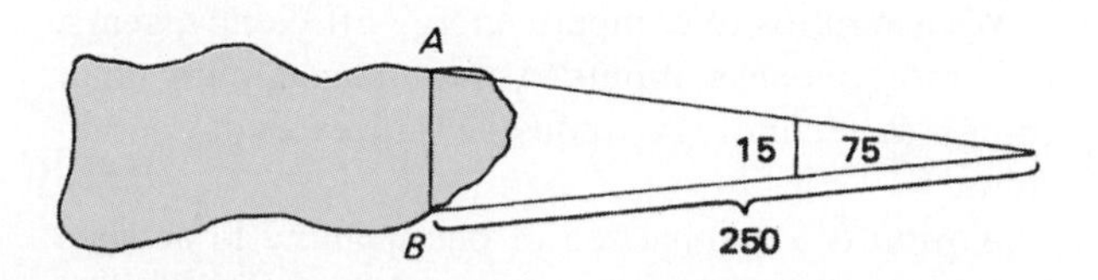

Figure 12–1

S–12–21. A blueprint lists the scale as $^1/_4$ inch to 6 inches. Determine the height of a doorway that measures 4 inches on the blueprint.

S–12–22. A copper wire is 360 feet long and has a resistance of 3.82 ohms. What is the resistance of 1000 feet of this wire?

S–12–23. The larger of two gears in a mesh makes 35 revolutions per minute. The smaller gear has a diameter of 38 centimeters and makes 180 revolutions per minute. What is the diameter of the larger gear?

S–12–24. Shag shoots a score of 195 on a rifle range, using 25 rounds. What would his score be if he fired 35 rounds?

S–12–25. Two speedboats start at the same time from the same place and travel in opposite directions. The ratio of their rates is 1:1.5. In 3 hours they are 60 miles apart. What is the rate of speed of the faster boat?

12.1 UNDERSTAND THE DEFINITION AND BASIC OPERATIONS WITH RATIOS

Frequently one quantity is compared to another quantity. For example, compare the gasoline consumption of a small car to that of a large car. If a small car uses 1 gallon of gasoline to travel 30 miles and a large car uses 1 gallon of gasoline to travel 15 miles, we say the small car travels twice as far as the large car on 1 gallon of gasoline. The ratio of miles per gallon is 2 to 1. We could also say that the ratio of miles traveled is 30 to 1 for the small car and 15 to 1 for the large car.

We use ratios to compare energy efficiency, gears, cost-effectiveness, inputs to outputs, and many other things in technology, industry and everyday situations.

A *ratio* is a comparison of one quantity to another. Similar comparisons are made when quantities are stated in fractions. Ratios compare like quantities. As the saying goes, apples must be compared to apples and oranges to oranges—not apples to oranges.

Examples of ratios and how they may be stated are:

$$\frac{15}{60} = 15{:}60 \text{ or } 15 \text{ is to } 60 = \frac{1}{4}$$

$$\frac{144}{12} = 144{:}12 \text{ or } 144 \text{ is to } 12 = \frac{12}{1}$$

$$\frac{3 \text{ hours}}{45 \text{ minutes}} = \frac{\frac{3 \text{ hours}}{1}\left(\frac{60 \text{ minutes}}{1 \text{ hour}}\right)}{45 \text{ minutes}} = \frac{180 \text{ minutes}}{45 \text{ minutes}} = \frac{4}{1}$$

When two quantities are compared, they must be in the same units. Ratios may be simplified by dividing equal quantities into each quantity. When a ratio is inverted, an *inverse ratio* results.

To simplify ratios, they must be in the same units. If they are not in the same units, convert the quantities to like units. Generally, this will require conversion to smaller units. Reducing should be only to whole numbers, not to decimals or fractions.

Example 12–1: Reduce a Simple Ratio

PROBLEM: What is the ratio of 12 to 48?

SOLUTION: The ratio of 12 to 48 should be written in a fractional form and then reduced by dividing the numerator and denominator by 12.

$$12{:}48 = \frac{12 \div 12}{48 \div 12} = \frac{1}{4}$$

Example 12–2: Reduce a Ratio with Values

PROBLEM: Write the ratio of 50 cents to \$3.

SOLUTION: Convert \$3 to cents by multiplying \$3 by 100, because there are 100 cents in 1 dollar. Then reduce to lowest terms by dividing the numerator and denominator by 50.

$$50 \text{ cents}{:}300 \text{ cents} = \frac{50 \div 50}{300 \div 50} = \frac{1}{6}$$

Example 12–3: Reduce a Ratio with Fractions

PROBLEM: What is the ratio of $5\frac{1}{2}$ to $\frac{1}{2}$?

SOLUTION: Write the ratio as a fraction.

$$5\frac{1}{2} \text{ to } \frac{1}{2} = \frac{5\frac{1}{2}}{\frac{1}{2}} = 5\frac{1}{2} \div \frac{1}{2}$$

Complete the indicated division.

$$5\frac{1}{2} \div \frac{1}{2} = \frac{11}{2} \div \frac{1}{2} = \frac{11}{\cancel{2}_{1}} \times \frac{\cancel{2}^{1}}{1} = \frac{11}{1}$$

Example 12–4: Reduce a Ratio with Time

PROBLEM: What is the ratio of 6 hours to 45 minutes?

SOLUTION: Convert 6 hours to minutes.

$$\left(\frac{6 \cancel{\text{hours}}}{1}\right)\frac{60 \text{ minutes}}{1 \cancel{\text{hour}}} = 360 \text{ minutes}$$

Set up the ratio and reduce to lowest terms.

$$\frac{360}{45} = \frac{\cancel{360}^{40}}{\cancel{45}_{5}} = \frac{\cancel{40}^{8}}{\cancel{5}_{1}} = \frac{8}{1}$$

EXERCISE 12–1 SIMPLIFY THE RATIOS

Simplify the following ratios by reducing to lowest terms.

12–1. $\frac{4}{8}$

12–2. $\frac{8}{24}$

12–3. $\frac{32}{24}$

12–4. $\frac{48}{12}$

12–5. $\frac{2.4}{0.3}$

12–6. $\frac{6\frac{1}{2}}{\frac{1}{2}}$

12–7. 7 to 42

12–8. 96 to 12

12–9. $\frac{5}{12}$ to $\frac{8}{15}$

12–10. 6.5 to 0.5

12–11. $\frac{\text{3 minutes}}{\text{1 hour}}$

12–12. 8 centimeters to 40 centimeters

12–13. 18 games to 2 games

12–14. 5 nickels to 4 dimes

12–15. 6 feet to 144 inches

12–16. 5 days to 6 weeks

12–17. $10 to 25 cents

12–18. 300 turns to 37.5 turns

12–19. a^2b^2 to ab

12–20. 6 feet 6 inches to 3 yards

12.2 UNDERSTAND PROPORTIONS AND THEIR SOLUTIONS

When one ratio is equal to another ratio, a proportion is formed. A *proportion* is a statement of equality between two ratios.

There are several ways a proportion may be stated. Two of the more common are:

Proportional form: $2:3 = 8:12$

Fraction form: $\frac{2}{3} = \frac{8}{12}$

A proportion has four terms. Using the example above, 2 is the *first term,* 3 is the *second term,* 8 is the *third term,* and 12 is the *fourth term.* The product of the first and fourth terms is called the product of the *extremes* (outside terms), and the product of the second and third terms is called the product of the *means* (inside terms). The product of the means equals the product of the extremes.

Occasionally, only three of the four terms of a proportion are known. To determine the fourth term, find the product of the extremes and set it equal to the product of the means. This process is often referred to as setting the cross products equal to each other, or *cross multiplication.* Once the products have been found, solve for the unknown.

Example 12–5: Solve a Linear Proportion

PROBLEM: Solve the proportion $X:8 = 6:24$ for the unknown.

SOLUTION: Find the product of the means and the product of the extremes.

$$X:8 = 6:24 \qquad (X)(24) = (8)(6)$$

(8 and 6: means; X and 24: extremes)

Divide both sides by 24.

$$(X)(24) = (8)(6)$$

$$\frac{\overset{1}{\cancel{24}}X}{\underset{1}{\cancel{24}}} = \frac{(\overset{1}{\cancel{8}})(\overset{2}{\cancel{6}})}{\underset{\underset{1}{\cancel{3}}}{\cancel{24}}}$$

$$X = 2$$

Then check the results as shown:

$$2:8 = 6:24 \qquad (2)(24) = (8)(6)$$

$$48 = 48$$

(8 and 6: means; 2 and 24: extremes)

Example 12–6: Solve a Fractional Proportion

PROBLEM: Solve the proportion for the unknown.

$$\frac{2}{3} = \frac{K}{12}$$

SOLUTION: Find the cross products of the fractional proportion. This operation is called *cross multiplying.*

$$\frac{2}{3} = \frac{K}{12}$$

$$(2)(12) = (3)(K)$$

Solve for K by dividing both sides by 3.

$$\frac{(2)(\overset{4}{\cancel{12}})}{\underset{1}{\cancel{3}}} = \frac{(\overset{1}{\cancel{3}})(K)}{\underset{1}{\cancel{3}}}$$

$$8 = K$$

Check the value as follows:

$$\frac{2}{3} = \frac{8}{12}$$

$$(2)(12) = (8)(3)$$

$$24 = 24$$

Example 12–7: Solve a Fractional Proportion

PROBLEM: Solve the proportion for the unknown.

$$\frac{30}{5} = \frac{10}{X}$$

SOLUTION: Find the cross products of the fractional proportion.

$$\frac{30}{5} = \frac{10}{X}$$

$$(30)(X) = (10)(5)$$

Solve for X by dividing both sides by 30.

$$\frac{(\overset{1}{\cancel{30}})(X)}{\underset{1}{\cancel{30}}} = \frac{(\overset{1}{\cancel{10}})(5)}{\underset{3}{\cancel{30}}}$$

$$X = \frac{5}{3} \text{ or } 1.\overline{6}$$

Check the results as follows:

$$(30)(X) = (10)(5)$$

$$(\overset{10}{\cancel{30}})\left(\frac{5}{\underset{1}{\cancel{3}}}\right) = (10)(5)$$

$$50 = 50$$

EXERCISE 12–2 SOLVE THE PROPORTIONS

Solve and check the following proportions.

12–21. $\frac{5}{8} = \frac{j}{24}$

12–22. $\frac{3}{7} = \frac{X}{49}$

12–23. $\frac{80}{r} = \frac{60}{5}$

12–24. $\frac{60}{360} = \frac{k}{45}$

12–25. $\frac{10}{s} = \frac{0.4}{36}$

12–26. $\frac{y}{15} = \frac{40}{3}$

12–27. $\frac{6}{x} = \frac{27}{36}$

12–28. $\frac{0.3}{24} = \frac{0.2}{a}$

12–29. $4:2 = 13:x$

12–30. $16:y = 64:8$

12–31. $x:40 = \frac{1}{4}:2$

12–32. $y:12 = 9:27$

12–33. $112:128 = 42:x$

12–34. $36:9 = 28:x$

12–35. $\frac{x - 4}{20} = \frac{14}{70}$

12–36. $\frac{9}{x} = \frac{x}{1}$

12–37. $\frac{36}{4x} = \frac{9x}{49}$

12–38. $\frac{x}{4.3} = \frac{19.7}{21.1775}$

12–39. $\frac{45}{5y} = \frac{9y}{64}$

12–40. $\frac{0.400}{0.014} = \frac{0.08}{p}$

12.3 UNDERSTAND AND PERFORM THE APPLICATIONS OF RATIOS AND PROPORTIONS

Many occupations require solving problems involving ratios and proportions. The following examples relate to some of the applications in industry.

Example 12–8: Solve a Rate/Time Proportion

PROBLEM: If a copying machine produces 48 copies in 30 seconds, how many seconds will it take to produce 240 copies?

SOLUTION: Set up the proportion in fractions and let X represent the time it will take to produce 240 copies.

$$\frac{48}{30} = \frac{240}{X}$$

Solve the proportion; remember that cross products are equal.

$$\frac{48}{30} = \frac{240}{X}$$

$$(48)(X) = (30)(240)$$

Divide each side of the equation by 48.

$$\frac{\overset{1}{\cancel{(48)}}(X)}{\underset{1}{\cancel{48}}} = \frac{(30)\overset{5}{\cancel{(240)}}}{\underset{1}{\cancel{48}}}$$

$$X = 150$$

It will take 150 seconds to produce 240 copies. Check your results as shown.

$$\frac{48}{30} = \frac{240}{150}$$

$$(48)(150) = (30)(240)$$

$$7200 = 7200$$

Example 12–9: Solve a Lever Proportion

PROBLEM: A man wishes to lift a rock out of a shallow hole. He estimates the rock weighs 420 pounds. If a 12-foot steel bar is used, where will the fulcrum need to be placed if the man applies 180 pounds of pressure?

SOLUTION: Recall the principle of the lever.

W_1 ↓ d_1 ▲ d_2 ↓ W_2

$$(W_1)(d_1) = (d_2)(W_2)$$

Then draw a sketch of the problem.

420 lb (rock) ↓ d_1 ▲ $12 - d_1$ ↓ 180 lb (man)

W_1 = weight to the left (420 lb)
d_1 = distance to the left (d_1)
W_2 = weight to the right (180 lb)
d_2 = distance to the right ($12 - d_1$)

Using the principle of the lever, write the problem as shown, where d_1 is the distance to the fulcrum from the rock and $12 - d_1$ is the distance of the man from the fulcrum.

$$(420)(d_1) = (12 - d_1)(180)$$

Then solve for d_1.

$$420d_1 = 12(180) - 180d_1$$
$$420d_1 + 180d_1 = (12)(180)$$
$$600d_1 = (12)(180)$$

$$\frac{\overset{1}{\cancel{600}}d_1}{\underset{1}{\cancel{600}}} = \frac{(\overset{1}{\cancel{12}})(\overset{18}{\cancel{180}})}{\underset{\underset{5}{\cancel{50}}}{\cancel{600}}}$$

$$d_1 = \frac{18}{5}$$

$$d_1 = 3.6$$

Thus the fulcrum should be 3.6 feet from the rock. Check as shown.

$$(420)(3.6) = (12 - 3.6)(180)$$
$$1512 = (8.4)(180)$$
$$1512 = 1512$$

Example 12–10: Solve a Construction Scale Proportion

PROBLEM: A brick wall measures $3\frac{1}{2}$ inches on a drawing. How long is the wall if the scale of the drawing is $\frac{1}{4}$ inch = 25 feet?

SOLUTION: Let x = the actual length. Set up the proportion.

$$\frac{\text{Actual size}}{\text{Scale size}} = \frac{\text{actual length}}{\text{drawing length}}$$

$$\frac{25 \text{ ft}}{\frac{1}{4} \text{ in.}} = \frac{x}{3\frac{1}{2} \text{ in.}}$$

Note: Convert the fractions to decimals.

$$\frac{25 \text{ ft}}{0.25 \text{ in.}} = \frac{x}{3.5 \text{ in.}}$$

Solve for x.

$$\frac{25}{0.25} = \frac{x}{3.5}$$

$$\frac{(25)(3.5)}{0.25} = x$$

$$x = 350 \text{ ft}$$

Check as shown.

$$\frac{25}{0.25} = \frac{350}{3.5}$$
$$(25)(3.5) = (350)(0.25)$$
$$875 = 875$$

Example 12–11: Solve a Circle Proportion

PROBLEM: Two pulleys are connected by a belt. The diameter of the larger pulley is 60 centimeters and the diameter of the smaller pulley is 45 centimeters. If the larger pulley rotates at 1500 revolutions per minute (rpm), how fast does the smaller pulley rotate?

SOLUTION: Sketch the pulleys (see Figure 12–2) and set up a proportion. Remember that pulley and gears are to be set up as an inverse ratio.

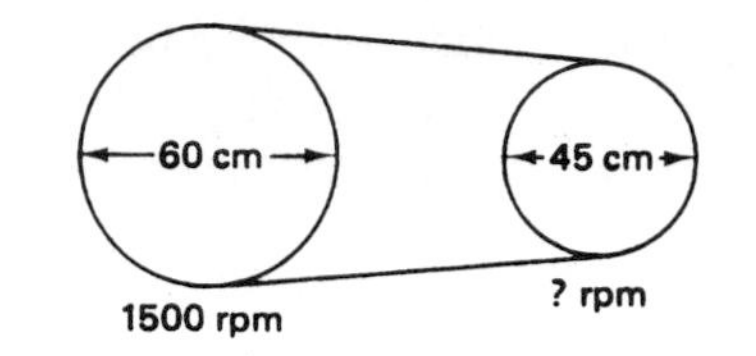

Figure 12–2

$$\frac{\text{Diameter of larger}}{\text{Diameter of smaller}} = \frac{\text{revolutions per minute of smaller}}{\text{revolutions per minute of larger}}$$

Let X be the revolutions per minute of the smaller pulley and set up the problem.

$$\frac{60}{45} = \frac{X \text{ rpm}}{1500}$$

Solve the ratio for X.

$$\frac{60}{45} = \frac{X \text{ rpm}}{1500}$$
$$(60)(1500) = (45)(X \text{ rpm})$$

$$\frac{(60)(1500)}{45} = \frac{(\overset{1}{\cancel{45}})(X \text{ rpm})}{\underset{1}{\cancel{45}}}$$

$$2000 = X \text{ rpm}$$

Check as shown.

$$\frac{60}{45} = \frac{2000}{1500}$$

$$(60)(1500) = (2000)(45)$$

$$90{,}000 = 90{,}000$$

Note: Another way to state the pulley or gear principle is as follows: The product of the diameter of the larger pulley times the revolutions per minute of the larger pulley equals the product of the diameter of the smaller pulley times the revolutions per minute of the smaller pulley. This is stated as $(D)(RPM) = (d)(rpm)$, where the capital letters indicate the larger pulley and the lowercase letters the smaller pulley.

EXERCISE 12–3 SOLVE APPLIED RATIO AND PROPORTION PROBLEMS

Work the following problems that are similar to applied problems in industry and daily living.

12–41. A baker sells 3 doughnuts for $0.99. How much will $1\frac{1}{2}$ dozen doughnuts cost?

12–42. A laborer is paid at a rate of $420 for every 3 days that she works. How much will she earn in 21 days?

12–43. A drawer is $1\frac{1}{2}$ feet wide. In a drawing the drawer is 1 inch wide. Determine the scale ratio of the actual width of the drawer to the scale width of the drawer in the drawing.

12–44. If 12 workers can complete a job in 75 days, how many workers would be required to complete the job in 10 days? (This is an application of an inverse ratio. See Example 12–11.)

12–45. The scale on a map is 1 inch = 12 miles. How many miles is it from one town to another if the distance measures $7\frac{1}{4}$ inches on the map?

12–46. In Arkansas, the ratio of harvested land to the total area of the state is 1:4.8. If the total area of the state is 53,225 square miles, how many square miles are harvested land?

12–47. A machine can produce 60,000 nails in 12 hours. At this rate, how long would it take to produce 160,000 nails?

12–48. The population of a certain country is 50,000,000. The area of the country is 215,000 square miles. Determine the population density (number of people per square mile).

12–67. A painter uses 1 quart of paint to cover 225 square feet of concrete wall. How many gallons would she need to cover 1800 square feet?

12–68. Thunder is heard 48 seconds after lightning flashes 10 miles away. When would a person 6 miles away hear the thunder?

12–69. What will it cost to serve 99 people ground beef that cost $1.89 per pound if 5 pounds will serve 20 people?

12–70. The maximum height reached by a bullet fired from a certain gun is proportional to the square of the time it takes to reach that height. If a bullet reaches a height of 5 feet 0.01 second after it is fired, what height would a bullet reach in half a second?

THINK TIME

Ratios and proportions may be used to solve a wide variety of problems. For example, they may be used to find screw thread depth, horsepower, mechanical efficiency of engines, taper and diameter, speed of pulleys and gears, strength of materials, velocity, density, and pressure. Think of ways in which you can use the principles of ratio and proportion to solve problems.

One of the common uses of ratio is to determine the portion of one's home that can be used as a business expense for tax purposes. Follow the example given below.

PROBLEM: Medora owns her home and uses one room exclusively as an office for her service station business. In this room she keeps her accounts and makes out required reports; at night she uses it to receive calls for emergency road service. She determines that the size of the room is 120 square feet. The total square footage of her home is 1440. If her electric, fuel, and water bills total $3672 for 1 year, how much of this can be allowed for the office as business expenses?

SOLUTION: First, set up a proportion.

$$\frac{120 \text{ sq ft}}{1440 \text{ sq ft}} = \frac{X}{\$3672}$$

Find the product of the means and the product of the extremes and divide to determine the amount allowed.

$$(12)(\$3672) = (X)(1440)$$

$$\frac{(120)(\$3672)}{1440} = X$$

$$X = \$306$$

Check as shown.

$$\frac{120}{1440} = \frac{\$306}{\$3672}$$

$$\$440{,}640 = \$440{,}640$$

PROCEDURES TO REMEMBER

1. To simplify a ratio:
   (a) Convert to like terms.
   (b) Write as a fraction.
   (c) Reduce to lowest terms.
2. To solve a proportion:
   (a) Set up the proportion, converting to like terms if necessary.
   (b) Determine the product of the means and the product of the extremes.
   (c) Solve for the unknown.
   (d) Check by substituting the result for the unknown and cross multiply.

CHAPTER SUMMARY

1. A *ratio* is a comparison of one quantity to another.
2. Ratios always compare like items.
3. When two quantities are compared, they must be like units.
4. A *true ratio* is found by reducing to the lowest term.
5. When a ratio is inverted, an *inverse ratio* results.
6. Simplification of ratios should be to whole numbers, not to decimals or fractions.
7. A *proportion* is a statement of equality between two ratios.

8. Two common ways of expressing a proportion are:
   (a) Proportional form (3:4 = 9:12)
   (b) Fractional form $\left(\frac{3}{4} = \frac{9}{12}\right)$
9. A proportion has four terms. In the example 6:8 = 18:24, 6 is the *first term,* 8 is the *second term,* 18 is the *third term,* and 24 is the *fourth term.*
10. The product of the first and fourth terms is called the product of the *extremes* (outside terms).
11. The product of the second and third terms is called the product of the *means* (inside terms).
12. The product of the means equals the product of the extremes.

CHAPTER TEST

Write in fractional form and simplify.

T–12–1. 4 inches to 12 inches

T–12–2. 7 to 63

T–12–3. 5 feet 8 inches to 6 feet 2 inches

T–12–4. $\dfrac{6\frac{1}{2}}{\frac{7}{8}}$

T–12–5. $\frac{x}{7} = \frac{1.5}{1}$

T–12–6. $\frac{1.5}{2} = \frac{3.5}{p}$

T–12–7. $\frac{E}{10.5} = \frac{9}{31.5}$

T–12–8. $\frac{84}{20} = \frac{21}{N}$

T–12–9. $j:8 = 18:j$

T–12–10. $18:21 = M:63$

T–12–11. What number is to 28 as 5 is to 12?

T–12–12. A trucker finds that he uses 63.5 gallons of fuel to drive 762 miles. How many gallons will he need to drive 4575 miles?

T–12–13. Burlap material sells at a rate of $5.88 for 3 yards. How much will $6\frac{1}{2}$ yards cost?

T–12–14. Dividends of $0.375 per share are paid every 90 days. How many days would it take to accumulate a total of $375.00 on 200 shares?

T–12–15. What is the ratio of two gears if one gear has 105 teeth and the other has 15 teeth?

T–12–16. Joel shoots a score of 164 on a rifle range using 30 rounds. What would his score be if he fired 45 rounds?

T–12–17. A jet plane travels 3972 miles in 6 hours. At this rate, how far can the plane travel in $4\frac{1}{2}$ hours?

T–12–18. A florist plants 475 tulip bulbs, of which 456 survive for sale. The following year the florist wishes to have 600 tulips for sale. How many tulips will need to be planted at the same survival rate?

T–12–19. If there are 160 pounds of nickel in 1 ton of alloy, how much nickel would be required to make 4500 pounds of the alloy?

T–12–20. A copying machine produces 35 copies in 20 seconds. How many copies will it produce in 5 minutes?

T–12–21. Find the missing dimension of the A-frame cottage shown in Figure 12–5.

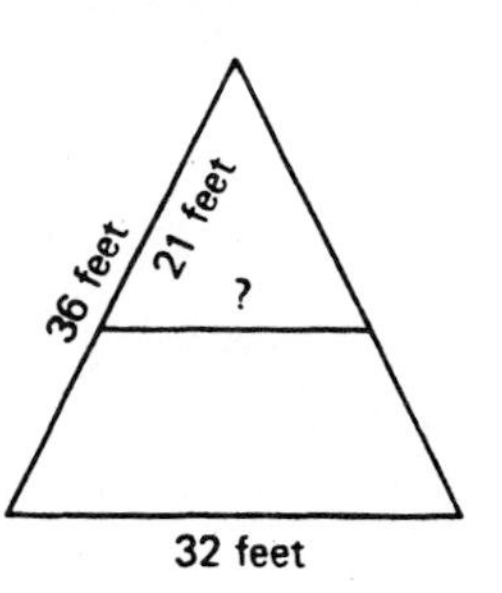

Figure 12–5

T–12–22. The scale on a drawing is $\frac{1}{8}$ inch = 3 feet. Determine the length in feet of a rafter if it measures $1\frac{3}{8}$ inches on the drawing.

T–12–23. A 5-inch pulley makes 1800 revolutions per minute and drives a larger pulley at 300 revolutions per minute. What is the diameter of the larger pulley?

T–12–24. A photograph has a length of 16 inches and width of 11 inches. An enlargement is made having a length of 45 inches. Determine the width of the enlargement.

T–12–25. A chef uses 22 pounds of flour to bake the cakes he serves during the day. He generally serves cake to about 160 people. He decides to cut the price he charges for a piece of cake. Now 240 people order cake each day. How much flour does he use?

13

GRAPHS AND LINEAR EQUATIONS

OBJECTIVES

After studying this chapter, you will be able to:

13.1. Understand the basic definitions and properties of the rectangular coordinate system.

13.2. Use various methods of graphing linear equations.

13.3. Understand slope, slope-intercept, finding the equation of a line, and graphing.

SELF-TEST

S–13–1. The point where the x- and y-axes intersect in a rectangular coordinate segment is called the ________________.

S–13–2. Graph the following points and label them: A (4, 5), B (−3, 0), C (−2, −8).

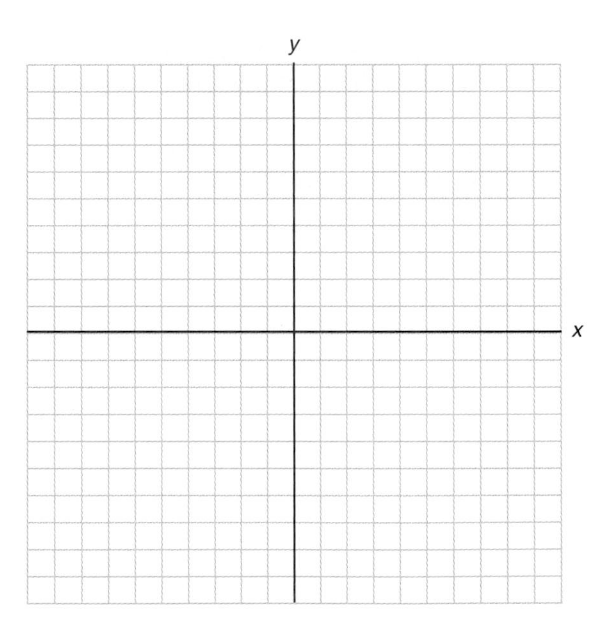

S–13–3. Given the following points, write the coordinates:

X (_____, _____)
Y (_____, _____)
Z (_____, _____)

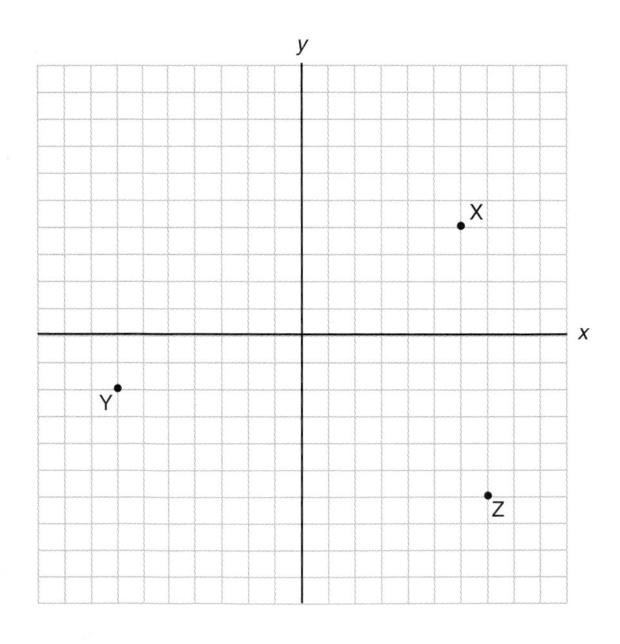

S–13–4. Given $y = -5x + 12$, find the y-value of the ordered pair (2, _____).

S–13–5. Given $2x - 4y = 8$, find the x-value of the ordered pair (______, 1).

S–13–6. Graph $y = x + 3$ by plotting the points.

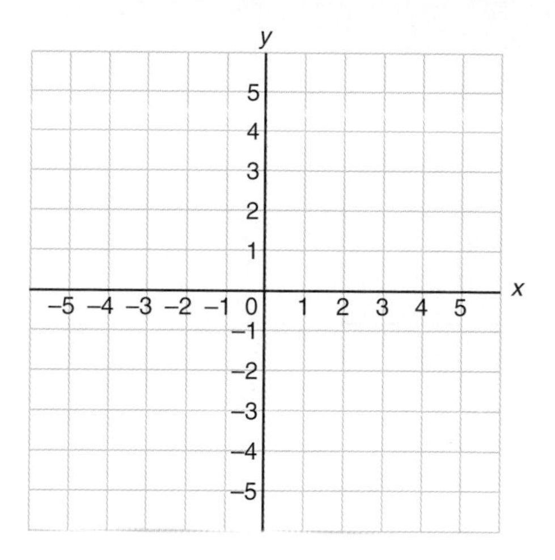

S–13–7. Graph the equation $y = 2x - 3$ by plotting the points.

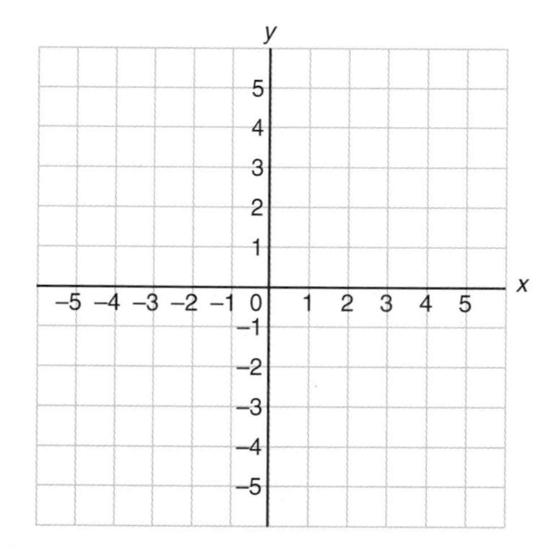

S–13–8. Graph $4x - 2y = 6$ by plotting the points.

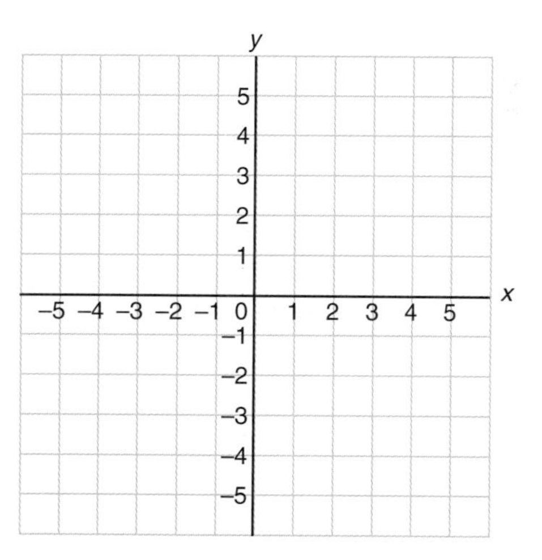

S–13–9. Graph the equation $x = -3$.

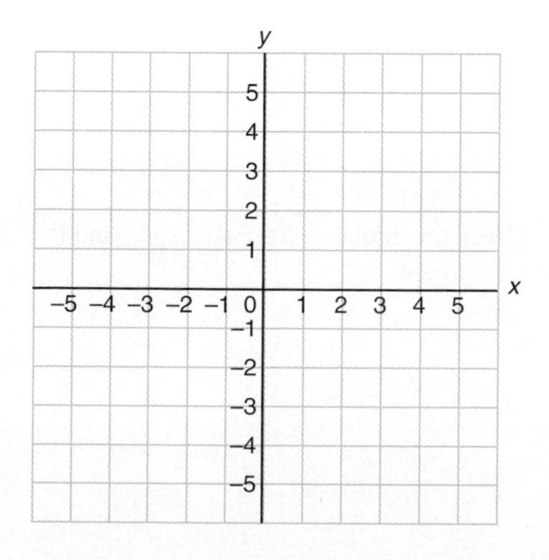

S–13–10. Graph $3x + y = 6$ using the intercept method.

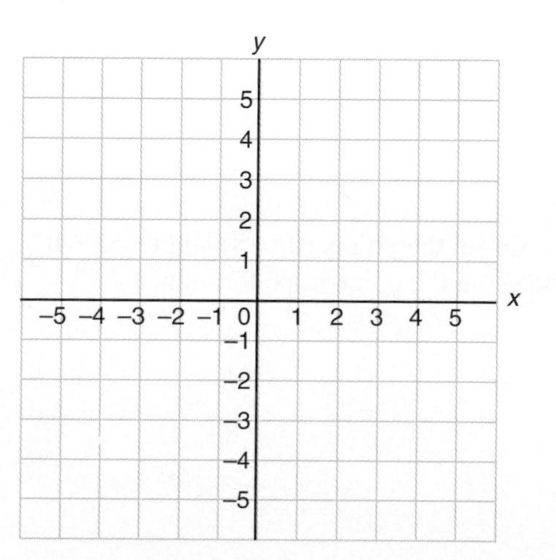

S–13–11. Graph $5x - 4y = 20$ using the intercept method.

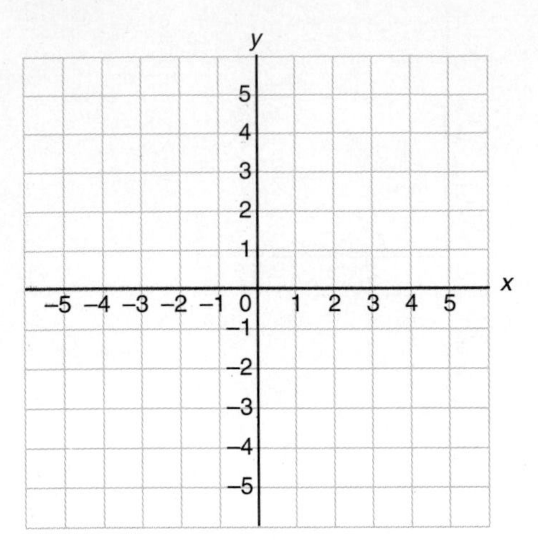

S–13–12. Graph $y = \frac{3}{4}x + 4$ using any method.

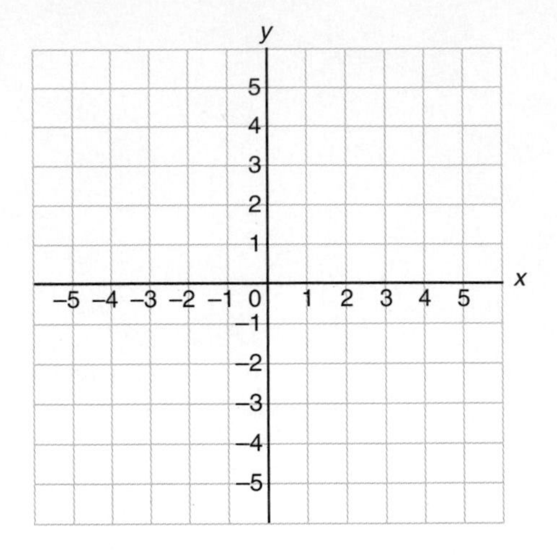

S–13–13. Graph $3x - 2y = 12$ using any method.

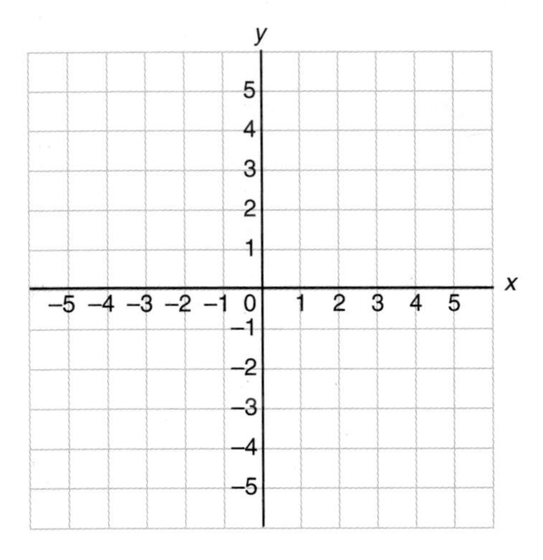

S–13–14. Given $3x - 5y = 1$, find the slope and y-intercept: $m =$ _____, $b =$ _____.

S–13–15. Given $5y - 3x + 8 = 0$, find the slope and y-intercept: $m =$ _____, $b =$ _____.

S–13–16. Given the points $(2, -3)$ and $(1, -5)$, find the slope using the point-slope formula.

S–13–17. Given the points $(0, -8)$ and $(-5, -10)$, find the slope using the point-slope formula.

S–13–18. Given the equation $5x - 4y = 12$, find the slope and the y-intercept.

S–13–19. Given $m = \frac{2}{3}$ and $b = -4$, graph the equation.

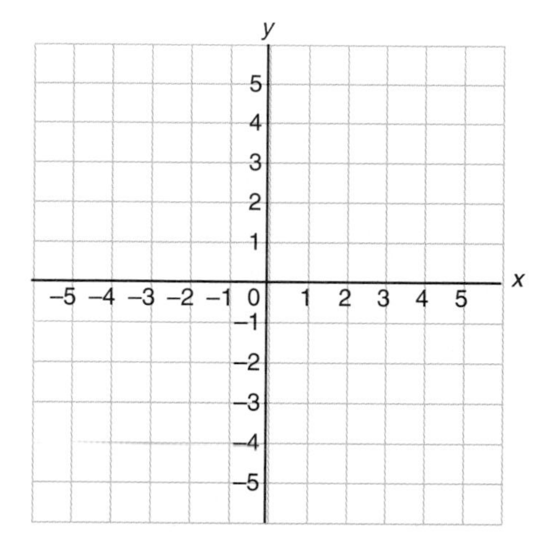

S–13–20. Given $m = -\frac{4}{3}$ and $b = +2$, graph the equation.

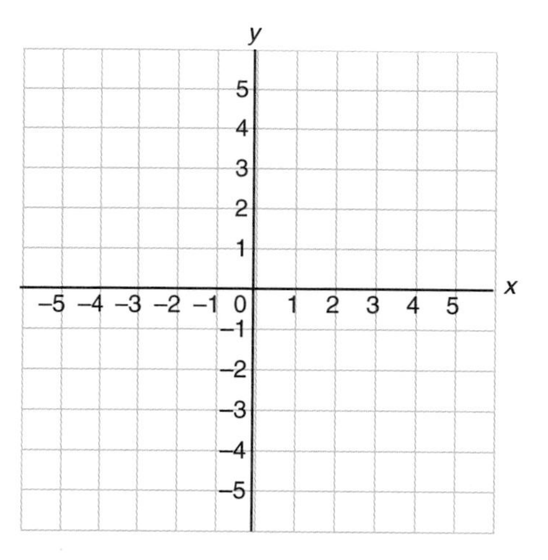

13.1 UNDERSTAND THE BASIC DEFINITIONS AND PROPERTIES OF THE RECTANGULAR COORDINATE SYSTEM

Many ideas can be better understood when a picture is used to show the concept. A graph can be considered a picture of certain conditions.

As shown in Chapter 7, pictographs, bar, line, and circle graphs are examples of pictures that describe particular conditions. An equation can be converted into a "picture" or graph.

The "canvas" of graphing equations is called the rectangular coordinate system. This is sometimes called the Cartesian coordinate system, named after René Descartes (1596–1650), a French mathematician who is credited with its development. However, it is generally known as the rectangular coordinate system. Some mathematicians consider the rectangular coordinate system the connection between algebra and geometry because we can graph (geometry) an equation (algebra).

Graphing has been around for thousands of years—some say that the Egyptians stretched ropes using pairs of numbers to survey land on the banks of the Nile River. Thus the rectangular coordinate system develops from perpendicular lines, two lines drawn at right angles, as shown in Figure 13–1.

Additional properties of the rectangular coordinate system are shown in Figure 13–2 (see p. 214).

One line is known as the x-axis or the *horizontal axis,* and the other line is the y-axis or *vertical axis*. The point of intersection of the lines is called the *origin.*

The intersection of two lines perpendicular to each other forms what is called the *rectangular coordinate system.* The two lines are known as the *coordinate axes.*

Rectangular Coordinate System

y
4
3
2
1
origin
–4 –3 –2 –1 0 1 2 3 4
x
–1
–2
–3
–4

Figure 13–1

They form a *plane,* which is divided into four regions known as *quadrants.* Note the quadrants are named in a counterclockwise direction, I, II III, and IV.

The axes are divided and numbered as shown in Figure 13–2. On the *x-axis,* to the right of the origin, the numbers are positive; to the left of the origin, the numbers are negative. On the *y-axis* above the origin, the values are positive; below the origin the values are negative.

Each point in the plane has an "address," which is identified by a pair of numbers known as an *ordered pair.*

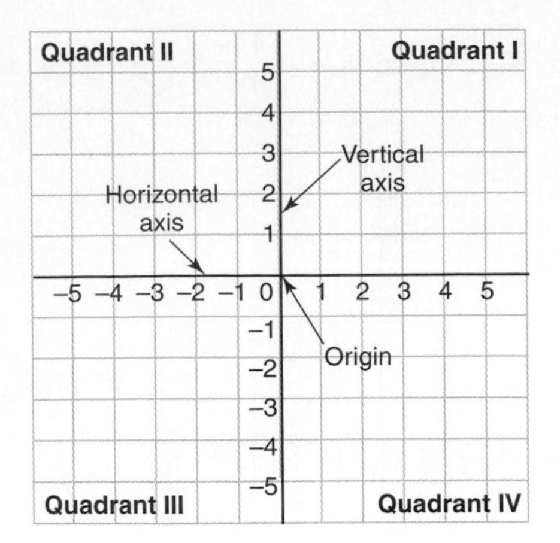

Figure 13–2

For example, the address of the ordered pair (3, 4) would be as shown in Figure 13–3.

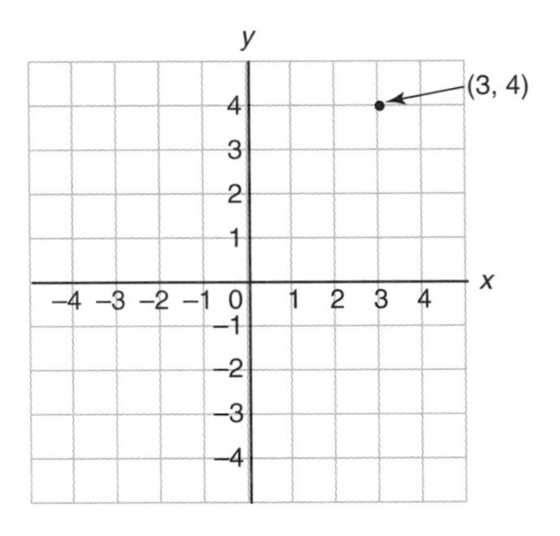

Figure 13–3

The first number is the x-value, the *abscissa* or the horizontal distance, which is +3 units to the right of the origin. The second number is the y-value, the *ordinate* or the vertical distance, which is 4 units above the origin.

Example 13–1: Plotting Points

PROBLEM: Plot the following points: (+4, +4), (−2, 3), and (−5, −4).

SOLUTION: To plot (+4, +4) start at the origin, move 4 units to the right on the x-axis, and then move up 4 units in the y-coordinate.

To plot (−2, 3) start at the origin, move 2 units to the left on the x-axis, and then move up 3 units in the y-coordinate.

To plot (−5, −4) start at the origin, move 5 units to the left on the x-axis, then move down 4 units in the y-coordinate. See Figure 13–4 for these illustrations.

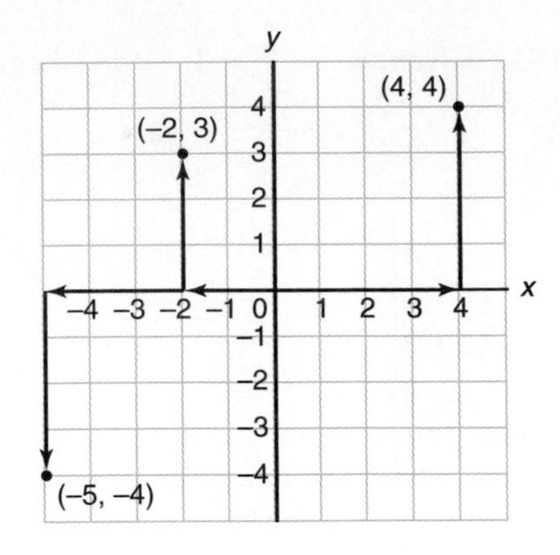

Figure 13–4

Example 13–2: Find the Coordinate of the Given Points

PROBLEM: Find the coordinates, given points A, B, and C in Figure 13–5.

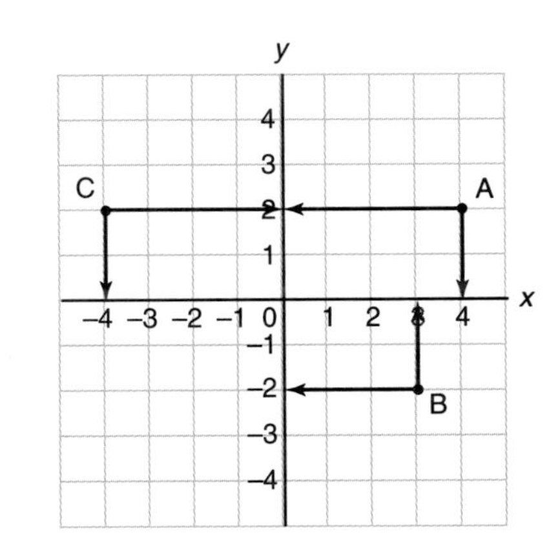

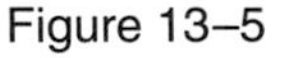
Figure 13–5

SOLUTION: Through point A, draw a vertical line to the x-axis, and note the value; then draw a horizontal line through A to the y-axis, and note the value. Thus the "address" of point A is (4, 2).

Through point B, draw a vertical line to the x-axis, and note the value; then draw a horizontal line through B to the y-axis, and note the value. Thus the address of B is the ordered pair (3, −2).

Through point C, draw a vertical line to the x-axis, and note the value; then draw a horizontal line through B to the y-axis, and note the value. Thus the ordered pair of point C is (−4, 2).

EXERCISE 13–1 PLOTTING AND LOCATING COORDINATES

Plot the given points for exercises 13–1 to 13–5 and label in Figure 13–6.

13–1. A (5, 2)

13–2. B (−3, 6)

13–3. C (9, −8)

13–4. D (−6, −2)

13–5. E (−2, −8)

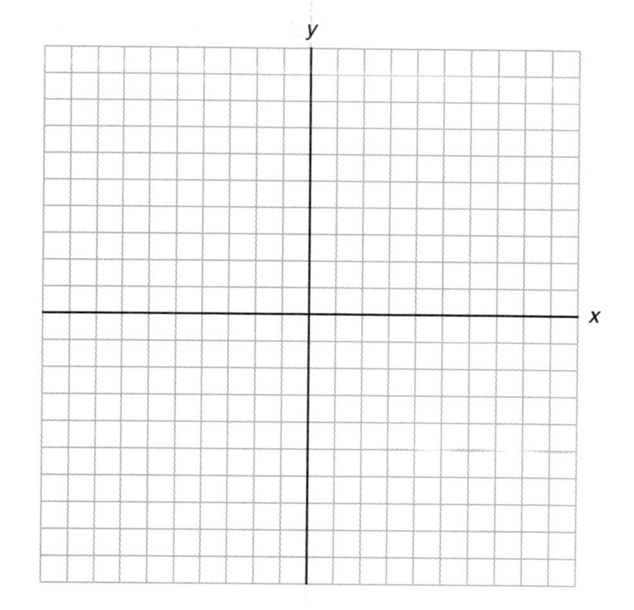

Figure 13–6

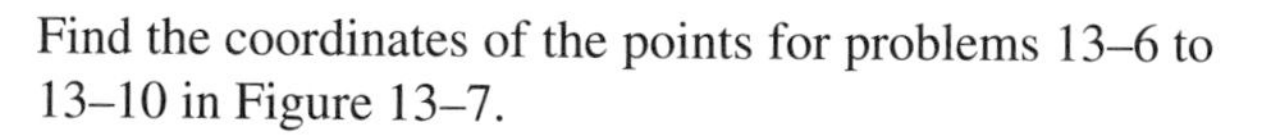

Find the coordinates of the points for problems 13–6 to 13–10 in Figure 13–7.

13–6. J (_____, _____)

13–7. K (_____, _____)

13–8. L (_____, _____)

13–9. M (_____, _____)

13–10. N (_____, _____)

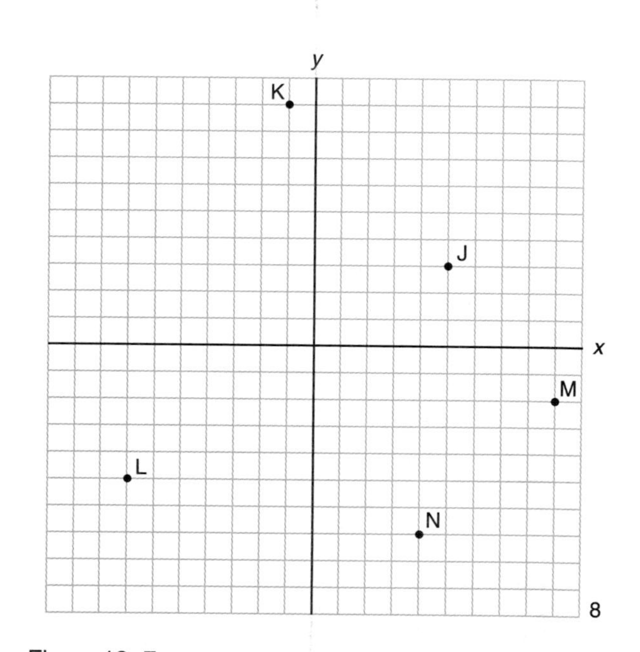

Figure 13–7

13.2 USE VARIOUS METHODS OF GRAPHING LINEAR EQUATIONS

The graph of an equation is "a picture" in the rectangular coordinate system that shows the ordered pairs that satisfy the equation. Plotting these ordered pairs is known as *graphing the equation.*

For example, given the equation $x + y = 5$, if $x = 2$, then

$$x + y = 5$$
$$2 + y = 5$$
$$y = 5 - 2$$
$$y = 3$$

So the ordered pair (2, 3) satisfies $x + y = 5$. Also, if $x = 1$, then

$$x + y = 5$$
$$1 + y = 5$$
$$y = 5 - 1$$
$$y = 4$$

So the ordered pair (1, 4) satisfies $x + y = 5$. There are several ordered pairs that will satisfy the equation. If $x = -3$, then

$$x + y = 5$$
$$-3 + y = 5$$
$$y = 5 + 3$$
$$y = 8$$

So the ordered pair $(-3, 8)$ also satisfies $x + y = 5$.

Therefore, if the ordered pairs (2, 3), (1, 4), and $(-3, 8)$ are plotted, the line satisfies the equation $x + y = 5$, as shown in Figure 13–8.

Another graphing procedure is to solve the equation for y and substitute values for x to obtain ordered pairs that satisfy the equation. This procedure is shown in Example 13–3.

Example 13–3: Graphing an Equation

PROBLEM: Graph the equation $-4x + y = -8$.

SOLUTION: Solve the equation for y, and substitute various values for x to obtain ordered pairs that satisfy

$$-4x + y = -8$$
$$y = +4x - 8$$

x	y	$y = 4x - 8$	*ordered pair*
0	−8	$y = 4(0) - 8 = 0 - 8 = -8$	(0, −8)
2	0	$y = 4(2) - 8 = 8 - 8 = 0$	(2, 0)
4	8	$y = 4(4) - 8 = 16 - 8 = 8$	(4, 8)
1	−4	$y = 4(1) - 8 = 4 - 8 = -4$	(1, −4)

Plot as shown in Figure 13–9; the ordered pairs satisfy the equation $-4x + y = -8$.

Another graphing procedure is known as the intercept form of graphing. This procedure is done by setting $x = 0$ in an equation to obtain a value for y. Then let $y = 0$ to obtain a value for x. These ordered pairs will indicate where the line of the graph crosses the x- and y-axes; thus, this is the intercept form of graphing. Since a line is determined by two points, the two intercepts will determine the graph of an equation. The definition of intercept is where the line crosses the x-axis or the y-axis.

The intercept form of graphing will be illustrated in Example 13–4.

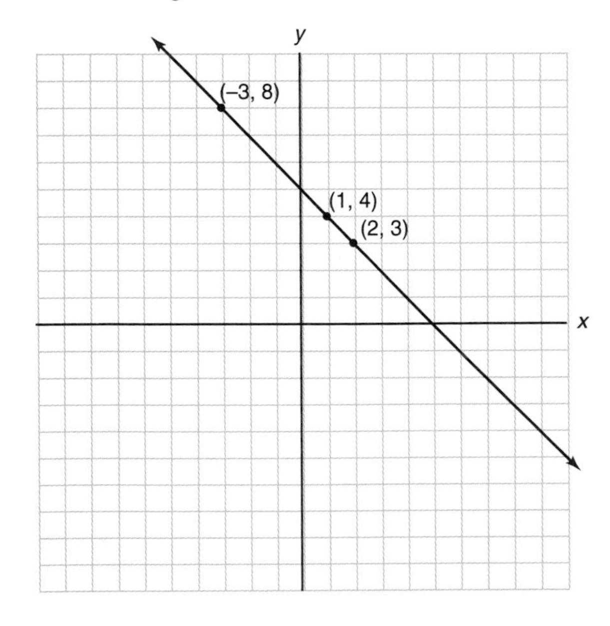

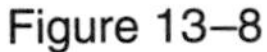
Figure 13–8

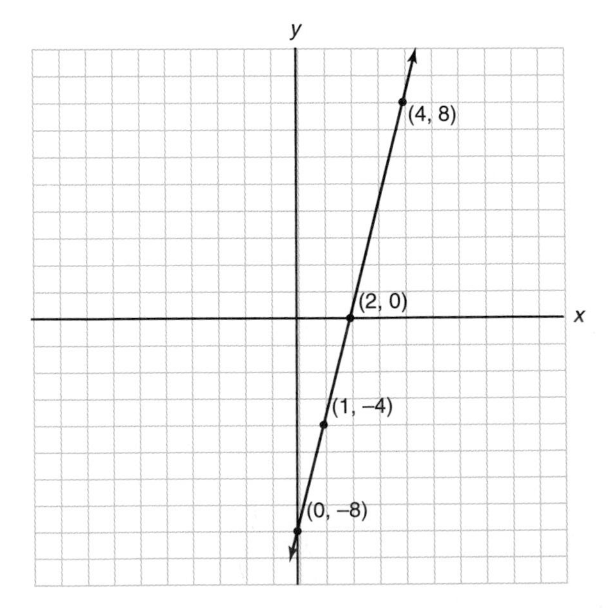

Figure 13–9

Example 13–4: Graphing an Equation: The Intercept Method

PROBLEM: Graph $3x - 4y = 12$.

SOLUTION: Set $y = 0$ to obtain a value for x, which is the x-intercept; then set $x = 0$ to obtain a value for y which is the y-intercept.

So given $3x - 4y = 12$, let $y = 0$.

$$3(x) - 4(0) = 12$$
$$3x - 0 = 12$$
$$\frac{3x}{3} = \frac{12}{3}$$
$$x = 4$$

Thus the ordered pair (4, 0) is the x-intercept. Then set $x = 0$ in the equation.

$$3x - 4y = 12$$
$$3(0) - 4y = 12$$
$$-4y = 12$$
$$\frac{-4y}{-4} = \frac{12}{-4}$$
$$y = -3$$

So (0, −3) is the y-intercept. Figure 13–10 shows the graph of the equation $3x - 4y = 12$ using the intercepts (4, 0) and (0, −3).

The intercept form of graphing will not graph equations of the type $ax + by = 0$, since the graph of those equations crosses the origin. That type of equation may be graphed using the plotting method. Solve the equation for one unknown, and substitute values as shown in Example 13–5.

Example 13–5: Graph an Equation of the Type: $Ax + By = 0$

PROBLEM: Graph the equation $-3x + y = 0$.

SOLUTION: Solve the equation for y; then substitute values in the table for x to obtain ordered pairs that satisfy the equation. So

$$-3x + y = 0$$
$$y = +3x$$

x	$y = 3x$	y	*ordered pair*
0	$y = 3(0) = 0$	0	(0, 0)
3	$y = 3(3) = 9$	9	(3, 9)
−3	$y = 3(-3) = -9$	−9	(−3, −9)
−2	$y = 3(-2) = -6$	−6	(−2, −6)

Plot the ordered pairs and draw the line to obtain the graph of $-3x + y = 0$ as show in Figure 13–11.

A special type graph results from an equation $Ax + By = C$ when either A or B equals zero. For example, if $Ax + By = 4$ and $A = 1$ and $B = 0$, the equation $1x + 0y = 4$ results. This indicates that for any value of y the x-value is 4. Thus this is a line parallel to the y-axis through $x = 4$. Example 13–6 illustrates graphing horizontal and vertical lines.

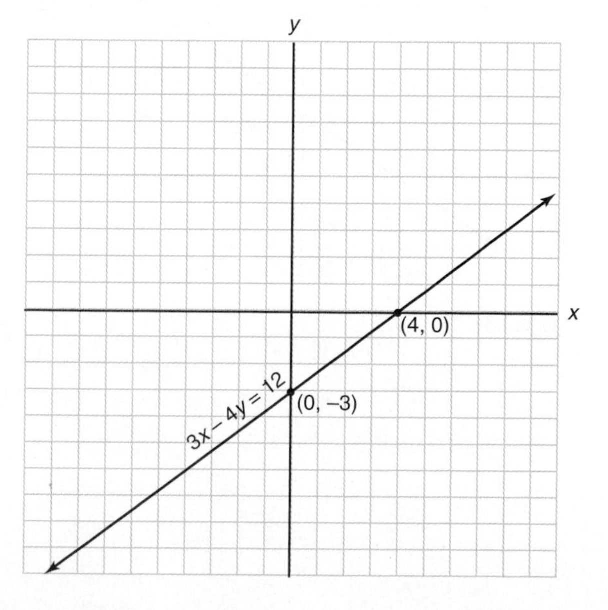

Figure 13–10

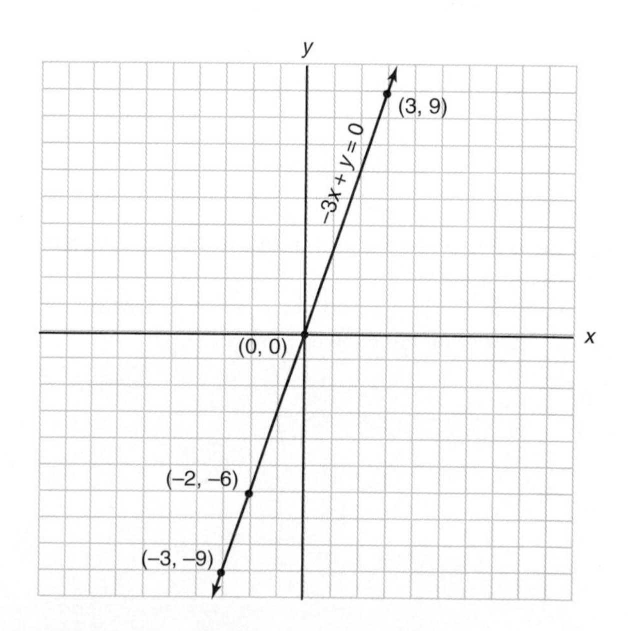

Figure 13–11

Example 13–6: Graphing Horizontal and Vertical Line

PROBLEM: Graph $x = 4$ and $y = -2$.

SOLUTION: The graph of x equaling any quantity is a vertical line parallel to the y-axis passing through the value of x. The graph of $x = 4$ is a vertical line passing through $x = 4$ as shown in Figure 13–12. The graph of y equaling any quantity is a horizontal line parallel to the x-axis passing through the value of y. The graph of $y = -2$ is a horizontal line passing through $y = -2$, as shown in Figure 13–12.

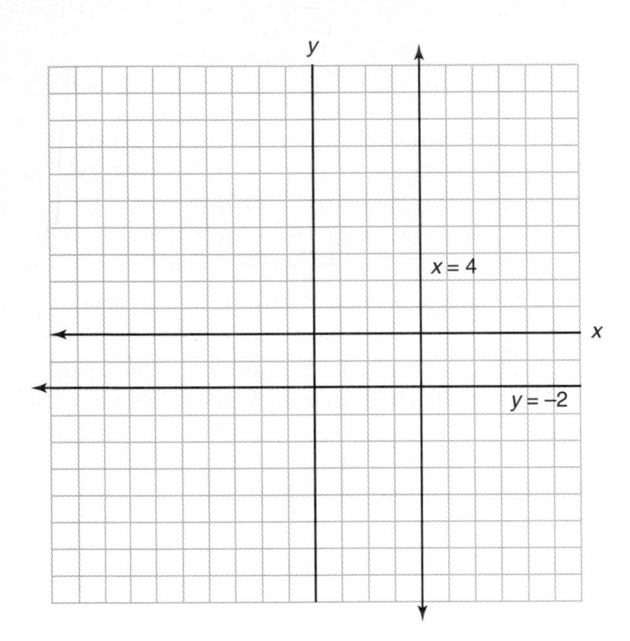

Figure 13–12

Example 13–7: Graphing an Equation with a Fraction

PROBLEM: Graph the equation $-\frac{2}{3}x + y = -4$.

SOLUTION: Solve the equation for y; then substitute values in the table for x to obtain ordered pairs that satisfy the equation.

Since the denominator of the fraction is 3, select multiples of 3 (which are easy to evaluate).

Thus

$$-\frac{2}{3}x + y = -4$$

$$y = \frac{2}{3}x - 4$$

then substitute values of x

x	$y = \frac{2}{3}x - 4$	y	*ordered pair*
0	$y = \frac{2}{3}(0) - 4 = -4$	-4	$(0, -4)$
3	$y = \frac{2}{\cancel{3}}\left(\frac{\cancel{3}}{1}\right) - 4 = -2$	-2	$(3, -2)$
9	$y = \frac{2}{\cancel{3}}\left(\frac{\overset{3}{\cancel{9}}}{3}\right) - 4 = +2$	$+2$	$(9, +2)$
-3	$y = \frac{2}{\cancel{3}}\left(\frac{\overset{-1}{\cancel{3}}}{1}\right) - 4 = -6$	-6	$(-3, -6)$

Plot the ordered pairs to show the graph of $-\frac{2}{3}x + y = -4$, as shown in Figure 13–13.

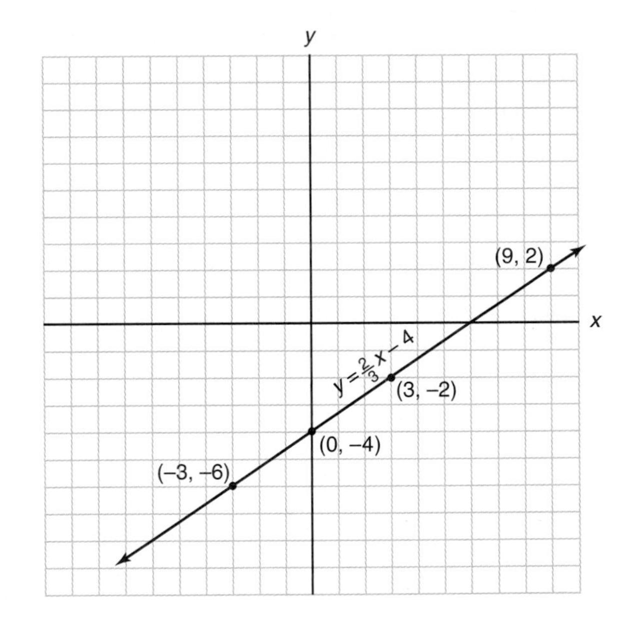

Figure 13–13

EXERCISE 13–2 GRAPHING EQUATIONS

Graph exercises 13–11 to 13–15 by solving for y and plotting.

13–11. $x + y = 4$

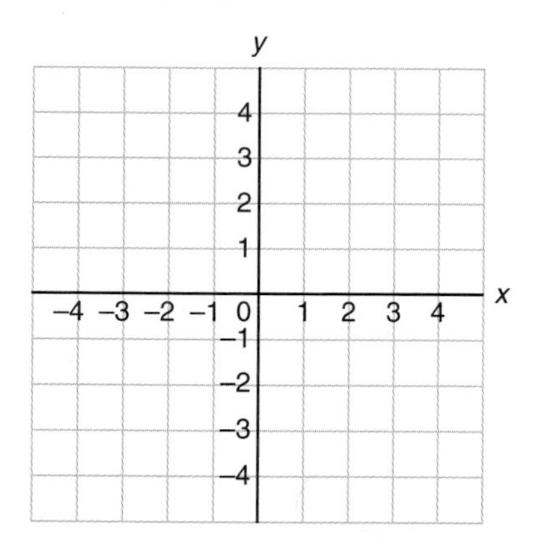

13–12. $x + 2y = 4$

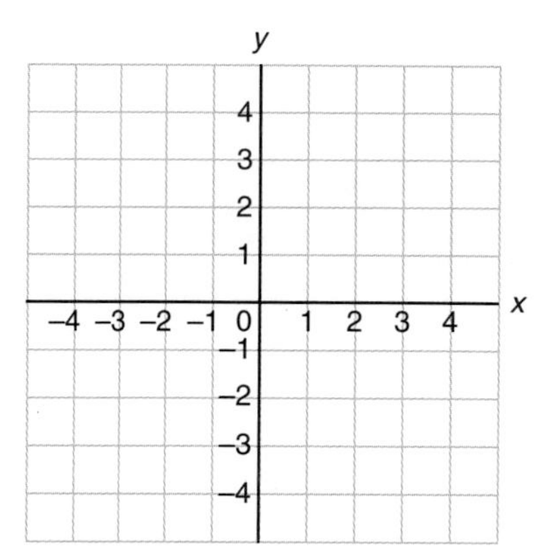

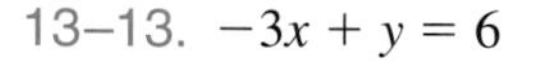

13–13. $-3x + y = 6$

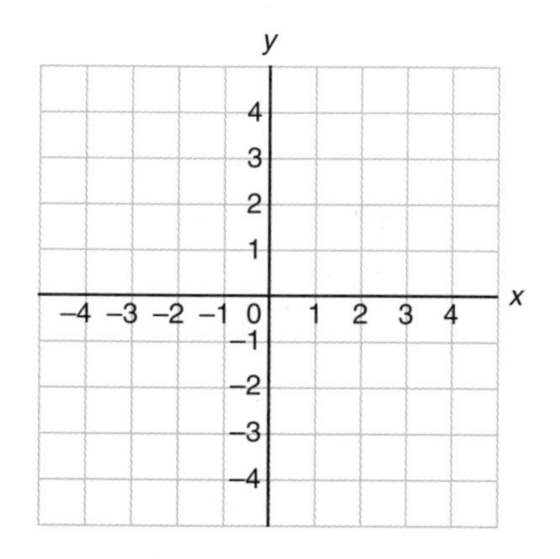

13–14. $-6x + 2y = 6$

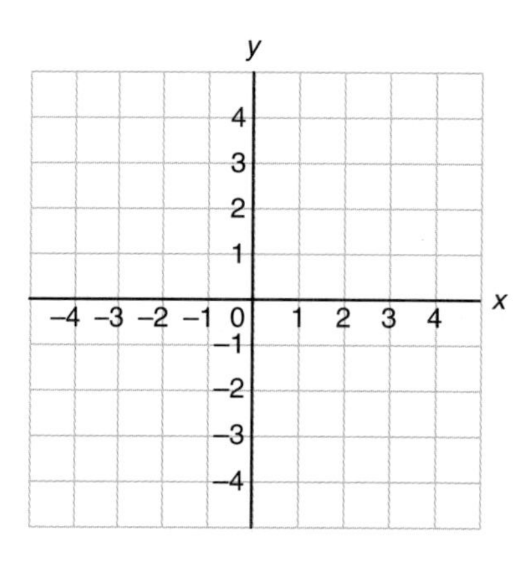

13–15. $2x - y = 7$

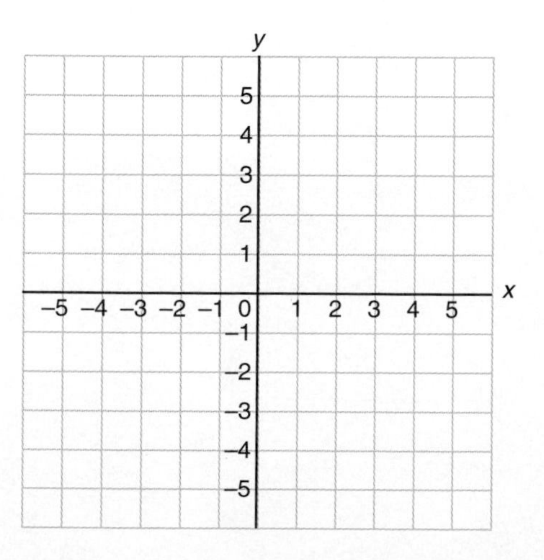

Graph exercises 13–16 to 13–20 using the intercept method.

13–16. $3x + 4y = 12$

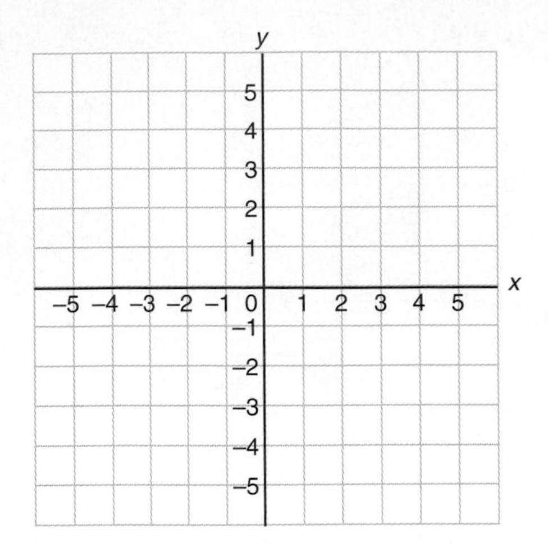

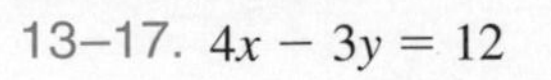

13–17. $4x - 3y = 12$

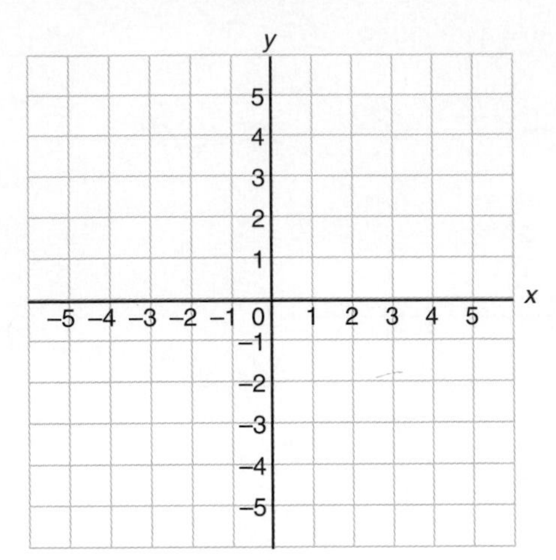

13–18. $x - y = 6$

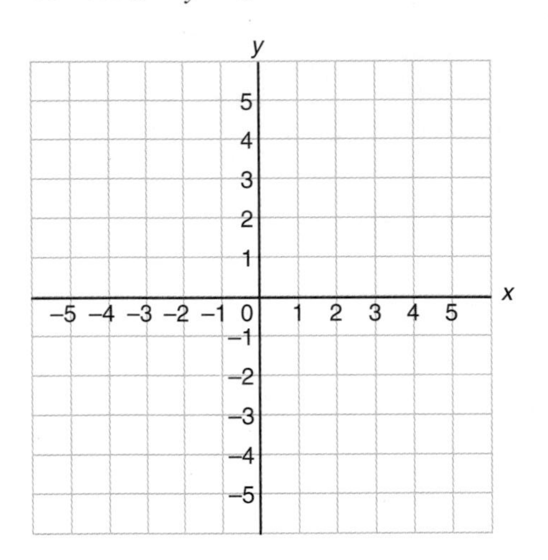

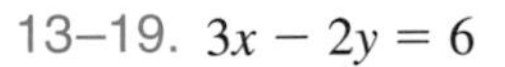

13–19. $3x - 2y = 6$

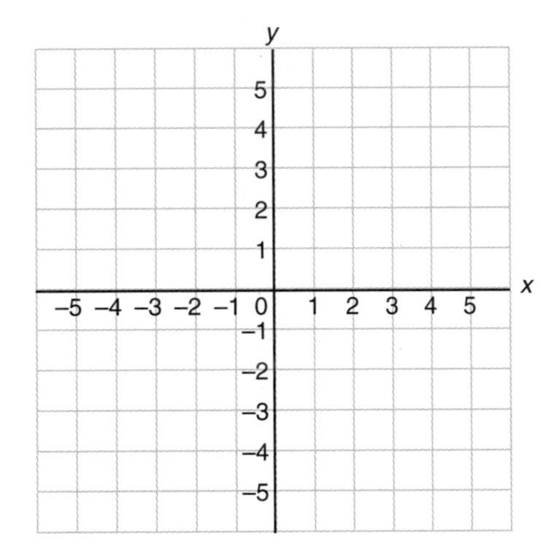

13–20. $x - 2y = 0$

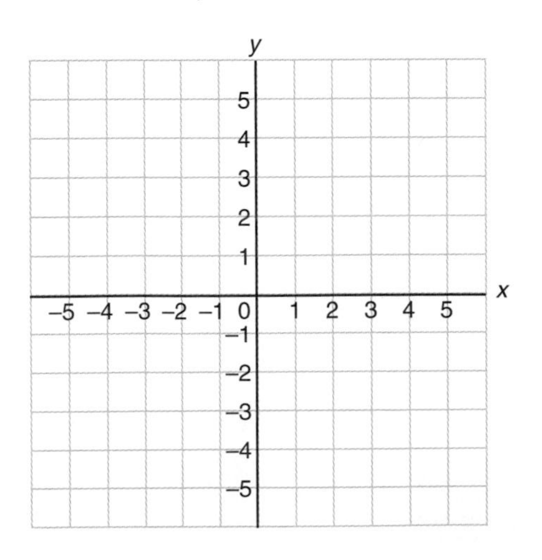

Graph exercises 13–21 to 13–25 using the most efficient method.

13–21. $x = -3$ and $y = 5$

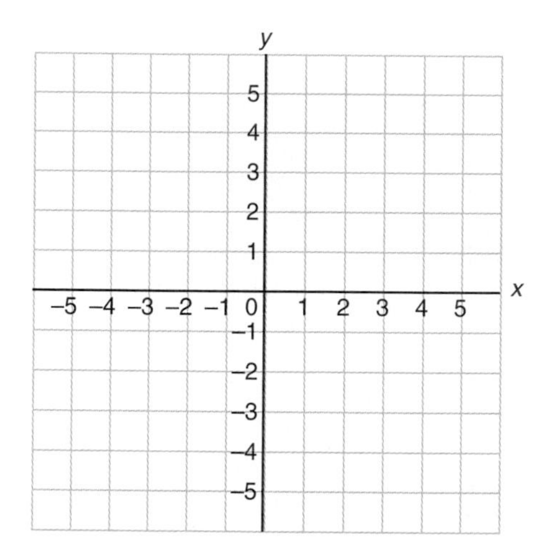

13–22. $4y - 6 = 0$

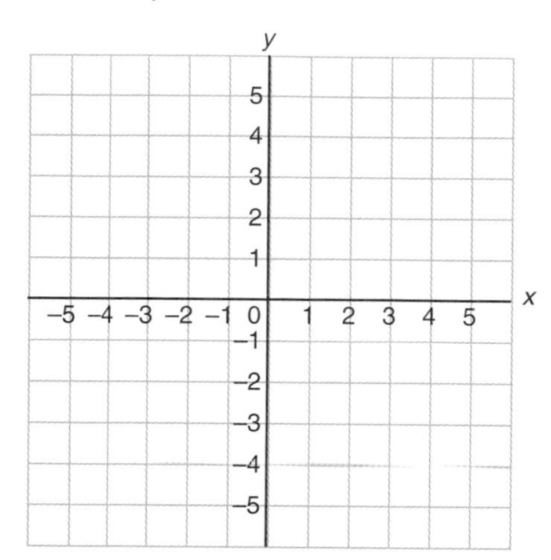

13–23. $2x - 5y = 10$

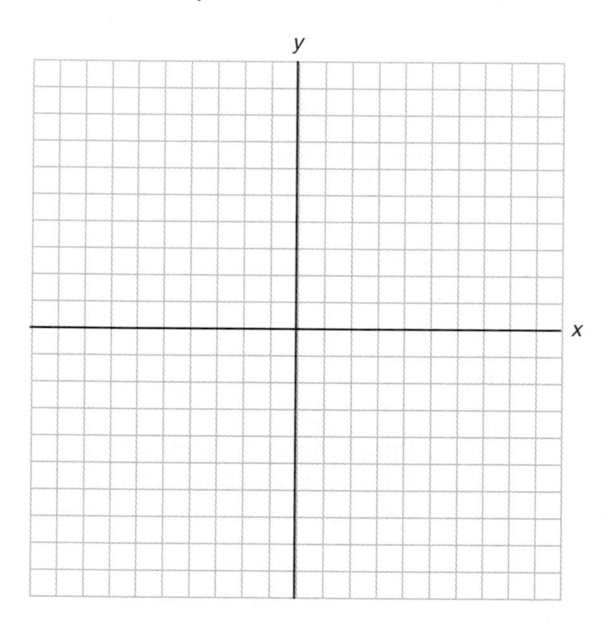

13–24. $y = -\frac{3}{4}x + 2$

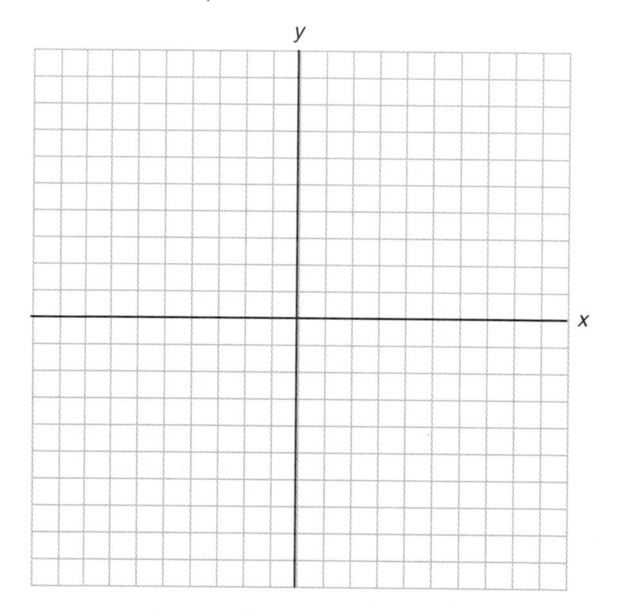

13–25. $y = -\frac{1}{2}x + 5$

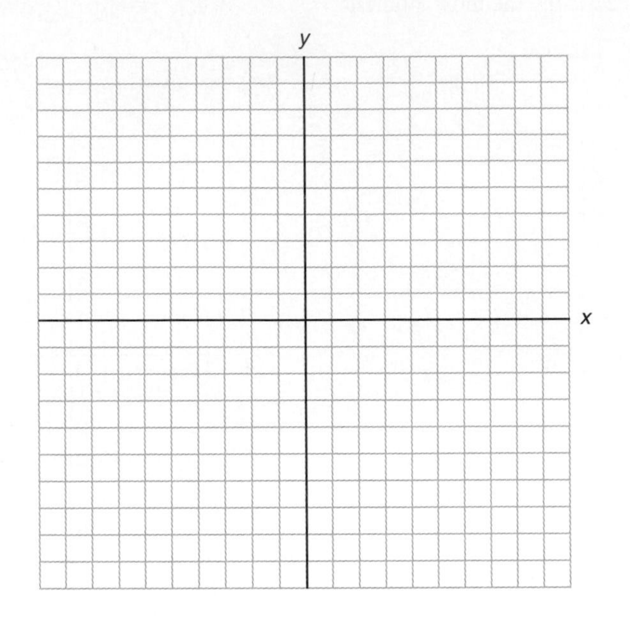

13.3 UNDERSTAND SLOPE, SLOPE-INTERCEPT, FINDING THE EQUATION OF A LINE, AND GRAPHING

Properties of Slope

The word *slope* is used to describe the slant of a hill or the pitch of a roof. How fast one goes down a ski hill depends on the slope of the hill. To best visualize slope, think of the roof of a building (not a flat-top building). The pitch of the roof is sometimes referred to as the slope of the roof. By definition, pitch is rise over run. This is shown in the following illustration.

$$\text{pitch} = \frac{\text{rise}}{\text{run}}$$

The slope of a line is defined as the change in the vertical distance compared to the change in the horizontal distance. In a rectangular coordinate system, the slope of a line is defined as the "change" in y compared to the "change" in x. The letter m is used to describe slope.

$$\text{slope} = m = \frac{\text{change in } y}{\text{change in } x} = \frac{\text{rise}}{\text{run}}$$

The slope of a line in a rectangular coordinate system can be found by selecting two points, and then finding the change in y over the change in x.

Parallel lines have the same slope. The properties and applications of slope will be shown in Examples 13–8 through 13–20.

Example 13–8: Find the Positive Slope of a Line

PROBLEM: Given the line in Figure 13–14 with P_1 and P_2, find the slope.

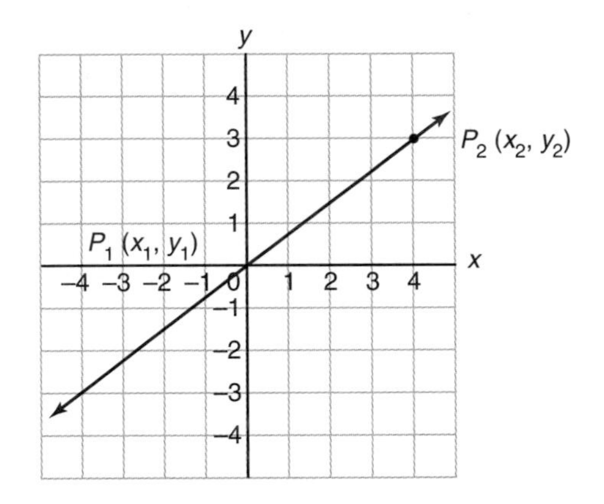

Figure 13–14

SOLUTION: Draw a perpendicular from P_2 to the x-axis, as in Figure 13–15. Count the change in the y-value. Then count the change in the x-value from P_1 to P_2.

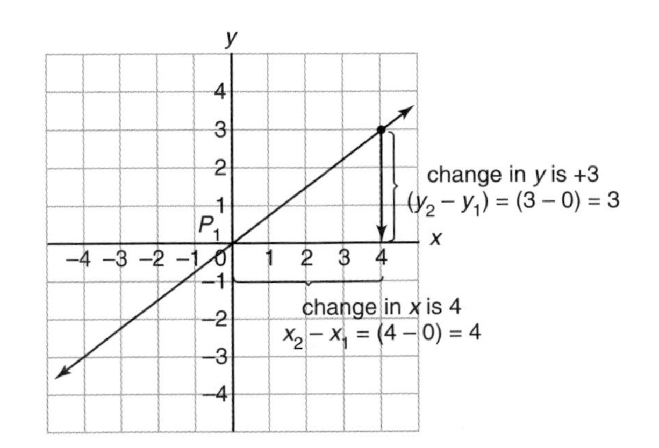

Figure 13–15

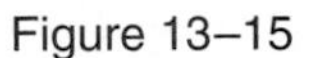

Thus the slope of the given line is:

$$\text{slope} = m = \frac{\text{change in } y}{\text{change in } x} = \frac{3-0}{4-0} = \frac{3}{4}$$

Example 13–9: Find the Negative Slope of a Line

PROBLEM: Given the line in Figure 13–16 with P_1 and P_2, find the slope.

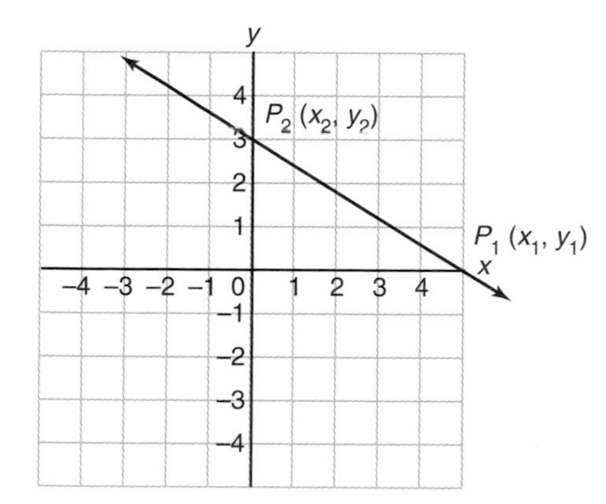

Figure 13–16

SOLUTION: Recall that slope is found by determining the ratio of the change of y from y_1, to y_2 to the change in x from x_1 to x_2.

Thus Figure 13–17 shows the change in x is -5 and the change in y is 3.

$$\text{slope} = m = \frac{\text{change in } y}{\text{change in } x} = \frac{(3, -0)}{(0, -5)} = \frac{3}{-5} \text{ or } \frac{-3}{5}$$

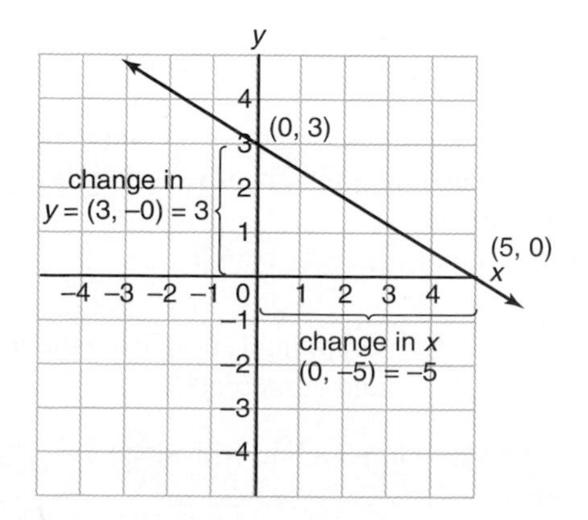

Figure 13–17

A line may have a positive or negative slope. *Note:* If the y-values increase from left to right, the slope is *positive.* If the y-values decrease from left to right, the slope is negative.

Slope Formula

The slope of any line may be determined given two points on the line. The formula for determining slope of a line given two points is:

$$\text{slope} = m = \frac{\text{change in } y}{\text{change in } x} = \frac{y_2 - y_1}{x_2 - x_1}$$

or

$$m = \frac{y_2 - y_1}{x_2 - x_1}$$

This formula may be used to find the slope without actually graphing the line.

Example 13–10: Find the Slope Given Two Points

PROBLEM: Find the slope of a line, given P_1 (2, 1) and P_2 (4, 5).

SOLUTION: Use the two-point slope formula $m = \frac{y_2 - y_1}{x_2 - x_1}$. Recall that P_1 means (x_1, y_1) and P_2 means (x_2, y_2). Thus $x_1 = 2$, $y_1 = 1$, $x_2 = 4$, and $y_2 = 5$, since P_1 (2, 1) and P_2 (4, 5).

So

$$m = \frac{y_2 - y_1}{x_2 - x_1} = \frac{5-1}{4-2} = \frac{4}{2} = \frac{2}{1}$$

Note: When slopes result in a digit they are written over 1, not as a single digit.

Example 13–11: Find the Slope Given Two Points

PROBLEM: Find the slope of a line through P_1 (4, −5) and P_2 (8, −8).

SOLUTION: Use the two-point slope formula, $m = \frac{y_2 - y_1}{x_2 - x_1}$. Recall that P_1 indicates (x_1, y_1) and P_2 indicates (x_2, y_2). Therefore, $x_1 = 4$, $y_1 = -5$, $x_2 = 8$, and $y_2 = -8$.

Therefore

$$m = \frac{y_2 - y_1}{x_2 - x_1} = \frac{-8 - (-5)}{8 - 4} = \frac{-8 + 5}{8 - 4} = \frac{-3}{4}$$

Slope-Intercept Formula

The two-point formula for slope, $m = \frac{y_2 - y_1}{x_2 - x_1}$, can be manipulated to a form known as slope-intercept form. Given the line as shown in Figure 13–18, $x_1 = 0$, $y_1 = b$, $x_2 = x$, and $y_2 = y$. Substitute these values into the two-point slope formula and solve for y.

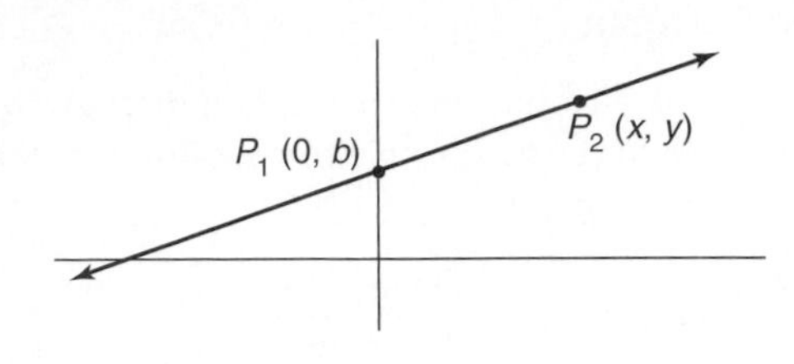

Figure 13–18

Given $\quad m = \dfrac{y_2 - y_1}{x_2 - x_1}$

Substitute $\quad m = \dfrac{y - b}{x - 0}$

Multiply by x $\quad m = \dfrac{y - b}{x}$

$$x(m) = \left(\frac{y - b}{x}\right)x$$

Solve for y

$$mx = y - b$$
$$mx + b = y$$
$$y = mx + b$$

Note: The b value is where the line crosses the y-axis, and m is the slope of the line.

This formula is known as the slope-intercept formula: $y = mx + b$. Thus any equation may be converted to $y = mx + b$, and the slope and y-intercept will be shown.

Example 13–12: Find the Slope and y-intercept Given an Equation

PROBLEM: What is the slope and y-intercept of $y = 5x - 4$?

SOLUTION: Recall the slope-intercept formula, $y = mx + b$. Given $y = 5x - 4$, the slope $= m = {}^5/_1$, and the y-intercept is -4. Thus the coordinates of the y-intercept are $(0, -4)$.

Example 13–13: Find the Slope and y-intercept of a Given Equation

PROBLEM: Given the equation $3x + 2y = -12$, determine the slope and y-intercept.

SOLUTION: Recall the slope-intercept form is $y = mx + b$. Solve the equation for y, making certain the numerical coefficient of y is 1.

Given: $\quad 3x + 2y = -12$

Move $3x$ to the right side $\quad 2y = -3x - 12$

Divide each term by 2 $\quad 2y = -3x - 12$

$$\frac{2y}{2} = \frac{-3x}{2} - \frac{12}{2}$$

$$y = -\frac{3}{2}x - 6$$

Thus $m = -{}^3/_2$, and the y-intercept is -6 or $(0, -6)$.

Example 13–14: Find the Slope and Intercept, Given a Complex Equation

PROBLEM: Given the equation ${}^3/_4\,x - 2y = 16$, find the slope and y-intercept.

SOLUTION: Convert the equation to the slope-intercept form by solving for y. Recall the numerical coefficient of y must be 1.

Given: $\quad \dfrac{3}{4}x - 2y = 16$

move $\dfrac{3}{4}x$ to the right side $\quad -2y = -\dfrac{3}{4}x + 16$

multiply each term by $-\dfrac{1}{2}$ $\quad -\dfrac{1}{2}(-2y) = -\dfrac{1}{2}\left(-\dfrac{3}{4}x\right) + -\dfrac{1}{2}(16)$

$$y = \frac{3}{8}x - 8$$

The equation is now in the form of $y = mx + b$.

Thus $m = {}^3/_8$, and the y-intercept is -8 or $(0, -8)$.

Example 13–15: Given the Slope and y-intercept, Write the Equation

PROBLEM: Given a slope of 3 and a y-intercept of -4 $(0, -4)$, write the equation of the line.

SOLUTION: Using the slope-intercept form $y = mx + b$, substitute the given values $m = 3$ and the y-intercept equals -4.

Thus the equation is $y = 3x - 4$.

Example 13–16: Graphing Lines Given the Slope and y-intercept

PROBLEM: Graph a line with slope $m = {}^3/_4$ and a y-intercept of -2.

SOLUTION: Remember, the y-intercept is where the line crosses the y-axis. The coordinates of the y-intercept are $(0, -2)$. Then recall that slope is

$$m = \frac{\text{change of } y}{\text{change of } x} = \frac{3}{4}$$

Therefore, start at $(0, -2)$; go to the right 4 units, and then up 3 units; repeat this procedure several times and connect the points to determine the line (see Figure 13–19).

Thus the equation $y = {}^{3}\!/_{4}x - 2$.

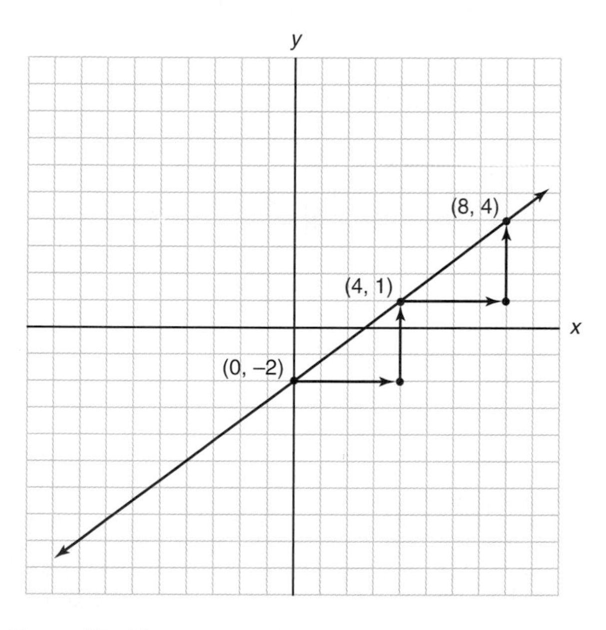

Figure 13–19

Example 13–17: Graph a Fractional Equation in Slope-Intercept Form

PROBLEM: Graph $y = -{}^{2}\!/_{5}x + 6$

SOLUTION: Since the equation is in slope-intercept form, the slope and the y-intercept are given. Thus with the equation $y = -{}^{2}\!/_{5}x + 6$, the slope is $-{}^{2}\!/_{5}$ and the y-intercept is $+6$ or $(0, +6)$. Therefore, start at $(0, +6)$ and move to the right 5 units, and then down 2 units (-2); repeat this procedure and draw the line through the points, or move 5 units left and up 2 units (see Figure 13–20).

Example 13–18: Graph an Equation Using the Slope-Intercept Form

PROBLEM: Graph $3x + 4y = 12$ using slope-intercept form.

SOLUTION: Convert the equation to $y = mx + b$ form to obtain the slope and y-intercept.

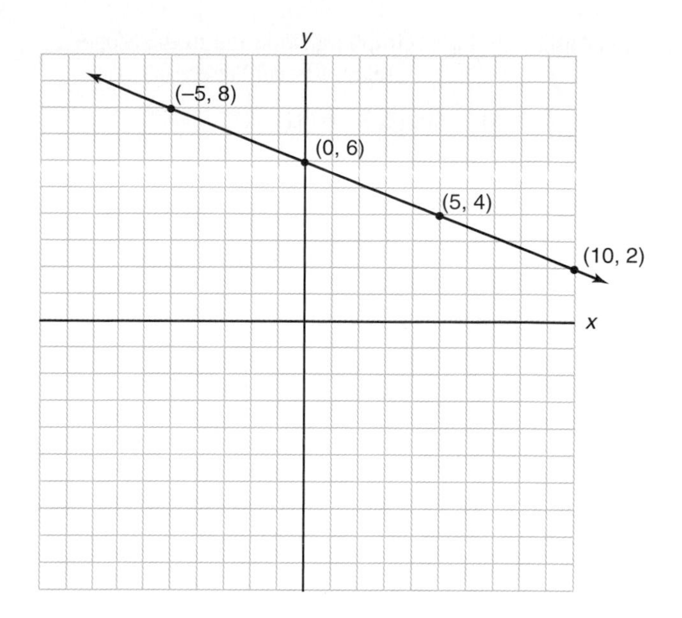

Figure 13–20

First solve for y $\quad 3x + 4y = 12$

Subtract $3x$ from each side $\quad 4y = -3x + 12$

Divide each term by 4 $\quad \frac{4y}{4} = \frac{-3x}{4} + \frac{12}{4}$

$$y = -\frac{3}{4}x + 3$$

Therefore, the $y = mx + b$ form where $m = -{}^{3}\!/_{4}$ and the y-intercept is 3. Thus start at $(0, 3)$, move 4 units to the right and down 3 units; repeat this procedure, and draw the line through the points or move 4 units to the left and up three (see Figure 13–21).

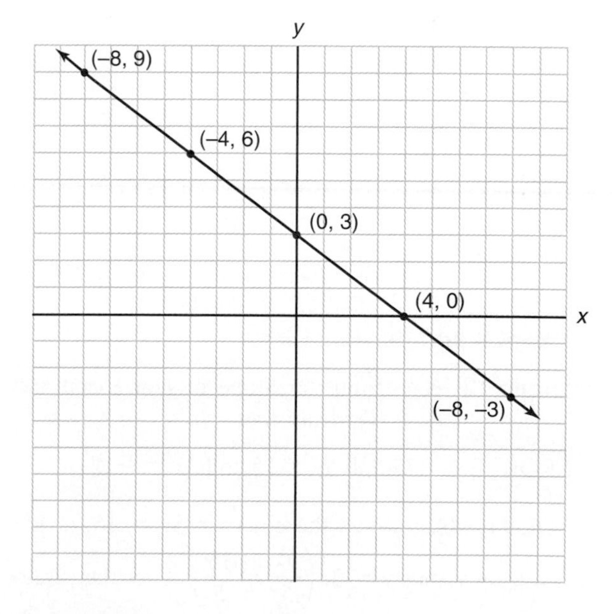

Figure 13–21

Example 13–19: Graph an Equation in the Slope-Intercept Form Where $b = 0$

PROBLEM: Graph $3y = 5x$.

SOLUTION: Convert the equation into $y = mx + b$ form by dividing each term by 3.

$$3y = 5x$$

$$\text{Divide each term by 3} \quad \frac{3y}{3} = \frac{5x}{3}$$

$$y = \frac{5}{3}x$$

Thus we have $y = mx + b$ form, so the slope is $5/3$; since there is no b value, the line goes through the origin. Therefore, starting at the origin (0, 0), move 3 units to the right and up 5 units; repeat this procedure and draw the line (see Figure 13–22).

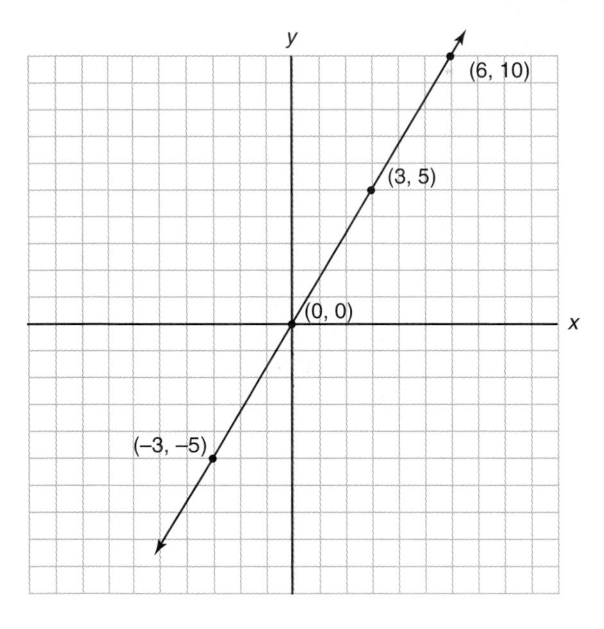

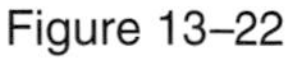

Figure 13–22

Horizontal and Vertical Lines

Example 13–20: Find the Slope of Horizontal and Vertical Lines

PROBLEM: Graph $5y = 15$ and $3x = -12$.

SOLUTION: Given the equation $5y = 15$, divide each side by 5; thus $y = 3$; $\left(\frac{5y}{5} = \frac{15}{5}\right)$. Since there is no x term, the line is parallel to the x-axis. The slope is 0, since there is no change in y, divided by the x change. This is shown in Figure 13–23.

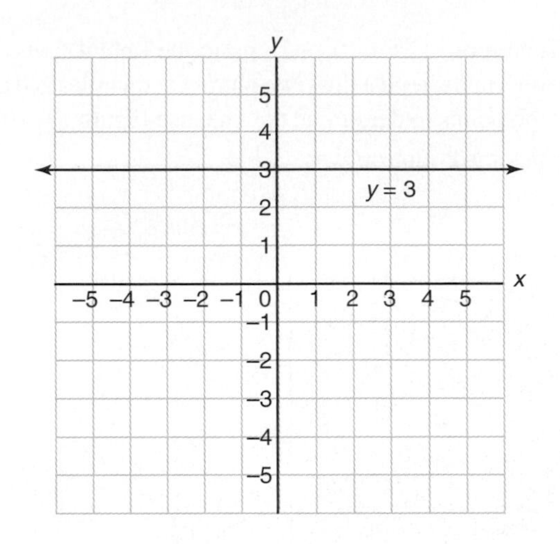

Figure 13–23

Given the equation $3x = -12$, divide each side by 3 $\left(\frac{3x}{3} = \frac{-12}{3}\right)$; thus $x = -4$. Since there is no y term, the line is parallel to the y-axis. The slope is undefined because there is 0 change in the x-value, and division by 0 is undefined. This is shown in Figure 13–24.

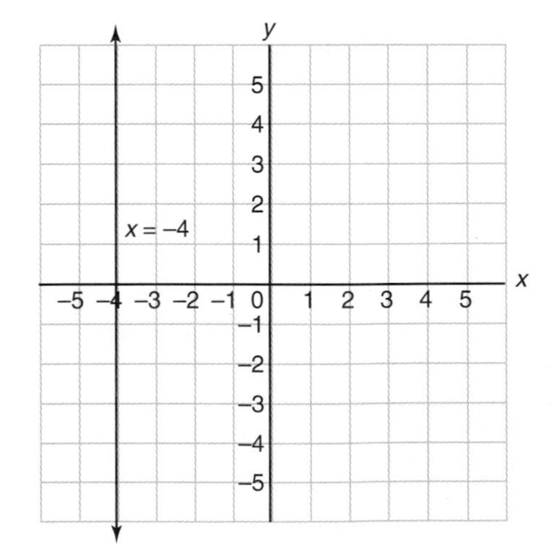

Figure 13–24

EXERCISE 13–3 SLOPE, SLOPE-INTERCEPT, FINDING EQUATIONS, AND GRAPHING

Find the slope of the lines given in the figures for exercises 13–26 to 13–31.

13–26. ______________

13–28. ______________

13–30. ______________

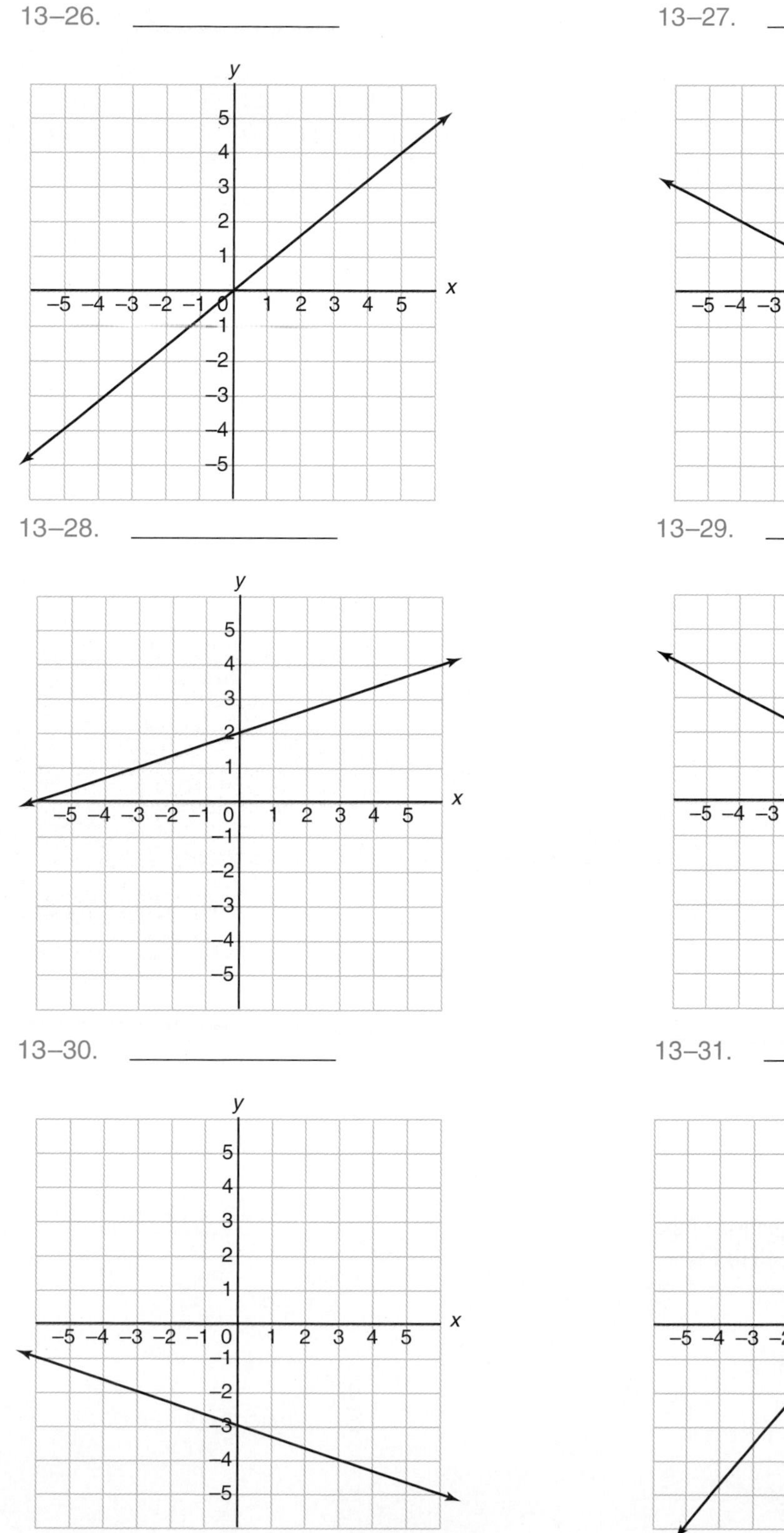

13–27. ______________

13–29. ______________

13–31. ______________

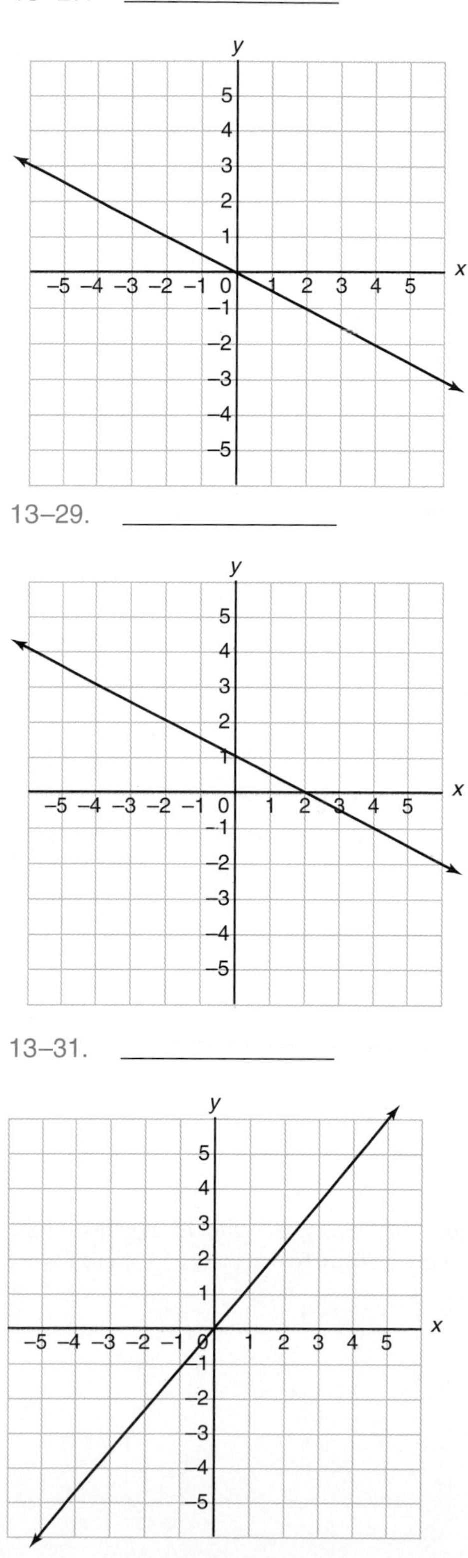

Find the slope for exercises 13–32 to 13–37, given two points.

13–32. (4, 3) and (0, 0) $m =$ ____

13–33. (0, 0) and (6, 8) $m =$ ____

13–34. (5, 2) and (2, 6) $m =$ ____

13–35. (5, 0) and (−2, 1) $m =$ ____

13–36. (21, −31) and (+5, −15) $m =$ ____

13–37. (−10, −8) and (−4, −2) $m =$ ____

Given the equation, find the slope and intercept for exercises 13–38 to 13–42.

13–38. $y = 6x - 4$ $m =$ ____ $b =$ ____

13–39. $y = \frac{3}{4}x + 9$ $m =$ ____ $b =$ ____

13–40. $9x - 3y = -18$ $m =$ ____ $b =$ ____

13–41. $5x + 5y = 30$ $m =$ ____ $b =$ ____

13–42. $\frac{2}{3}x + \frac{1}{3}y = 12$ $m =$ ____ $b =$ ____

Given the slope and y-intercept, write the equation of the line in slope-intercept form for exercises 13–43 to 13–46.

13–43. $m = 4$ $b = 4$

13–44. $m = -1$ $b = -7$

13–45. $m = \frac{3}{5}$ $b = -5$

13–46. $m = -\frac{5}{2}$ $y = 2$

Graph the following equations given the slope and y-intercept form for exercises 13–47 to 13–58.

13–47. $m = 1$ $y = 1$

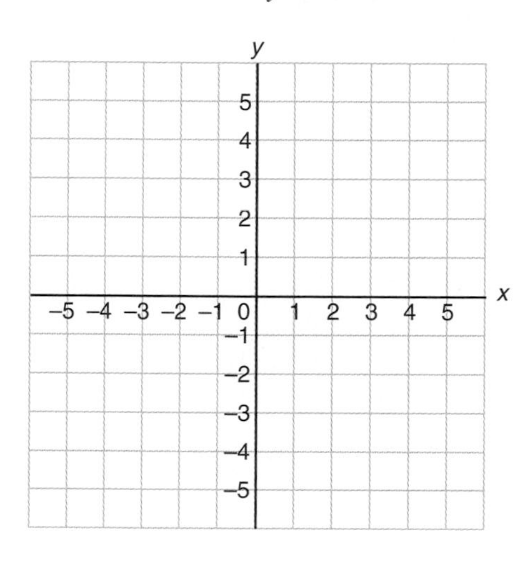

13–48. $m = 3$ $y = -3$

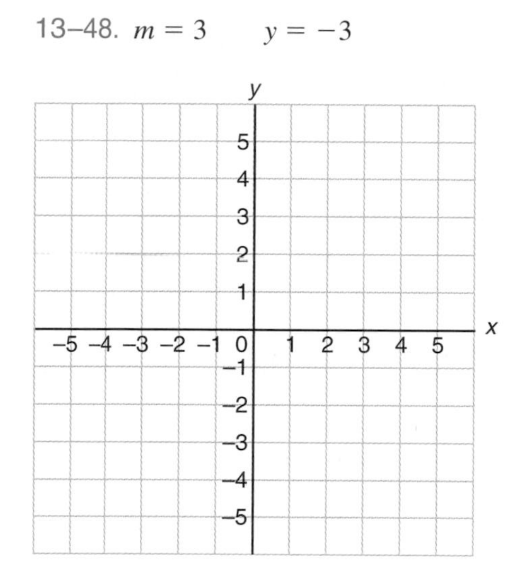

13–49. $m = \frac{2}{3}$ $y = 6$

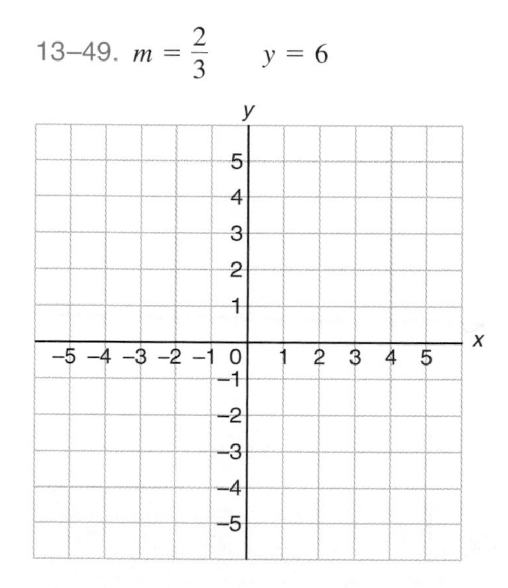

13–50. $m = \frac{5}{3}$ $y = -3$

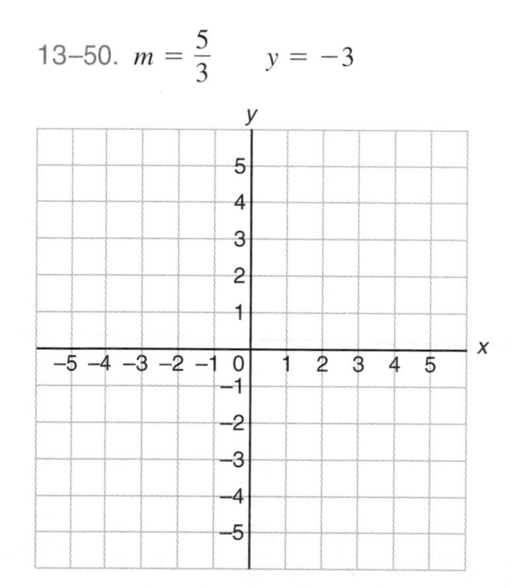

13–51. $-3x = -6y$

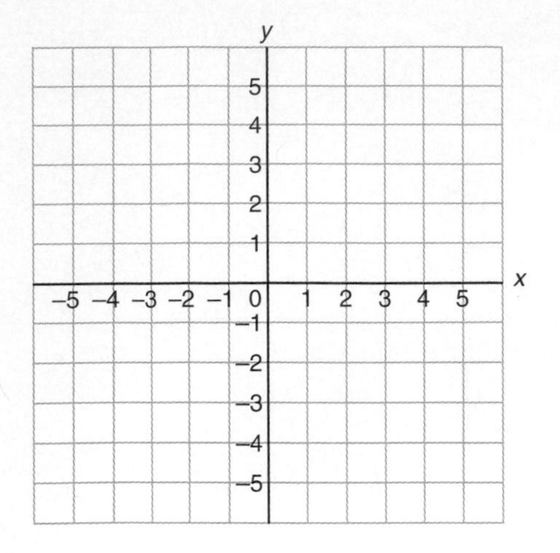

13–52. $\frac{3}{5}x = -\frac{4}{5}y$

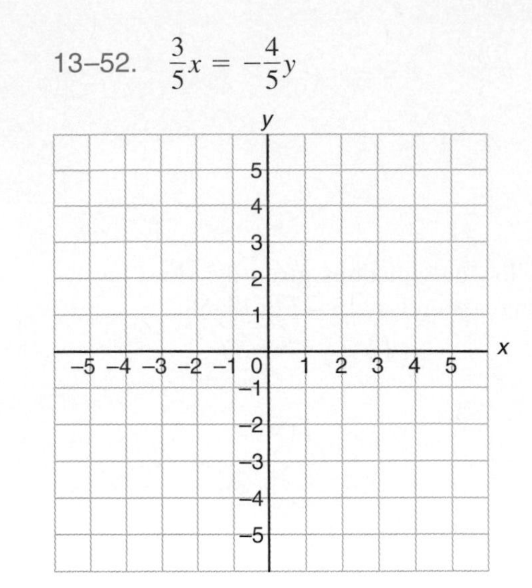

13–53. $3x + 2y = 6$

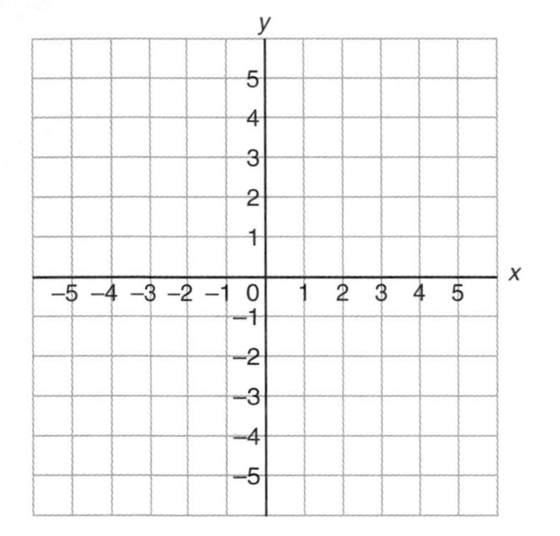

13–54. $2y = -8x + 5$

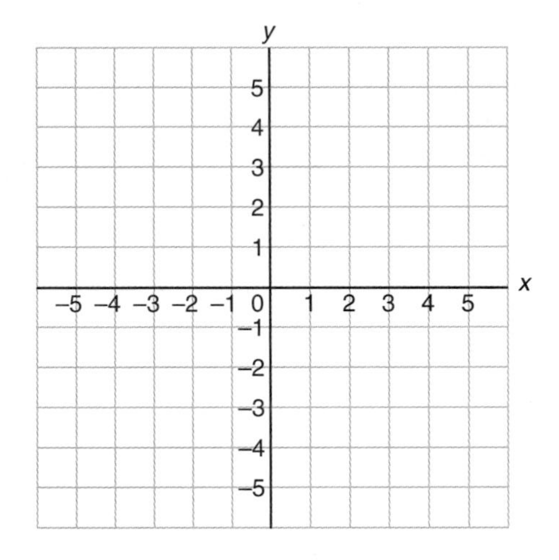

13–55. $3x + 6 = 0$

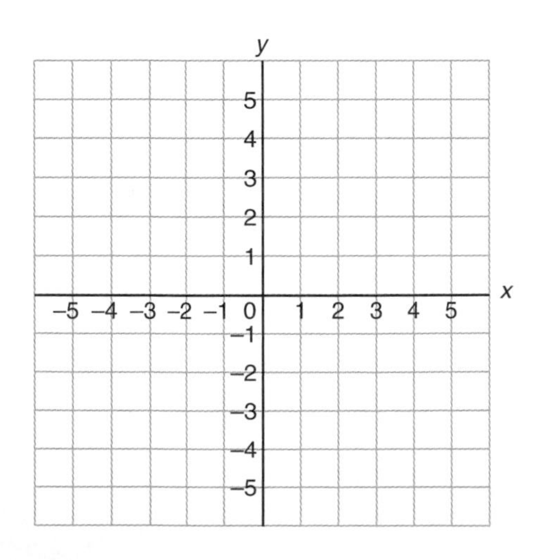

13–56. $2y = 7$

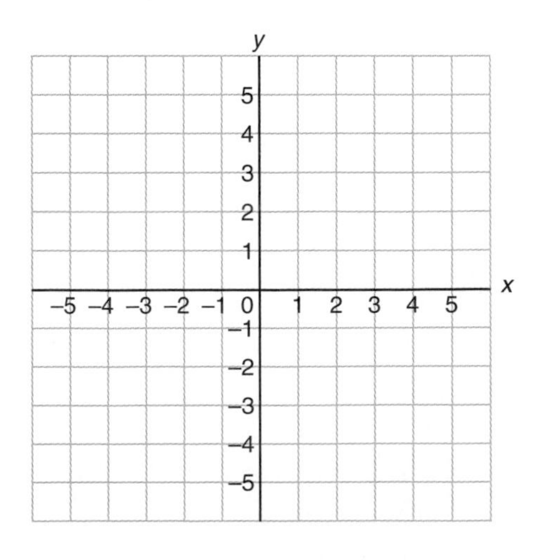

13–57. $6y - 18 + x = 0$

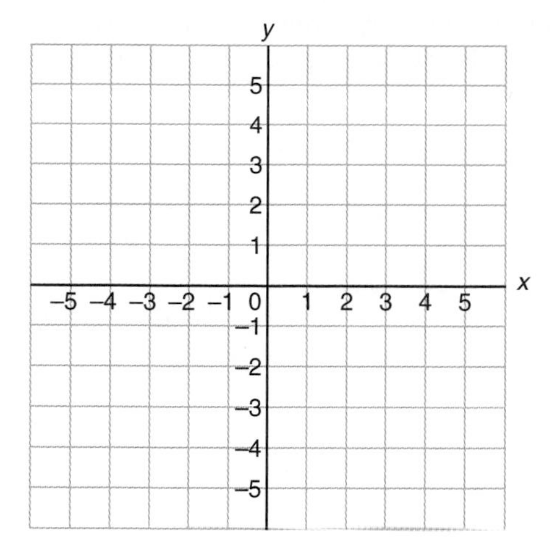

13–58. $-\frac{3}{4}x - \frac{6}{8}y + 2 = \frac{8}{4}$

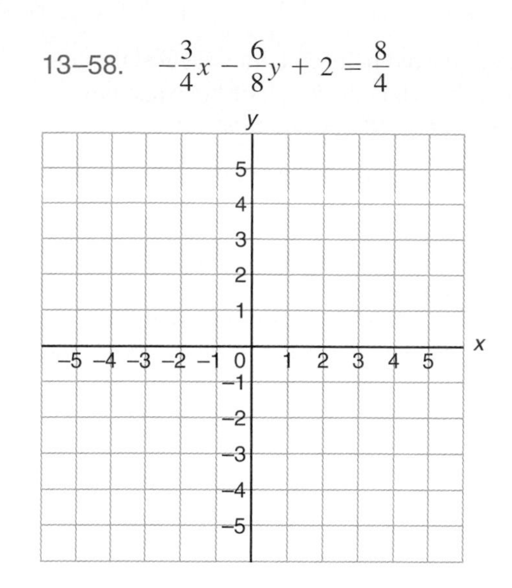

THINK TIME

Given the equations $x - 2y + 8 = 8$ and $x + 2y = 4$, if these equations are graphed on the same coordinate system, will they intersect and, if so, at what point?

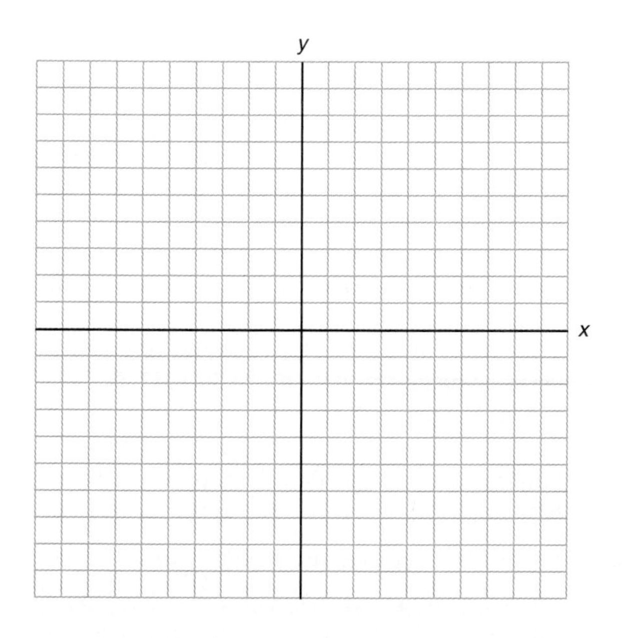

PROCEDURES TO REMEMBER

To plot a point given the x- and y-coordinates:

1. Start at the origin with the x-value and move to the right for positive values and to the left for negative values.
2. From this point, for the y-value, move up for positive and down for negative values.
3. Then label the point (x, y) according to the values of x and y.

To find coordinates of a point given the graph of the point:

1. Through the point, draw a vertical line to the x-axis and a horizontal line to the y-axis.
2. Then count the x-value from the origin to where the vertical line crossed the x-axis and count the y-value from where the horizontal line crossed the y-axis.
3. Thus the "address" of the point (x_1, y_1).

To graph an equation using the ordered-pair procedure:

1. Solve the equation for one unknown.
2. Substitute values for one unknown and solve for the other to determine ordered pairs that satisfy the equation.

3. Plot these ordered pairs and draw the line through the points, which is the graph of the equation.
4. Label the graph with the equation.

To graph an equation using the intercept procedure:

1. Given the equation set $y = 0$, this coordinate will indicate the x-intercept, where the line (graph) crosses the x-axis.
2. Then set the $x = 0$; this coordinate will indicate the y-intercept, where the line (graph) crosses the y-axis.
3. Draw the line through the two intercepts. This will be the graph of the equation.
4. Label the graph with the equation.

To graph an equation of the $Ax + By = 0$ form:

1. Solve the equation for y.
2. Substitute values for x to obtain ordered pairs that satisfy the equation.
3. Plot the ordered pairs and draw the line through the ordered pairs, which will be the graph of the equation.
4. Label the graph with the equation.

To graph horizontal and vertical lines:

1. For an $x = a$ equation draw a line parallel to the y-axis through a.
2. For a $y = b$ equation, draw a line parallel to the x-axis through b.

To find the slope given the graph of the line:

1. Draw perpendicular lines from a point on the line to the x- and y-axes.
2. Determine the length of x and y from the origin.
3. Place the y-value over the x-value, which is the slope.

To find the slope given two points on the line:

1. Use the formula $m = \dfrac{y_2 - y_1}{x_2 - x_1}$.
2. Substitute the values of x_1, x_2, y_1, and y_2 into the formula.
3. Simplifying this will indicate the slope of the line.

To find the slope and y-intercept given an equation:

1. Given an equation, solve the equation for y.
2. Thus the coefficient of x is the slope and the b-value is the y-intercept.

To write an equation given the slope and the y-intercept:

1. Use the slope-intercept formula $y = mx + b$.
2. Substitute the given values for slope and y-intercept.

To graph lines given the slope and y-intercept:

1. Graph the y-intercept.
2. From the y-intercept, plot the x-value—right for positive and left for negative.
3. Then, from this point, plot the x-value—up for positive and down for negative. Then draw the line for the equation.

CHAPTER SUMMARY

1. The *rectangular coordinate system* is formed by two perpendicular number lines or axes, which are used to graph ordered pairs of numbers. The rectangular coordinate system is divided into *four quadrants* by the *horizontal* and *vertical* axes. The *origin* is the point where the perpendicular lines meet.
2. The location of a point in the rectangular coordinate system is determined by *ordered pairs* known as coordinates of the point. The ordered pair is given by (x, y), where x is the x-value and y is the y-value.
3. *Graphing* a linear equation is done by plotting several points (ordered pairs) that satisfy the equation. *Graphing* an equation can also be done by finding the *intercepts,* where the line of the equation crosses the x- and y-axes.
4. The graph of a *horizontal line* is parallel to the x-axis. The graph of a *vertical line* is parallel to the y-axis.
5. *Slope* of a line is the change in the vertical distance compared to the change in the horizontal distance or the change in the rise over the run. A line that slants upward from left to right has a *positive slope.* A line that slants downward from left to right has a *negative slope.* The *formula* for finding the slope given two points is: $m = \dfrac{y_2 - y_1}{x_2 - x_1}$ where m equals slope. The formula $y = mx + b$ is called the *slope-intercept formula,* where m equals the slope and b equals the y-intercept.

CHAPTER TEST

T–13–1. A rectangular coordinate system is divided by x- and y-axes into four ________________.

T–13–2. Graph the following points and label them: P (7, 3), Q (0, −7), R (−4, −2).

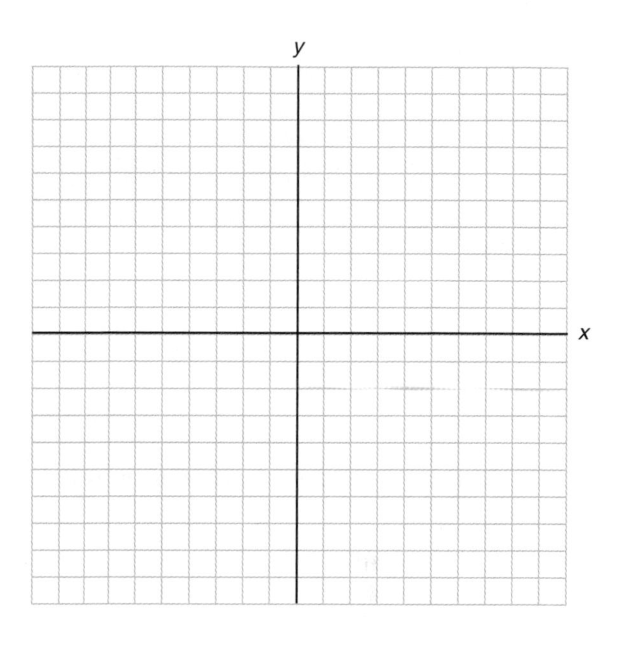

T–13–3. Given the points on the graph, write the coordinates: J (_____, _____), K (_____, _____), L (_____, _____).

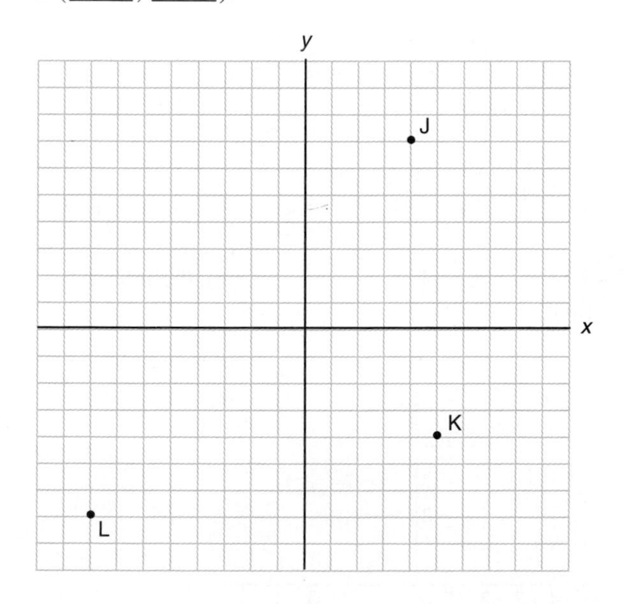

T–13–4. Given the equation $y = 3x - 8$, find the y-value of the ordered pair (4, _____).

T–13–5. Given $3x - 6y = 12$, find the x-value of the ordered pair (_____, 2).

T–13–6. Graph $y = 2x - 4$ by plotting the points.

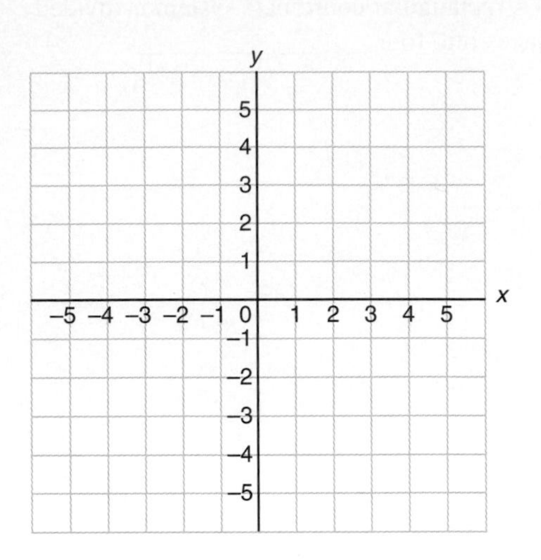

T–13–7. Graph $y = -3x - 2$ by plotting the points.

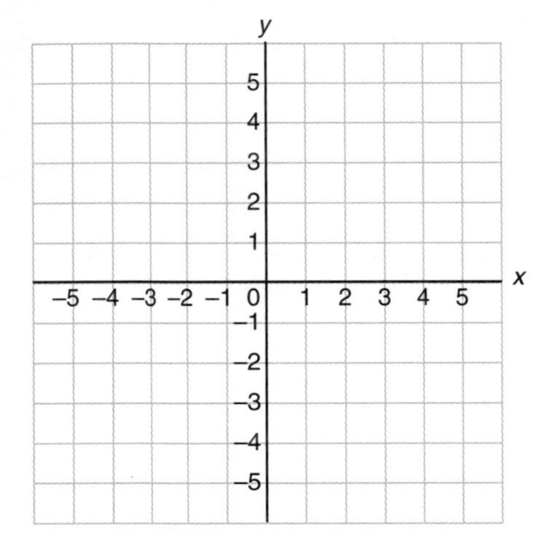

T–13–8. Graph $5x - 3y = -15$ by plotting the points.

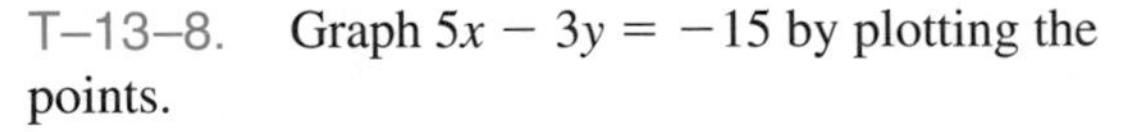

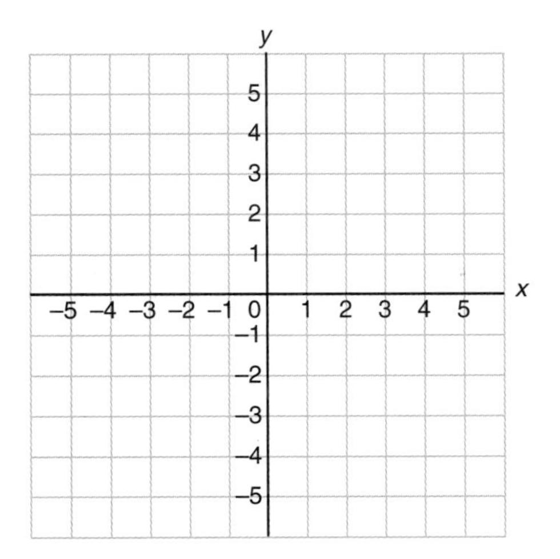

T–13–9. Graph $y = -5.5$.

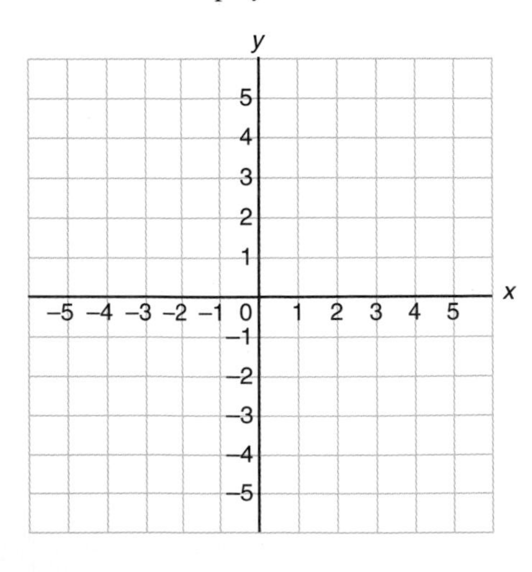

T–13–10. Graph $4x + y = -5$ using the intersect method.

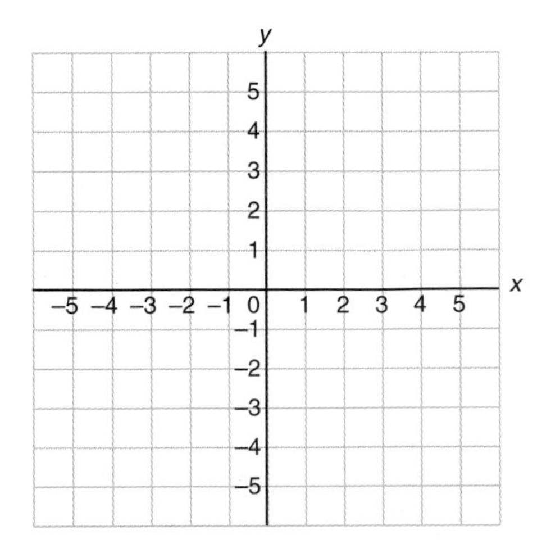

T–13–11. Graph $3x - 6y = 18$ using the intercept method.

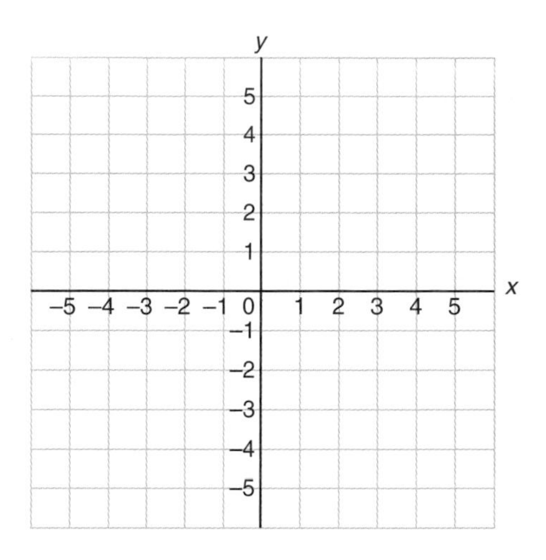

T–13–12. Graph $y = \frac{3}{8}x - 4$ using any method.

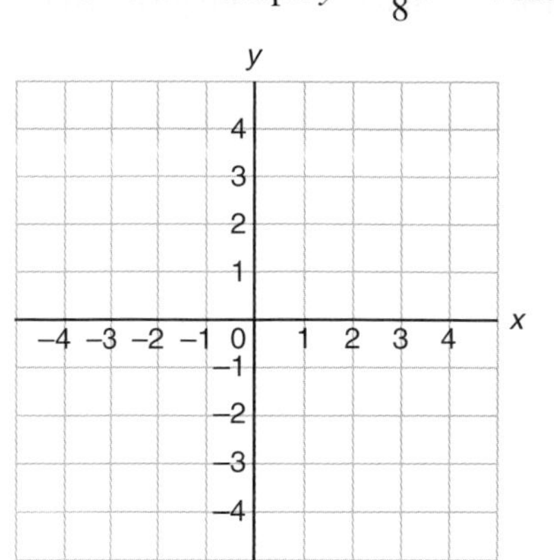

T–13–13. Graph $6x - 3y = 12$ using any method.

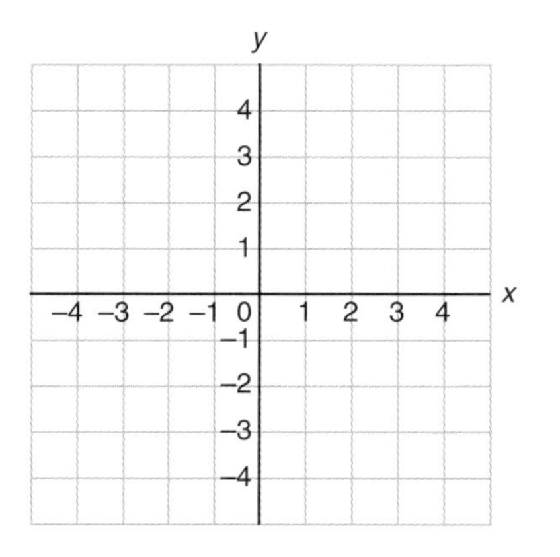

T–13–14. Given $4x - 20y = 40$, what is the slope and y-intercept? $m =$ _____, $b =$ _____.

T–13–15. Given $5.5y - 11x - 44 = 0$, what is the slope and y-intercept? $m =$ _____, $b =$ _____.

T–13–16. Given the two points (4, 8) and (12, 16), find the slope using the point-slope formula.

T–13–17. Given the two points $(-8, 0)$ and $(12, -30)$, find the slope using the point-slope formula.

T–13–18. Given the equation $7x - 21y = 28$, find the slope and the y-intercept.

S–14–7. How many degrees would the needle of the compass rotate if you turned from north to south?

S–14–8. Three fourths of a complete revolution gives an angle of __________ degrees.

S–14–9. Minutes used to measure angles are subdivided into __________.

S–14–10. The symbol for a straight line is __________.

S–14–11. A triangle with all sides of equal length is called an __________ triangle.

S–14–12. What angle do the hands of a clock make at 7 o'clock?

S–14–13. Measure with a protractor angle *NOP* shown in Figure 14–1.

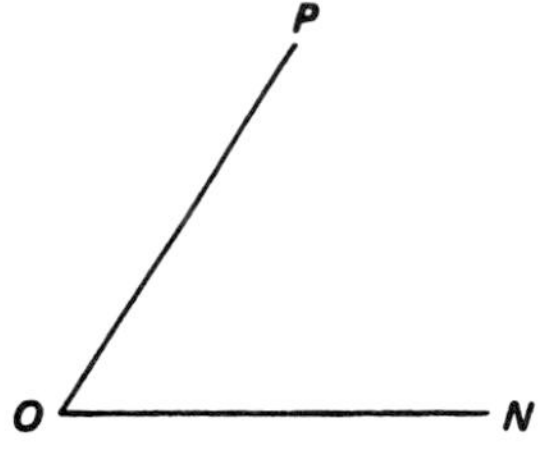

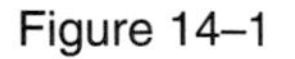
Figure 14–1

S–14–14. Measure with a protractor angle *J* shown in Figure 14–2.

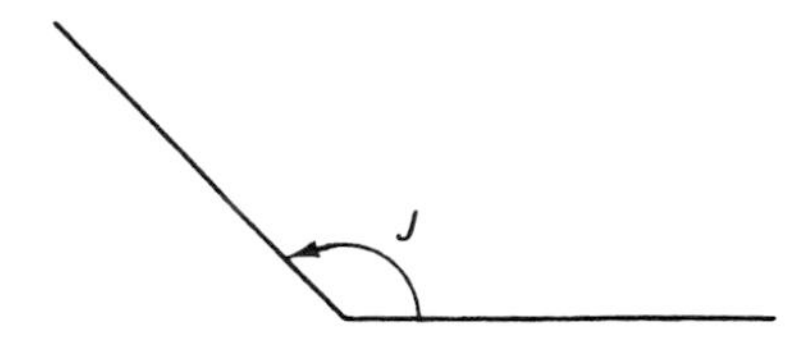

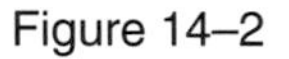
Figure 14–2

S–14–15. Construct a 73° angle with a protractor.

S–14–16. Construct a 153° angle with a protractor.

S–14–17. Find the complement of a 36° angle.

S–14–18. Find the supplement of a 67° angle.

S–14–19. How many degrees are there in one-fourth of a right angle?

S–14–20. Measure $\angle WOZ$ in Figure 14–3.

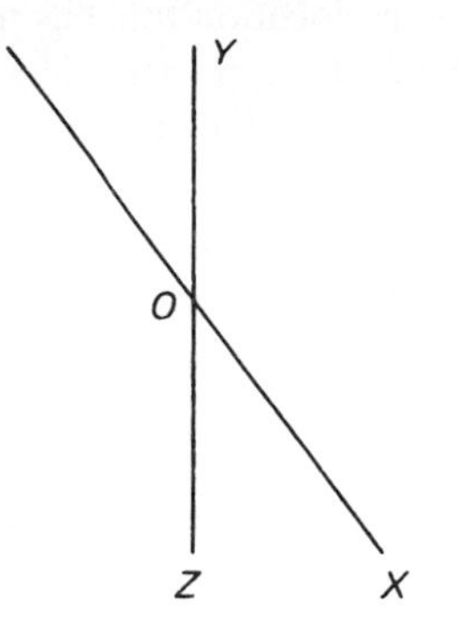

Figure 14–3

S–14–21. Measure $\angle WOY$ in Figure 14–3.

S–14–22. Angles WOZ and ZOX are called __________ angles.

S–14–23. Measure $\angle PON$ in Figure 14–4.

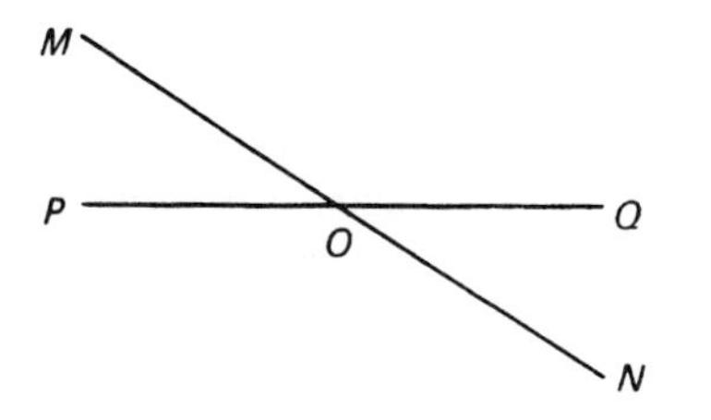

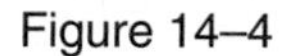

Figure 14–4

S–14–24. Measure $\angle POM$ in Figure 14–4.

S–14–25. Measure $\angle POQ$ in Figure 14–4.

14.1 UNDERSTAND GEOMETRY: ITS HISTORY, DEFINITIONS, AND FIGURES

Geometry is an ancient branch of mathematics. Its roots can be traced back five or six thousand years to the Egyptian and Babylonian civilizations. The precise construction of the pyramids is a monument to the use of geometry and skill of the Egyptians. These skills were also used to reestablish boundary lines that washed away during the flooding of the Nile River.

The first known geometer was a Greek mathematician, Euclid, who lived about 300 B.C. Euclid gathered the geometric principles, combined them with his own concepts, and wrote the first geometry book. This book included geometric principles that have changed very little over the centuries. In fact, Euclid's book could be used today with few changes. It is a rarity that any book of science or mathematics could be useful for such a long period. Euclid wrote a nearly perfect geometry book even though he worked with the simplest of tools and equipment to make his calculations. His knowledge and skills of geometry make our work and life easier.

The three basic properties of geometry are point, line, and plane. A *point* is a position or location that has no size or dimension. Generally, a point is represented by a dot with a capital letter beside it (.*P*). A *line* may be defined as a series or set of points extending indefinitely in opposite directions and may form a straight, broken, or curved line. A line does not have width or depth. It has length only and is generally identified by a lowercase letter beside it:

$\overleftrightarrow{a}$ straight line; $\overset{\wedge\!\wedge}{\mathrm{b}}$ broken line; $\overset{\frown}{\mathrm{c}}$ curved line.

A *straight line* is the shortest distance between two points. Generally, when the word *line* is used, it means a straight line. The symbol for a straight line is $\overleftrightarrow{b}$. This symbol indicates that straight line *b* extends indefinitely in both directions. It is understood that a line extends indefinitely; thus the arrows are not generally added. Two lines that never meet, no matter how far they are extended, are called *parallel lines* and are identified by the symbol ||. A *plane* is a flat surface that has two dimensions, length and width. The sides of a box are plane surfaces.

Two types of geometry are plane and solid. *Plane geometry* deals with two dimensions: length and width. *Solid geometry* includes three dimensions: length, width, and height. Plane geometry is like two-dimensional art such as a painting, whereas solid geometry may be likened to a three-dimensional art sculpture. The basic definitions given above will be of value when you consider the applied geometric concepts and forms.

As stated previously, a plane is a flat surface that extends indefinitely in all directions. A *plane figure* is a figure drawn on a flat surface. A *plane closed figure* is a figure that encloses an area or surface. Any plane figure can be described by line segments, points, and angles. Observe the plane closed figures illustrated in Figure 14–5. Note that they may be drawn with straight or curved lines.

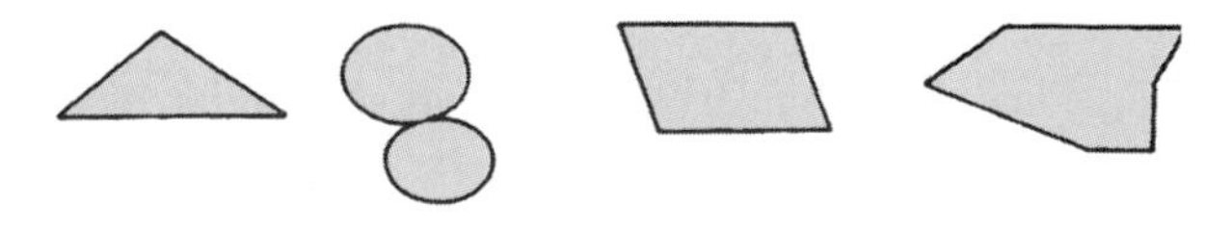

Figure 14–5

There are names for certain plane figures. A *polygon* is a plane figure formed by straight-line segments. A *triangle* is any polygon formed by three straight-line segments. A *right triangle* is a triangle containing a right angle. An *equilateral triangle* is a triangle with all three sides of equal length. An *isosceles triangle* is a triangle with two equal sides. Observe the illustration of the triangles in Figure 14–6.

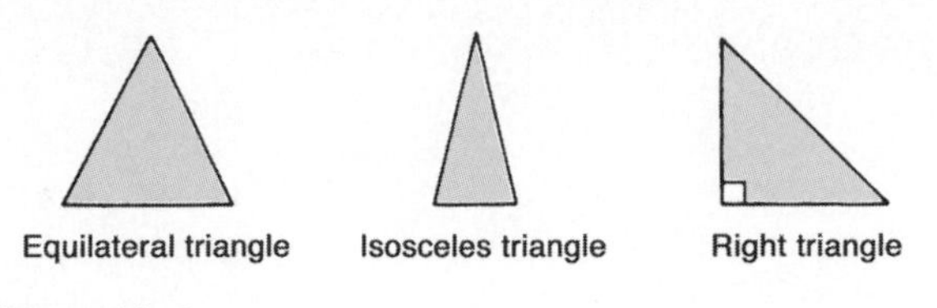

Figure 14–6

Four-sided polygons are formed with parallel lines and have special names. A *parallelogram* is a four-sided plane figure whose opposite sides are parallel. A *rectangle* is a parallelogram containing all right angles. A *rhombus* is a parallelogram with all sides of equal length. A *square* is a parallelogram with all equal sides and all right angles. Observe the illustrations of parallelograms in Figure 14–7.

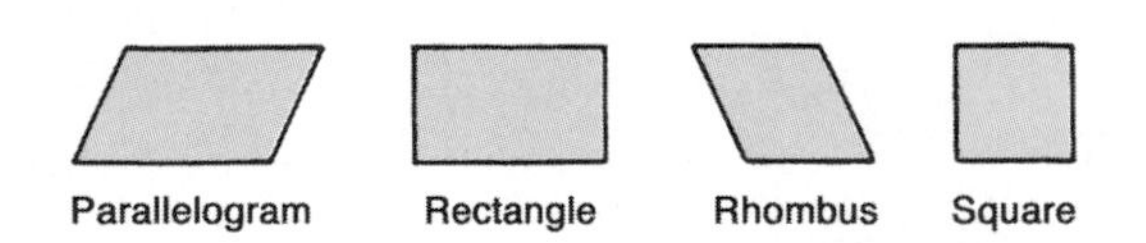

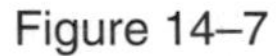

Figure 14–7

Observe the drawing of a block shown in Figure 14–8. Because it has length, width, and height, it is called a *rectangular solid.* A solid is a figure that has depth, length, and width. Letters indicate the corners of the block.

To draw a rectangular solid, first draw the front rectangle, then draw parallel lines from each of the four corners of the rectangle. Add the perpendicular and horizontal lines to complete the rectangular solid. (Dotted lines may be used for the invisible lines as shown in Figure 14–8.)

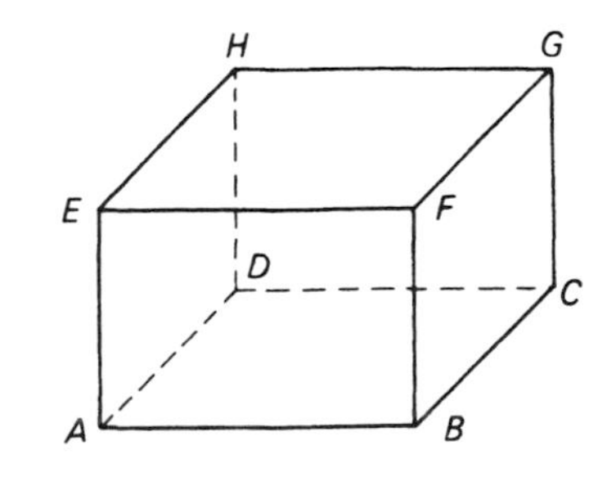

Figure 14–8

The line from point *A* to point *B* is called a *line segment* because it has *endpoints A* and *B.* The symbol for line segment *AB* is $\overline{AB}$. This means that line segment *AB* is only a portion of a line that extends indefinitely in both directions. Line segment *AB* is referred to as the edge of a rectangular solid.

Lines $\overline{AB}$, $\overline{BF}$, $\overline{FE}$, and $\overline{EA}$ form a plane or a *face* of the rectangular solid. The plane is called *ABFE*. Point *F*, where the lines *EF*, *BF*, and *GF* meet and where planes *ABFE*, *BCGF*, and *EFGH* meet, is called a *vertex* of the rectangular solid.

Example 14–1: Sketch a Rectangular Solid

PROBLEM: Sketch a rectangular solid, letter it, and determine its planes, line segments, and vertices.

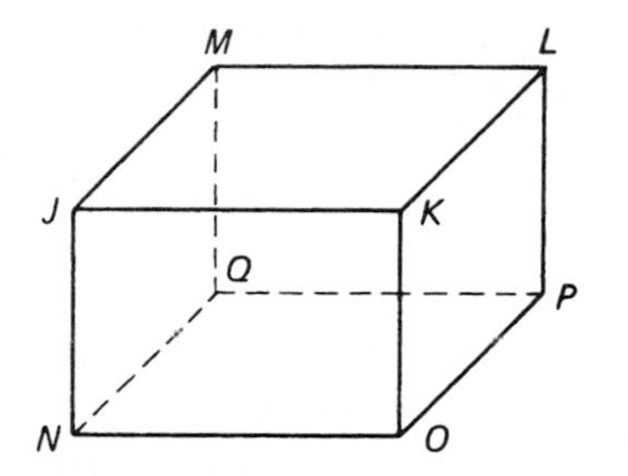

Figure 14–9

SOLUTION: First, draw the rectangular solid. Second, letter the rectangular solid. See Figure 14–9. Third, determine the planes. The planes are *JKLM, JKON, JMQN, KLPO, LMQP,* and *NOPQ*. Fourth, determine the line segments. The line segments are *JK, JN, JM, KL, KO, LM, LP, MQ, NO, NQ, OP,* and *PQ*. Fifth, indicate the eight vertices. The vertices are

J, formed by planes *JKLM, JKON, JMQN*
K, formed by planes *JKLM, JKON, KIPO*
L, formed by planes *JKLM, LMQP, KLPO*
M, formed by planes *JKNO, JMQN, LMQP*
N, formed by planes *JKNO, JMQN, LMQP*
O, formed by planes *JKNO, KLPO, NOPQ*
P, formed by planes *LMQP, KLPO, NOPQ*
Q, formed by planes *JMNQ, LMQP, NOPQ*

EXERCISE 14–1 GEOMETRIC DEFINITIONS AND PROPERTIES

Complete the following exercises to help learn the basic geometric definitions.

14–1. List three types of parallelograms.

14–2. List three types of triangles.

14–3. Draw and label a line segment.

14–4. Draw and label a rectangular solid.

14–5. What are the three dimensions of a solid geometry?

14–6. A triangle with all three sides of equal length is called an __________ triangle.

14–7. Draw the symbol for parallel lines.

14–8. Draw the symbol for straight line *f*.

14–9. A plane closed figure formed by straight line segments is called a __________.

14–10. A triangle containing a right angle is called __________.

14.2 UNDERSTAND THE DEFINITIONS, PROPERTIES, AND TYPES OF ANGLES

When two lines intersect (cross) each other, angles are formed. The point of intersection is called the *vertex* of the angles and is indicated by a capital letter. See Figure 14–10. Lines *AB* and *CD* intersect at point *O*. Point *O* is the vertex

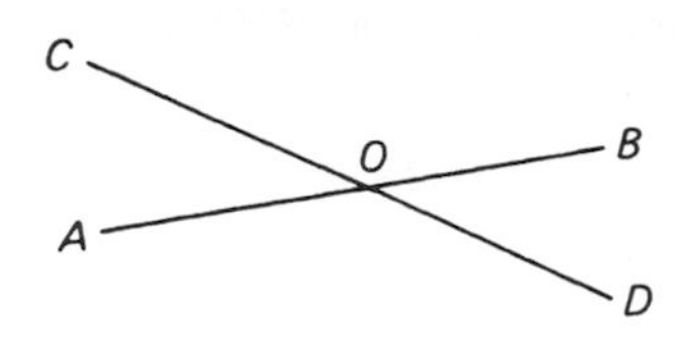

Figure 14–10

of the angles. Each angle is indicated by the angle symbol ($\angle$) and three letters that make up the angle. The angles for the illustration are $\angle COB$, $\angle COA$, $\angle AOD$, and $\angle BOD$. Note the middle letter of each angle is always the vertex of the angle. Angles may also be indicated by a lowercase letter inside an angle ($\angle$). Angles may also be formed by rotating a line from another line, provided they have a common endpoint. These lines are called *rays.* By definition, a ray is part of a line that begins at a given point and goes in one direction without ending. The symbol for a ray is $\overrightarrow{JK}$. See Figure 14–11. Angle *LJK* is formed by rotating ray *JK* in a counterclockwise direction from ray *JL*. Un-

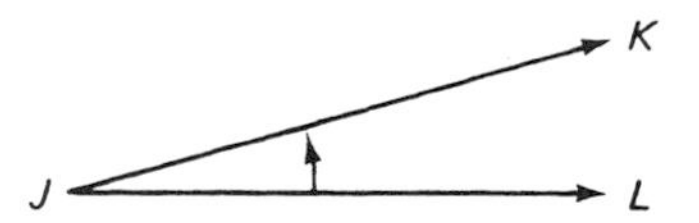

Figure 14–11

less otherwise indicated, angles are generated in a counterclockwise direction as indicated by the curved arrow between the rays *JL* and *JK*. If a ray is rotated completely around another ray from a common vertex, it will make a complete revolution. A complete revolution is divided into 360 parts called *degrees.* See Figure 14–12.

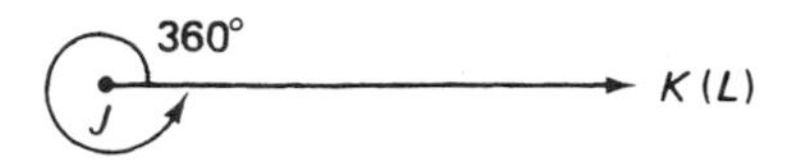

Figure 14–12

When ray *JK* is rotated completely around from ray *JL* as indicated by the curved arrow, it has made a complete revolution of 360°. Degrees are indicated by the symbol °. Thus 35 degrees is written as 35°. For more accurate measurement, each degree is divided into 60 parts called *minutes* and is indicated by the symbol ′. Each minute is divided into 60 parts, called *seconds,* and is indicated by the symbol ″. When a line or ray rotates 180° or half of a complete revolution, it forms a *straight angle.* See Figure 14–13. When a line or ray rotates 90° or a quarter of a complete revolution, it forms a *right angle.* The

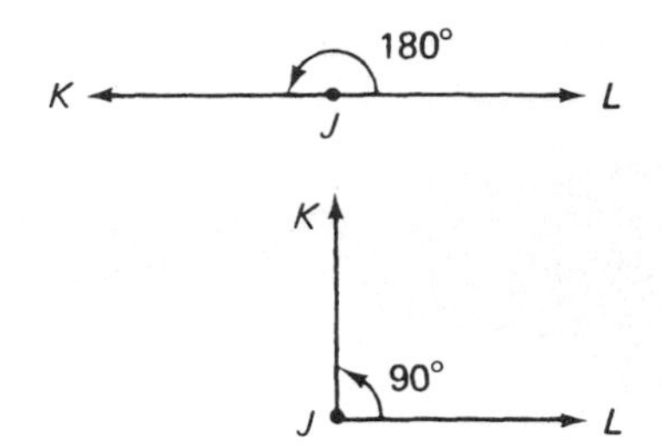

Figure 14–13

symbol for a right angle is ∟. Lines or rays that form a right angle are said to be *perpendicular* to each other. The symbol for perpendicular is $\perp$.

An angle greater than 90° and less than 180° is called an *obtuse angle.* An angle less than 90° is called an *acute angle.* Two angles whose sum is 90° are called *complementary angles.*

A simple illustration of angles is the hands of a clock. Each hour, the minute hand of a clock moves 360°. If the point where the hands of the clock are attached is considered to be the vertex of an angle and the hands are considered as sides or rays of the angle, an infinite number of angles would be formed. Since a clock has 12 numbers, the degrees between numbers, $\frac{360°}{12} = 30°$. If a clock has indications for minutes, the degrees between each minute would be $\frac{360°}{60} = 6°$.

Example 14–2: Clock Angles

PROBLEM: What angle is formed by the hands of a clock at four o'clock?

SOLUTION: Draw a sketch of a clock indicating four o'clock. See Figure 14–14. First, count the number of hours between twelve o'clock and four o'clock. The number is 4.

Figure 14–14

Second, multiply the number of hours (4) times the degrees between numbers on the face of the clock (30°) to arrive at the angle indicated by the hands of the clock (120°): $4 \times 30° = 120°$.

EXERCISE 14–2 DEFINITIONS AND PROPERTIES OF ANGLES

Complete the following exercises concerning geometric definitions.

14–11. Sketch and label two intersecting line segments and their point of intersecton.

14–12. Sketch three lines and have them intersect at three different points. Label the points of intersection *A*, *B*, and *C*.

14–13. How many degrees are there in a half revolution?

14–14. How many degrees are there in the complement of a 68° angle?

14–15. How many degrees are there in the supplement of a 58° angle?

14–16. The complement of a 36° angle contains how many degrees?

14–17. How many degrees are there in the supplement of a 118° angle?

14–18. How many degrees are there in an acute angle?

14–19. How many degrees are there in a straight angle?

14–20. How many degrees are there in a right angle?

14–21. How many degrees are in a fourth of a right angle?

14–22. The number of degrees in a circle is equal to __________ straight angles and __________ right angles.

14–23. How many degrees are there in an obtuse angle?

14–24. Name the angle shown in Figure 14–15.

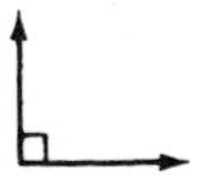

Figure 14–15

14–25. What type of angle is shown in Figure 14–16?

J O B

Figure 14–16

14–26. What type of angle is shown in Figure 14–17?

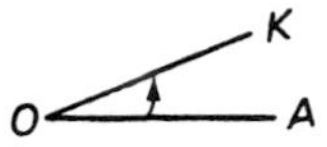

K O A

Figure 14–17

14–27. What angle will the hands of a clock form at two o'clock?

14–28. What central angle will the hands of a clock form at five o'clock?

14–29. What obtuse angle will the hands of a clock form at eight o'clock?

14–30. If you were driving due north and made a 90° turn to the right, what would be your new direction?

14.3 MEASURING ANGLES

Angles are constructed and measured using a *protractor.* Study the protractor illustrated in Figure 14–18. The protractor may be divided into 10° and 1° increments.

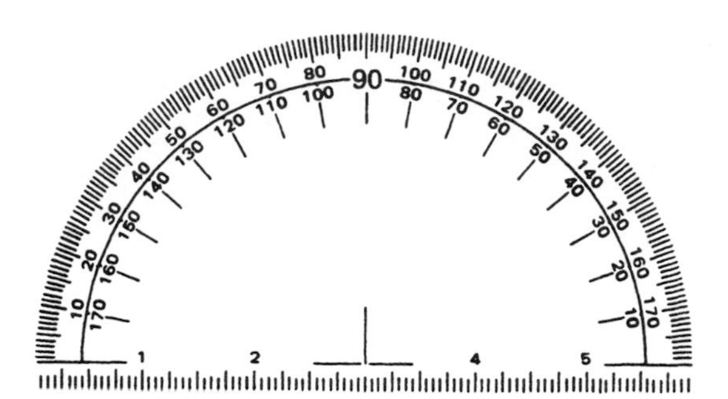

Figure 14–18

Example 14–3: Measure an Acute Angle

PROBLEM: Measure a given acute angle using a protractor.

SOLUTION: First, place the protractor on the acute angle as illustrated in Figure 14–19. The flat edge of the protractor should be placed on ray *JL,* with the midpoint of the flat edge directly on point *J.* Second, read the inside degrees from right to left. Thus *KJL* is 52°.

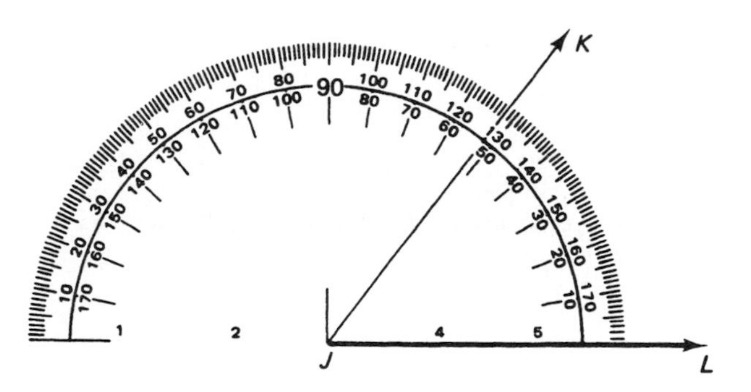

Figure 14–19

Example 14–4: Measure an Obtuse Angle

PROBLEM: Measure a given obtuse angle using a protractor.

SOLUTION: First, place the protractor on the obtuse angle as illustrated in Figure 14–20. The flat edge of the protractor should be placed on ray *CB,* with the midpoint of the flat edge directly on point *C.* Second, read the inside degrees from right to left. Thus ∠*KJL* is 133°.

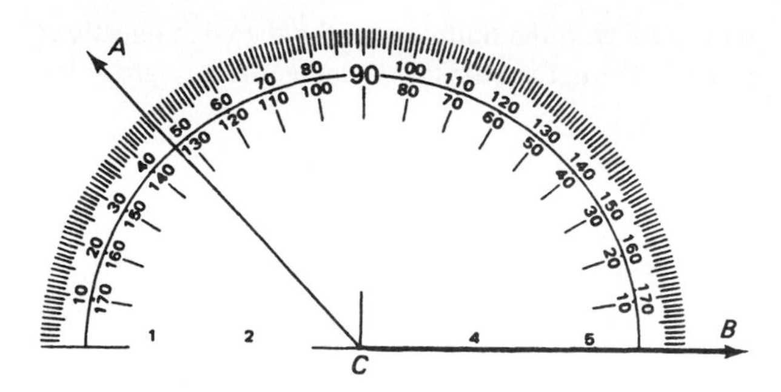

Figure 14–20

Example 14–5: Construct an Acute Angle

PROBLEM: Construct an angle of 37°.

SOLUTION: First, draw a "working" ray, *FG,* and assign the vertex of the angle as point *F* on working ray *FG* as shown in Figure 14–21. Second, place the protractor on ray *FG* with the midpoint of the flat edge directly at point *F.* Third, read the inside degrees from right to left and make a point at 37°. Fourth, draw in ray *FH* to form ∠*HFG.*

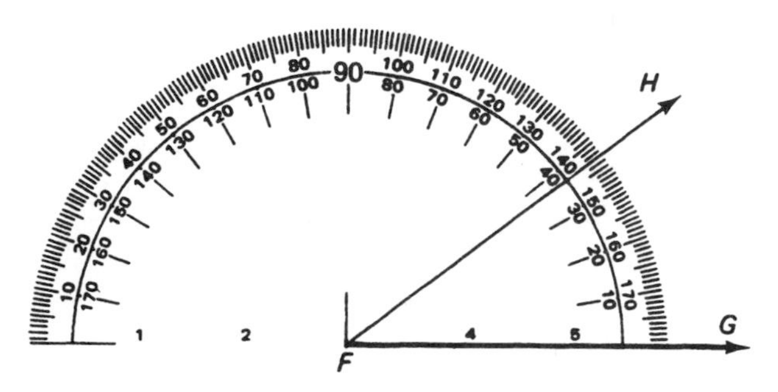

Figure 14–21

Example 14–6: Construct an Obtuse Angle

PROBLEM: Construct an angle of 162°.

SOLUTION: First, draw, a working ray, *ST,* and assign the vertex of the angle at point *S* on working ray *ST* as shown in Figure 14–22. Second, place the protractor

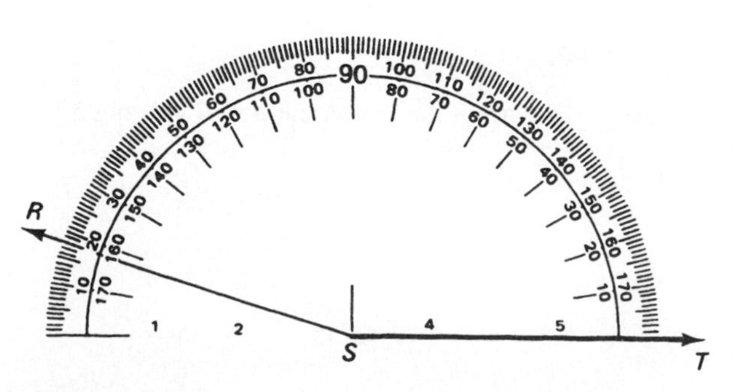

Figure 14–22

on ray *ST* with the midpoint of the flat edge directly at point *S*. Third, read the inside degrees from right to left and make a point at 162°. Fourth, draw in ray *SR* to form $\angle RST$.

EXERCISE 14–3 MEASURING ANGLES

Using a protractor, measure the degrees of the angles and draw the angles as indicated.

14–31. In Figure 14–23, $\angle AOB$ = _____.

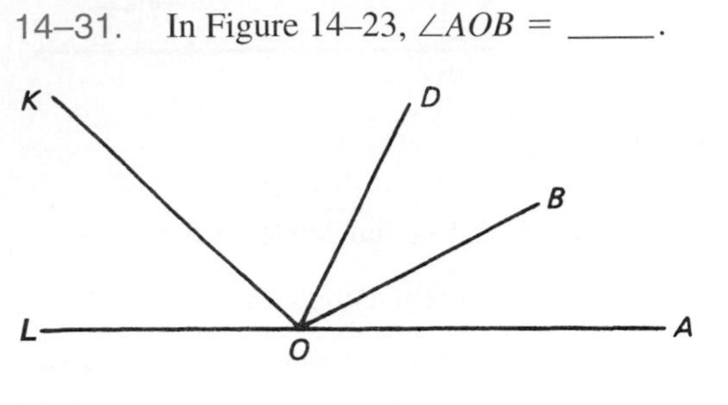

Figure 14–23

14–33. In Figure 14–23, $\angle AOK$ = _____.

14–35. In Figure 14–25, $\angle PQR$ = _____.

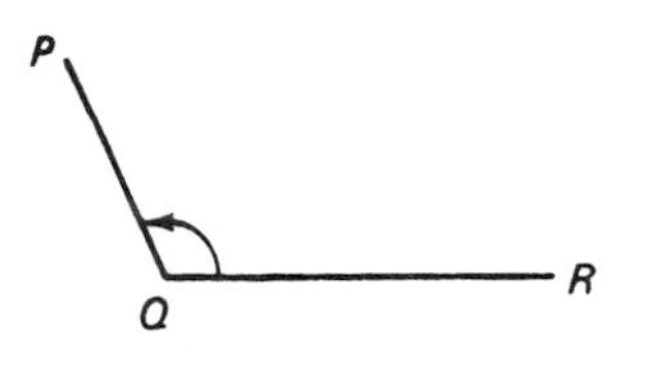

Figure 14–25

14–37. Construct $\angle KOB$ as 48° in Figure 14–27.

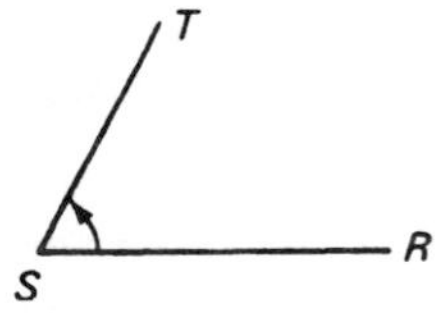

O B

Figure 14–27

14–39. Construct the supplement of a 51° angle.

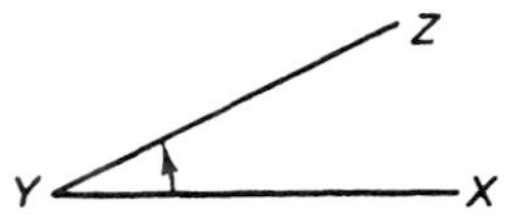

14–32. In Figure 14–23, $\angle BOD$ = _____.

14–34. In Figure 14–24, $\angle RST$ = _____.

T S R

Figure 14–24

14–36. In Figure 14–26, $\angle XYZ$ = _____.

Z Y X

Figure 14–26

14–38. Construct $\angle ROT$ as 163° on Figure 14–28.

O T

Figure 14–28

14–40. In Figure 14–29, $\angle AOC$ = _____.

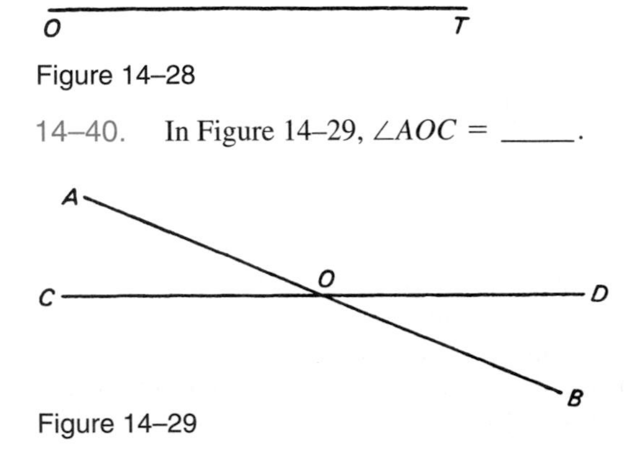

Figure 14–29

THINK TIME

The number of possible geometric figures is infinite. The names for geometric figures are determined by the number of sides or size of angles. Sometimes the figure is named by the number of sides used to construct a planed closed figure. A *polygon* is a geometric figure that has three or more sides. A *pentagon* is a five-sided figure. A *hexagon* is a six-sided figure. A *septagon* is a seven-sided figure. An *octagon* is an eight-sided figure. If all sides and angles are equal, the polygon is called a *regular polygon.* For example, a square is a regular polygon.

Some of the more common figures and their descriptions have been indicated in the following chart. Complete the chart to use as a reference.

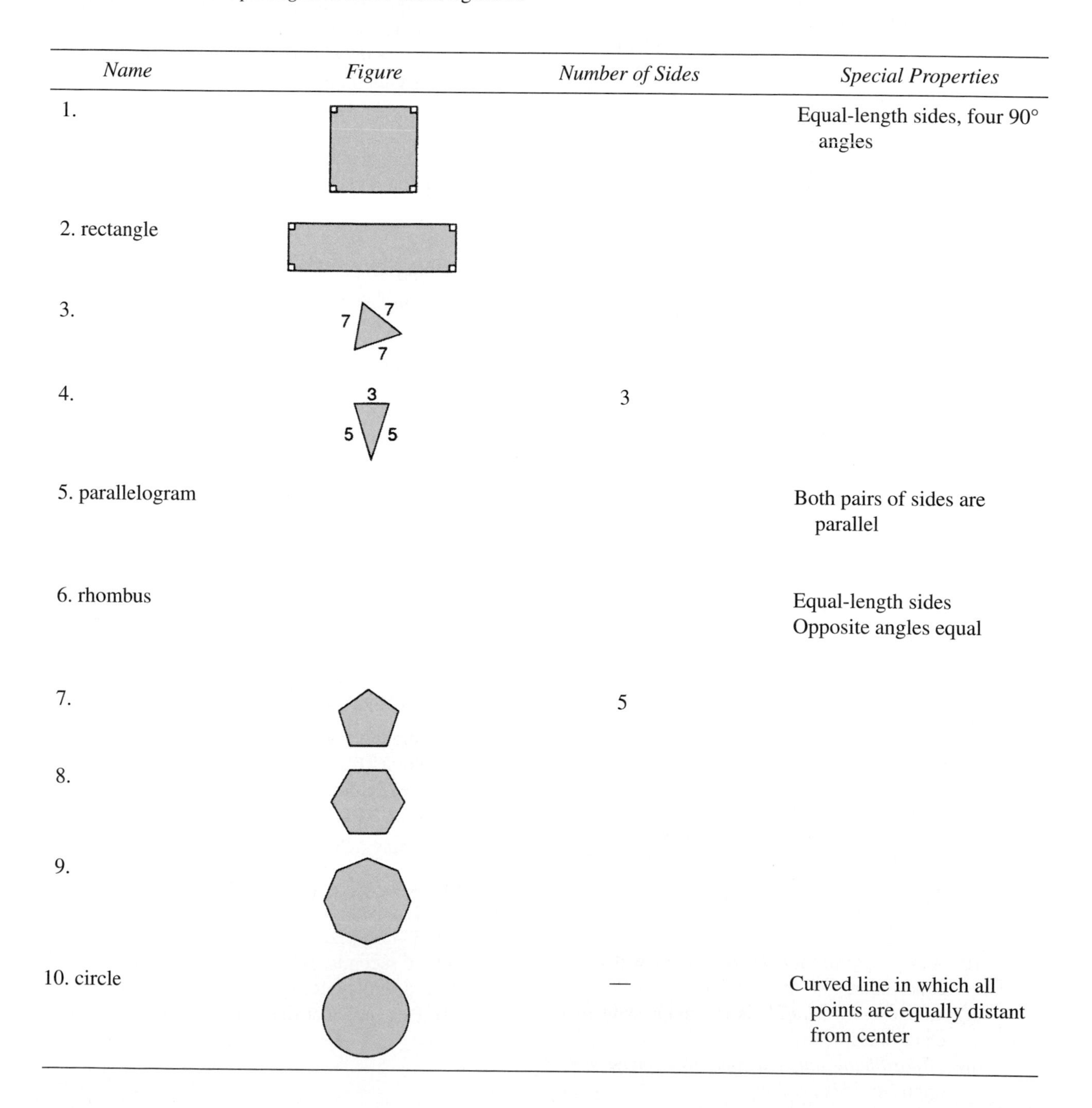

Name	*Figure*	*Number of Sides*	*Special Properties*
1.			Equal-length sides, four 90° angles
2. rectangle			
3.	7, 7, 7		
4.	3, 5, 5	3	
5. parallelogram			Both pairs of sides are parallel
6. rhombus			Equal-length sides Opposite angles equal
7.		5	
8.			
9.			
10. circle		—	Curved line in which all points are equally distant from center

PROCEDURES TO REMEMBER

1. To measure an angle, place the flat edge of a protractor on one of the rays or sides of the angle so that the midpoint of the flat edge of the protractor is directly on the vertex. Read the inside degrees of the protractor for angles less than 90°. Read the outside degrees for angles greater than 90°.
2. To construct an angle, draw a working ray and vertex, and place the flat edge of the protractor on the working ray or side with the midpoint of the flat edge of the protractor on the vertex. Read the degrees and make a mark at the given number of degrees. Draw a line from the vertex through the mark to form the angle.

CHAPTER SUMMARY

1. The following geometric terms and definitions are essential in solving geometry problems:
   - (a) *Point* is a position or location that has no size or dimension.
   - (b) A *line* is a series or set of points which, when connected, extend indefinitely in opposite directions or may be arranged to form a straight, broken, or curved line.
   - (c) A *straight line* is the shortest distance between two points.
   - (d) A *plane* is a flat surface that has length and width but no depth or height.
   - (e) *Plane geometry* involves two dimensions: length and width.
   - (f) *Solid geometry* involves three dimensions: length, width, and height.
   - (g) A *plane figure* is any figure that can be drawn on a flat surface.
   - (h) A *plane closed figure* is any plane figure that encloses an area.
   - (i) A *polygon* is a plane closed figure described by straight-line segments and angles.
   - (j) A *triangle* is a polygon with three straight-line segments and angles.
   - (k) A *right triangle* is a triangle that contains a right angle.
   - (l) An *equilateral triangle* is a triangle with three equal sides.
   - (m) An *isosceles triangle* is a triangle with two equal sides.
   - (n) A *parallelogram* is a four-sided plane with opposite sides parallel.
   - (o) A *rectangle* is a parallelogram with all right angles.
   - (p) A *square* is a parallelogram with equal sides and all right angles.
   - (q) A *rhombus* is a parallelogram with equal sides.
   - (r) A *rectangular solid* is a geometric figure with length, width, and height; the six face planes are rectangular.
   - (s) A *line segment* is a portion of a line indicated by endpoints.
   - (t) A *vertex* of a solid is the point of intersection of three planes.
   - (u) A *ray* is a part of a line that begins at a given point and goes in one direction.
   - (v) A *degree* is a unit of angular measure equal to $\frac{1}{360}$ of a revolution. There are 360° in one complete revolution.
   - (w) A *straight angle* contains 180°.
   - (x) A *right angle* contains 90°.
   - (y) An *obtuse angle* is an angle greater than 90° and less than 180°.
   - (z) *Complementary angles* are two angles whose sum is 90°.
   - (aa) An *acute angle* is an angle less than 90°.
   - (ab) *Supplementary angles* are two angles whose sum is 180°.
   - (ac) A *protractor* is an instrument used to measure and construct angles.
2. The following geometric symbols are important for use in working problems in geometry:
   - (a) Curved line *a:* $\frown{a}$
   - (b) Broken line *b:* $\overset{\wedge\!\wedge}{b}$
   - (c) Straight line *c:* $\overleftrightarrow{c}$ or $\overline{c}$
   - (d) Point *D:* . D
   - (e) Parallel lines: ||
   - (f) Line segment *AB:* $\overline{AB}$
   - (g) Angle *ABC:* $\angle ABC$ or $\angle B$
   - (h) Ray *JK:* $\overrightarrow{JK}$
   - (i) 276 degrees: 276°
   - (j) 53 minutes: 53′
   - (k) 16 seconds: 16″
   - (l) A right angle: ∟
   - (m) Perpendicular lines: ⊥

CHAPTER TEST

T–14–1. The symbol ∟ means __________ angle.

T–14–2. The symbol $\overrightarrow{AB}$ means __________.

T–14–3. Write the symbol for line segment *LK*.

T–14–4. Write the symbol for line *n*.

T–14–5. How long is a ray?

T–14–6. If you were flying south and made a 90° turn to the right, in what direction would you be going?

T–14–7. The symbol for perpendicular lines is __________.

T–14–8. An instrument used to measure degrees is called a __________.

T–14–9. Degrees used to measure angles are subdivided into __________.

T–14–10. Angles are labeled by three capital letters. The middle letter indicates the __________ of the angle.

T–14–11. A triangle with two equal length sides is a (an) __________ triangle.

T–14–12. What degree of angle would the hands of a clock make at five o'clock?

T–14–13. Measure with a protractor angle *r* shown in Figure 14–30.

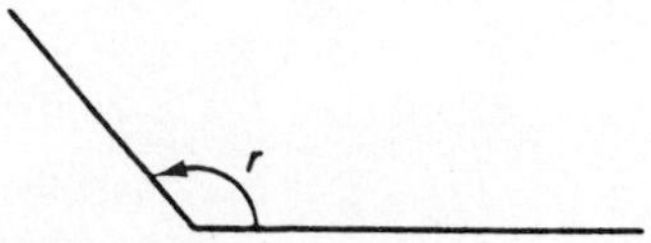

Figure 14–30

T–14–14. Measure with a protractor angle *PQR* shown in Figure 14–31.

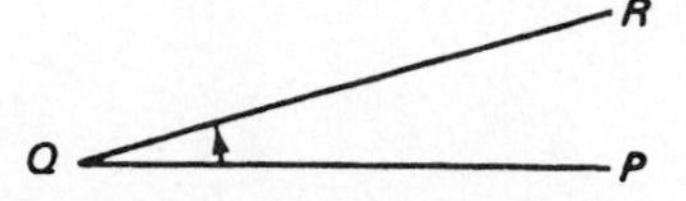

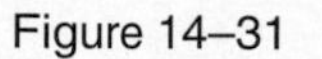
Figure 14–31

T–14–15. Construct a 39° angle with a protractor.

T–14–16. Construct a 121° angle with a protractor.

T–14–17. What is the complement of a 26° angle?

T–14–18. Determine the supplement of a 46° angle.

T–14–19. How many degrees are there in one third of a straight angle?

T–14–20. A position or location that has no size or dimension is called a(an) __________.

T–14–21. The point of intersection of three planes or angles is called __________.

T–14–22. A triangle with equal-length sides is called a(an) __________.

T–14–23. Measure $\angle JOL$ in Figure 14–32.

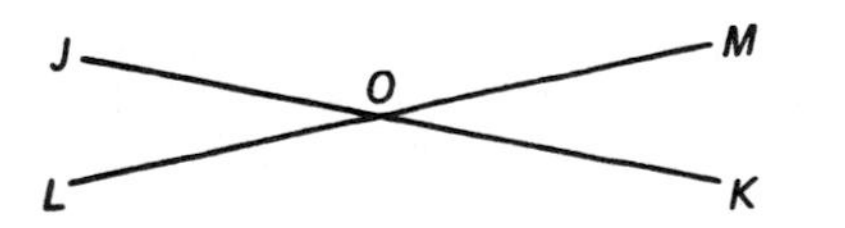

Figure 14–32

T–14–24. Measure $\angle MOJ$ in Figure 14–32.

T–14–25. Angles *MOJ* and *JOL* are called __________ angles.

15

PERIMETERS AND AREAS OF PLANE GEOMETRIC FIGURES

OBJECTIVES

After studying this chapter, you will be able to:

15.1. Understand definitions and basic operations of perimeters and areas.
15.2. Understand definitions and properties of circles.
15.3. Understand definitions and properties of complex figures.

SELF-TEST

The following test will measure your ability to find the perimeter, circumference, or area of a plane geometric figure. Successful completion of these problems shows that you know how to find perimeters and areas and you are ready to complete more advanced geometry problems. Problems you are unable to solve will show areas you need to study.

S–15–1. What type of geometric figure is Figure 15–1?

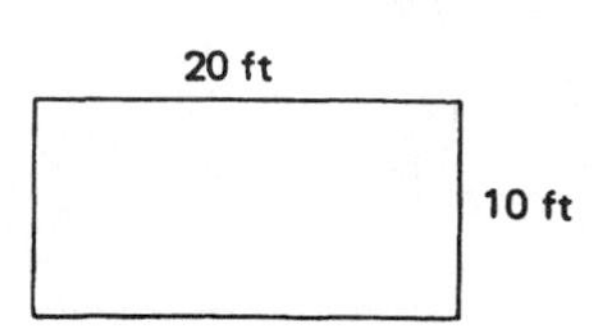

Figure 15–1

S–15–2. Find the perimeter of Figure 15–1.

S–15–3. Find the area of Figure 15–1.

S–15–4. Figure 15–2 is what type of geometric figure?

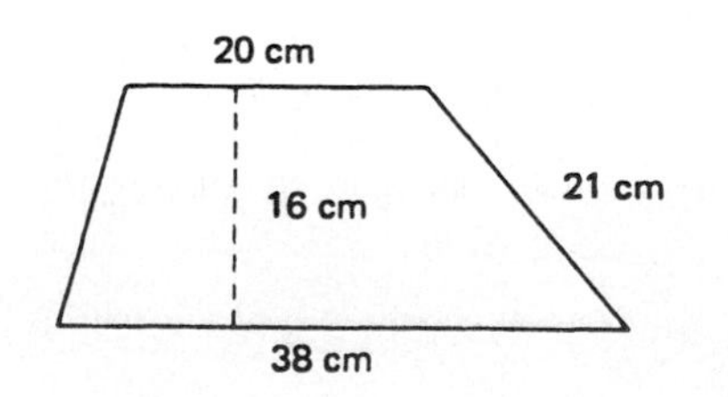

Figure 15–2

S–15–5. Find the perimeter of Figure 15–2.

S–15–6. Determine the area of Figure 15–2.

S–15–7. What type of geometric figure is Figure 15–3?

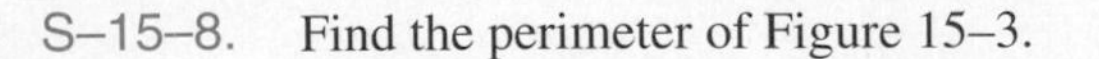

S–15–8. Find the perimeter of Figure 15–3.

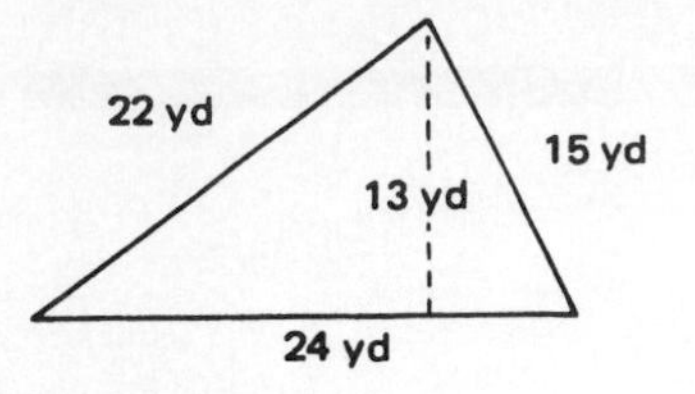

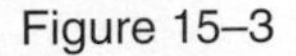

Figure 15–3

S–15–9. What is the altitude of Figure 15–3?

S–15–10. Calculate the area of Figure 15–3.

S–15–11. What is the diameter of Figure 15–4?

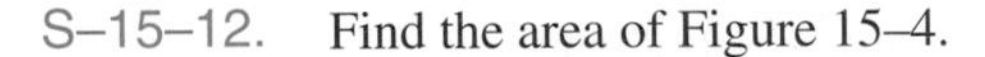

S–15–12. Find the area of Figure 15–4.

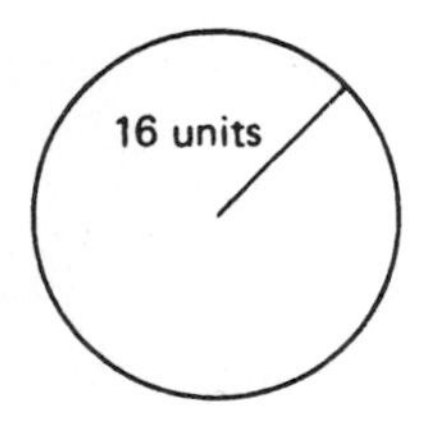

Figure 15–4

S–15–13. Find the circumference of Figure 15–4.

S–15–14. Find the amount of fencing needed to fence the garden area represented by Figure 15–5.

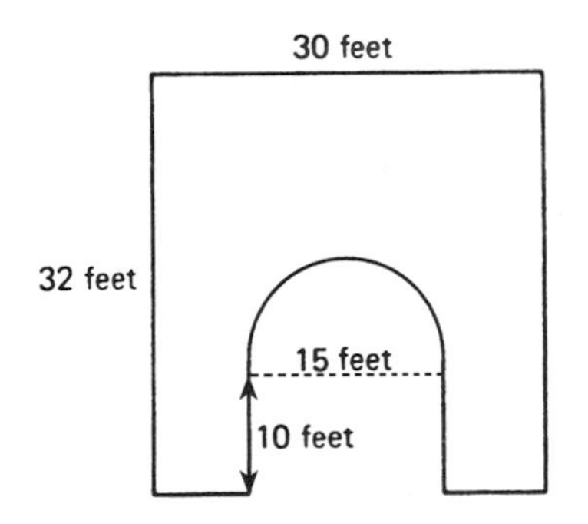

Figure 15–5

S–15–15. Find the area of the garden shown in Figure 15–5.

S–15–16. Find the inside circumference of the sidewalk around the fountain illustrated by Figure 15–6.

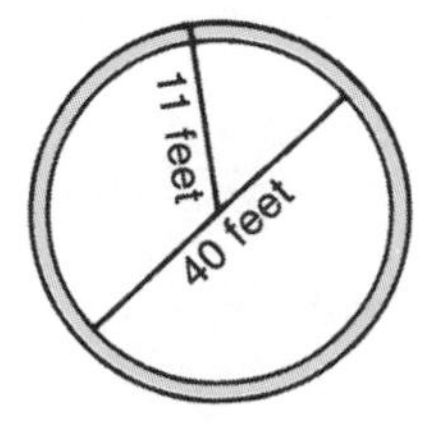

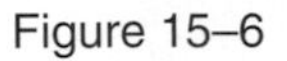

Figure 15–6

S–15–17. Find the outside circumference of the sidewalk shown in Figure 15–6.

S–15–18. What is the area of the sidewalk shown in Figure 15–6?

S–15–19. Calculate the perimeter of an ice-skating rink represented by Figure 15–7.

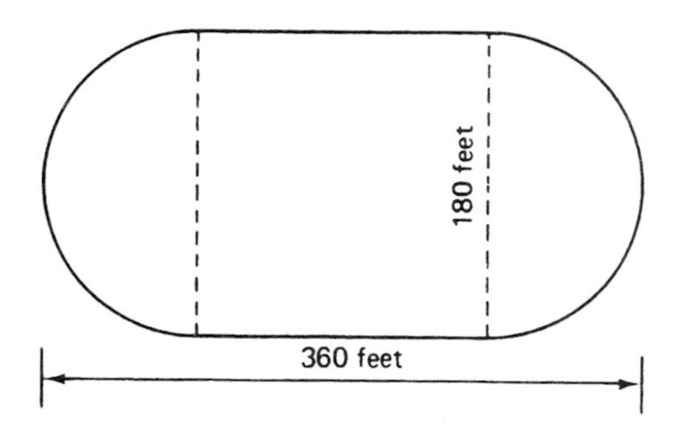

Figure 15–7

S–15–20. What is the area of the ice-skating rink shown in Figure 15–7?

15.1 DEFINITIONS AND BASIC OPERATIONS OF PERIMETERS AND AREAS

The use of perimeters, circumferences, and areas of geometric figures is a daily experience for many people. However, geometric principles are seldom considered when a slice of bread is spread with peanut butter, a pie crust is rolled out, a wall is painted, or a tire size is selected. Yet in each case geometric problems are solved either directly or indirectly.

Formulas used to find perimeters, circumferences, and areas can be useful in saving both time and money at home and work. Knowing how to use formulas to find areas is of great value when, for example, carpeting a room, buying a lot, or determining the amount of lawn fertilizer to purchase. The ability to determine perimeters, circumferences, and areas is necessary for many crafts and trades.

Perimeter is the distance around a geometric figure or the sum of the length of the sides. It is stated in linear measure such as inches, feet, miles, centimeters, and kilometers. The perimeter of a circle is called the *circumference* of the circle. *Area* is the amount of surface of a geometric figure.

The procedures for finding perimeter, circumference, and area can best be illustrated through the use of examples. Study each of the examples, noting the formulas used and the procedures followed.

Example 15–1: Find the Perimeter of a Square

PROBLEM: Find the perimeter of the square shown in Figure 15–8.

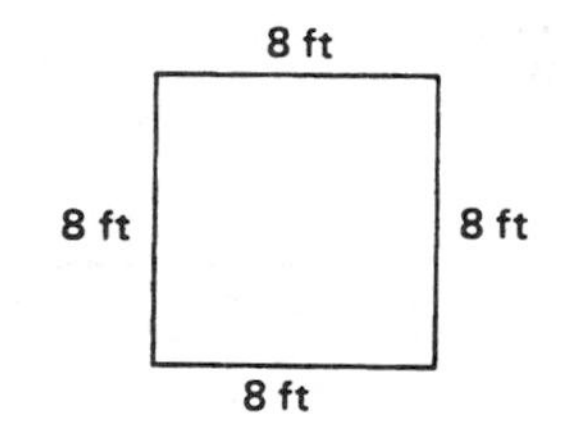

Figure 15–8

SOLUTION: The perimeter of the square can be found by adding the lengths of the four sides. Thus

$$\begin{aligned} \text{perimeter} &= 8 \text{ ft} + 8 \text{ ft} + 8 \text{ ft} + 8 \text{ ft} \\ &= 32 \text{ ft} \end{aligned}$$

The formula for determining the perimeter of a square is

$$P = s + s + s + s \text{ or } P = 4s$$

Example 15–2: Find the Perimeter of a Rectangle

PROBLEM: Find the perimeter of the rectangle shown in Figure 15–9.

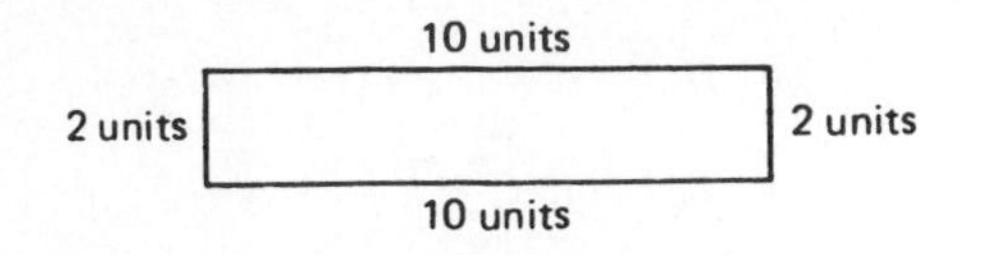

Figure 15–9

SOLUTION: The perimeter of the rectangle is the sum of the lengths of the four sides. Thus

$$\begin{aligned} \text{perimeter} &= 10 \text{ units} + 2 \text{ units} + 10 \text{ units} + 2 \text{ units} \\ &= 24 \text{ units} \end{aligned}$$

The formula for determining the perimeter of a rectangle is

$$\begin{aligned} \text{perimeter} &= \text{length} + \text{width} + \text{length} + \text{width} \\ P &= l + w + l + w \\ P &= 2l + 2w \\ P &= 2(l + w) \end{aligned}$$

Example 15–3: Find the Perimeter of a Triangle

PROBLEM: Find the perimeter of the triangle shown in Figure 15–10.

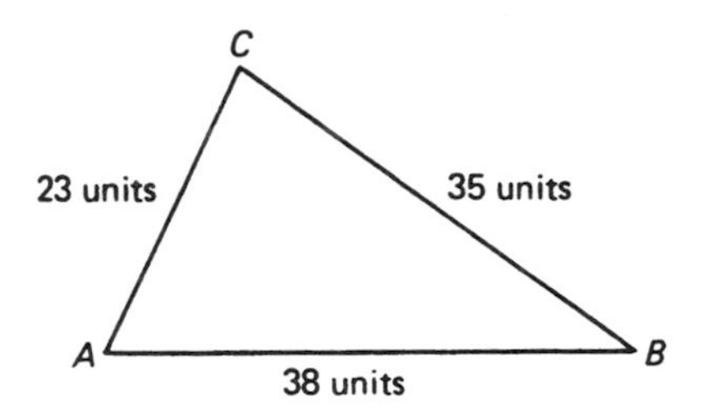

Figure 15–10

SOLUTION: The perimeter of the triangle is the sum of the lengths of the three sides. Thus

$$\begin{aligned} \text{perimeter} &= 23 \text{ units} + 35 \text{ units} + 38 \text{ units} \\ P &= 96 \text{ units} \end{aligned}$$

The formula for determining the perimeter of a triangle is

$$\text{perimeter of a triangle} = \text{side } a + \text{side } b + \text{side } c$$

$$P = a + b + c$$

Example 15–4: Find the Area of a Square

PROBLEM: Find the area shown in Figure 15–11.

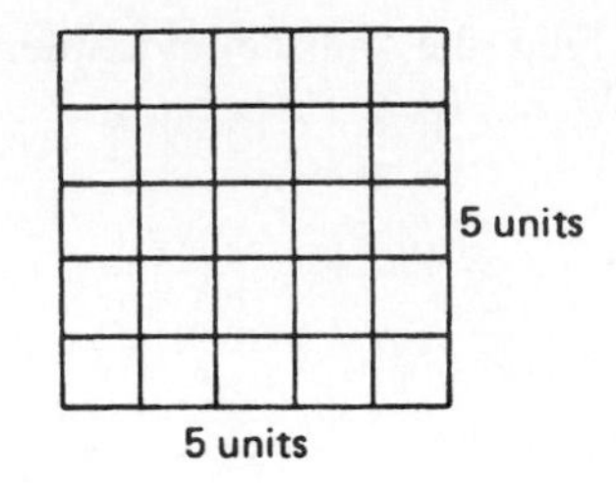

Figure 15–11

SOLUTION: One way to find the area of a square is to count each of the individual smaller squares. An easier way is to count the number of units in one column or side and then multiply this number by the number of rows. Thus

$$\begin{aligned} \text{area} &= 5 \text{ units} \times 5 \text{ units} \\ &= 25 \text{ square units} \end{aligned}$$

The formula for determining the area of a square is

$$\begin{aligned} \text{area} &= \text{side} \times \text{side} \\ A &= s^2 \end{aligned}$$

It is important that area always be calculated in like units; thus the area should be in square inches, square yards, square meters, and so on.

Example 15–5: Find the Area of a Rectangle

PROBLEM: Find the area of the rectangle shown in Figure 15–12.

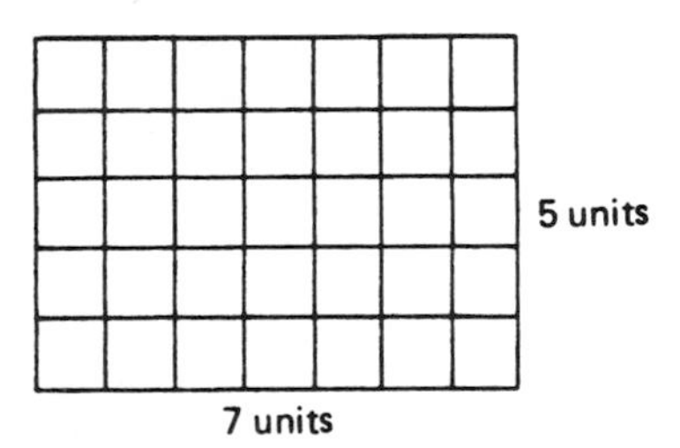

Figure 15–12

SOLUTION: The area of a rectangle can be found by multiplying the number of square units in one row (7) by the number of columns (5). Thus

$$\begin{aligned} \text{area} &= 7 \text{ units} \times 5 \text{ units} \\ A &= 35 \text{ square units} \end{aligned}$$

The formula for finding the area of a rectangle is

$$\begin{aligned} \text{area} &= \text{length} \times \text{width} \\ A &= lw \end{aligned}$$

Example 15–6: Find the Area of a Parallelogram

PROBLEM: Find the area of the parallelogram shown in Figure 15–13.

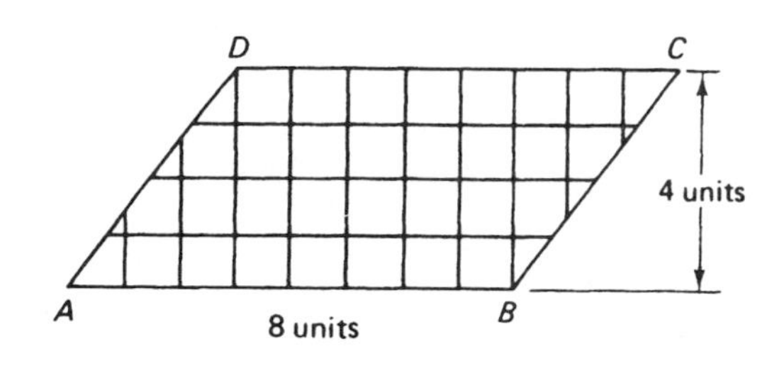

Figure 15–13

SOLUTION: One way to determine the area of the parallelogram is to move triangle ADE to the right side BC to form rectangle $DE(E)C$ (see Figure 15–14) and

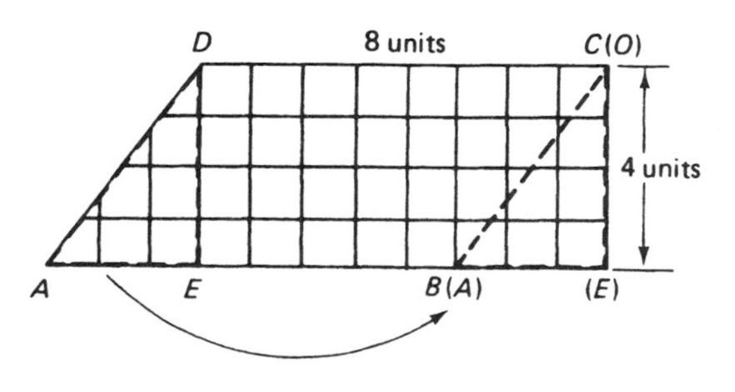

Figure 15–14

then multiply length DC times width DE. An easier way is to multiply the length DC times the height DE. Thus

$$\text{area} = 8 \text{ units} \times 4 \text{ units}$$
$$= 32 \text{ square units}$$

The formula for finding the area of a parallelogram is

$$\text{area} = \text{base (length)} \times \text{height (altitude)}$$
$$A = bh$$

Example 15–7: Find the Area of a Triangle

PROBLEM: Find the area of the triangle shown in Figure 15–15.

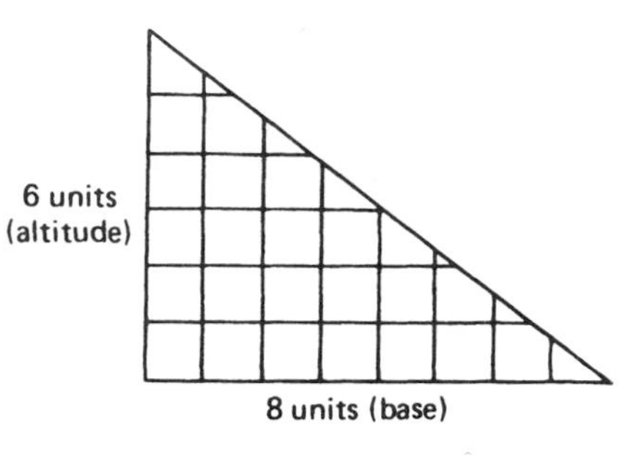

Figure 15–15

SOLUTION: One way to find the area of the triangle is to form a rectangle as illustrated in Figure 15–16.

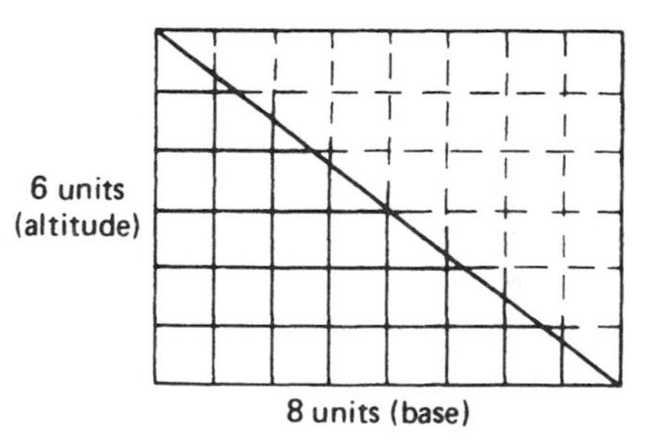

Figure 15–16

As shown in the sketch, the triangle is half of the area of the rectangle. Thus

$$\text{area} = \frac{8 \text{ units (base)} \times 6 \text{ units (altitude)}}{2}$$
$$= 24 \text{ square units}$$

The formula for determining the area of any triangle is

$$\text{area} = \frac{\text{base} \times \text{altitude}}{2}$$

$$A = \frac{ba}{2} \quad \text{or} \quad A = \frac{1}{2}bh \text{ or } 0.5\,bh$$

EXERCISE 15–1 BASIC PERIMETERS AND AREAS

Solve the following problems for the given quantity.

15–1. Find the perimeter of the square shown in Figure 15–17.

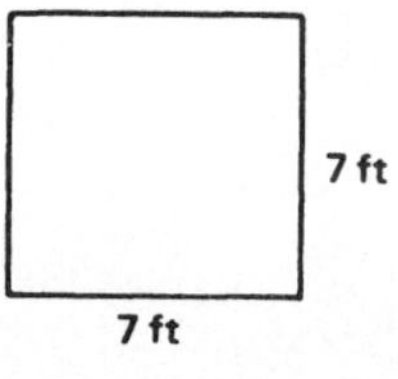

Figure 15–17

15–2. Find the area of the square shown in Figure 15–17.

15–3. Find the area of the square shown in Figure 15–18.

15–4. Find the perimeter of the square shown in Figure 15–18.

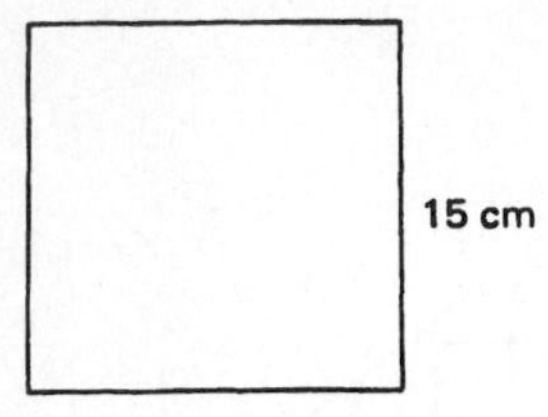

Figure 15–18

15–5. Find the area of the rectangle shown in Figure 15–19.

15–6. Find the perimeter of the rectangle shown in Figure 15–19.

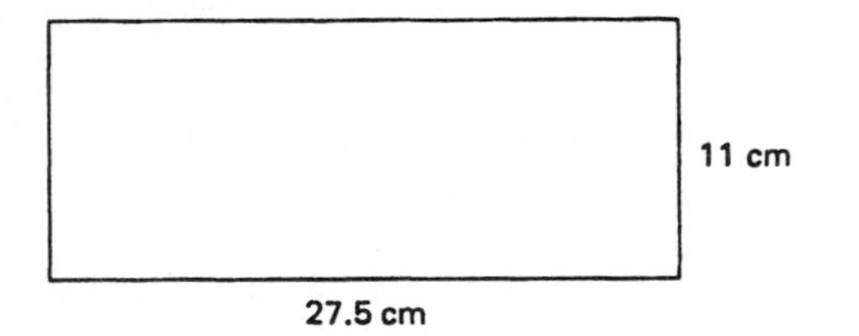

Figure 15–19

15–7. Find the perimeter of the rectangle shown in Figure 15–20.

15–8. Find the area of the rectangle shown in Figure 15–20.

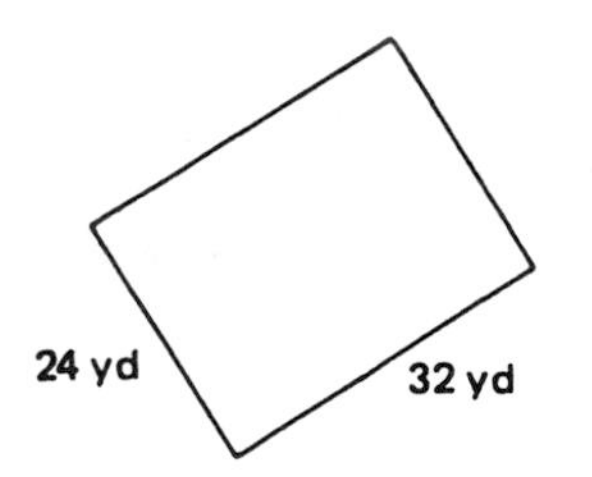

Figure 15–20

15–9. Calculate the area of the right triangle shown in Figure 15–21.

15–10. Find the perimeter of the right triangle shown in Figure 15–21.

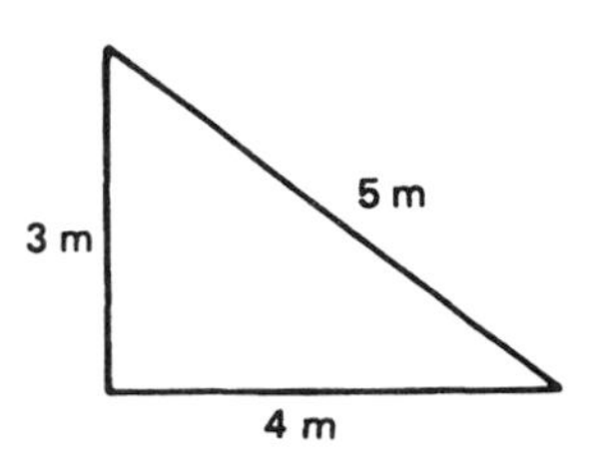

Figure 15–21

15–11. Find the area of the triangle shown in Figure 15–22.

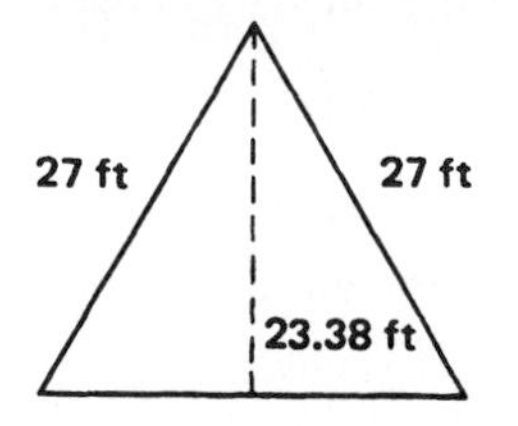

Figure 15–22

15–12. Find the perimeter of the triangle shown in Figure 15–22.

15–13. Find the perimeter of the triangle shown in Figure 15–23.

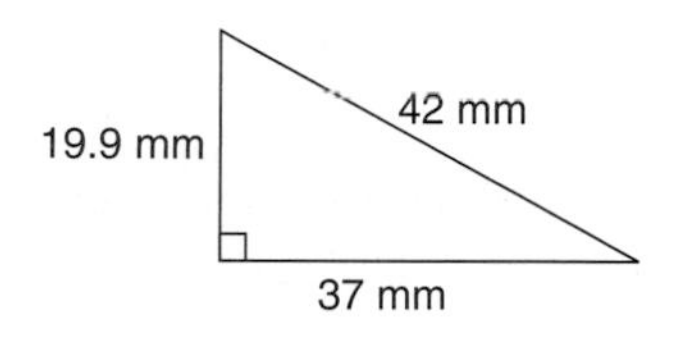

Figure 15–23

15–14. Find the area of the triangle shown in Figure 15–23.

15–15. What is the altitude of the triangle shown in Figure 15–24?

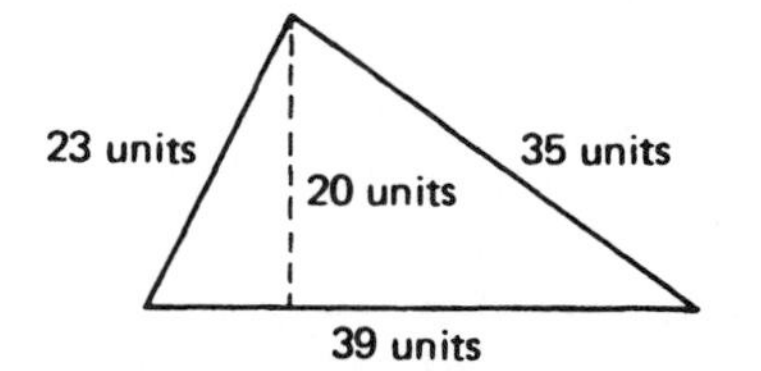

Figure 15–24

15–16. What is the perimeter of the triangle shown in Figure 15–24?

15–17. Find the area of the triangle shown in Figure 15–24.

15–18. What is the altitude of triangle shown in Figure 15–25?

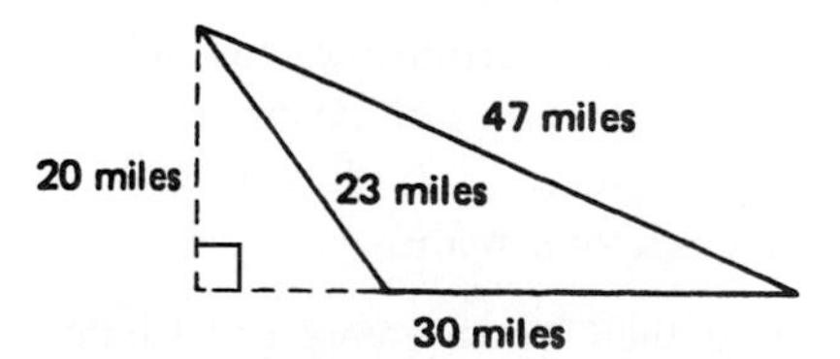

Figure 15–25

15–19. Find the perimeter of the triangle shown in Figure 15–25.

15–20. Determine the area of the triangle shown in Figure 15–25.

15.2 DEFINITIONS AND PROPERTIES OF CIRCLES

A *circle* is a plane curve with a set of points that are the same distance from a fixed point. The fixed point is called the *center* of the circle. The perimeter or distance around a circle is called the *circumference* of the circle. Figure 15–26 shows a circle, its center, and its circumference.

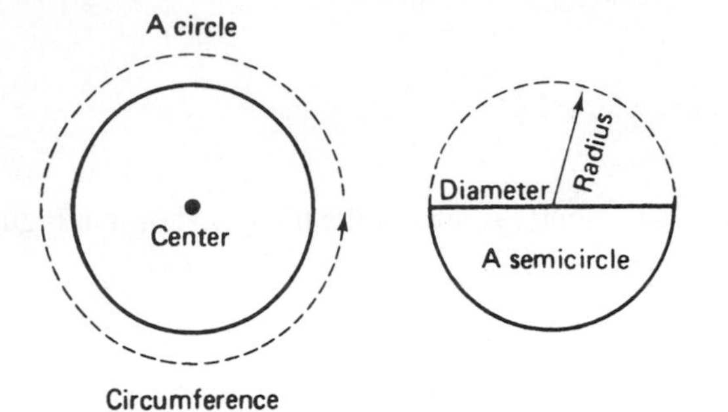

Figure 15–26

The *diameter* of a circle is a line drawn from one side of a circle, through the center, to the opposite side of the circle. The diameter divides the circle into two *semicircles*. Any line drawn from the center of a circle to the circumference of the circle is called a *radius*. The radius is half of the diameter or a circle. Study the illustration showing diameter, radius, and semicircle.

The *circumference* of a circle can be found by multiplying the diameter of the circle times 3.14. The special number, 3.14, is called *pi* (pronounced "pie") and its symbol is the Greek letter π. The approximate numerical value for π is 3.14. This number is the ratio of the circumference of a circle to its diameter. A more exact value for π is 3.141592653. For calculation purposes, $\pi = 3.14$ or 3.142 is used, depending on the accuracy desired. π was first determined by the ancient Egyptians as they developed a formula for finding the circumference of a circle. Most calculators have a π key, which gives π as 3.1415927.

The *area* of a circle can be determined by multiplying the radius of the circle $\times$ the radius of the circle $\times$ π.

Study the following examples for finding the circumference and area of a circle or semicircle.

Example 15–8: Find the Circumference of a Circle

PROBLEM: Find the circumference of the circle shown in Figure 15–27.

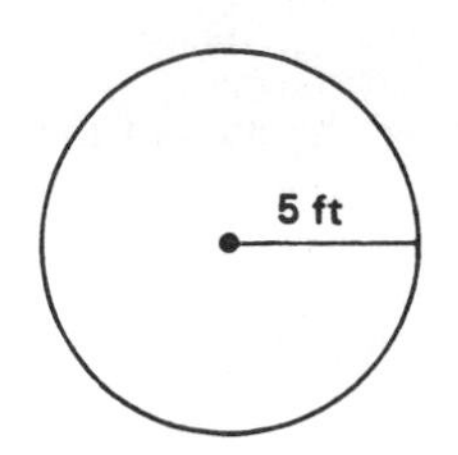

Figure 15–27

SOLUTION: The circumference of a circle is the distance around the circle. The circumference is found by multiplying the diameter (2 $\times$ the radius) by π, thus:

$$\begin{aligned}\text{circumference} &= 2(5\text{ ft}) \times 3.14\\ &= 31.4\text{ ft}\end{aligned}$$

The formula for determining the circumference of a circle is

$$\text{circumference} = 2 \times \text{the radius} \times \pi$$
$$C = 2r\pi$$

or

$$\text{circumference} = \text{diameter} \times \pi$$
$$C = d\pi$$

Example 15–9: Find the Circumference of a Semicircle

PROBLEM: Find the circumference of the semicircle shown in Figure 15–28.

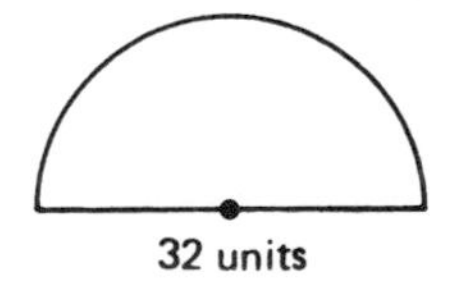

Figure 15–28

SOLUTION: The circumference of a semicircle is half the circumference of the whole circle plus the length of the diameter. First, determine the circumference of the whole circle.

$$\begin{aligned}C &= d\pi\\ &= 32\text{ units} \times 3.14\\ &= 100.48\text{ units}\end{aligned}$$

Second, divide the circumference (100.48 units) by 2.

$$\frac{100.48}{2} = 50.24\text{ units}$$

Third, add the diameter (32 units) to half of the circumference (50.24 units).

$$50.24\text{ units} + 32\text{ units} = 82.24\text{ units}$$

The formula for determining the circumference of a semicircle is

$$\text{circumference of a semicircle} = \frac{\pi \times \text{diameter}}{2} + \text{diameter}$$
$$C = \frac{\pi d}{2} + d$$

Example 15–10: Find the Area of a Circle Given the Radius

PROBLEM: Find the area of the circle shown in Figure 15–29.

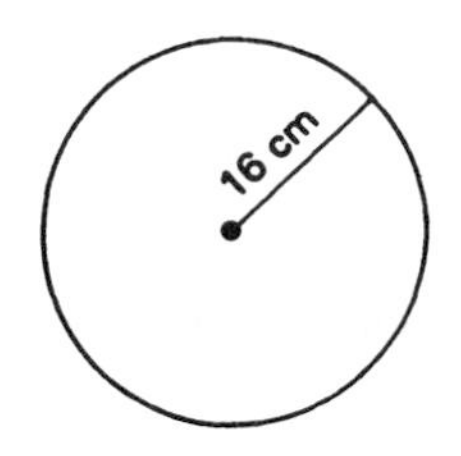

Figure 15–29

SOLUTION: The area of a circle is found by multiplying π × the radius × the radius. Thus

$$\begin{aligned} \text{area} &= 3.14 \times 16 \text{ cm} \times 16 \text{ cm} \\ &= 803.8 \text{ cm}^2 \end{aligned}$$

The formula for determining the area of a circle using the radius is

$$\begin{aligned} \text{area} &= \pi \times \text{radius} \times \text{radius} \\ A &= \pi r^2 \end{aligned}$$

Example 15–11: Find the Area of a Circle Given the Diameter

PROBLEM: Find the area of the circle shown in Figure 15–30.

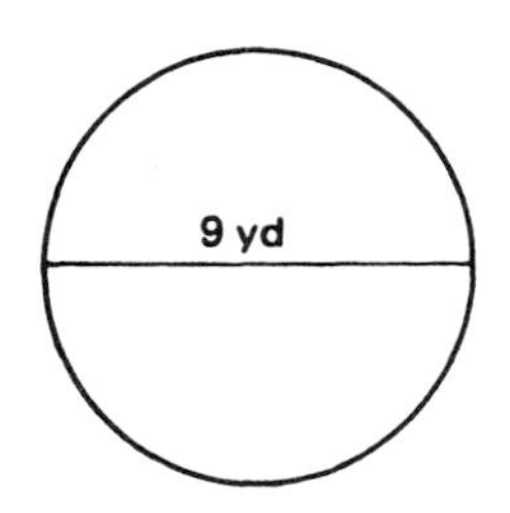

Figure 15–30

SOLUTION: The area of a circle, given the diameter, is determined by dividing the diameter by 2 to obtain the radius, and then multiplying π × the radius × the radius. First, determine the radius.

$$9 \text{ yd} \div 2 = 4.5 \text{ yd}$$

Second, use the formula for determining area using the radius.

$$\begin{aligned} \text{Area} &= 3.14 \times 4.15 \text{ yd} \times 4.5 \text{ yd} \\ &= 63.585 \text{ sq yd} \end{aligned}$$

The formula for determining the area of a circle using the diameter is:

$$\begin{aligned} \text{area} &= \pi \times \left(\frac{\text{diameter}}{2}\right)^2 \\ A &= \pi\left(\frac{d}{2}\right)^2 \\ &= \frac{\pi d^2}{4} \end{aligned}$$

This could be written $A = \dfrac{3.14159d^2}{4}$, which could be written $A = 0.7854d^2$, being $3.14159 \div 4$ can be rounded to 0.7854. (Note where the numbers 0.7854 are found on a calculator.) Thus another formula to find the area of a circle is $A = 0.7854d^2$.

Example 15–12: Find the Area of a Circle using $A = 0.7854d^2$

PROBLEM: Find the area of a 25-unit circle.

SOLUTION: The area of a circle, given the diameter, is found by multiplying 0.7854 by the diameter squared. A 25-unit circle means that the circle has a diameter of 25 units.

$$\begin{aligned} A &= 0.7854d^2 \\ &= 0.7854 \times 25 \text{ units} \times 25 \text{ units} \\ &= 490.875 \text{ square units} \end{aligned}$$

Example 15–13: Find the Area of a Semicircle

PROBLEM: Find the area of the semicircle shown in Figure 15–31.

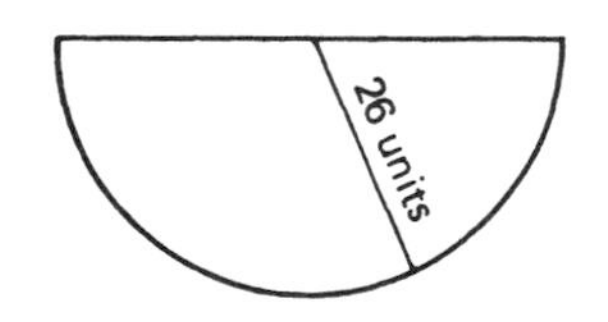

Figure 15–31

SOLUTION: The area of a semicircle is half the area of the whole circle. First, determine the area of the whole circle.

$$\begin{aligned} \text{Area} &= 3.14 \times 26 \text{ units} \times 26 \text{ units} \\ \text{Area} &= 2122.64 \text{ square units} \end{aligned}$$

15.3 DEFINITIONS AND PROPERTIES OF COMPLEX FIGURES

The basic formulas for finding the perimeter and area of a simple geometric figure can be used for more complex geometric figures. This is done by dividing the more complex figures into simple figures. The following examples illustrate how some of the changes are made and the formulas that are involved.

Example 15–14: Find the Area of a Trapezoid

PROBLEM: Find the area in Figure 15–38.

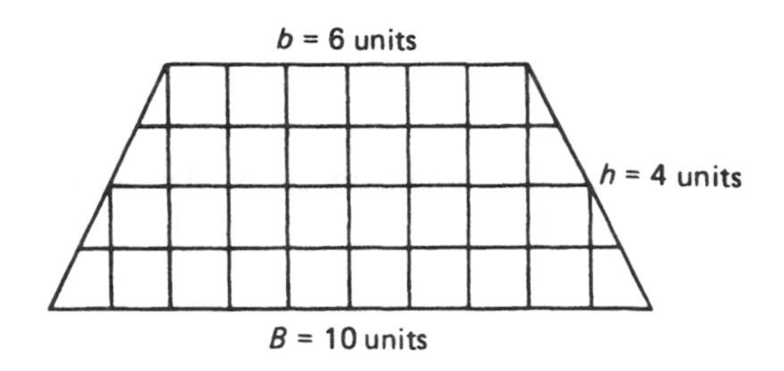

Figure 15–38

SOLUTION: The figure shown in Figure 15–38 is called a trapezoid. The upper base, b, and the lower base, B, are parallel to each other and the figure is closed. One way to determine the area of the trapezoid is to cut off the small triangles from each end of the trapezoid and reattach them to the upper corners to form a rectangle. See Figure 15–39. Note that the length of the newly formed rectangle is 8 units, or the average of the lengths of the upper base (6) and the lower base (10).

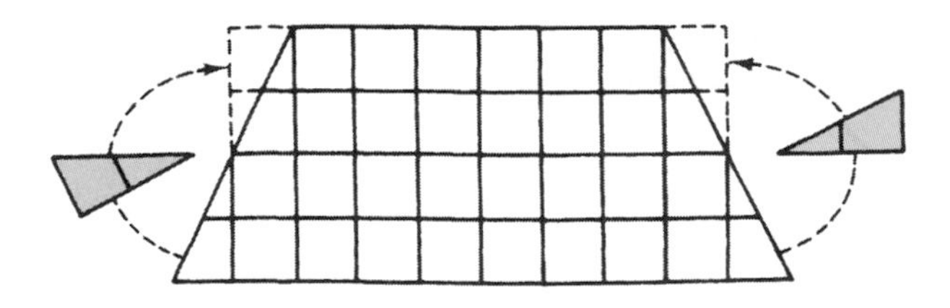

Figure 15–39

The height of the rectangle is 4 units. Thus

$$\begin{aligned} \text{area} &= \frac{10 \text{ units} + 6 \text{ units}}{2} \times 4 \text{ units} \\ &= \frac{16}{2} \text{ units} \times 4 \text{ units} \\ &= 8 \text{ units} \times 4 \text{ units} \\ &= 32 \text{ units} \end{aligned}$$

The formula for determining the area of a trapezoid is

$$\text{area} = \frac{\text{upper base} + \text{lower base}}{2} \times \text{height}$$

$$A = \left(\frac{B + b}{2}\right)h$$

or

$$A = 0.5(B + b)h$$

Example 15–15: Find the Area of a Hexagon

PROBLEM: Find the area in Figure 15–40.

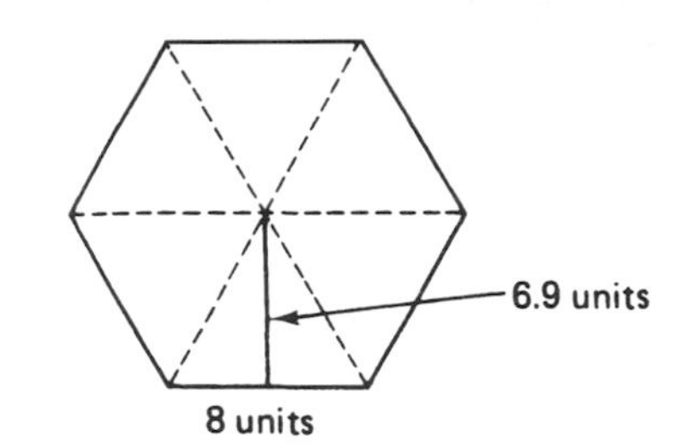

Figure 15–40

SOLUTION: The given figure is a hexagon. A *regular hexagon* is a polygon with six equal sides and angles, as shown in the sketch. Note that it can be divided into six equilateral triangles. Use the following procedures to determine the area of the given hexagon. First, find the area of one of the six equilateral triangles. Use the formula for finding the area of a triangle, $A = \frac{bh}{2}$.

$$\begin{aligned} A &= \frac{8 \text{ units} \times 6.9 \text{ units}}{2} \\ &= \frac{55.2 \text{ square units}}{2} = 27.6 \text{ square units} \end{aligned}$$

Then multiply the area of one triangle (27.6 square units) by the number of triangles (6) to determine the area of the hexagon.

$$A = 27.6 \text{ square units} \times 6 = 165.6 \text{ square units}$$

The formula for determining the area of a hexagon is

$$\text{area} = 6\left(\frac{\text{base} \times \text{height}}{2}\right)$$

$$A = 6\left(\frac{bh}{2}\right) \quad \text{or} \quad 3bh$$

Example 15–16: Applied Carpet Area

PROBLEM: Find the cost of carpeting the floor plan shown in Figure 15–41. The cost of carpeting is \$22 per yard.

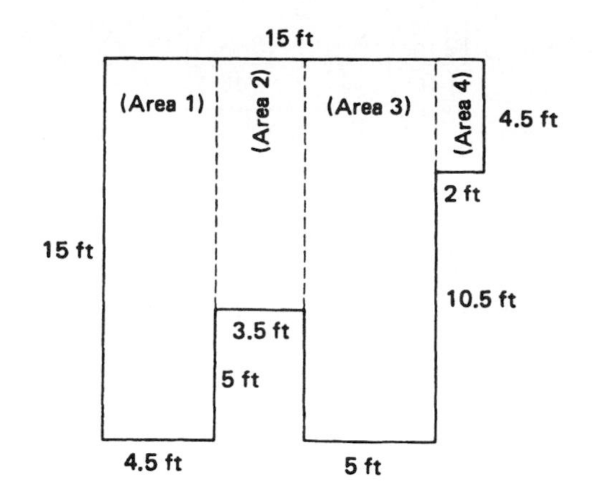

Figure 15–41

SOLUTION: First, divide the floor plan into rectangles as indicated by the dashed lines. Then find the area of each rectangle.

$$\begin{aligned}\text{Area 1} &= 15 \text{ ft} \times 4.5 \text{ ft} = 67.5 \text{ sq ft}\\ \text{Area 2} &= 10 \text{ ft} \times 3.5 \text{ ft} = 35.0 \text{ sq ft}\\ \text{Area 3} &= 15 \text{ ft} \times 5.0 \text{ ft} = 75.0 \text{ sq ft}\\ \text{Area 4} &= 4.5 \text{ ft} \times 2 \text{ ft} = 9 \text{ sq ft}\end{aligned}$$

Find the total square footage of the four areas.

$$\begin{aligned}\text{Area 1} &= 67.5 \text{ sq ft}\\ \text{Area 2} &= 35.0 \text{ sq ft}\\ \text{Area 3} &= 75.0 \text{ sq ft}\\ +\ \text{Area 4} &= 9.0 \text{ sq ft}\\ \hline \text{Total area} &= 186.5 \text{ sq ft}\end{aligned}$$

Since 9 square feet = 1 square yard, multiply as shown to convert square feet to square yards.

$$\left(\frac{186.5 \text{ sq ft}}{1}\right)\left(\frac{1 \text{ sq yd}}{9 \text{ sq ft}}\right) = 20.7 \text{ sq ft}$$

The amount of carpet needed is 20.7 square yards. Round up 20.7 square yards to the next full yard. Thus 20.7 square yards becomes 21 square yards needed. To find the cost of the carpeting, multiply the number of square yards (21) by the cost per square yard ($22).

$$\text{Cost} = \left(\frac{21 \text{ sq yd}}{1}\right)\left(\frac{\$22}{\text{sq yd}}\right) = \$462 \text{ for 21 sq yd}$$

Example 15–17: Applied Brick Cost Example

PROBLEM: Find the cost of bricks to build a wall 38 feet long and 4 feet high. Each brick with mortar covers an area of $8\frac{1}{2}$ inches by $2\frac{3}{4}$ inches and its cost is $0.98.

SOLUTION: Draw a sketch of the wall and the brick to gain a better understanding of what you are to do. See Figure 15–42.

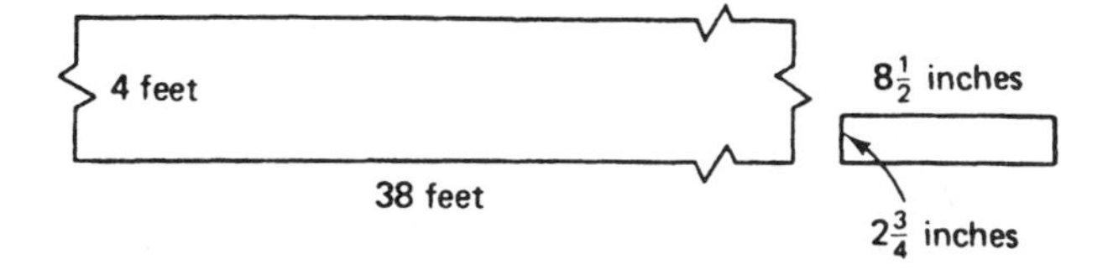

Figure 15–42

Convert all measurements to inches to eliminate possible confusion.

$$\left(\frac{38 \text{ ft}}{1}\right)\left(\frac{12 \text{ in.}}{1 \text{ ft}}\right) = 456 \text{ in.}$$

$$\left(\frac{4 \text{ ft}}{1}\right)\left(\frac{12 \text{ in.}}{1 \text{ ft}}\right) = 48 \text{ in.}$$

Find the area of the wall using the formula

$$\begin{aligned}A &= lw\\ &= (456 \text{ in.})\,(48 \text{ in.})\\ &= 21{,}888 \text{ sq in.}\end{aligned}$$

Then convert the mixed fractions to decimal fractions to simplify the process.

$$8\frac{1}{2} \text{ in.} = 8.5 \text{ in.}$$

$$2\frac{3}{4} \text{ in.} = 2.75 \text{ in.}$$

Find the area of one brick using the formula

$$\begin{aligned}A &= lw\\ &= (8.5 \text{ in.})\,(2.75 \text{ in.})\\ &= 23.375 \text{ sq in.}\end{aligned}$$

To find the number of bricks needed, divide the area of the wall (21,888 square inches) by the area of one brick (23.375 square inches).

$$\left(\frac{21{,}888 \text{ sq in.}}{1}\right)\left(\frac{1 \text{ brick}}{23.375 \text{ sq in.}}\right) = 936.38 \text{ bricks}$$

Round the decimal fraction to the next complete whole number. Thus 936.36 bricks = 937 bricks. To find the total cost of the bricks, multiply the total number of bricks (937) by the cost per brick ($0.98).

$$\text{Cost} = \left(\frac{937 \text{ bricks}}{1}\right)\left(\frac{\$0.98}{1 \text{ brick}}\right) = \$918.26$$

This problem could be done more efficiently by using a calculator as follows:

$$\text{cost of bricks} = \left(\frac{\begin{array}{c}\text{total area of wall}\\ \text{in square inches}\end{array}}{\begin{array}{c}\text{area of one brick}\\ \text{in square inches}\end{array}}\right)(\text{cost per brick})$$

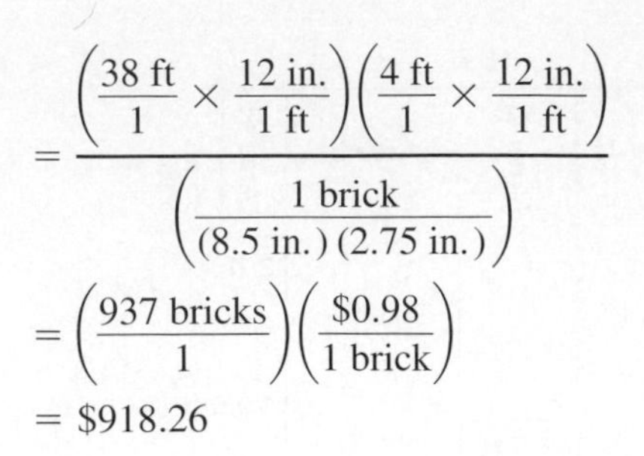

$$= \frac{\left(\frac{38\text{ ft}}{1} \times \frac{12\text{ in.}}{1\text{ ft}}\right)\left(\frac{4\text{ ft}}{1} \times \frac{12\text{ in.}}{1\text{ ft}}\right)}{\left(\frac{1\text{ brick}}{(8.5\text{ in.})(2.75\text{ in.})}\right)}$$

$$= \left(\frac{937\text{ bricks}}{1}\right)\left(\frac{\$0.98}{1\text{ brick}}\right)$$

$$= \$918.26$$

EXERCISE 15–3 PERIMETERS AND AREAS OF COMPLEX FIGURES

Solve the following problems concerning perimeters, circumferences, and areas to further develop your skills.

15–41. What kind of geometric figure is Figure 15–43?

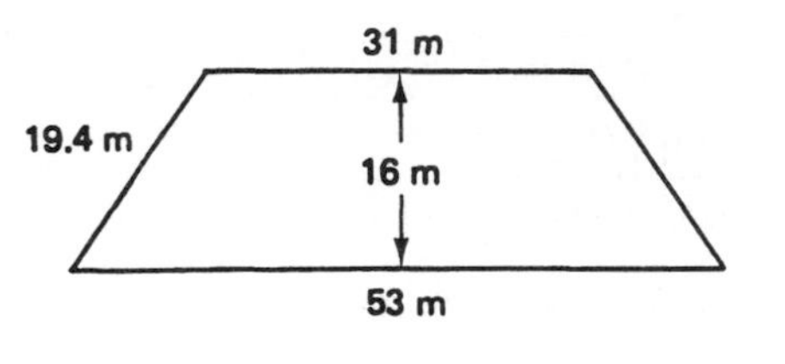

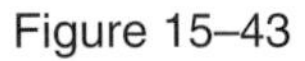

Figure 15–43

15–42. Find the perimeter of Figure 15–43.

15–43. Find the area of Figure 15–43.

15–44. Find the perimeter of Figure 15–44.

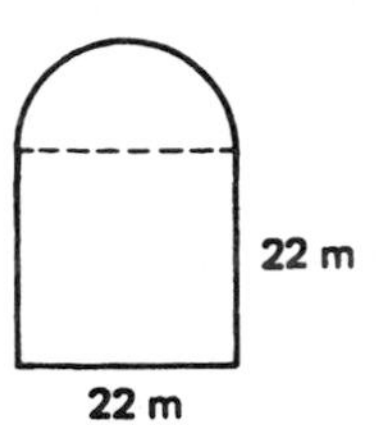

Figure 15–44

15–45. Find the area of Figure 15–44.

15–46. What kind of geometric figure is Figure 15–45?

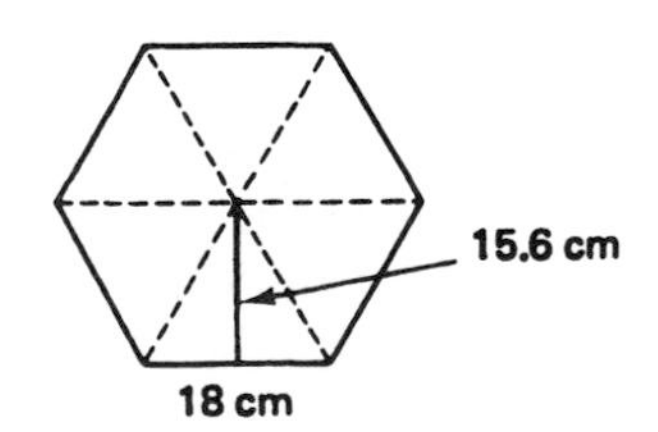

Figure 15–45

15–47. Find the perimeter of one of the triangles in Figure 15–45.

15–48. Find the perimeter of Figure 15–45.

15–49. Find the area of one of the triangles in Figure 15–45.

15–50. Calculate the area of Figure 15–45.

15–51. Find the area of Figure 15–46.

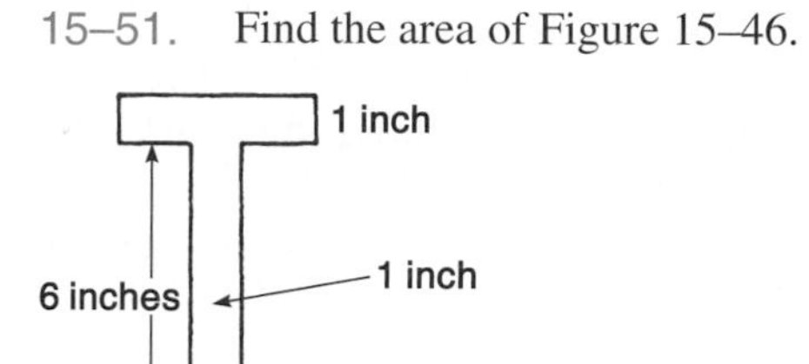

Figure 15–46

15–52. Find the perimeter of Figure 15–46.

15–53. Find the perimeter of the triangle in Figure 15–47.

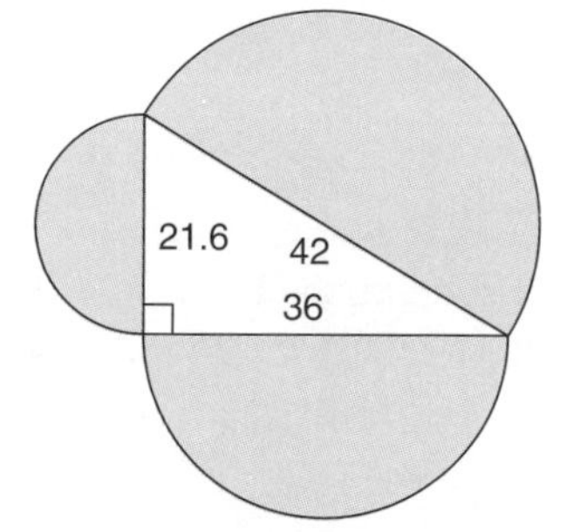

Figure 15–47

15–54. Find the perimeter of Figure 15–47.

15–55. Find the area of the triangle in Figure 15–47.

15–56. Find the area of the shaded portion of Figure 15–47.

15–57. Calculate the area of Figure 15–47.

15–58. Find the perimeter of Figure 15–48.

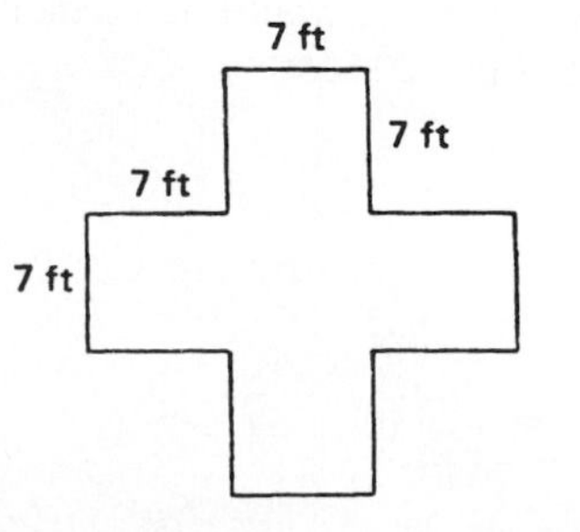

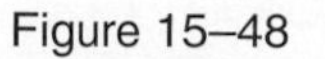

Figure 15–48

15–59. Find the area of Figure 15–48.

15–60. What would be the cost of painting Figure 15–48 be if the cost of paint and labor is \$0.12 per square foot?

15–61. Calculate the area of Figure 15–49.

15–62. If Figure 15–49 represents a wheat field, determine the yield per acre if 157.7 bushels were harvested from the given field (Round to the nearest tenth). (160 square rods equals 1 acre.)

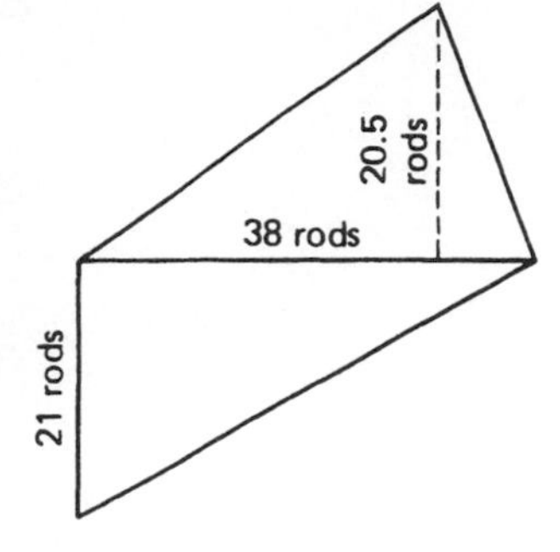

Figure 15–49

15–63. What is the cost of sodding the playing area of the football field represented by Figure 15–50 at the cost of \$4.25 per square yard?

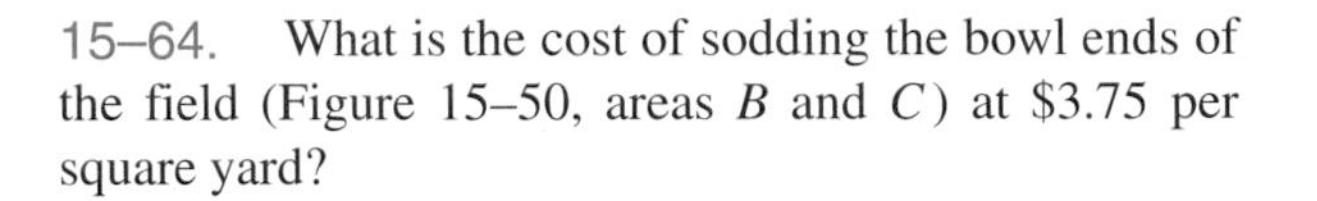

15–64. What is the cost of sodding the bowl ends of the field (Figure 15–50, areas *B* and *C*) at \$3.75 per square yard?

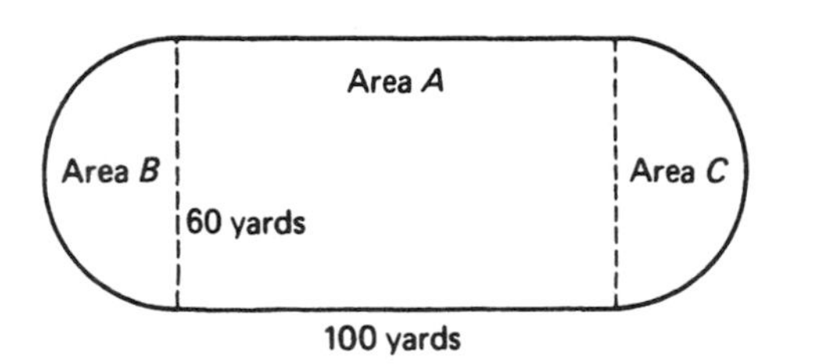

Figure 15–50

15–65. How many feet of fencing would be needed to fence in the entire area of Figure 15–50?

15–66. Find the cost of carpeting a 32-foot circular stage if the installed cost per yard is \$34.95 per square yard.

15–67. How many feet of baseboard molding would be needed for the floor plan represented by Figure 15–51?

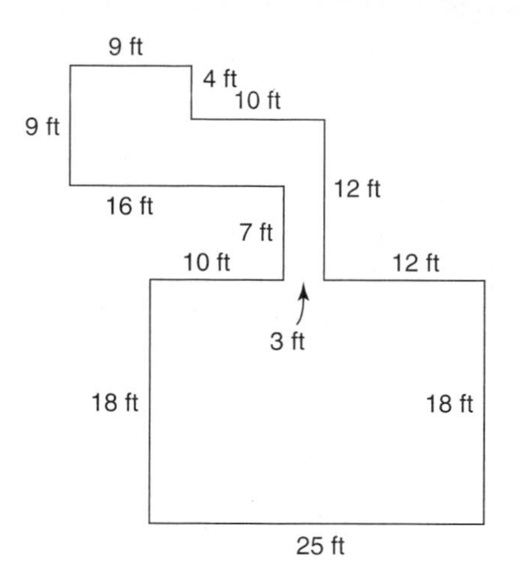

Figure 15–51

15–68. What is the total area in square feet represented by Figure 15–51?

15–69. What is the cost of carpeting the rooms represented by Figure 15–51 at \$28 per square yard? (1 square yard = 9 square feet.)

15–70. What is the wall area of a room represented by Figure 15–52?

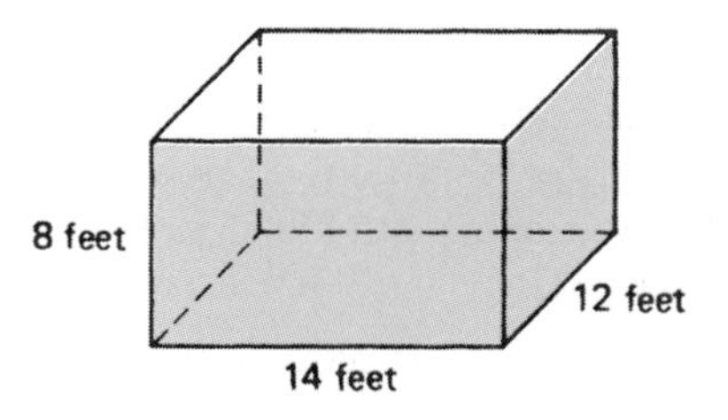

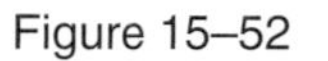

Figure 15–52

15–71. What is the area of the ceiling of the room shown in Figure 15–52?

15–72. Find the cost of installing hardwood parquet flooring in 6-inch squares at \$0.89 per 6-inch square, shown in Figure 15–52.

15–73. How many gallons of paint are needed to paint the room shown in Figure 15–52 (walls and ceiling) if 1 gallon covers 550 square feet?

15–74. Using the information from problem 15–73, find the cost of painting the room (Figure 15–52) if the paint costs \$24.95 per gallon and \$8.95 per quart.

15–75. How many 2-inch by 2-inch tiles would be needed to cover the floor of a bathroom that is $7\frac{1}{2}$ feet by $8\frac{1}{2}$ feet?

15–76. A landscape contractor wishes to sod an area illustrated by Figure 15–53. How much sod, in sqaure yards will be needed?

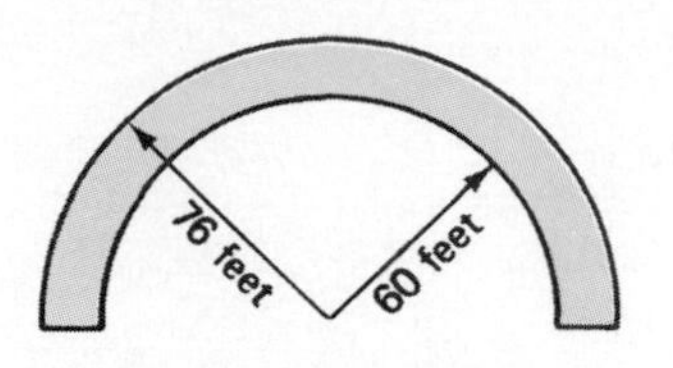

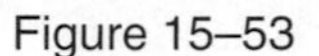
Figure 15–53

15–77. Find the cost of fencing Figure 15–53 if the cost is $1.29 per foot.

15–78. A carpenter has accepted the job of paneling a wall 12 feet by 72 feet. How many panels will he need if each panel is 4 feet by 8 feet?

15–79. A nurseryman finds that he needs 36 square feet of ground space to grow a spruce tree. How many spruce trees could he grow on a plot of land 200 yards by 300 yards?

15–80. What is the swimming area of a 35-foot-diameter pool?

THINK TIME

Formulas for finding the perimeter, circumference, or area of a plane geometric figure are difficult to remember. To refresh your memory from time to time, you may want to construct a chart for reference. Complete the chart below. When you finish, cover portions of the chart with a blank piece of paper to check your recall.

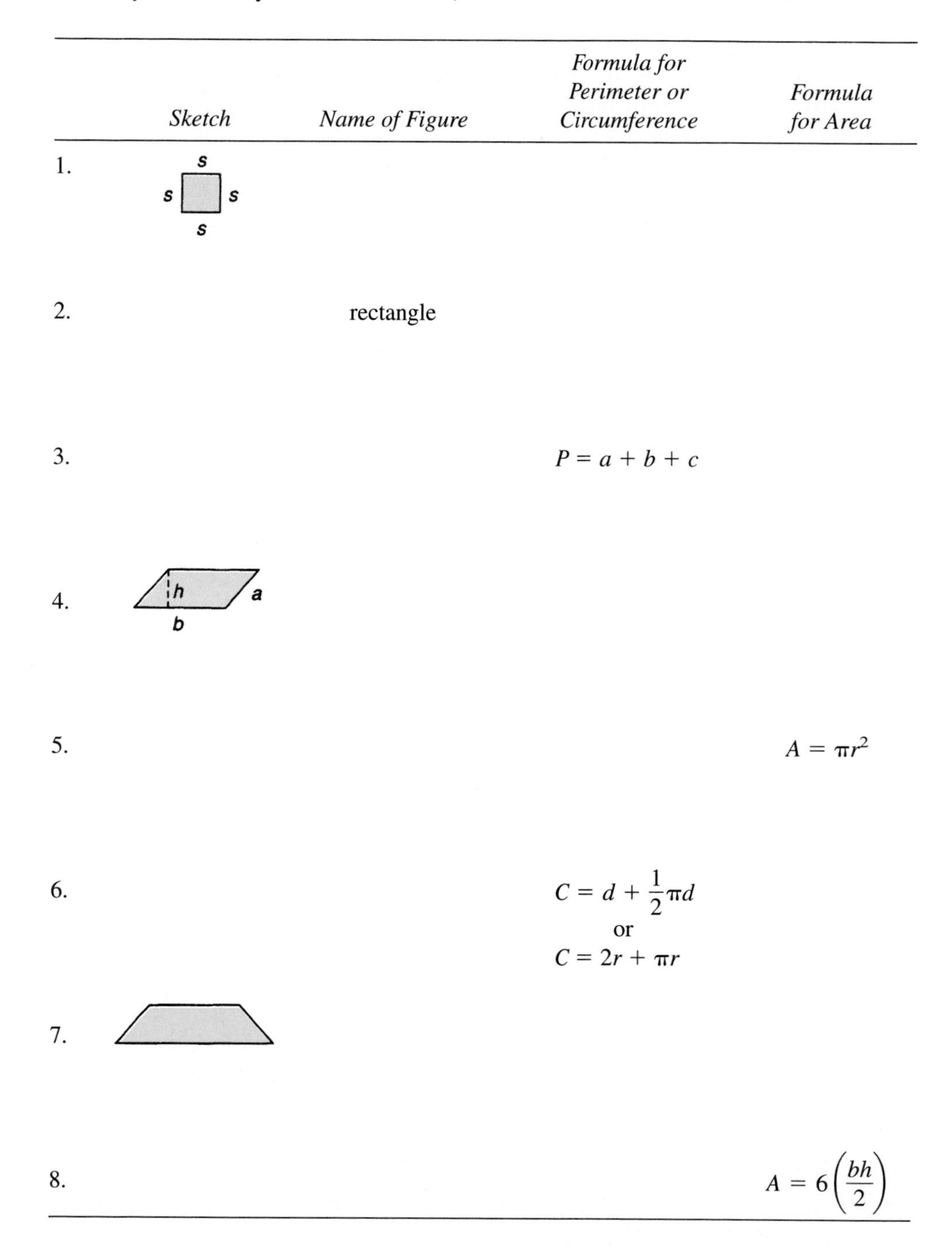

	Sketch	*Name of Figure*	*Formula for Perimeter or Circumference*	*Formula for Area*
1.	s, s, s, s			
2.		rectangle		
3.			$P = a + b + c$	
4.	h, a, b			
5.				$A = \pi r^2$
6.			$C = d + \frac{1}{2}\pi d$ or $C = 2r + \pi r$	
7.				
8.				$A = 6\left(\frac{bh}{2}\right)$

PROCEDURES TO REMEMBER

1. When determining the perimeter or the area, use only like units of measurement.
2. Convert mixed fractions to decimals, where possible, when finding the area of a plane geometric figure.
3. Divide or change complex plane geometric figures into simple figures, where possible, and apply basic formulas for finding perimeter, circumference, or area.

CHAPTER SUMMARY

1. The following are definitions concerning plane geometric figures:
   (a) *Perimeter* is the distance around a plane geometric figure and is stated in linear measure such as inches, miles, centimeters, and meters.
   (b) *Circumference* is the distance around a circle.
   (c) *Area* is the amount of surface of a plane geometric figure and is measured in square inches, square yards, square meters, and so on.
   (d) A *circle* is a set of points that are the same distance from a fixed point called the *center*.
   (e) The *radius* of a circle is the distance from the center of the circle to the circumference.
   (f) The *diameter* of a circle is a line drawn from the circumference of a circle, through the center, to the opposite circumference.
   (g) A *semicircle* is half of a circle.
   (h) *Pi* is a special number, equal to approximately 3.14, which is the ratio for the circumference of a circle to its diameter. The symbol for pi is the Greek letter π. For calculation purposes, $\pi = 3.14$ or 3.142 is used, depending on the accuracy desired. The calculator value of π is 3.1415927.

2. The following formulas are essential to solving geometric figures.

Figure	*Perimeter (Circumference)*	*Area*
Square	$P = 4s$	$A = s^2$
Rectangle	$P = 2l + 2w$	$A = lw$
	$P = 2(l + w)$	
Triangle	$P = a + b + c$	$A = \frac{1}{2}bh$
Parallelogram	$P = 2a + 2b$	$A = bh$
	$P = 2(a + b)$	
Circle	$C = d\pi$	$A = \pi r^2$
	$C = 2\pi r$	$A = \frac{\pi d^2}{4}$
Semicircle	$C = d + \frac{1}{2}\pi d$	$A = \frac{\pi r^2}{2}$
	$C = 2r + \pi r$	
Trapezoid	$P = 2a + B + b$	$A = \left(\frac{B + b}{2}\right)h$
Hexagon	$P = 6s$	$A = 6\left(\frac{bh}{2}\right)$

CHAPTER TEST

T–15–1. Figure 15–54 is what type of geometric figure?

814 km

814 km

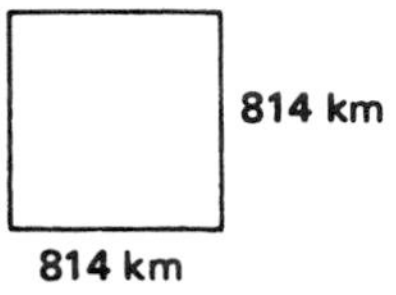

Figure 15–54

T–15–2. Determine the area of Figure 15–54.

T–15–3. Find the perimeter of Figure 15–54.

T–15–4. What type of geometric figure is Figure 15–55?

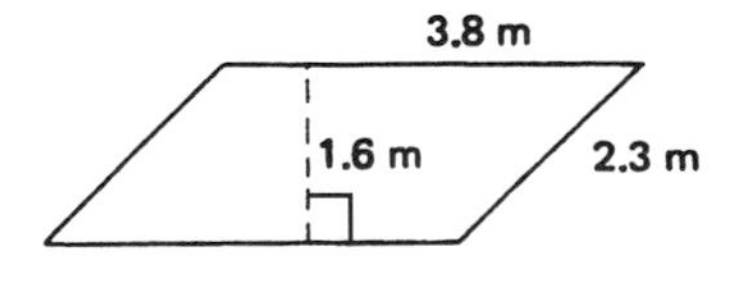

Figure 15–55

T–15–5. What is the altitude of Figure 15–55?

T–15–6. What is the perimeter of Figure 15–55?

T–15–7. Find the area of Figure 15–55.

T–15–8. What is the altitude of Figure 15–56?

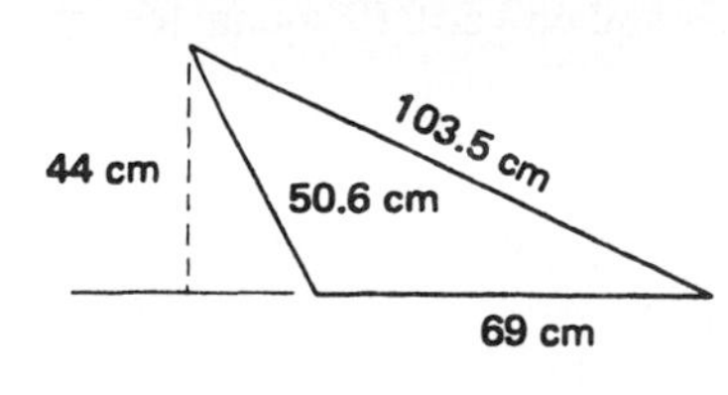

Figure 15–56

T–15–9. Find the perimeter of Figure 15–56.

T–15–10. What is the area of Figure 15–56?

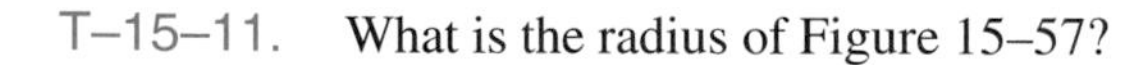

T–15–11. What is the radius of Figure 15–57?

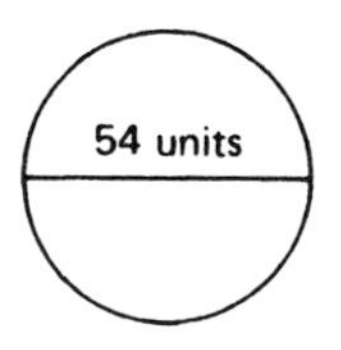

Figure 15–57

T–15–12. Calculate the area of Figure 15–57.

T–15–13. What is the circumference of Figure 15–57?

T–15–14. What kind of geometric figure is Figure 15–58?

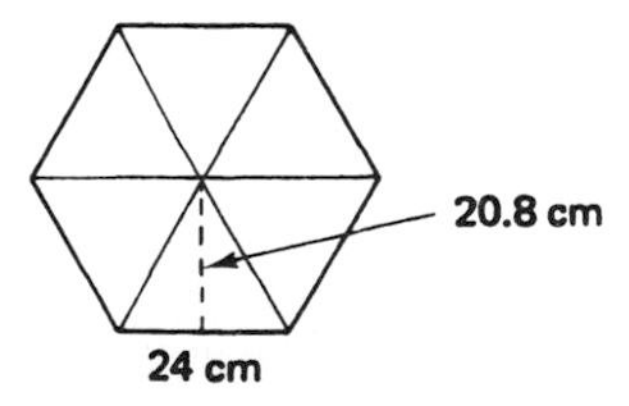

Figure 15–58

T–15–15. Find the perimeter of Figure 15–58.

T–15–16. Calculate the area of Figure 15–58.

T–15–17. Find the cost of carpeting the area around the swimming pool illustrated by Figure 15–59. The cost of carpeting is $34.95 per square yard. (9 square feet = 1 square yard.)

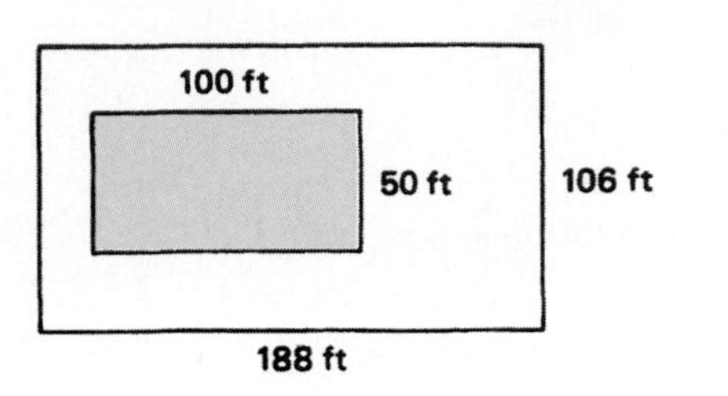

Figure 15–59

T–15–18. Two heating ducts are joined together as illustrated by Figure 15–60. Find what the diameter of the single opening (c) would have to be if the area of the openings is to be equal to the area of the openings $a + b$.

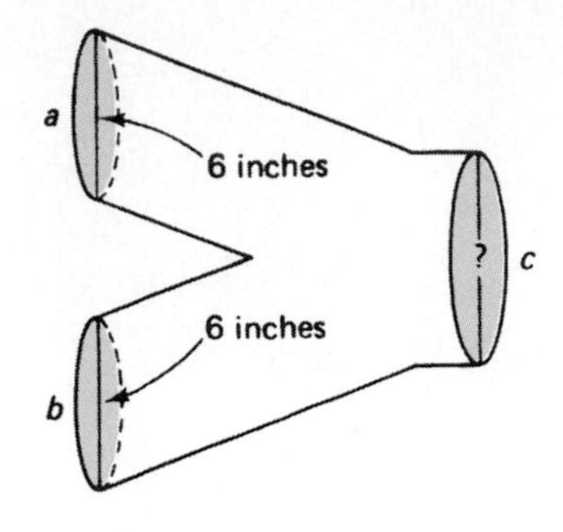

Figure 15–60

T–15–19. How many sections of acoustical ceiling tile will be needed to cover the ceiling illustrated by Figure 15–61? The tile sections are 2 feet by 4 feet.

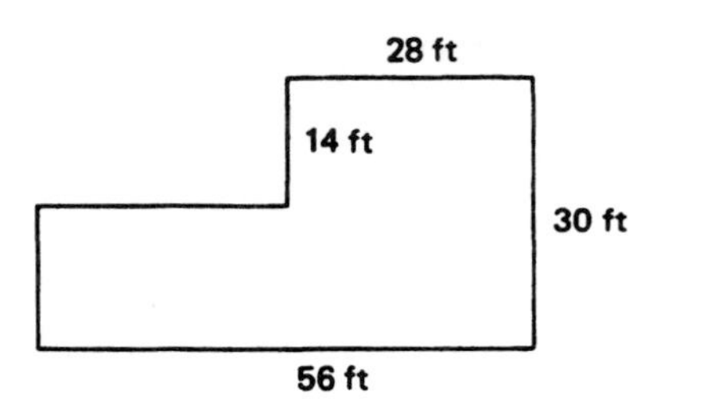

Figure 15–61

T–15–20. How many gallons of paint are needed to paint the walls and ceiling of the storage room illustrated by Figure 15–62? Each gallon of paint covers 520 square feet.

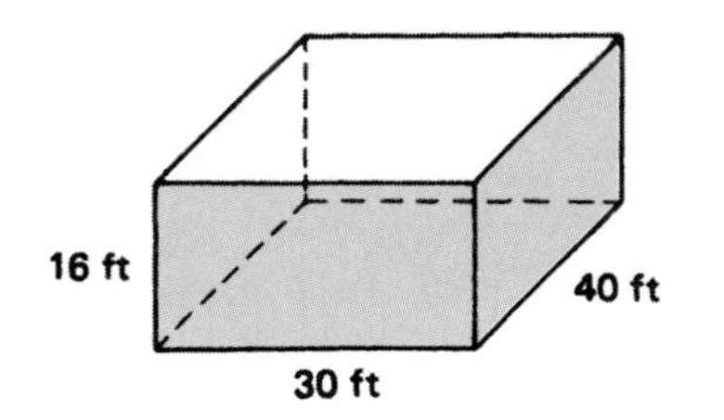

Figure 15–62

16

SURFACE AREAS AND VOLUMES OF GEOMETRIC FIGURES

OBJECTIVES

After studying this chapter, you will be able to:

16.1. Understand definitions and properties of surface areas of geometric solids.
16.2. Understand definitions and properties of surface areas of circular solids.
16.3. Understand definitions and properties of surface areas of pyramids and spheres.
16.4. Understand definitions and properties of volumes of geometric solids.
16.5. Use geometric solid applications.

SELF-TEST

This test will determine your ability to find lateral surface areas, total surface areas, and volumes of solid geometric figures. Successful completion of this self-test will indicate your ability to find areas and volumes of solid figures. Problems that you are unable to solve will indicate areas that will need further study.

S–16–1. Find the volume of the prism shown in Figure 16–1.

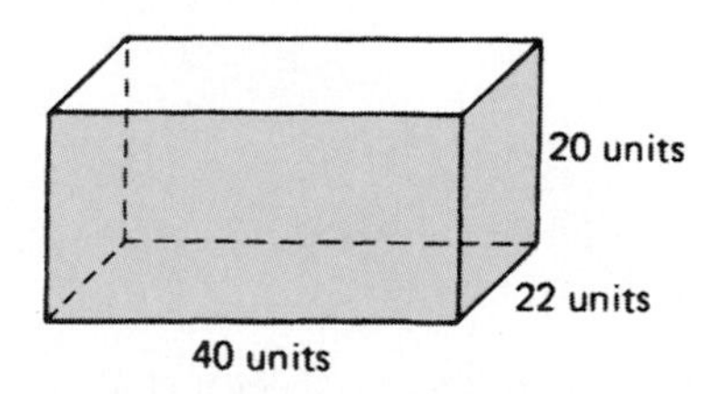

Figure 16–1

S–16–2. Find the total surface area of a 15-inch cube.

S–16–3. Find the lateral surface area of a rectangular solid 3 cm wide, 4 cm long, and 30 cm high.

S–16–4. Find the lateral surface area of the cylinder shown in Figure 16–2.

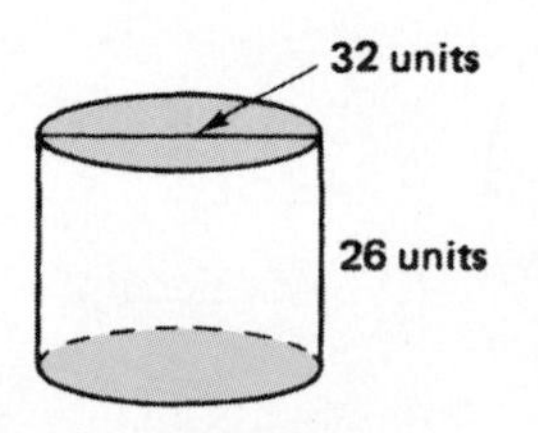

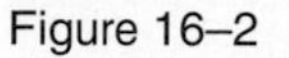

Figure 16–2

S–16–5. Find the volume of the cone shown in Figure 16–3.

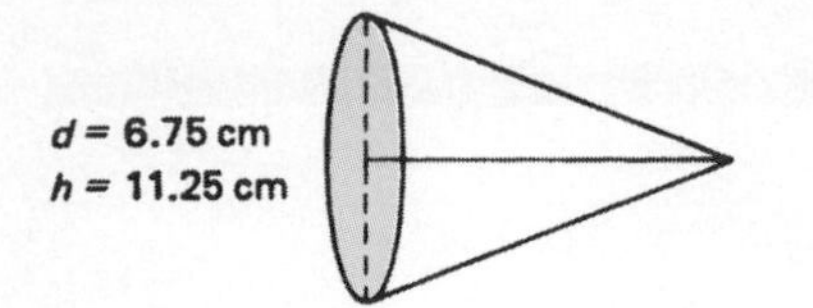

Figure 16–3

S–16–6. Find the volume of the cylinder shown in Figure 16–4.

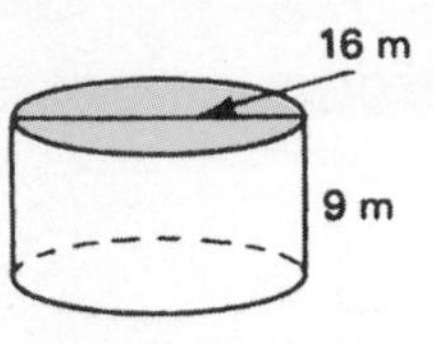

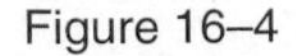
Figure 16–4

S–16–7. Find the total surface area of the cylinder shown in Figure 16–5.

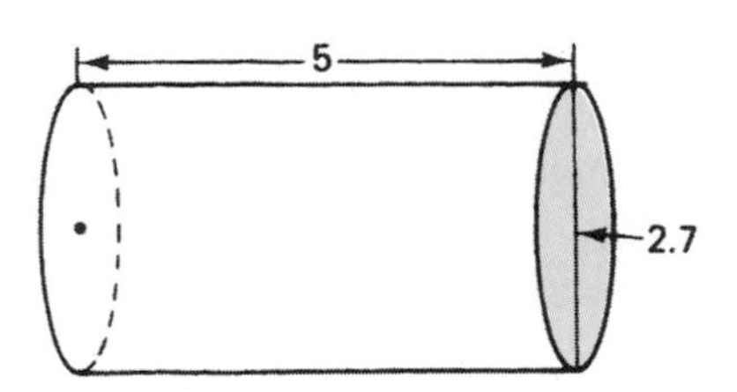

Figure 16–5

S–16–8. Find the lateral surface area of the cylinders shown in Figure 16–6.

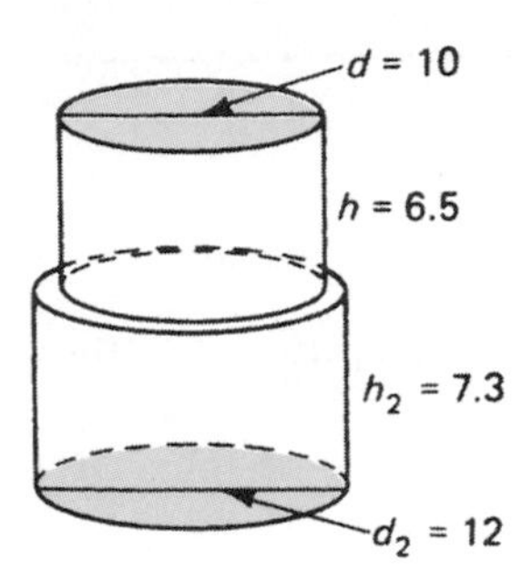

Figure 16–6

S–16–9. Find the volume of a 24-inch-diameter sphere.

S–16–10. Determine the volume of the pyramid shown in Figure 16–7.

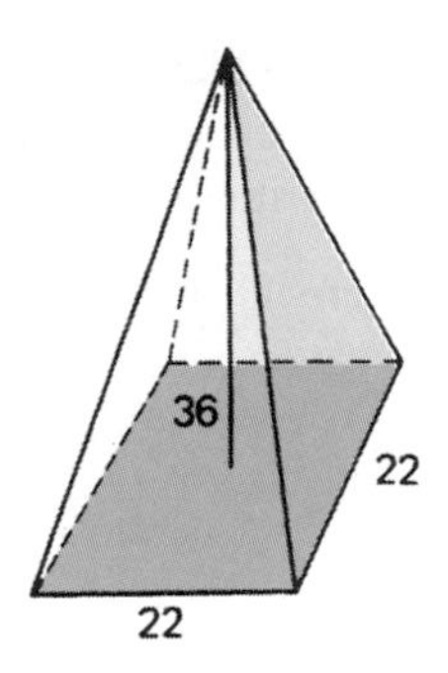

Figure 16–7

S–16–11. How many gallons will be held by the tank shown in Figure 16–8? (1 cubic foot = 7.5 gallons.)

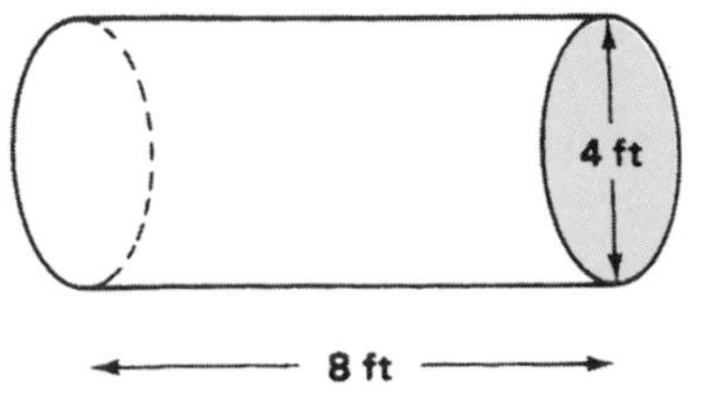

Figure 16–8

S–16–12. How many square feet of sheet metal are needed to construct the duct (does not include end pieces) illustrated in Figure 16–9? (Add 15% for seams.)

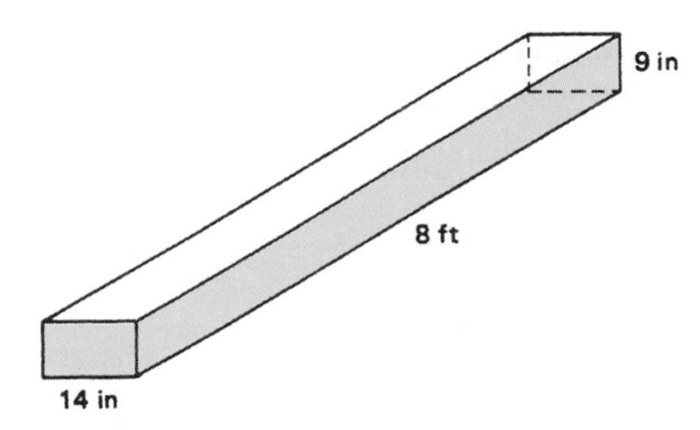

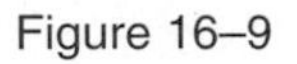
Figure 16–9

S–16–13. Find the capacity of the cone portion of the funnel shown in Figure 16–10.

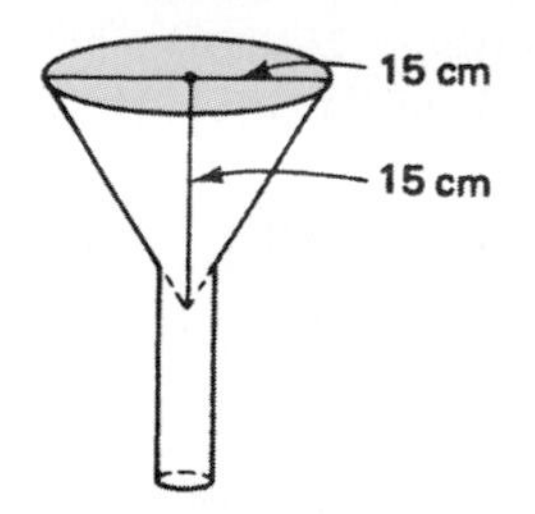

Figure 16–10

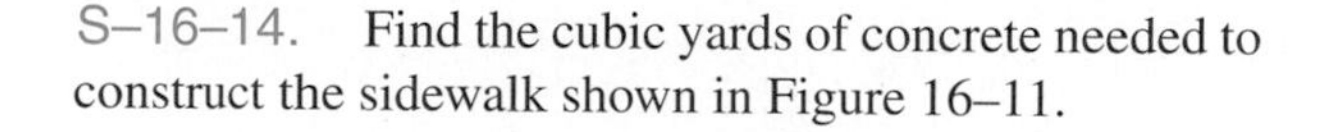
S–16–14. Find the cubic yards of concrete needed to construct the sidewalk shown in Figure 16–11.

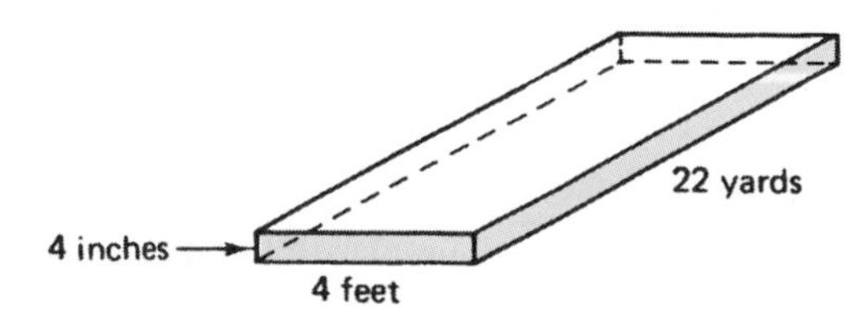

Figure 16–11

S–16–15. Determine the cost of painting a 50-foot-diameter spherical water tower at $1.89 per square yard.

16.1 DEFINITIONS AND PROPERTIES OF SURFACE AREAS OF GEOMETRIC SOLIDS

Every object, whether it is as large as the earth or as small as an atom, has surface area and volume. The surface area and volume of geometric solids are determined by using formulas involving length, width, and height. Automobile manufacturers and buyers are concerned about "the cubic inches" or volume of compression of an engine. A smart shopper notes the content or volume of a container as well as its cost. An air-conditioning salesperson determines the volume of space to be air-conditioned. Many occupations require the skills to find the surface area and volume of geometric solids.

The formulas for finding the surface area and volume of geometric solids are related to formulas used to find perimeter and area of plane figures. Skill with plane figure formulas will help in finding the areas and volumes of solid figures.

Surface Area: Prisms

A *prism* is a solid geometric figure with parallel edges and uniform cross sections. Figure 16–12 illustrates various prisms.

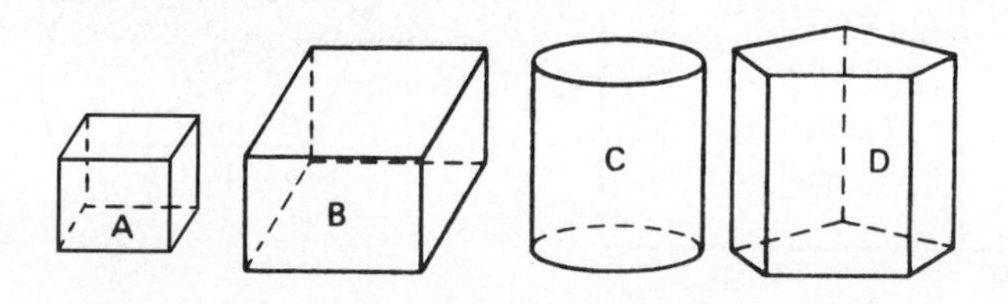

Figure 16–12

A *face* is any plane surface of a solid figure. A *lateral face* is a side of a geometric figure. The *bases* of a prism are the top face and bottom face, generally called the top and bottom. The *lateral edge* of a prism is where two sides of a solid intersect. Study Figure 16–13.

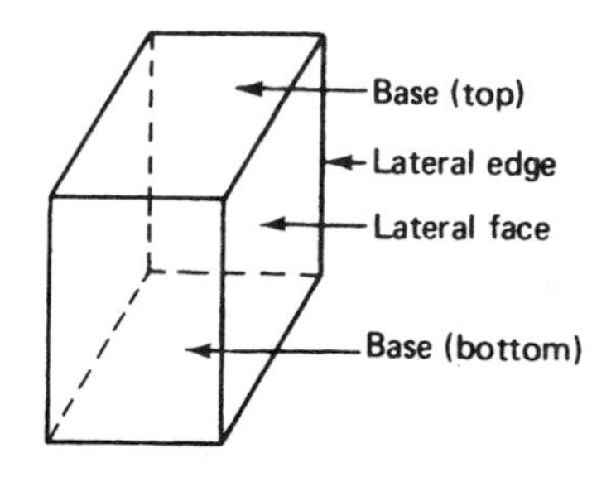

Figure 16–13

All prisms are classified by the shape of their base. A *right prism* has lateral faces that are perpendicular to the base. A prism whose sides are not perpendicular to its base is called an *oblique prism*.

Figure 16–12A is a *right square prism* because it has a square for a base. Figure 16–12B and 16–13 are *right rectangular prisms* because they have rectangular bases. Figure 16–12C represents a special type of prism called a *cylinder*. Figure 16–12D is a *right pentagonal prism* because of its pentagon-shaped base.

Lateral Surface Areas

The *lateral surface area* of a solid geometric figure is the area or surface of the sides of the figure, excluding the area of the top and bottom. The shaded portion of Figure 16–14 indicates the lateral surface.

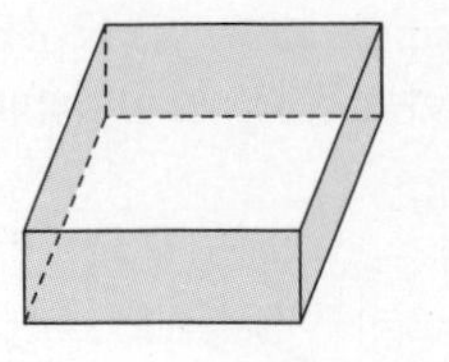

Figure 16–14

The *total surface area* of a geometric solid includes the area of the top and bottom of the figure as well as the area of the sides. To find the lateral surface of a prism multiply the perimeter of the prism by its height.

Example 16–1: Find the Lateral Surface of a Solid

PROBLEM: Find the lateral surface (area) of Figure 16–15.

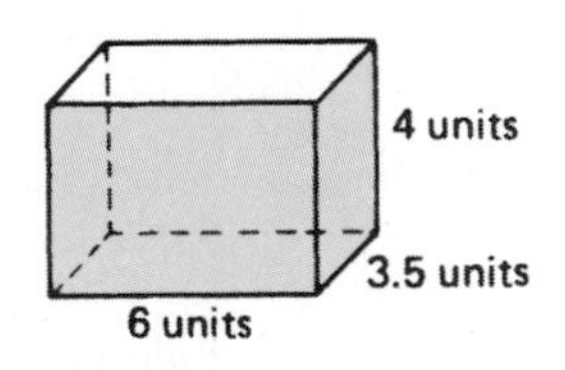

Figure 16–15

SOLUTION: First, determine the type of figure shown in Figure 16–15. Since the base of the figure is a rectangle and its faces are perpendicular to its base, Figure 16–15 is a right rectangular prism.

Then find the lateral area by finding the area of each of the four sides and adding these areas together. See Figure 16–16.

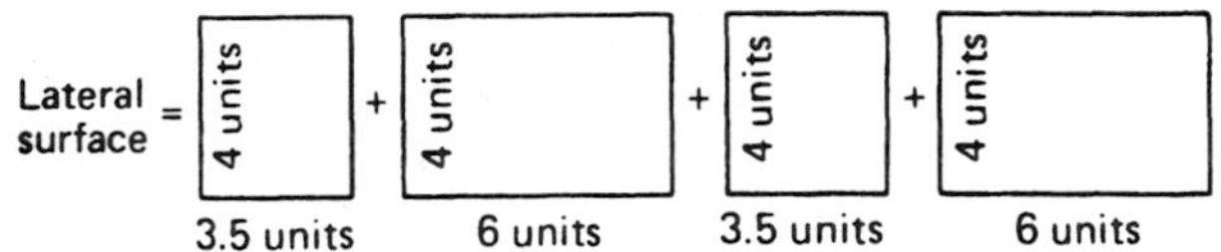

Figure 16–16

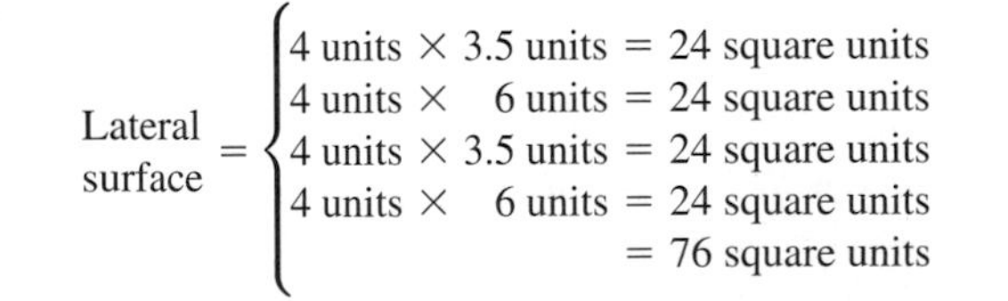

$$\text{Lateral surface} = \begin{cases} 4 \text{ units} \times 3.5 \text{ units} = 24 \text{ square units} \\ 4 \text{ units} \times 6 \text{ units} = 24 \text{ square units} \\ 4 \text{ units} \times 3.5 \text{ units} = 24 \text{ square units} \\ 4 \text{ units} \times 6 \text{ units} = 24 \text{ square units} \\ \qquad\qquad = 76 \text{ square units} \end{cases}$$

An easier way to find the lateral surface is to use the formula

$$\text{lateral surface (area)} = \text{perimeter } (2l + 2w) \times \text{height } (h)$$
$$A = p \times h$$
$$= [2(6) + 2(3.5)]\,4$$
$$= 76 \text{ square units}$$

Example 16–2: Find the Total Surface of a Prism

PROBLEM: Find the total surface area of Figure 16–17.

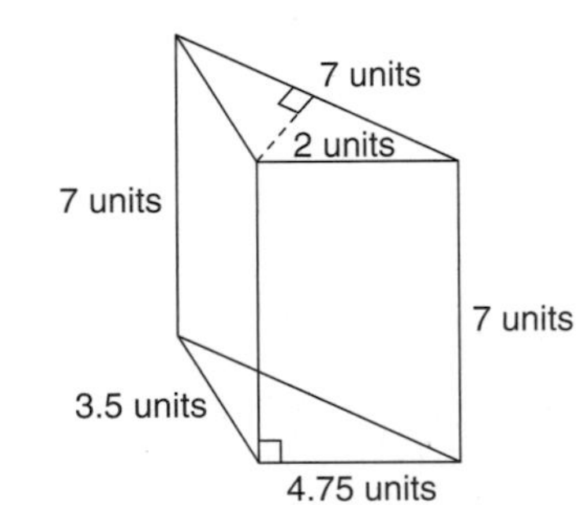

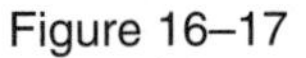

Figure 16–17

SOLUTION: Determine the type of figure shown in Figure 16–17. Since the base of the figure is a triangle and its faces are perpendicular to its base, the figure is a triangular prism.

Then find the surface areas. The total surface area may be found by finding the areas of each of the faces, the top, and the bottom and then adding these areas together. See Figure 16–18.

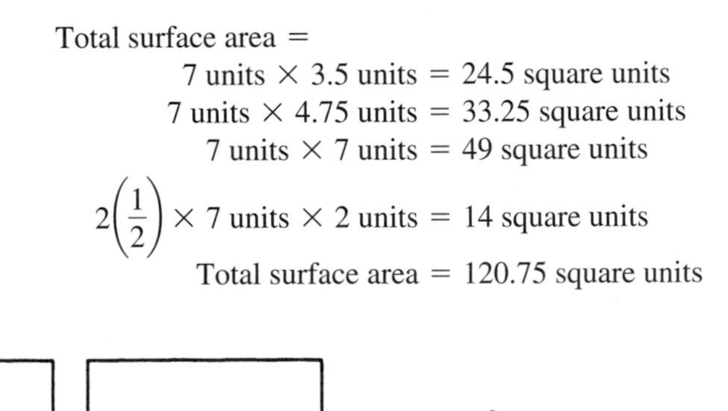

Total surface area =

$$7 \text{ units} \times 3.5 \text{ units} = 24.5 \text{ square units}$$
$$7 \text{ units} \times 4.75 \text{ units} = 33.25 \text{ square units}$$
$$7 \text{ units} \times 7 \text{ units} = 49 \text{ square units}$$
$$2\left(\frac{1}{2}\right) \times 7 \text{ units} \times 2 \text{ units} = 14 \text{ square units}$$
$$\text{Total surface area} = 120.75 \text{ square units}$$

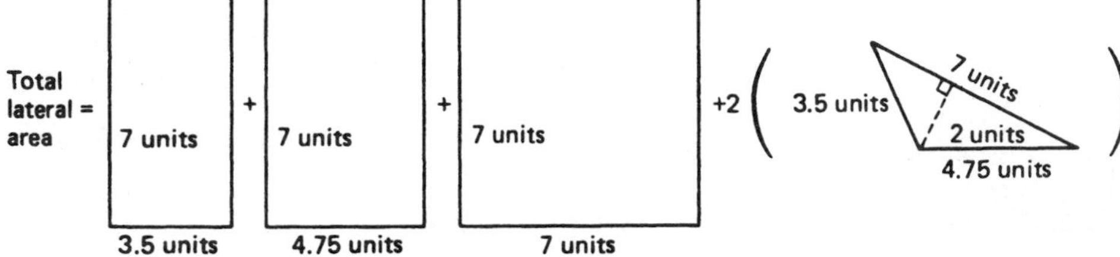

Figure 16–18

The formula for determining the total surface area of a right triangular prism is

$$\text{total surface area} = \text{perimeter } (a + b + c) \times \text{height} + \frac{2\,(\text{base} \times \text{height})}{2}$$

$$A = p \times h + bh$$

$$\begin{aligned} A &= (3.5 + 4.75 + 7)(7) \\ &\quad + 2\,[0.5(7)\,(2)] \\ &= (15.25)(7) + (2)(7) \\ &= 106.75 + 14 \\ &= 120.75 \text{ sq units} \end{aligned}$$

EXERCISE 16–1 SURFACE AREAS AND VOLUMES OF SOLIDS

Find the lateral or total surface areas for the following as indicated.

16–1. What type of prism is Figure 16–19?

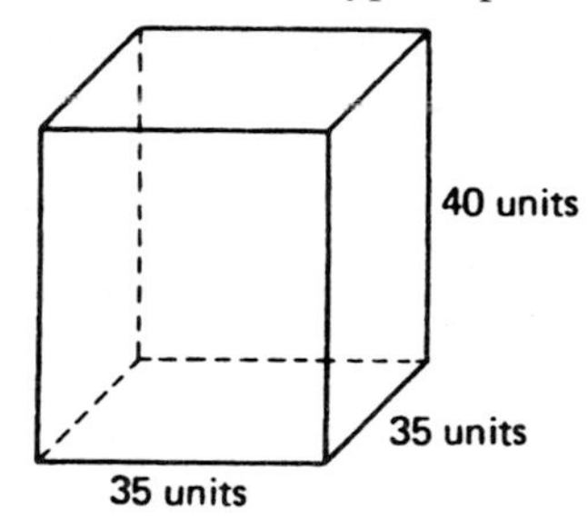

Figure 16–19

16–2. What is the lateral area of Figure 16–19?

16–3. Find the total surface area of Figure 16–19.

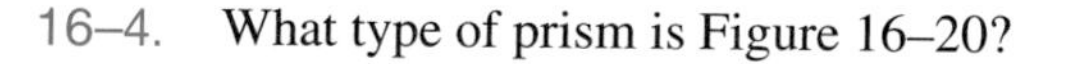

16–4. What type of prism is Figure 16–20?

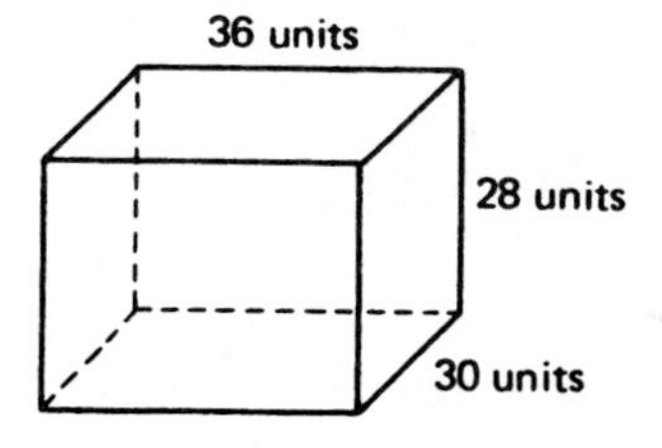

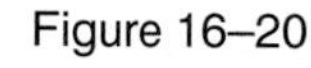

Figure 16–20

16–5. Find the lateral area of Figure 16–20.

16–6. Determine the total surface area of Figure 16–20.

16–7. What type of prism is Figure 16–21?

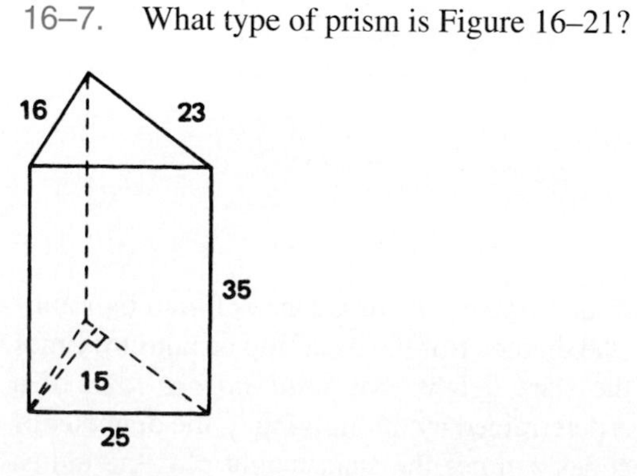

Figure 16–21

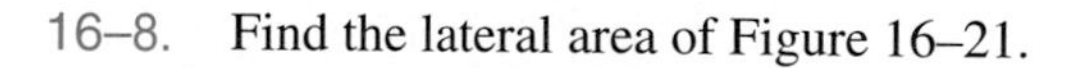

16–8. Find the lateral area of Figure 16–21.

16–9. Determine the total surface area of Figure 16–21.

16–10. What type of prism is Figure 16–22?

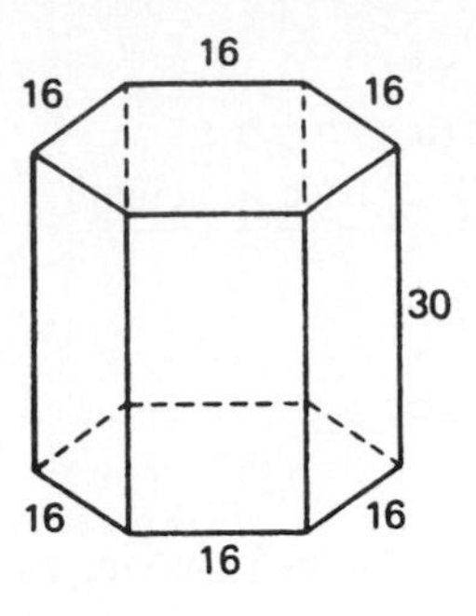

Figure 16–22

16–11. Find the lateral area of Figure 16–22.

16–12. Determine the total surface area of Figure 16–22.

16.2 DEFINITIONS AND PROPERTIES OF SURFACE AREAS OF CIRCULAR SOLIDS

Cylinders are used in various ways in our daily lives. Many products are packaged in cylinders. Automobile engines are described by the number of cylinders. A *cylinder* is a solid geometric figure with equal, parallel, circular bases.

The *lateral surface area* of a cylinder is the surface area of the side of the cylinder, excluding the area of the top and bottom. The shaded portion of Figure 16–23 indicates the lateral surface area. The *total surface area* of a cylinder includes the area of the top and bottom plus the area of the side. The area of the shaded portion plus the top and bottom areas equal the total lateral surface of the cylinder, as shown by Figure 16–23.

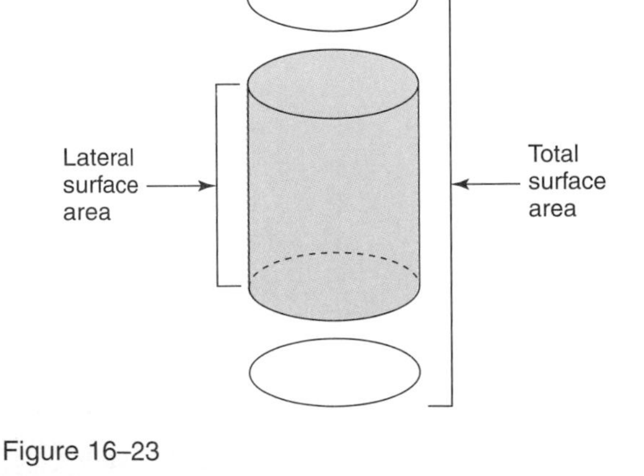

Figure 16–23

A *cone* is a solid geometric figure with one circular base whose side tapers evenly to a point. The *lateral surface area* of a cone is the surface area of the side of the cone, excluding the base area. The *total surface area* of a cone includes the area of the base and the area of the side. Figure 16–24 indicates the lateral surface. The lateral area plus the shaded portion of the base equals the total surface area of the cone.

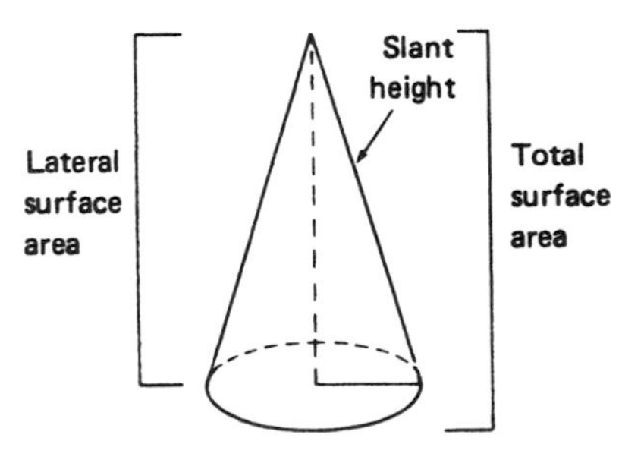

Figure 16–24

The *lateral surface area* of a cone is found by multiplying $\frac{1}{2}$ the diameter of the base (top or bottom) times π times the slant height. The *total surface area* of a cylinder is determined by multiplying $\frac{1}{2}$ the diameter of the base times π times the slant height plus the radius squared times π.

Example 16–3: Find the Total Surface Area of a Cylinder

PROBLEM: Determine the total surface area of the cylinder represented by Figure 16–25, with a diameter of 12 meters and a height of 20 meters.

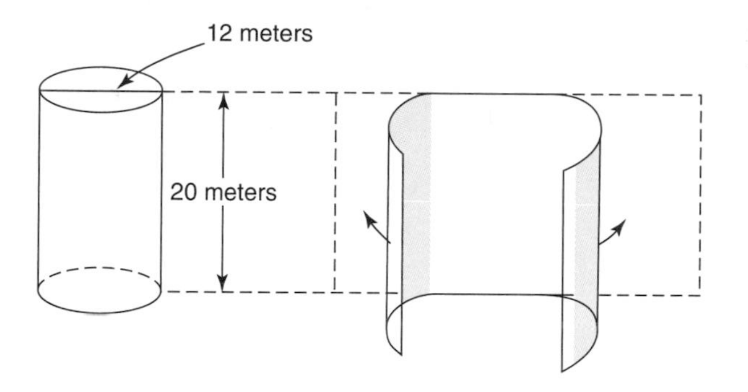

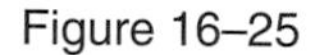
Figure 16–25

SOLUTION: Observe the cylinder; it could be "unrolled" to form a rectangle. If this were done, the lateral surface area of a cylinder could be found by multiplying the length times the width.

Then find the lateral surface area by multiplying the diameter (12 meters) × π (3.142) × the height (20 meters):

$$\begin{aligned}\text{lateral surface area} &= 12\text{ m} \times 3.142 \times 20\text{ m}\\ &= 754.08\text{ m}^2\end{aligned}$$

Find the area of the bases (top and bottom). The bases are two equal circles; thus use the formula for finding the area of a circle. Multiply the radius times the radius times π (3.142) times 2, the number of bases. (Recall that the radius is $\frac{1}{2}$ of the diameter.)

$$\begin{aligned}\text{Area of bases} &= \frac{1}{2}(12\text{ m}) \times \frac{1}{2}(12\text{ m}) \times 3.142 \times 2\\ &= 226.224\text{ m}^2\end{aligned}$$

Then add the lateral surface area and the surface area of the bases to find the total surface area.

$$\begin{aligned}\text{Total surface area} &= \text{lateral surface area}\\ &\quad + \text{area of bases}\\ &= 754.08\text{ m}^2 + 226.224\text{ m}^2\\ &= 980.304\text{ m}^2\end{aligned}$$

Note the formulas used to solve this problem:

$$\text{Lateral surface area of a cylinder} = d\pi h \text{ or } 2r\pi h$$

$$\begin{aligned}\text{Total surface area of a cylinder} &= d\pi h + 2\left(\frac{d}{2}\right)^2\pi\\ &= 2r\pi h + 2\pi r^2\end{aligned}$$

Example 16–4:

PROBLEM: Find the square feet of material needed to make the wind tunnel shown in Figure 16–26. The cone-shaped tunnel is 10 feet long and the radius is 2.5 feet.

SOLUTION: Observe the drawing of the cone in Figure 16–26. Note the height, h = 10 feet, the radius of the base r = 2.5 feet, and the slant height, s, is unknown.

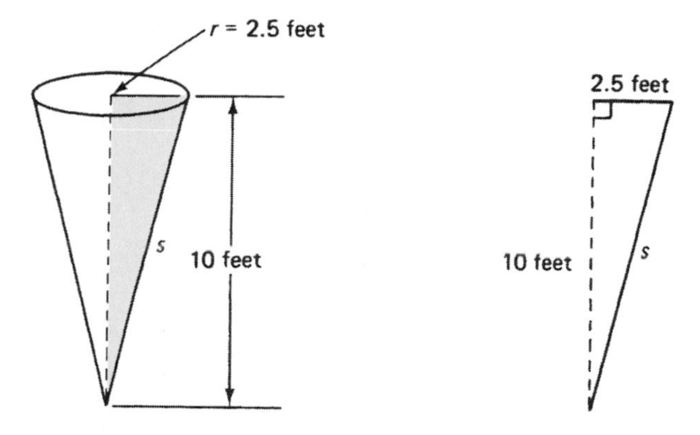

Figure 16–26

Find the slant height, s. Note in Figure 16–26 that a right triangle has been formed. Thus the formula for finding the hypotenuse of a right triangle can be used to find the slant height s.

$$\begin{aligned}\text{Slant height} &= \sqrt{(\text{radius})^2 + (\text{height})^2}\\ &= \sqrt{(2.5)^2 + (10)^2}\\ &= \sqrt{6.25 + 100}\\ &= \sqrt{106.25}\\ &= 10.3\text{ ft (to the nearest tenth)}\end{aligned}$$

The lateral area of a right circular cone equals the circumference of the base times one-half the slant height or the radius times the slant height times π.

$$\begin{aligned}\text{Lateral surface area} &= \text{r} \times \text{sl} \times \pi\\ &= 2.5\text{ ft} \times 10.3\text{ ft} \times 3.142\\ &= 80.9065\text{ sq ft}\end{aligned}$$

Find the total surface area of the cone.

$$\begin{aligned}\text{Total surface area} &= \text{lateral surface area}\\ &\quad + \pi r^2 \text{ (the top)}\\ &= 80.9065\text{ sq ft}\\ &\quad + 19.6375\text{ sq ft}\\ &= 100.544\text{ sq ft}\end{aligned}$$

Thus 101 square feet of material would be needed to make the wind tunnel cone. More material would be needed for seams and waste in the fabrication process.

The following formulas were used to solve this problem:

Lateral surface area of a cone = $r\pi \times$ slant height

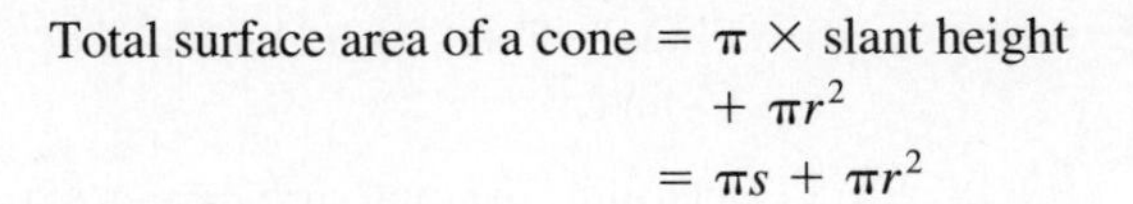

Total surface area of a cone = $\pi \times$ slant height $+ \pi r^2$
$= \pi s + \pi r^2$

EXERCISE 16–2 SURFACE AREAS AND PROPERTIES OF CIRCULAR SOLIDS

16–13. Figure 16–27 is what type of figure?

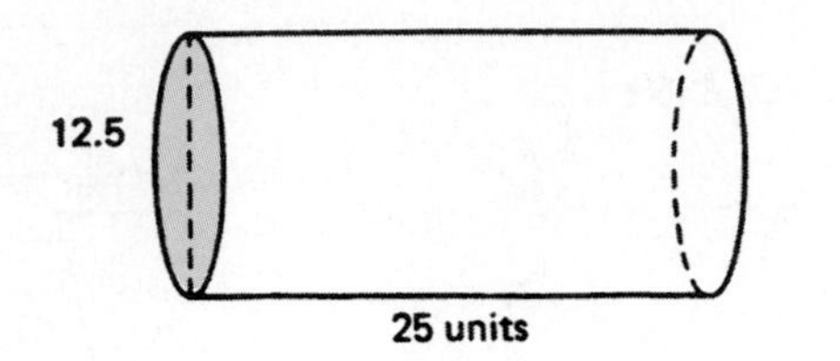

Figure 16–27

16–14. Find the lateral surface area of Figure 16–27.

16–15. Find the total surface area of Figure 16–27.

16–16. What type of solid is Figure 16–28?

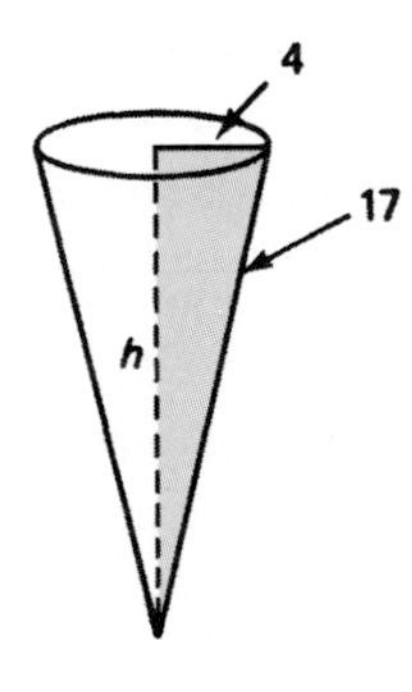

Figure 16–28

16–17. Find the diameter of Figure 16–28.

16–18. Find the height of Figure 16–28.

16–19. Find the lateral surface area of Figure 16–28.

16–20. Find the total surface of Figure 16–28.

16–21. What is the radius of Figure 16–29?

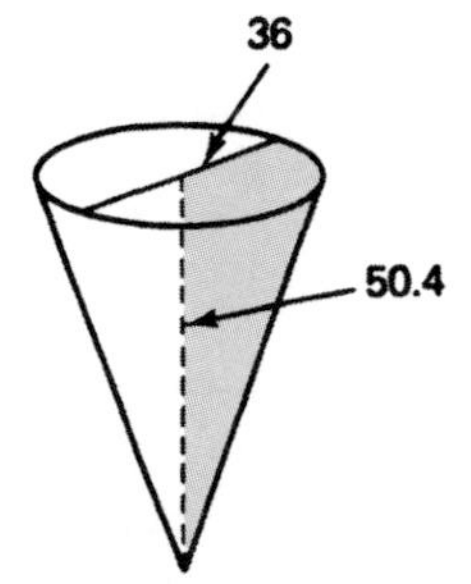

Figure 16–29

16–22. What is the slant height of Figure 16–29?

16–23. Determine the lateral surface area of Figure 16–29.

16–24. Find the total surface area of Figure 16–29.

16.3 DEFINITIONS AND PROPERTIES OF SURFACE AREAS OF PYRAMIDS AND SPHERES

Another type of solid figure, similar to the cone, is the pyramid. The names of pyramids are determined by the shape of their base. A *right pyramid* is a pyramid with a regular polygon base whose sides form equal isosceles triangles. Study the drawings of three different types of pyramids: square pyramid (Figure 16–30), rectangular pyramid (Figure 16–31), and hexagonal pyramid (Figure 16–32); the first two are right pyramids. Note the names of the various parts of the pyramids.

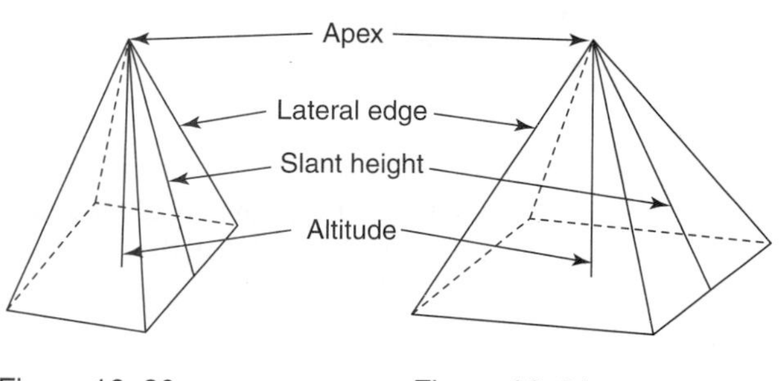

Figure 16–30

Figure 16–31

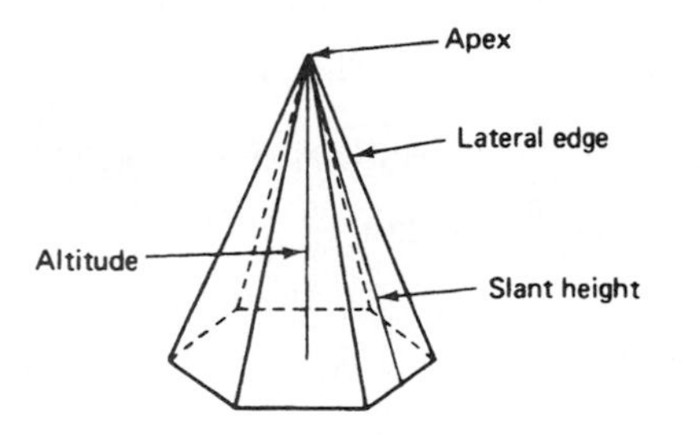

Figure 16–32

A *sphere* is a solid figure with all points on the surface the same distance from the center. A unique property of the sphere is that it has a greater volume in proportion to its surface area than any other solid geometric figure. A *hemisphere* is a half sphere. Study the sketches of the sphere (Figure 16–33) and hemisphere (Figure 16–34). Note the various parts of the sphere and hemisphere.

The *lateral surface area* of a pyramid is found by adding the area of each of the face triangles. The *total surface area* of a pyramid is found by adding the lateral surface area and the base area. The *total surface area* of a sphere is found by the formula $4\pi r^2$ or πd^2. The *lateral surface area* of a hemisphere is found by the formula $2\pi r^2$ or $\frac{\pi d^2}{2}$. The *total surface area* of a hemisphere is determined by the formula $3\pi r^2$ or $0.75\ \pi d^2$.

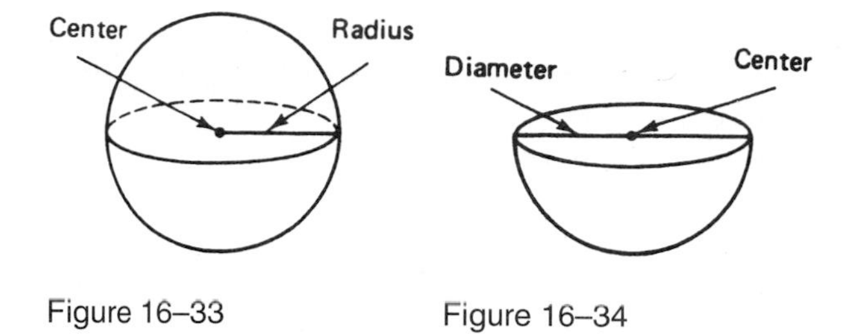

Figure 16–33

Figure 16–34

Example 16–5: Find the Lateral and Total Surface Area of a Pyramid

PROBLEM: Find the lateral and total surface areas of the pyramid shown in Figure 16–35.

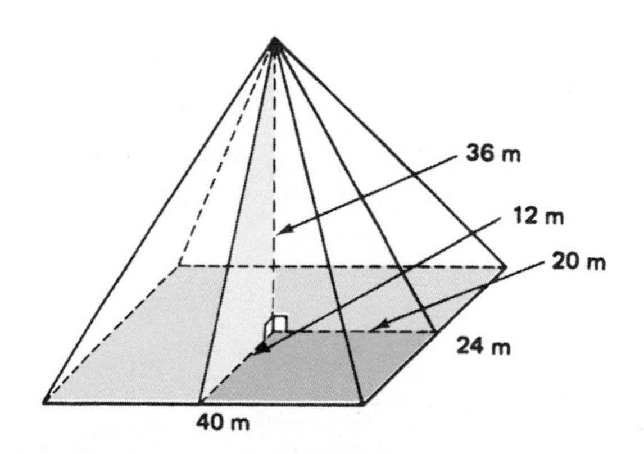

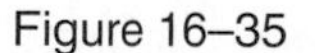
Figure 16–35

SOLUTION: Note that the figure is a rectangular pyramid because its base is a rectangle and all faces are isosceles triangles. Then find the slant height of two different faces. See Figure 16–36. Find the lateral surface area of the pyramid by using the formula for the surface area of a triangle. The lateral surface area of a pyramid is found by adding the surface areas of the four triangles.

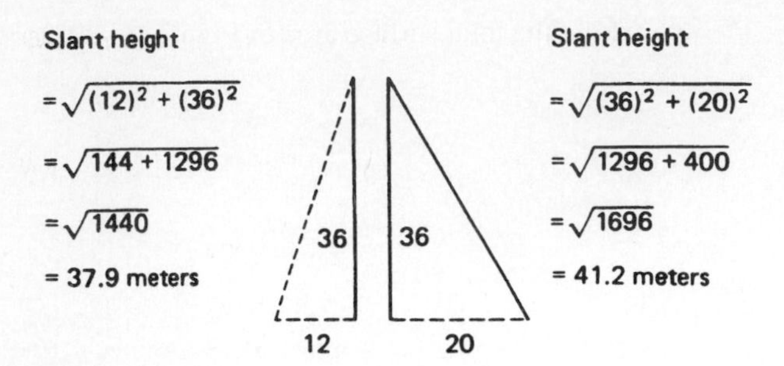

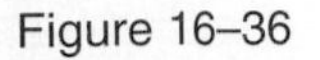
Figure 16–36

Lateral surface area

$$= 2\left[\frac{\text{base (40 m)} \times \text{slant height (37.9 m)}}{2}\right]$$

$$+ 2\left[\frac{\text{base (24 m)} \times \text{slant height (41.2 m)}}{2}\right]$$

$$= \overset{1}{\cancel{2}}\left(\frac{40\text{ m} \times 37.9\text{ m}}{\underset{1}{\cancel{2}}}\right) + \overset{1}{\cancel{2}}\left(\frac{24\text{ m} + 41.2\text{ m}}{\underset{1}{\cancel{2}}}\right)$$

$$= 1516\text{ m}^2 + 988.8\text{ m}^2$$
$$= 2504.8\text{ m}^2$$

Then find the total surface area of the pyramid.

$$\begin{aligned}\text{Total surface area} &= \text{lateral surface area} \\ &\quad + \text{base surface area} \\ &= 2504.8\text{ m}^2 + (40\text{ m})(24\text{ m}) \\ &= 2504.8\text{ m}^2 + 960\text{ m}^2 \\ &= 3464.8\text{ m}^2\end{aligned}$$

Note the formulas used to solve this problem:

$$\text{Slant height} = \sqrt{(\text{altitude})^2 + \left(\frac{1}{2}\text{ base}\right)^2}$$

$$\text{sh} = \sqrt{a^2 + \left(\frac{b}{2}\right)^2}$$

$$\begin{aligned}\text{Lateral surface area} &= 2\left(\frac{1}{2}\text{ base} \times \text{slant side 1}\right) \\ &\quad + 2\left(\frac{1}{2}\text{ base} \times \text{slant side 2}\right) \\ &= 2\left(\frac{b}{2}\text{ sh}_1\right) + 2\left(\frac{b}{2}\text{ sh}_2\right) \\ &= b(\text{sh}_1) + b(\text{sh}_2)\end{aligned}$$

$$\text{Total surface area} = \text{lateral surface area} + \text{base area}$$

Example 16–6: Determine the Surface Area of a Sphere

PROBLEM: Find the surface area of the sphere shown in Figure 16–37.

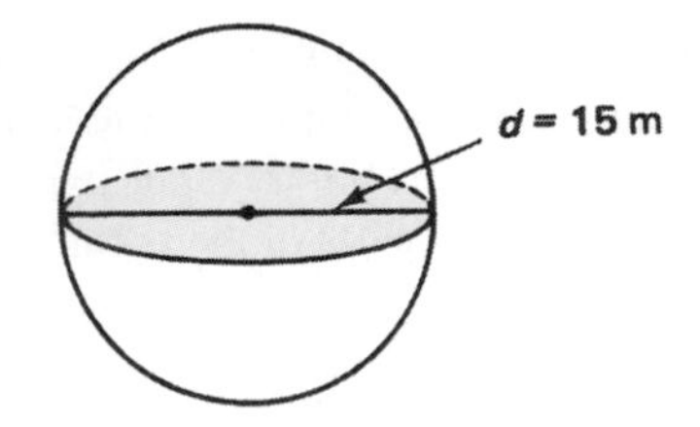

Figure 16–37

SOLUTION: The lateral surface area and the total surface area of a sphere are the same. Recall the formula for determining the total surface area of a sphere: $4\pi r^2$ or πd^2.

$$\begin{aligned}\text{Total surface area} &= \pi d^2 \\ &= (3.142)(15\text{ m})(15\text{ m}) \\ &= 706.95\text{ m}^2\end{aligned}$$

Note the lateral surface area of a hemisphere is found by the formula $\frac{\pi d^2}{2}$ or $2\pi r^2$.

EXERCISE 16–3 SURFACE AREAS AND PROPERTIES OF PYRAMIDS AND SPHERES

Find the lateral and total surface areas; as called for in the following problems.

16–25. What type of solid figure is Figure 16–38?

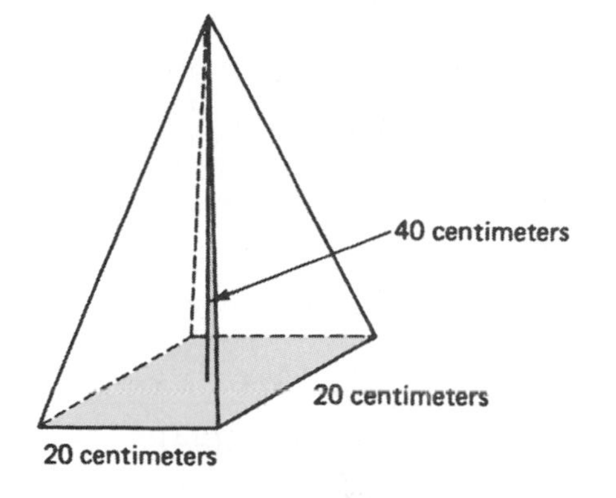

Figure 16–38

16–26. Find the slant height of Figure 16–38.

16–27. Find the lateral surface area of Figure 16–38.

16–28. Find the total surface area of Figure 16–38.

16–29. What is the total surface area of Figure 16–39?

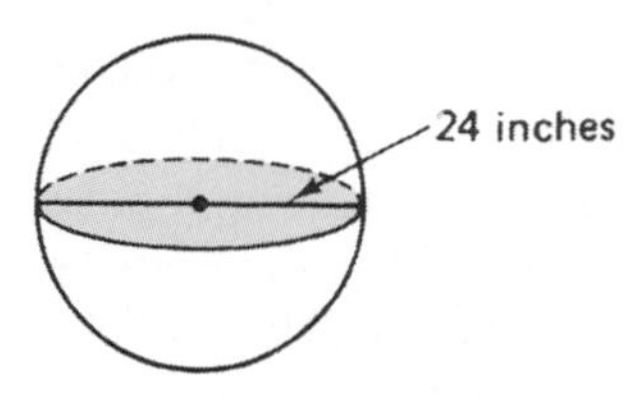

Figure 16–39

16–30. Determine the lateral surface area of the hemisphere of Figure 16–39.

16–31. What kind of solid figure is Figure 16–40?

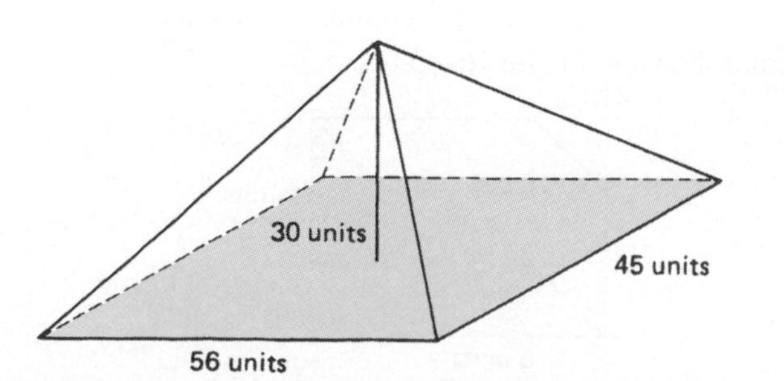

Figure 16–40

16–32. Find the slant height on the 45-unit side of Figure 16–40.

16–33. Find the slant height on the 56-unit side of Figure 16–40.

16–34. Find the lateral surface area of Figure 16–40.

16–35. Find the total surface area of Figure 16–40.

16–36. What type of figure is Figure 16–41?

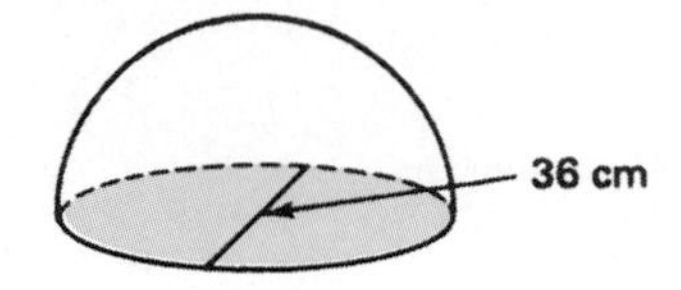

Figure 16–41

16–37. What is the radius of Figure 16–41?

16–38. What is the altitude of Figure 16–41?

16–39. Find the lateral surface area of Figure 16–41.

16–40. Determine the total surface area of Figure 16–41.

16.4 DEFINITIONS AND PROPERTIES OF VOLUMES OF GEOMETRIC SOLIDS

Volume is the amount of content or cubic space in a solid. The volume of a prism or cylinder is found by multiplying the area of the base times the altitude or height.

To better understand volume, study Figure 16–42. The cubes in the first layer are numbered from 1 to 16. Recall that a cube is a solid figure with all square faces; thus all edges are equal. If each cube is 1 cubic inch, the volume of the first layer would be 16 cubic inches. The volume of all three layers would be 16 cubic inches per layer times the number of layers (3), or a total of 48 cubic inches.

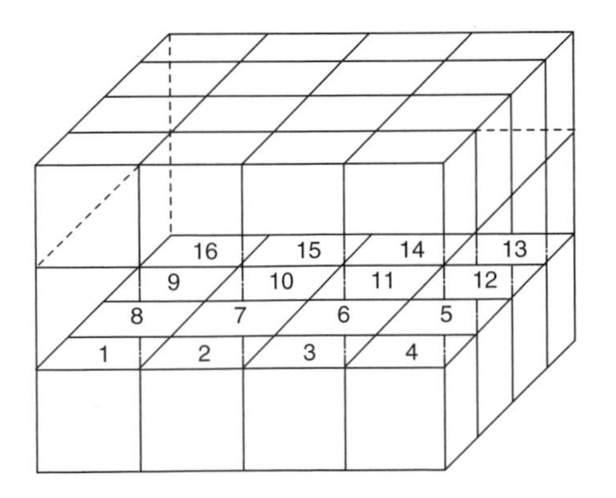

Figure 16–42

The volume of a pyramid or cone is found by multiplying the area of the base times the altitude or height times one third.

Example 16–7: Determine the Volume of a Rectangular Solid

PROBLEM: Find the volume of the rectangular solid shown in Figure 16–43.

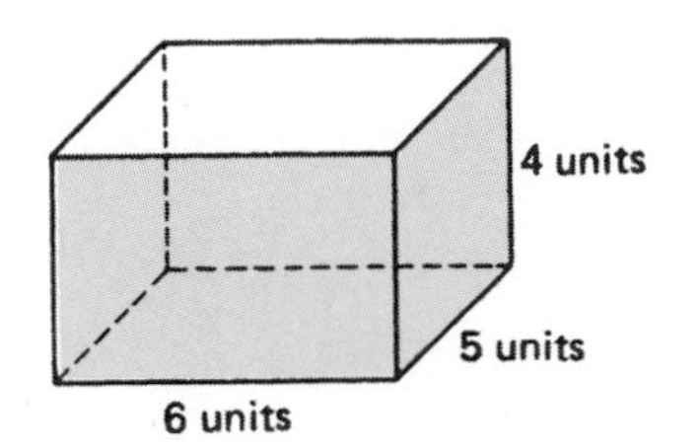

Figure 16–43

SOLUTION: Use the formula for finding the volume of a rectangular solid.

$$\begin{aligned}\text{Volume} &= \text{length} \times \text{width} \times \text{height}\\ &= (6 \text{ units})(5 \text{ units})(4 \text{ units})\\ &= 120 \text{ cubic units}\end{aligned}$$

Example 16–8: Determine the Volume of a Cylinder

PROBLEM: Find the volume of the cylinder shown in Figure 16–44.

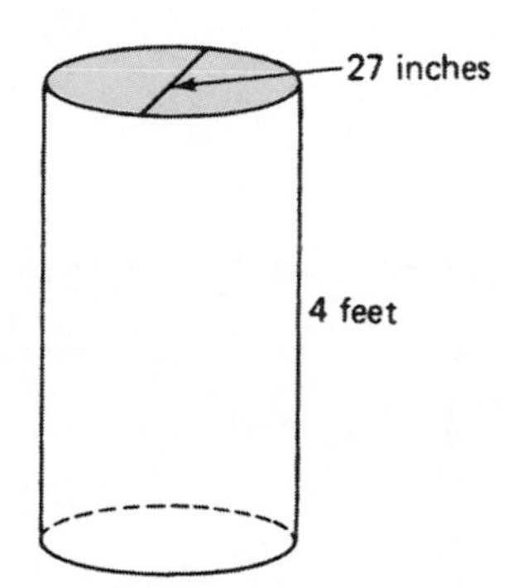

Figure 16–44

SOLUTION: First, convert to like units (12 in. = 1 ft).

$$\left(\frac{4 \not{ft}}{1}\right)\left(\frac{12 \text{ in.}}{1 \not{ft}}\right) = 48 \text{ in.}$$

The radius = $\frac{1}{2}$ (27 inches) = 13.5 inches. Then use the formula for finding the volume of a cylinder: $V = \pi r^2 h$.

$$\begin{aligned}\text{Volume} &= \text{area of base } (\pi r^2) \times \text{height or altitude}\\ &= (3.142)(13.5 \text{ in.})(13.5 \text{ in.})(48 \text{ in.})\\ &= 27{,}486.216 \text{ cu in.}\end{aligned}$$

Example 16–9: Find the Volume of a Cone

PROBLEM: Find the volume of the cone shown in Figure 16–45.

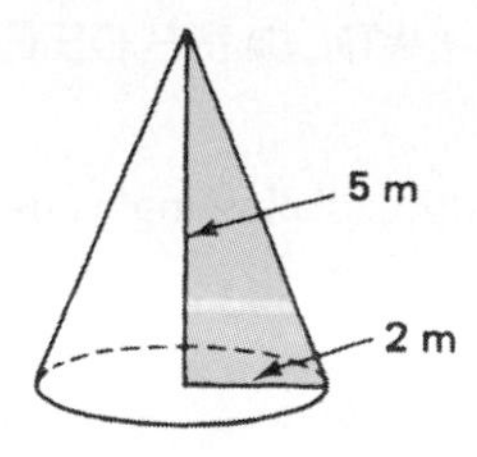

Figure 16–45

SOLUTION: Use the formula for finding the volume of a cone, $V = \dfrac{\pi r^2 h}{3}$.

$$\begin{aligned}\text{Volume} &= \frac{\pi r^2 h}{3}\\ &= \text{area of base} \times \text{altitude} \div 3\\ &= \frac{(3.142)(2)(2)(5)}{3}\\ &= \frac{(3.142)\ (20 \text{ m}^3)}{3}\\ &= 20.95 \text{ m}^3\end{aligned}$$

The formula for the volume of a pyramid is

Volume = area of the base × the height × one-third

Example 16–10: Find the Volume of a Sphere

PROBLEM: Find the volume of an 8-inch sphere.

SOLUTION: Use the formula for finding the volume of a sphere.

$$\begin{aligned}\text{Volume} &= 4 \times \pi \times \text{the radius cubed} \div 3\\ &= \frac{(4)(3.142)(4)(4)(4)}{3}\\ &= 268.12 \text{ in.}^3\end{aligned}$$

The formula for finding the volume of a sphere is $V = \dfrac{4\pi r^3}{3}$.

The formula for determining the volume of a hemisphere is $V = \dfrac{2\pi r^3}{3}$.

EXERCISE 16–4 VOLUMES OF GEOMETRIC SOLIDS

Find the volume of the following as indicated.

16–41. What type of solid is shown in Figure 16–46?

16–42. Find the volume of the first layer of the solid in Figure 16–46.

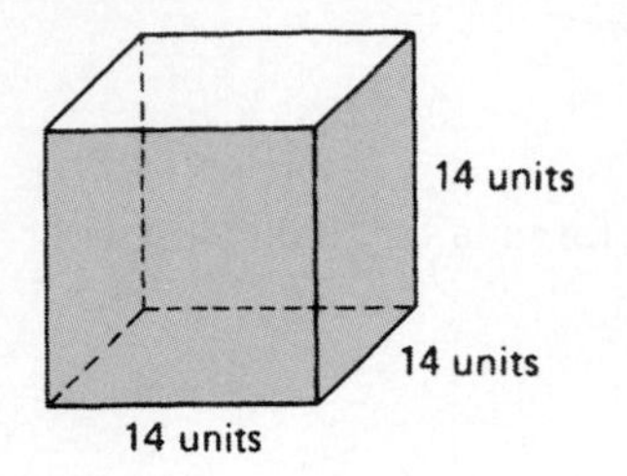

Figure 16–46

16–43. Find the volume of the solid in Figure 16–46.

16–44. What type of solid is Figure 16–47?

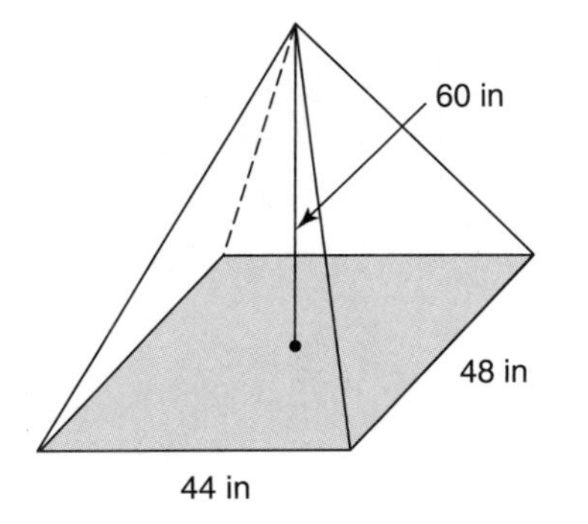

Figure 16–47

16–45. Find the area of the base of Figure 16–47.

16–46. Determine the volume of Figure 16–47.

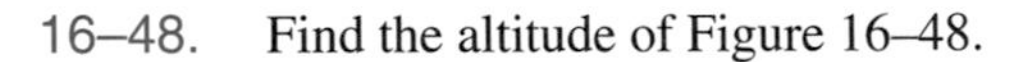

16–47. What type of solid is shown in Figure 16–48?

16–48. Find the altitude of Figure 16–48.

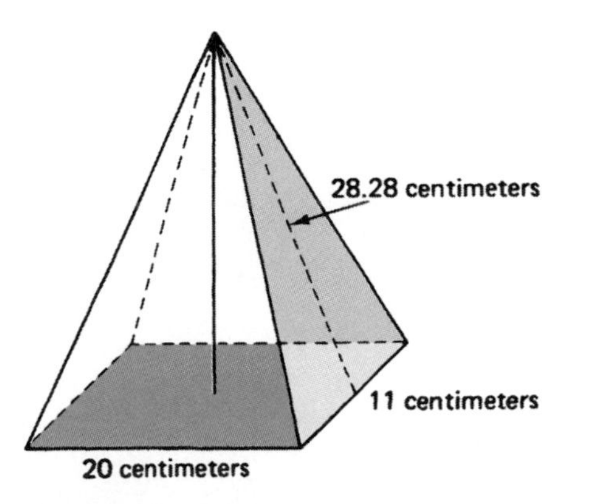

Figure 16–48

16–49. Find the area of the base of Figure 16–48.

16–50. Determine the volume of Figure 16–48.

16–51. What type of solid is shown in Figure 16–49?

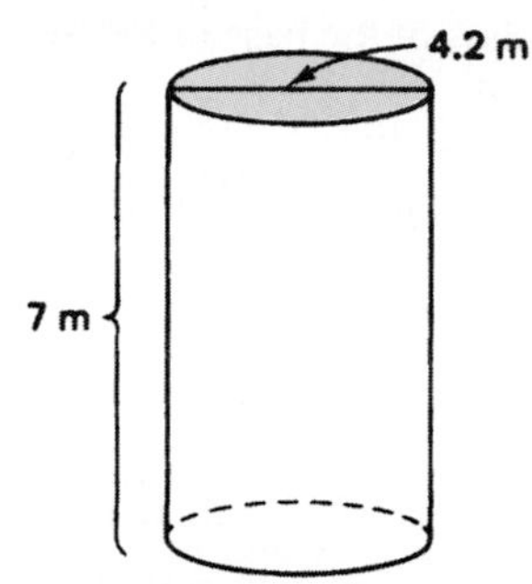

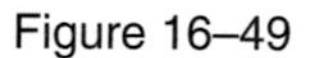
Figure 16–49

16–52. Find the area of the base of Figure 16–49.

16–53. Find the volume of Figure 16–49.

16–54. What type of solid is Figure 16–50?

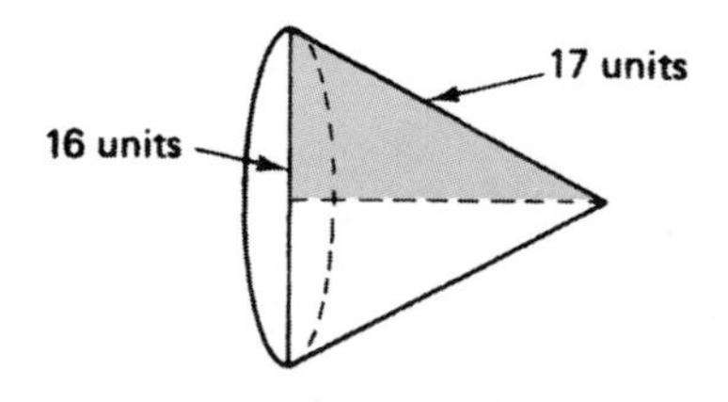

Figure 16–50

16–55. Find the area of the base of Figure 16–50.

16–56. Find the volume of Figure 16–50.

16–57. What type of solid is shown in Figure 16–51?

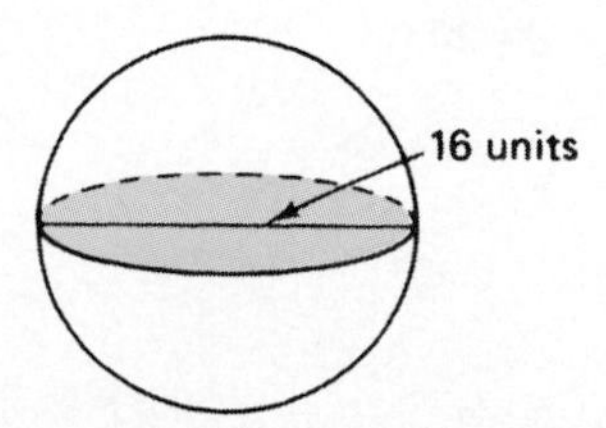

Figure 16–51

16–58. Find the volume of Figure 16–51.

16–59. What type of solid is shown in Figure 16–52?

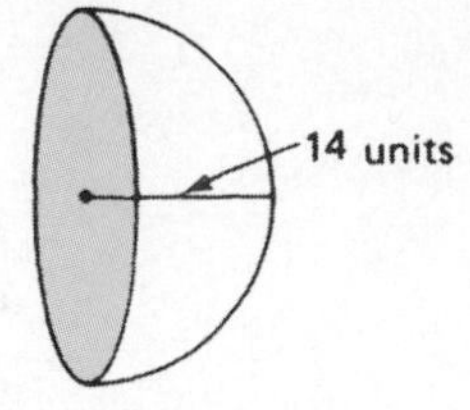

Figure 16–52

16.5 GEOMETRIC SOLID APPLICATIONS

The following examples illustrate applied problems. In each application, a formula for finding the volume of a solid is used. Sometimes more than one formula is used.

Example 16–11: Determine the Capacity of a Cylinder

PROBLEM: Find the capacity in gallons of the cylindrical tank shown in Figure 16–53.

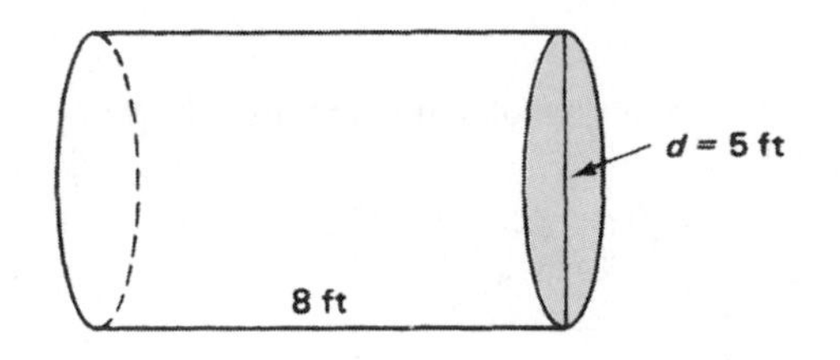

Figure 16–53

SOLUTION: Find the volume of the cylinder using the formula $V = \pi r^2 h$.

$$V = (3.142)(2.5\text{ ft})(2.5\text{ ft})(8\text{ ft})$$
$$= 157.1\text{ cu ft}$$

Then find the number of gallons contained in 157.1 cubic feet using the fact 1 cubic foot = 7.5 gallons.

$$\frac{157.1\ \cancel{\text{cu ft}}}{1} \times \frac{7.5\text{ gal}}{1\ \cancel{\text{cu ft}}} = 1178.2\text{ gal}$$

Example 16–12: Determine the Volume of a Cone

PROBLEM: Find the tons of coal in a pile with a circumference of 62.8 feet and a height of 9 feet shown in Figure 16–54.

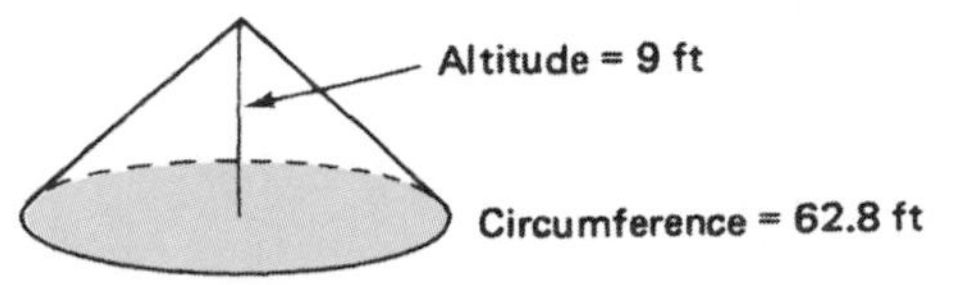

Figure 16–54

16–60. Find the volume of Figure 16–52.

SOLUTION: Find the radius of the base of a pile. Use the formula $c = d\pi$; thus $\frac{c}{\pi} = d$.

$$\text{diameter} = \frac{\text{circumference}}{\pi}$$
$$= \frac{62.8\text{ ft}}{\pi}$$
$$= 20\text{ ft}$$
$$\text{radius} = 10\text{ ft}$$

Then find the volume of the coal pile by using the formula for the volume of a cone.

$$V = \frac{\pi r^2 h}{3}$$
$$= \frac{(3.142)(10\text{ ft})(10\text{ ft})(9\text{ ft})}{3}$$
$$= 943\text{ cu ft}$$

To find the tons of coal, use the fact that 1 cubic foot = 0.0235 ton of coal.

$$\left(\frac{943\ \cancel{\text{cu ft}}}{1}\right)\left(\frac{0.0235\text{ ton}}{\cancel{\text{cu ft}}}\right) = 22.1605\text{ or }22.2\text{ tons of coal}$$

Example 16–13: Find the Surface Area of a Cylinder and a Sphere

PROBLEM: Find the amount of paint needed to paint the water tower shown in Figure 16–55.

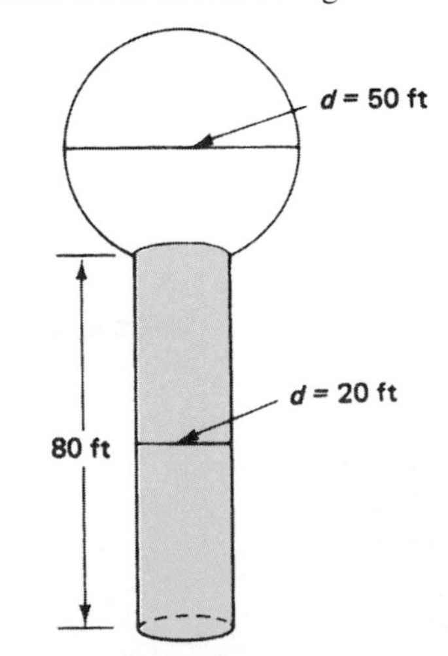

Figure 16–55

SOLUTION: Find the total surface area of the sphere using the formula πd^2.

$$\text{Total surface area} = (3.142)(50 \text{ ft})(50 \text{ ft}) = 7855 \text{ sq ft}$$

Then find the lateral surface area of the cylinder using the formula $LS = d\pi rh$.

$$\text{Lateral surface area} = (20 \text{ ft})(3.142)(80 \text{ ft}) = 5027 \text{ sq ft}$$

Determine the surface area that is under the sphere, inside the cylinder, where the sphere is attached to the cylinder, using the formula πr^2.

$$A = (3.142)(10 \text{ ft})(10 \text{ ft}) = 314.2 \text{ sq ft}$$

Determine the total surface area to be painted.

$$\begin{aligned} \text{Total area of sphere} &= 7855 \text{ sq ft} \\ + \text{ Total area of cylinder} &= 5027 \text{ sq ft} \\ \hline \text{Total area} &= 12{,}882 \text{ sq ft} \\ - \text{ Area under the sphere} &= 314 \text{ sq ft} \\ \hline \text{Area to be painted} &= 12{,}568 \text{ sq ft} \end{aligned}$$

The gallons of paint needed: 1 gallon of paint covers 550 square feet.

$$\left(\frac{12{,}568 \text{ sq ft}}{1}\right)\left(\frac{1 \text{ gal}}{550 \text{ sq ft}}\right) = 22.85 \text{ gal or } 23 \text{ gal}$$

EXERCISE 16–5 APPLIED GEOMETRIC SOLID PROBLEMS

Find the areas and volumes of the following applied problems.

16–61. How much fabric would be needed to cover a cylindrical wastebasket 18 inches high and 8 inches in diameter? Do not include the base.

16–62. Find the amount of stainless steel needed to make a tank 6.75 meters long and 6 meters in diameter. (Allow 5% for seams.)

16–63. Find the volume in cubic meters of the tank in problem 16–62.

16–64. Determine the weight of a brass solid 3 yards long, 1 inch thick, and $\frac{1}{2}$ foot wide. (Brass weighs 512 pounds per cubic foot.)

16–65. Find the capacity in gallons of a rectangular fish tank 12 inches by 14.5 inches by 48 inches. (1 gallon = 231 cubic inches.)

16–66. Find the capacity in gallons of a cylindrical gas tank 12 inches in diameter and 48 inches long. (1 gallon = 231 cubic inches.)

16–67. How many gallons of crude oil will a mile section of the Alaskan pipeline contain if the inside diameter is 48 inches? (1 cubic foot = 7.5 gallons.)

16–68. Find the capacity of the mill-dust collector shown in Figure 16–56.

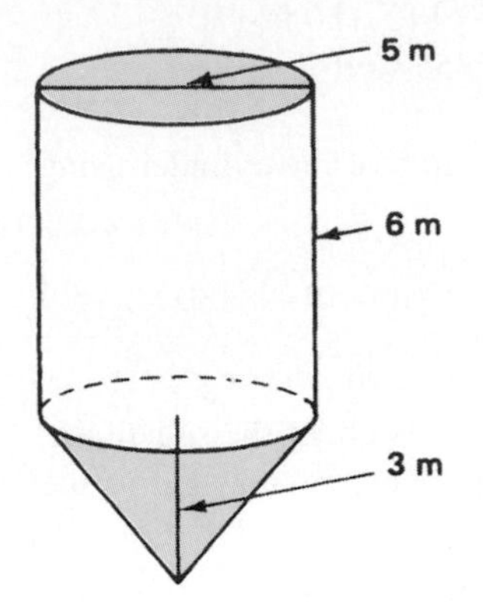

Figure 16–56

16–69. A stainless steel cylindrical milk tanker is 38 feet long and 5 feet in diameter. Determine the amount of stainless steel material needed to construct a tanker, allowing 12% for seams and waste.

16–70. Find the capacity in gallons of the milk tanker in problem 16–69. (1 cubic foot = 7.5 gallons.)

16–71. A swimming pool 50 meters long, 20 meters wide with an average depth of 3 meters is under construction. How many cubic meters must be excavated if an additional meter must be excavated for each side and the bottom?

16–72. Determine the square meters to be tiled if the inside and bottom of the pool (problem 16–71) are tiled.

16–73. A new house is built in the form of a hemisphere having a radius of 7 meters. If the air in the house is to be circulated every 90 seconds, how many cubic meters of air must be moved per hour?

16–74. A 50-foot length of $^5/_8$-inch garden hose contains how many quarts (to the nearest tenth)? (1 quart = 57.75 cubic inches.)

16–75. Find the monthly heating cost for a 6.5-foot-diameter circular hot tub filled to a depth of $2^1/_2$ feet. The cost to heat 1 gallon of water per month is $0.057. (1 cubic foot = 7.5 gallons.)

THINK TIME

To provide yourself with a simple review guide for geometric figures and the formulas for determining their areas, complete the following chart from the given sketches shown.

Sketch	*Name of Figure*	*Formula for Lateral Surface Area*	*Formula for Total Surface Area*	*Formula for Volume*
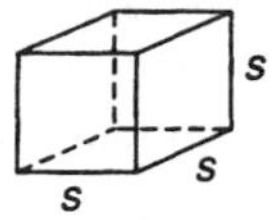 s, s, s				
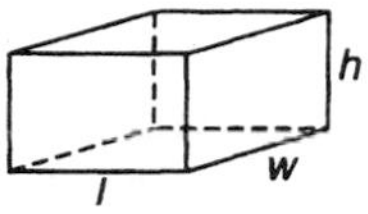 l, w, h				
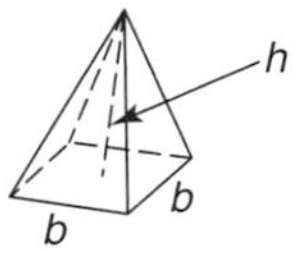 b, b, h				
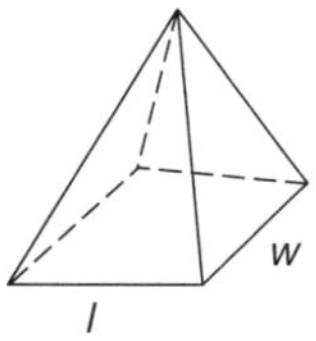 l, w				
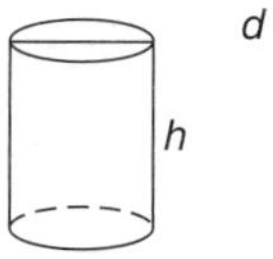 d, h				
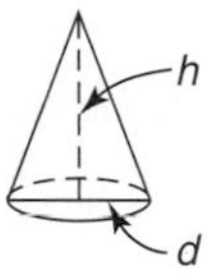 h, d				
 d				
 d				

PROCEDURES TO REMEMBER

1. Be certain to use only like units when finding an area or volume.
2. To find the area or volume of complex figures, change to simple figures where possible, and apply basic formulas for determining lateral surface area, total surface area, or volume.

CHAPTER SUMMARY

The following definitions are essential when working with geometric figures.

1. A *prism* is a solid geometric figure with parallel edges and uniform cross sections.
2. A *face* is any plane surface of a solid figure.
3. A *lateral face* of a prism is a side of a geometric figure.
4. The *bases* of a prism are the top face and the bottom face and are generally called the top and bottom of the solid.
5. The *lateral edge* of a prism is where two sides of a solid intersect.
6. A *right prism*'s faces are perpendicular to its base.
7. An *oblique prism*'s faces are not perpendicular to its base.
8. A *right square prism* has a square base.
9. A *right rectangular prism* has a rectangular base.
10. A *right pentagonal prism* has a pentagon-shaped base.
11. The *lateral surface area* of a solid geometric figure is the surface area of the sides not including the area of the top and bottom.
12. The *total surface area* of a solid geometric figure is the area of the top and bottom of the figure including the area of the sides.
13. A *right pyramid* is a pyramid with a regular polygon base and sides that form equal isosceles triangles.
14. A *sphere* is a solid figure with all points equal distances from the center.
15. A *hemisphere* is a half sphere.
16. *Volume* is the amount of cubic units or content in a solid figure.

CHAPTER TEST

T–16–1. Find the volume of a rectangular solid 2 centimeters by 5 centimeters by 8 centimeters.

T–16–2. Find the lateral surface area of Figure 16–57.

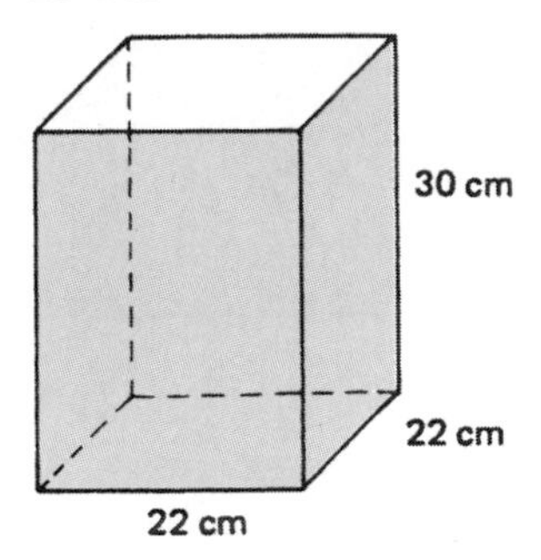

Figure 16–57

T–16–3. Determine the total surface area of a rectangular room 8 feet wide, 16 feet long, and 8 feet high.

T–16–4. Find the lateral surface area of Figure 16–58.

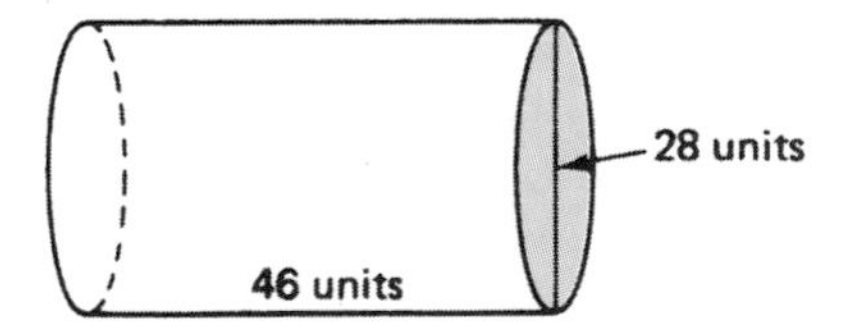

Figure 16–58

T–16–5. Find the volume in Figure 16–59.

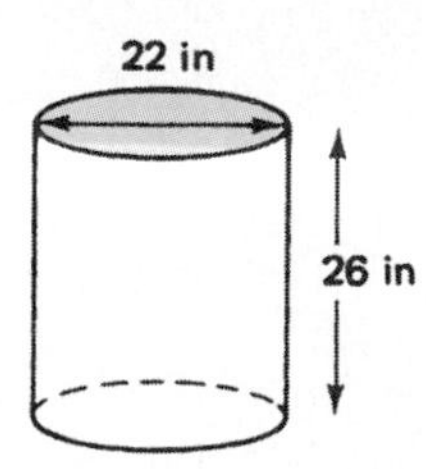

Figure 16–59

T–16–6. Find the lateral surface area of the cone in Figure 16–60.

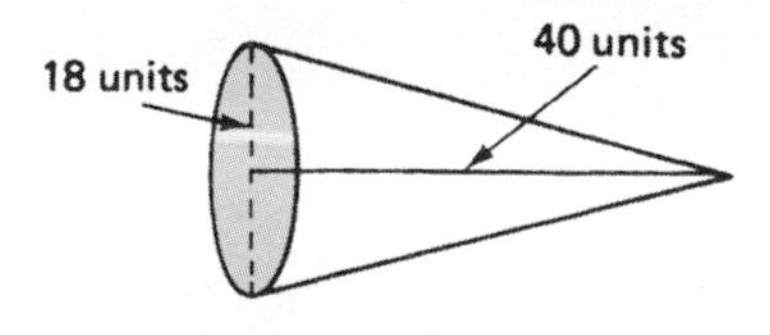

Figure 16–60

T–16–7. Determine the total surface area of the cylinders in Figure 16–61.

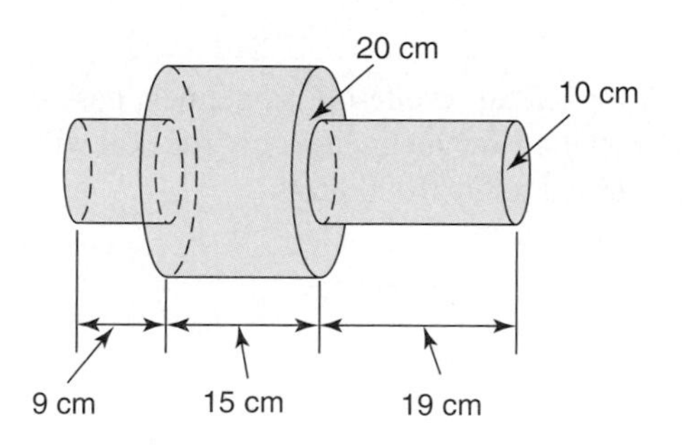

Figure 16–61

T–16–8. Determine the volume of the cone in Figure 16–62.

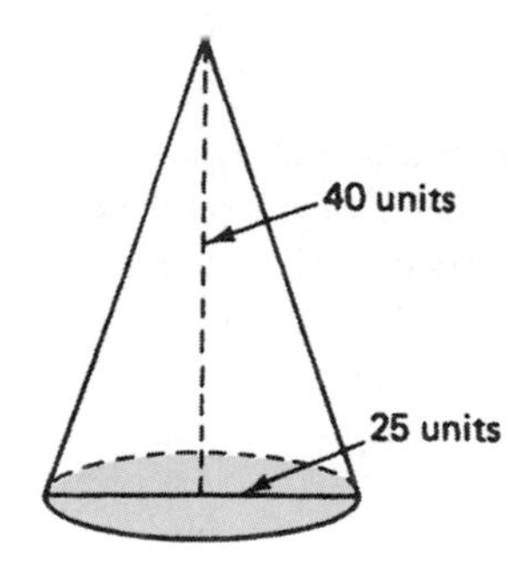

Figure 16–62

T–16–9. Find the surface area of the hemisphere in Figure 16–63.

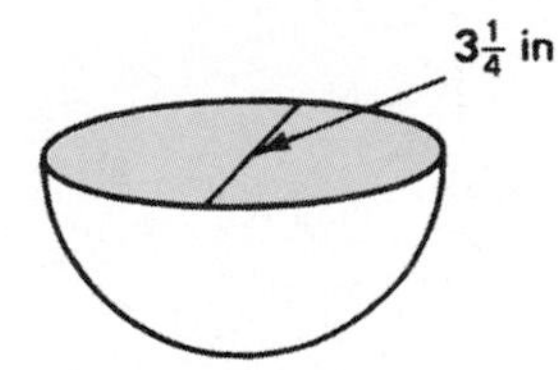

Figure 16–63

T–16–10. Determine the total surface area of the pyramid in Figure 16–64.

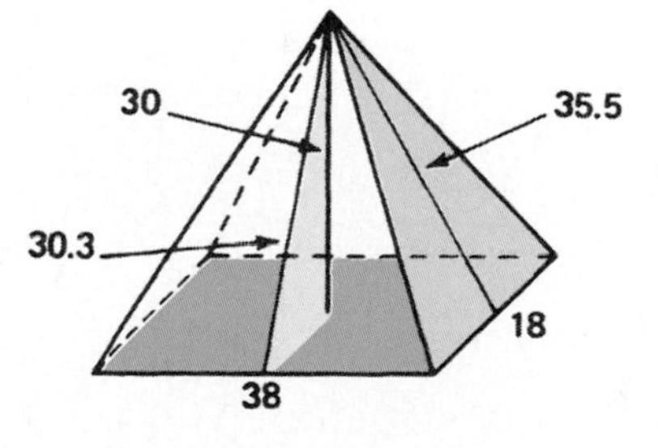

Figure 16–64

T–16–11. A stainless steel milk tanker is 36 feet long and 6 feet in diameter. How many gallons will it hold? (1 cubic foot = 7.5 gallons.)

T–16–12. Find the capacity of the sawdust collector shown in Figure 16–65.

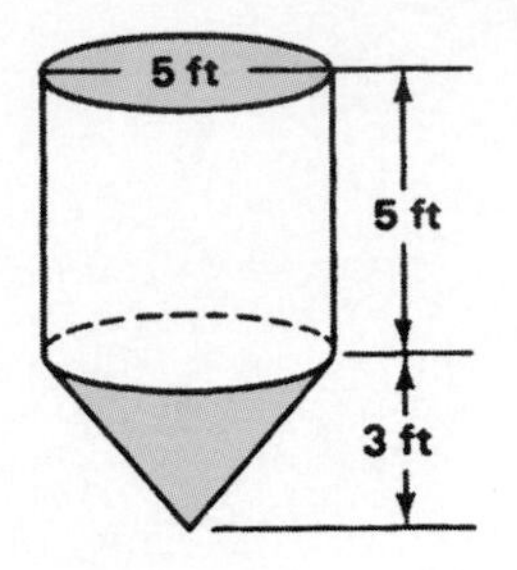

Figure 16–65

T–16–13. A cylindrical toothpaste container is 18 centimeters long and has a diameter of 3 centimeters. Find the number of cubic centimeters of toothpaste in the container.

T–16–14. A spherical crude-oil container has a diameter of 16 meters. How many liters of crude oil will the container hold if 1 m^3 = 1000 liters?

T–16–15. A pipeline in a refinery has 400 feet between shutoff valves. How many gallons of crude oil would be in the 400-foot section if the inner diameter of the pipe is 3 feet? (1 cubic foot = 7.5 gallons.)

17

APPLICATIONS OF METRIC AND U.S. MEASURE

OBJECTIVES

After studying this chapter, you will be able to:

17.1. Understand the basics of the metric system.
17.2. Find metric area.
17.3. Find metric volume.
17.4. Find metric weight.
17.5. Understand Temperature—Celsius/Fahrenheit.
17.6. Understand applications of the metric/U.S. measurement systems.

SELF-TEST

This test will measure your ability to make conversions in the metric system and the U.S. system, as well as your skill with applied measurement problems. You may use the conversion tables. Show your work so that possible areas of difficulty may be noted.

S–17–1. 521 cm = _________ m.

S–17–2. 5.5 kg = _________ g.

S–17–3. 5280 mL = _________ L.

S–17–4. 17,500 mm = _________ km.

S–17–5. 625 cm^2 = _________ sq in.

S–17–6. 64.255 m^2 = _________ ha.

S–17–7. 700 miles = __________ km.

S–17–8. 4 lb 3 oz = __________ g.

S–17–9. If top-grade hogs weigh 220 pounds, how many metric tons would 50 top-grade hogs weigh?

S–17–10. How many kilograms does a 260-pound defensive end weigh?

S–17–11. 32.3 kg = __________ g.

S–17–12. 5 lb = __________ kg.

S–17–13. 2.36 metric tons = __________ lb.

S–17–14. 5000 ml = __________ liters.

S–17–15. 5.55 dg = __________ g.

S–17–16. Find the weight of 500 milliliters of water?

S–17–17. 45°C = __________ °F.

S–17–18. 210°F = __________ °C.

S–17–19. During a thunderstorm, the temperature dropped from 86°F to 62°F. Find the temperature change on the Celsius scale.

S–17–20. A metal shaper makes a $^{3}/_{64}$-inch cut on a piece of brass. Find the size of the cut in millimeters.

S–17–21. Building codes in a city require the window glass in houses to be at least $^{3}/_{16}$ inch thick. If European windows are used, find the thickness of the glass in millimeters.

S–17–22. A ship loaded with 6600 tons (English) of wheat sails from Duluth, Minnesota to Yokohama, Japan. How many metric tons does the wheat weigh?

S–17–23. A classroom is 30 feet by 40 feet by 12 feet and exchanges air every 3 minutes. How many cubic meters of air is exchanged every hour?

S–17–24. A steel casting weighs 1.5 kilograms. Fifty holes are bored in the casting; 2.75 grams of steel is removed from each hole. Find the weight of the casting after the holes have been bored.

S–17–25. An airplane can fly 1 kilometer on 12 kilograms of fuel. How many metric tons of fuel will be used to fly nonstop from New York to Minneapolis, a distance of 1760 kilometers?

17.1 INTRODUCTION AND BASICS OF THE METRIC SYSTEM

Gabriel Mouton of France realized the need for a practical decimal system of measurement and developed a base 10 system of measurement in 1670. People liked Mouton's system, but it was not until 1790 that the French Academy endorsed Mouton's system. The French Academy adopted Mouton's system based on the decimal concept and a definite nonchanging value. This definite value was the distance from the North Pole to the equator. The distance was divided into 10 million parts which were called *meters.*

To ensure precise measurement standards throughout the world, the System International, or SI, international conference established the standards. Nearly all the nations of the world have since adopted the SI metric system. The United States is the only large nation that does not use the metric system. However, much international trade and business is done using the metric system of measure.

There are seven base units of measure in the SI metric system. They are meter (length), kilogram (mass or weight), kelvin (temperature), second (time), ampere (electricity), candela (luminous intensity), and mole (amount of substance). We will study four basic types of metric measure: kilogram, meter, liter, and Celsius degree.

Contrary to U.S. measurement, in the metric system a relationship exists between the basic units of measure and water. Study the relationship shown in Figure 17–1. This relationship between the basic units of measure makes the metric system very logical.

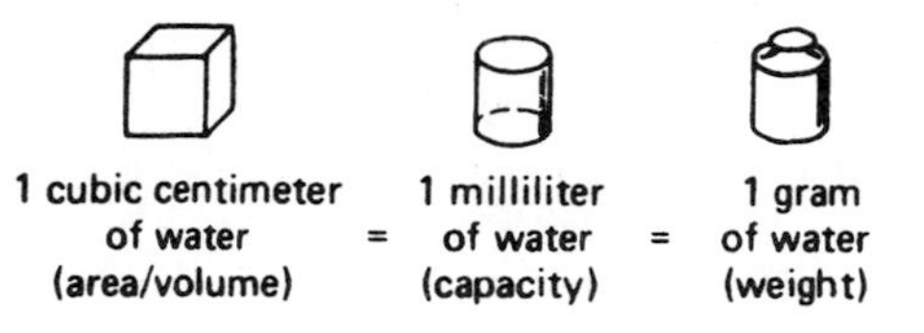

Figure 17–1

As noted before, the metric system is a base 10 or decimal measurement system. This means that each unit may be multiplied or divided by 10 to change a base unit into a more workable unit. Prefixes are added to the base units for identification of the units. When units are *larger* than the base unit, Greek prefixes are used. The most common of these prefixes are *kilo* = 1000, *hecto* = 100, and *deka* = 10. When units are *smaller* than the base unit, Latin prefixes are used. The most common of these prefixes are *deci* = 0.1, *centi* = 0.01, and *milli* = 0.001. Thus a kilometer = 1000 meters and a millimeter = 0.001 meter. Study Figure 17–2 to understand the relationship between metric prefixes and decimal numeration.

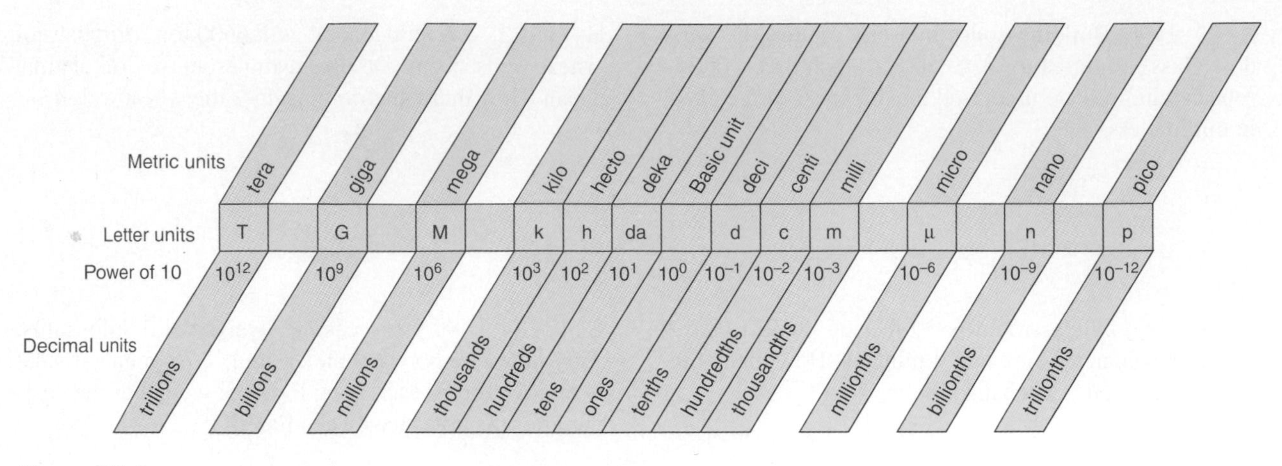

Figure 17–2

Movement of the decimal point to the left or right of a given quantity is all that is needed to change to the next-higher or next-lower unit in the metric system. Naturally, the new unit prefix should be indicated. Thus 89 meters = 890 decimeters. The procedure for making conversions will be shown in the examples.

The base unit of measure for length in the metric system is the *meter*. A meter may be approximated in the following manner (see Figure 17–3). If an average adult extends his hand out from his shoulder and turns his head away from the extended hand, the distance from the tip of his fingers to the tip of his nose is approximately 1 meter. The thickness of a dime is about the length of a millimeter.

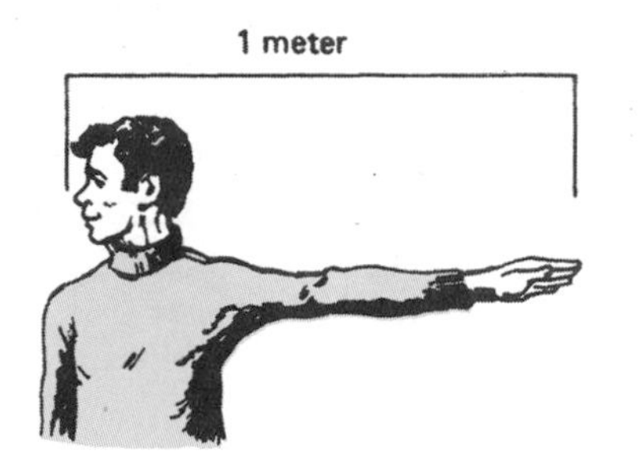

Figure 17–3

Like other base metric units, we can change the meter into more workable units by either multiplying or dividing by 10. Figure 17–4 shows the various meter units, their names, and the standard abbreviations.

It should be noted the dekameter, hectometer, and decimeter are rarely used, whereas the kilometer, meter, and millimeter are commonly used.

Changing from one unit to another may be accomplished simply by moving the decimal point. To better understand the procedure, study the examples.

Meter Units		*Comparison to Meter*		*Abbreviation*
1 kilometer	=	1000	meters	km
1 hectometer	=	100	meters	hm
1 dekameter	=	10	meters	dam
1 meter	=	1	meter	m
1 decimeter	=	0.1	meter	dm
1 centimeter	=	0.01	meter	cm
1 millimeter	=	0.001	meter	mm

Figure 17–4 Meter Table of Equivalents

Example 17–1: Convert Meters to Kilometers

PROBLEM: Change 1250 meters (m) into kilometers (km).

SOLUTION: Using the meter chart given, note that 1000 meters = 1 kilometer. Since 1000 meters = 1 kilometer, divide both sides by 1000 meters, thus:

$$\frac{1000\text{ m}}{1000\text{ m}} = \frac{1\text{ km}}{1000\text{ m}} \quad \text{and} \quad 1 = \frac{1\text{ km}}{1000\text{ m}}$$

Multiply 1250 meters by a fraction equal to 1. (Restudy Example 2–8, if necessary.) In this problem, $\frac{1\text{ km}}{1000\text{ m}}$ is the fraction equal to 1. Note that the number 1, representing the fraction, is used for illustration.

$$\frac{\overset{1.25}{\cancel{1250}}}{1} \times \frac{1\text{ km}}{\underset{1}{\cancel{1000}}\text{ m}} = 1.250\text{ km}$$

An easier procedure to change meters to kilometers is by dividing the given number of meters (1250) by 1000.

$$\frac{1250 \text{ m}}{1000} = 1.250 \text{ km}$$

The easiest procedure is to move the decimal point three places to the left.

km	**hm**	**dam**	**m**	**dm**	**cm**	**mm**
1000	100	10	1	0.1	0.01	0.001
3	2	1				

1250 m = 1 2 5 0 km (decimal moved 3 2 1)

1250 m = 1.25 km

Remember: To change a smaller unit to a larger unit, move the decimal point the appropriate number of spaces to the left.

Example 17–2: Convert Meters to Centimeters

PROBLEM: Convert 38.8 meters (m) to centimeters (cm).

SOLUTION: Using the meter table, note that 100 centimeters = 1 meter. Multiply 38.8 meters by one:

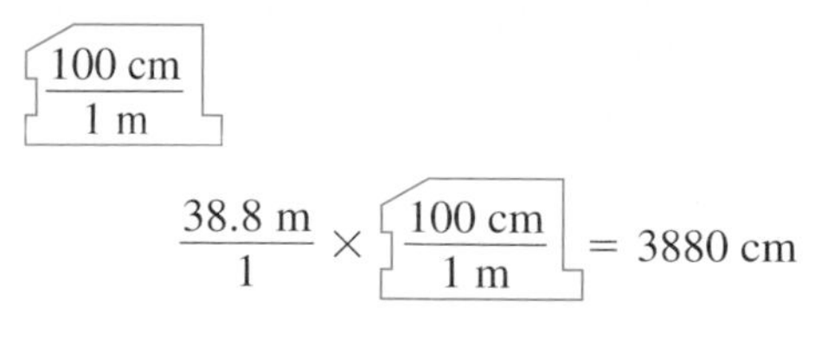

$$\frac{100 \text{ cm}}{1 \text{ m}}$$

$$\frac{38.8 \text{ m}}{1} \times \frac{100 \text{ cm}}{1 \text{ m}} = 3880 \text{ cm}$$

A more efficient procedure to change the meters into centimeters is by multiplying the given number of meters (38.8) by 100.

$$38.8 \text{ m} \times 100 = 3880 \text{ cm}$$

The easiest procedure is to move the decimal point two places to the right.

km	**hm**	**dam**	**m**	**dm**	**cm**	**mm**
1000	100	10	1	0.1	0.01	0.001
				1	2	

38.8 m = 3 8 8 0 cm (decimal moved 1 2)

38.8 m = 3880 cm

Remember: To change a larger unit into a smaller unit, move the decimal point the appropriate number of spaces to the right.

EXERCISE 17–1 METRIC CONVERSIONS—LENGTH

17–1. 15 m = __________ cm.

17–2. 7 dam = __________ m.

17–3. 2 mm = __________ m.

17–4. 200 cm = __________ mm.

17–5. 53 m = __________ dam.

17–6. 3005 mm = __________ km.

17–7. 2 cm = __________ dm.

17–8. 54 hm = __________ m.

17–9. 44 cm = __________ mm.

17–10. 5280 mm = __________ m.

17–11. 12.4 km = __________ m.

17–12. 91.5 mm = __________ m.

17–13. 213.6 dm = __________ mm.

17–14. 2215 mm = __________ hm.

17–15. 35 750 mm = __________ km.

17.2 METRIC AREA

Recall that area is determined by finding the number of square units contained within an object. Area is expressed in square units such as square meters, square decimeters, and square centimeters. Square units may be abbreviated: for example, a square centimeter is abbreviated as cm^2. The exponent of 2 indicates square units.

Common abbreviations for area are as follows: km^2, hm^2, m^2, cm^2, and mm^2.

The land measure in the metric system is known as a *hectare.* A hectare equals 10,000 square meters, or a square piece of land 100 meters by 100 meters. The following examples illustrate the use of areas in the metric system.

Example 17–3: Find the Area in Hectares

PROBLEM: Find the number of hectares (ha) in a rectangular plot of land, 1.6 kilometers by 1.2 kilometers.

SOLUTION: Convert kilometers to meters for each of the measurements, since a hectare is equal to 10,000 square meters.

$$1.6 \text{ km} = 1600 \text{ m}$$
$$1.2 \text{ km} = 1200 \text{ m}$$

Using the formula for finding area, $A = lw$, find the area of the plot of land.

$$A = 1600 \text{ m} \times 1200 \text{ m}$$
$$A = 1{,}920{,}000 \text{ m}^2$$

Multiply 1,920,000 square meters by 1, $\frac{1 \text{ ha}}{10{,}000 \text{ m}^2}$ to find the number of hectares.

$$\overset{192}{\cancel{1{,}920{,}000}} \times \frac{1 \text{ ha}}{\underset{1}{\cancel{10{,}000}} \text{ m}^2} = 192 \text{ ha}$$

Another procedure for converting square meters to hectares is to divide the square meters (1,920,000) by the number of square meters in a hectare (10,000), since 1 hectare = 10,000 square meters.

$$\frac{1{,}920{,}000 \text{ m}^2}{10{,}000 \text{ m}^2} = 192 \text{ ha}$$

EXERCISE 17–2 METRIC CONVERSIONS—AREA

Complete the following area problems.

17–16. 15 dam × 25 dam = __________ m².

17–17. 54.5 cm × 35.5 cm = __________ m².

17–18. 0.88 dm × 0.44 dm = __________ cm².

17–19. 35.1 km × 0.5 km = __________ ha.

17–20. 0.0055 km × 0.0065 km = __________ cm².

17–21. A shelf in a closet measures 182 centimeters by 30 centimeters. How many square meters of shelving will be needed for four shelves?

17–22. How many hectares are in a wheat farm that is 1.25 kilometers by 0.35 kilometers?

17–23. How many square meters would there be in a table top that is 214 centimeters by 107 centimeters?

17–24. What is the surface area in square centimeters of a cubic box that has a side length of 45 millimeters?

17–25. Find the cost of installing carpeting in a room that is 400 centimeters by 350 centimeters if carpet, pad, and installation cost $29.95 per square meter.

17.3 METRIC VOLUMES

Volume is determined by finding the number of cubic units in an object. The basic unit of measure for volume in the metric system is the cubic meter (m^3). The exponent 3 indicates cubic measure. A common usage of cubic measurement is found in health occupations, where the term "cc" is used as an abbreviation for cubic centimeters.

The most common unit of measure capacity in the metric system is the liter. The liter is used extensively throughout the world, as a unit of volume.

The words "capacity" and "volume" are used interchangeably to mean the same thing. Technically, *volume* is the amount of space within a container, and *capacity* is the amount of substance that a container will hold. The relationship between volume and capacity in the metric system may best be seen in Figure 17–5.

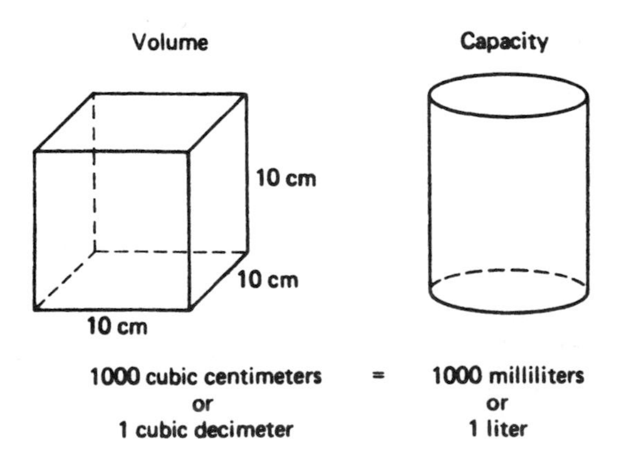

Figure 17–5

Figure 17–6 shows the relationship of a liter in liquid capacity to other units of liquid capacity. Note that the kiloliter, liter, and milliliter are the most commonly used measures.

Liquid Capacity						*Volume*
1 kiloliter (kL)	=	1000	liters	=	1000 dm^3	= 1.0 m^3
1 hectoliter (hL)	=	100	liters	=	100 dm^3	= 0.1 m^3
1 dekaliter (daL)	=	10	liters	=	10 dm^3	= 0.01 m^3
1 liter (L)	=	1	liter	=	1 dm^3	= 0.0001 m^3
1 deciliter (dL)	=	0.1	liter	=	100 cm^3	= 0.0001 m^3
1 centiliter (cL)	=	0.01	liter	=	10 cm^3	= 0.00001 m^3
1 milliliter (mL)	=	0.001	liter	=	1 cm^3	= 0.000001 m^3

Figure 17–6 Liter Table of Equivalents

Changing from one unit to another unit with a liter as the base unit can be accomplished by moving the decimal point to the appropriate number of spaces to the right or left. To change a larger unit (liter) to a smaller unit (deciliter), move the decimal point to the right the appropriate number of places (one). To change a smaller unit (liter) to a larger unit (dekaliter), move the decimal point to the left the appropriate number of places (one). Study the following examples that show the procedures.

Example 17–4: Convert Liters to Milliliters

PROBLEM: Convert 17 liters (L) to milliliters (mL).

SOLUTION: Using the liter chart, note that 1000 mL = 1 L. Multiply 17 liters by 1 or

$$\frac{1000 \text{ mL}}{1 \text{ L}}$$

$$\frac{17 \text{ L}}{1} \times \frac{1000 \text{ mL}}{1 \text{ L}} = 17{,}000 \text{ mL}$$

Another procedure for changing liters to milliliters is by multiplying the given number of liters (17) by 1000.

$$17 \text{ L} \times 1000 = 17{,}000 \text{ mL}$$

The simplest solution, however, is to move the decimal point three places to the right.

kL	hL	daL	L	dL	cL	mL
1000	100	10	1	0.1	0.01	0.001
				1	2	3

$$17\text{ L} = 17.\underbrace{0}_{1}\underbrace{0}_{2}\underbrace{0}_{3}\text{ mL}$$

$$17\text{ L} = 17000\text{ mL}$$

Example 17–5: Find the Volume in Liters

PROBLEM: Find the volume in cubic centimeters and liters of the aquarium shown in Figure 17–7.

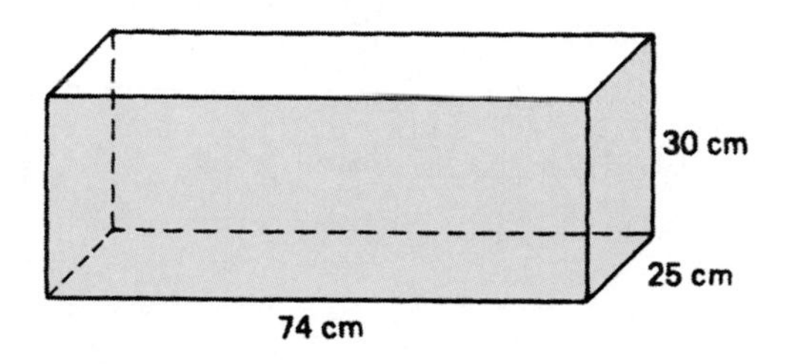

Figure 17–7

SOLUTION: Find the volume of the aquarium. Recall the formula for finding volume: volume (V) = length (l) × width (w) × height (h). Thus

$$\begin{aligned} V &= lwh \\ &= (74\text{ cm})(25\text{ cm})(30\text{ cm}) \\ &= 55{,}500\text{ cm}^3 \end{aligned}$$

Convert cubic centimeters to milliliters. Recall that 1 cubic centimeter = 1 milliliter. Thus

$$55{,}500\text{ cm}^3 = 55{,}500\text{ mL}$$

Then convert 55,500 milliliters to liters. (*Note:* 1000 milliliters = 1 liter.) Thus

$$55{,}500\text{ mL} = 55.\underbrace{5}_{3}\underbrace{0}_{2}\underbrace{0}_{1}\text{ L}$$

$$55{,}500\text{ mL} = 55.5\text{ L}$$

EXERCISE 17–3 METRIC CONVERSIONS—VOLUME

Solve the following as indicated, using a conversion table.

17–26. 77 L = __________ mL.

17–27. 25 cc = __________ cm^3.

17–28. 3.8 dm^3 = __________ cm^3.

17–29. 55 mL = __________ cL.

17–30. 0.66 cm^3 = __________ dm^3.

17–31. 8880 mL = __________ L.

17–32. $747{,}720 \text{ mm}^3 =$ __________ m^3.

17–33. $76.7 \text{ cc} =$ __________ cm^3.

17–34. $0.0075 \text{ dm}^3 =$ __________ mm^3.

17–35. 3836 dL = __________ kL.

17–36. $1.38 \text{ m}^3 =$ __________ mm^3.

17–37. 5001 L = __________ hL.

17–38. $0.0076 \text{ cm}^3 =$ __________ mm^3.

17–39. 3.27 mL = __________ L.

17–40. Find the capacity in liters of a gasoline tank 5.5 meters long, 2.5 meters wide, and 2 meters deep.

17–41. Find the volume in cubic centimeters of a soft-water container 2.1 meters long, 24 centimeters wide, and 26 millimeters deep.

17–42. Find the volume in cubic meters of a cube with one side 340 centimeters long.

17–43. An excavation for a building is to be 55 meters by 40 meters by 15 meters. Find the cost of excavation at $4.85 per cubic meter.

17–44. A plastic underlayment sheet for a bike path is 10 kilometers by 2 meters by 4 millimeters. How many cubic meters of plastic are in the sheet?

17–45. A swimming pool 25 meters long and 10 meters wide has an average depth of 1.5 meters. How many kiloliters of water will the pool hold?

17–46. How many 500-milliliter containers of honey could be filled from 100 hectoliters?

17–47. How many cubic meters of blacktop are needed to make a driveway 10 meters long, 5 meters wide, and 12 centimeters thick?

17–48. If rain fell at the rate of 2 millimeters per minute, how many minutes would it take to reach 0.1 meters?

17–49. Find the volume in liters of a rectangular water cooling tank 2 meters by 20 decimeters by 28 centimeters.

17–50. A patient was given 20 cubic centimeters of a medicine in 500 milliliters of water. What percent of the solution was medicine?

17.4 METRIC WEIGHT

Mass is defined as the quantity of material contained in a given body. *Weight* is the measure of the force of gravity on a given body. These are the technical definitions of mass and weight; however, they are frequently used to mean the same thing. For example, on the planet Earth, the terms "mass" and "weight" may be used interchangeably because the Earth's gravity affects objects in a similar manner. However, on the moon, a mass weighs about $\frac{1}{6}$ of its weight on the Earth. This is due to the moon's gravity, which is about $\frac{1}{6}$ of the Earth's gravity. For our discussion we assume that mass or weight may be used similarly.

The basic SI metric unit for measuring mass is the kilogram. An average-sized paper clip weighs about 1 gram. Thus a kilogram is about the weight of 1000 paper clips each of which weighs 0.001 of a kilogram.

The relationship between volume capacity and weight in the metric system can be seen in Figure 17–8. The following should be noted:

$$1 \text{ cm}^3 = 1 \text{ mL} = 1 \text{ g}$$
$$1000 \text{ cm}^3 = 1 \text{ L} = 1000 \text{ g}$$
$$1 \text{ dm}^3 = 1 \text{ L} = 1 \text{ kg}$$

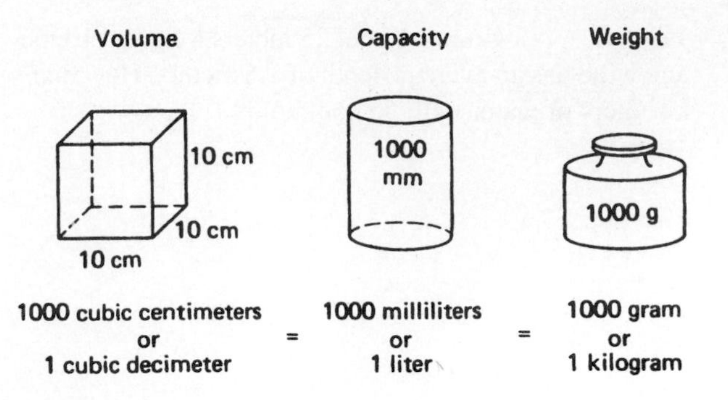

Figure 17–8

Figure 17–9 shows the relationship of a kilogram of weight to other units of weight, volume, and capacity in the metric system.

Weight	*Liquid Capacity*	*Volume*
1000 kilograms (kg) = 1 metric ton (t)	= 1 kL	= 1 m^3
1 kilogram (kg) = 1000 grams	= 1 L	= 1 dm^3
1 hectogram (hg) = 100 grams	= 0.1 L	= 100 cm^3
1 dekagram (dag) = 10 grams	= 0.01 L	= 10 cm^3
1 gram (g) = 1 gram	= 1 mL	= 1 cm^3
1 decigram (dg) = 0.1 gram	= 0.1 mL	= 1 mm^3
1 centigram (cg) = 0.01 gram	= 0.01 mL	= 0.1 mm^3
1 milligram (mg) = 0.001 gram	= 0.001 mL	= 0.01 mm^3

Figure 17–9 Metric System Equivalents

It should be noted that 1000 kilograms = 1 metric ton = 1 kiloliter = 1 cubic meter. The procedures for changing units in the metric weight system are the same as those shown previously for changing other metric units. Study the examples that show the procedures.

Example 17–6: Convert Milligrams to Grams

PROBLEM: Convert 2700 milligrams (mg) to grams (g).

SOLUTION: Using the kilogram table of equivalents, 1000 milligrams = 1 gram. Multiply 2700 by 1:

$$\frac{1\text{ g}}{1000\text{ mg}}.$$

$$\frac{\overset{2.7}{\cancel{2700}}\text{ mg}}{1} \times \frac{1\text{ g}}{\underset{1}{\cancel{1000}}\text{ mg}} = 2.7\text{ g}$$

We could change the milligrams into grams by dividing the milligrams (2700) by 1000.

$$\frac{2700\text{ mg}}{1000} = 2.7\text{ g}$$

The easiest way to change mg to g is to move the decimal point three places to the left.

kg	**hg**	**dag**	**g**	**dg**	**cd**	**mg**
1000	100	10	1	0.1	0.01	0.001
				3	2	1

2700 mg = 2.7 0 0 g (3 2 1)

2700 mg = 2.7 g

Example 17–7: Convert Cubic Centimeters to Kilograms

PROBLEM: What is the weight in kilograms of gasoline in a rectangular tank 1.3 meters long, 80 centimeters wide and 42 centimeters deep? (Gasoline weighs 0.8 times as much as water.)

SOLUTION: Change the measurements to centimeters: 1 m = 100 cm and 1 dm = 10 cm. Thus

1.3 m = 130 cm
80 cm = 80 cm
42 cm = 42 cm

Find the volume of the tank using the formula $V = lwh$.

$$V = (130\text{ cm})(80\text{ cm})(42\text{ cm})$$
$$V = 436{,}800\text{ cm}^3$$

Change cubic centimeters to kilograms using the formula 1 kg = 1000 cm^3.

$$\frac{436{,}800\text{ cm}^3}{1}\left(\frac{1\text{ kg}}{1000\text{ cm}^3}\right) = 436.8\text{ kg}$$

Multiply to find the weight of the gasoline, which is 0.8 times the weight of water.

(436.8 kg) (0.8 weight of gasoline) = 349.44 kg

Thus the tank contains 349.44 kilograms.

EXERCISE 17–4 METRIC CONVERSION—WEIGHT

The following problems will give you the opportunity to develop your skills with weight measurements. Use any tables you need to solve the problems.

17–51. Convert 3.8 kilograms to grams.

17–52. Convert 520 milligrams to grams.

17–53. Convert 81.3 decigrams to milligrams.

17–54. Convert 7.8 centigrams to grams.

17–55. Convert 57.3 dekagrams to milligrams.

17–56. 7886 mg = __________ g.

17–57. 85.6 dg = __________ g.

17–58. 45,000 g = __________ kg.

17–59. 800 dag = __________ kg.

17–60. 3804 kg = __________ metric tons.

17–61. 84 g = __________ cg.

17–62. 81 dg = __________ mg.

17–63. 8.08 kg = __________ cg.

17–64. 5.54 tons = __________ kg.

17–65. Find the sum of the following weights: 4.3 g, 4.9 cg, 781 mg, and 5.05 cg.

17–66. Find the sum of the following: 0.83 kg, 0.38 kg, 33.3 g, 5050 mg, and 4.4 dg.

17–67. What is the weight in kilograms of 34 castings that weigh 386 grams each?

17–68. An irregular-shaped plastic container weighed 1400 grams when empty. When filled with water it weighed 5.4 kilograms. Find the volume of the container in cubic centimeters.

17–69. If there are 8 grams of polluting substance per liter of sewage, how many kilograms of polluting matter remain per 700 kiloliters?

17–70. Find the weight in metric tons of 54 pieces of plastic weighing 85.6 kilograms each.

17–71. A chemist has a 2-liter beaker of an aqueous solution containing 550 grams of sugar. The air evaporates 300 cubic centimeters of the solution. What is the weight of the remaining solution?

17–72. A brick measures 5.8 centimeters by 9.5 centimeters by 20.2 centimeters. Find its weight in kilograms (to the nearest tenth) if bricks weigh 2.4 grams per cubic centimeter.

17–73. Determine the weight of the water in metric tons in a swimming pool 25 meters by 10 meters by 2 meters.

17–74. A beaker holds 500 milliliters of a salt solution containing 15 grams of salt. How much salt must be added to make it a 10% solution?

17–75. Determine the weight of a solid piece of brass 95 millimeters thick, 5 centimeters wide, and 3 meters long. (Brass weighs 8.2 kilograms per cubic decimeter.)

17.5 TEMPERATURE

The basic SI unit for measuring temperature is the *Kelvin* (K) scale. The Kelvin scale is used by scientists and is based on a scale from absolute zero where water freezes at 273.15°K and boils at 373.15°K. Since this scale is not very practical for general use, the Celsius (C) or centigrade scale is used in the metric system. The Celsius scale, based on the kelvin system, indicates water freezing at 0 degrees and boiling at 100 degrees. Thus 1 degree Celsius is equal to 1 degree Kelvin. Figure 17–10 shows the relationships of the three types of measurement scales.

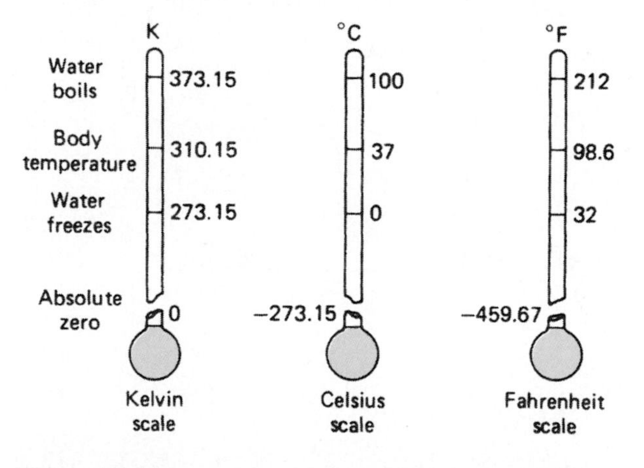

Figure 17–10

The United States may adopt the Celsius system of measuring temperature to replace the Fahrenheit temperature measurement. Thus it is necessary to be able to convert from Fahrenheit to Celsius and Celsius to Fahrenheit. The United States is the largest country in the world that does not use the Celsius temperature system. The following temperature conversion formulas will be helpful.

From Celsius to Fahrenheit: $F = 1.8C + 32°$

From Fahrenheit to Celsius: $C = \frac{5}{9}(F - 32°)$

The following examples show how the formulas are used.

Example 17–8: Convert Celsius to Fahrenheit

PROBLEM: Convert 40°C to °F

SOLUTION: Using the formula $F = 1.8C + 32°$, substitute 40° for C and solve the formula for the unknown, F.

$$\begin{aligned} F &= 1.8C + 32° \\ &= 1.8\,(40°) + 32 \\ &= 72° + 32° \\ &= 104° \end{aligned}$$

Example 17–9: Convert Fahrenheit to Celsius

PROBLEM: A chef has an oven with a Celsius thermostat. A recipe calls for a cake to be baked at 320°F. Find the Celsius setting.

SOLUTION: Use the formula $C = \frac{5}{9}(F - 32°)$. Substitute 320° for F and solve for the unknown, C.

$$\begin{aligned} C &= \frac{5}{9}(320° - 32°) \\ &= \frac{5}{\cancel{9}_{1}}\left(\frac{\cancel{288°}^{32°}}{1}\right) \\ &= 160° \end{aligned}$$

Thus the Celsius thermostat setting would be 160°.

EXERCISE 17–5 TEMPERATURE CONVERSIONS

Converting from one measurement to another is simple when the correct formulas are used. Use the formulas to convert the following temperatures.

17–76. 20°C = __________ °F.

17–77. 28°C = __________ °F.

17–78. 98.6°F = __________ °C.

17–79. 100°F = __________ °C.

17–80. 212°C = __________ °F.

17–81. 62°F = __________ °C.

17–82. 45°F = __________ °C.

17–83. −30°F = __________ °C.

17–84. 72°F = __________ °C.

17–85. 85°F = __________ °C.

17–86. Sleet occurs when the temperature is about 27°F. What is the Celsius equivalent to this temperature?

17–87. Wrought iron melts at 1550°C. What temperature is this on the Fahrenheit scale?

17–88. The temperature in the oxyacetylene process of welding sometimes reaches 5800°F. What is the temperature on the Celsius scale?

17–89. The temperature dropped from 20°F to −10°F in 5 hours. What would the change of temperature be in degrees Celsius?

17–90. At what temperature is the reading on the Celsius thermometer the same as the reading on the Fahrenheit thermometer?

17.6 CONVERTING BETWEEN THE METRIC AND U.S. SYSTEMS

As business and industry continue to become international, it is a valuable skill to understand and use the metric system. Thus it is essential to make conversions from one system to the other, although frequently measurements are given in both systems which is known as dual dimensioning. There is a need to be able to work efficiently with both systems and to convert measurements from one system to the other. Electronic calculators, conversion charts, and other devices aid in the conversion process.

You may find it necessary to convert measurements from one system to the other, so it is important for you to learn the few common approximate equivalents shown in Figure 17–11.

English to Metric		*Metric to English*	
1 inch	= 2.54 cm	1 cm	= 0.3937 inch
1 yard	= 0.9144 m	1 m	= 1.0936 yards
1 mile	= 1.609 km	1 km	= 0.6213 mile
1 ounce (dry)	= 28.349 g	1 g	= 0.0352 ounce
1 pound (dry)	= 0.4536 kg	1 kg	= 2.2046 pounds
1 ounce (liquid)	= 29.75 mL	1 mL	= 0.0338 ounce
1 quart (liquid)	= 0.946 L	1 L	= 1.0568 quarts

Figure 17–11 Common U.S. and Metric Equivalents

The basic conversions given, with your understanding of the metric and U.S. systems, should allow you to move from one system to the other. You may find it helpful to memorize some of the commonly used conversion factors. Additional conversion factors are stated in Figures 17–12 to 17–15.

Recall that this procedure is similar to the equivalence method from Chapter 6. The examples will give you an

English to Metric		*Metric to English*	
1 in.	= 2.54 cm	1 mm	= 0.03937 in.
	= 25.4 mm	1 cm	= 0.3937 in.
1 ft	= 30.48 cm		= 0.0328 ft
	= 0.3048 m	1 m	= 39.37 in.
1 yd	= 91.44 cm		= 3.2808 ft
	= 0.9144 m	1 km	= 0.62136 mile
1 mile	= 1609.344 m		= 3280.8 ft

Figure 17–12 Length Equivalents

English to Metric		*Metric to English*	
1 sq in.	= 6.4516 cm^2	1 cm^2	= 0.155 sq in.
1 sq ft	= 0.09290 m^2	1 m^2	= 10.7639 sq ft
1 sq yd	= 0.83613 m^2	1 hectare	= 2.471 acres
1 acre	= 0.4047 ha		

Figure 17–13 Area Equivalents

English to Metric		*Metric to English*	
1 cu in.	= 16.387 cm^3	1 cm^3	= 0.061 cu in.
1 cu ft	= 28.317 dm^3	1 dm^3	= 61.0237 cu in.
	= 0.0283 m^3		= 0.0353 cu ft
1 cu yd	= 0.7646 m^3	1 m^3	= 35.3147 cu ft
			= 1.3079 cu yd

Figure 17–14 Volume Equivalents

English to Metric		*Metric to English*	
1 pound	= 0.4536 kg	1 kilogram	= 2.2406 lb
1 short ton	= 907.2 kg	1 gram	= 0.0352 oz
1 long ton	= 1016 kg	1 metric ton	= 2204.6 lb

Figure 17–15 Weight and Mass Equivalents

opportunity to study the conversion procedures and to practice making conversions.

Example 17–10: Convert Yards to Meters

PROBLEM: How long is a 100-yard football field in meters?

SOLUTION: Using the conversion fact 1 yard = 0.9144 meter, multiply by $\dfrac{0.9144\text{ m}}{1\text{ yd}}$.

$$\left(\frac{100\ \cancel{\text{yd}}}{1}\right)\left(\frac{0.9144\text{ m}}{1\ \cancel{\text{yd}}}\right) = 91.44\text{ m}$$

Thus a 100-yard football field is 91.44 meters long.

Example 17–11: Convert Miles to Kilometers

PROBLEM: The speed limit is 65 miles per hour. What is this speed limit in kilometers per hour?

SOLUTION: Use the conversion fact 1 mile = 1.609 kilometers. Thus $\dfrac{1.609\text{ kmph}}{1\text{ mph}}$.

$$\left(\frac{65\ \cancel{\text{mph}}}{1}\right)\left(\frac{1.609\text{ kmph}}{1\ \cancel{\text{mph}}}\right) = 104.585\text{ kmph}$$

Thus 65 miles per hour is equal to 104.6 kilometers per hour.

Example 17–12: Convert Liters to Gallons

PROBLEM: The capacity of a gasoline tank in a German sportscar is 42 liters. How many gallons is this to the nearest tenth of a gallon?

SOLUTION: Using the conversion factor 1 liter = 1.0568 quarts and 4 quarts = 1 gallon, set up the procedures and solve for gallons.

$$\left(\frac{42\ \cancel{\text{L}}}{1}\right)\left(\frac{1.0568\ \cancel{\text{qt}}}{1\ \cancel{\text{L}}}\right)\left(\frac{1\text{ gal}}{4\ \cancel{\text{qt}}}\right) = 11.0964\text{ gal}$$

Thus 42 liters is equal to about 11.1 gallons.

Example 17–13: Convert Grams to Pounds

PROBLEM: A French chef is to prepare 14 servings of filet mignon for a party. Each filet should weigh about 250 grams. How many pounds of steak will the chef need to order from his suppliers?

SOLUTION: Change grams to kilograms by moving the decimal point three places to the left.

$$250\text{ g} = .\underbrace{2\ 5\ 0}_{3\ 2\ 1}\text{ kg}$$

$$250\text{ g} = 0.25\text{ kg}$$

Then find the total amount of meat needed in kilograms.

$$14\text{ servings} \times 0.25\text{ kg per serving} = 3.5\text{ kg of meat}$$

Using the conversion factor 1 kilogram = 2.2046 pounds, set up the procedure and solve for pounds.

$$\left(\frac{3.5\ \cancel{\text{kg}}}{1}\right)\left(\frac{2.2046\text{ lb}}{1\ \cancel{\text{kg}}}\right) = 7.7161\text{ lb of meat}$$

Thus the chef should order 7.72 pounds of meat from his suppliers to serve 14 people 250-gram filets.

EXERCISE 17–6 CONVERSIONS BETWEEN MEASUREMENT SYSTEMS

Conversion from metric to U.S. and U.S. to metric is not difficult when the correct facts are used and the proper mathematical operations are applied. Practice making conversions using the correct conversion formulas.

17–91. 1 ft = __________ cm.

17–92. 1 mile = __________ m.

17–93. 8.3 km = __________ miles.

17–94. 105 mm = __________ in.

17–95. 55 m = __________ yd.

17–96. 24 in. = __________ mm.

17–97. 66 sq yd = __________ m^2.

17–98. 48 cm^2 = __________ sq in.

17–99. 66.2 mm^2 = __________ sq in.

17–100. What is the size in centimeters of a 12-inch by 12-inch tile?

17–101. Find the area of a 12-inch by 12-inch tile in square centimeters.

17–102. $\frac{1}{2}$ gal = __________ L.

17–103. 25 cu in. = __________ cm^3.

17–104. 1 L = __________ oz.

17–105. 0.35 kg = __________ oz.

17–106. One-half kilogram of nails costs \$0.98. What is the cost per pound?

17–107. 25,000 cu in. = __________ cm^3.

17–108. A 12-ounce container of frozen orange juice is equal to how many ml?

17–109. How many cubic meters of earth will need to be excavated for a house 80 feet by 30 feet and 4 feet deep?

17–110. A contractor needs 150 cubic yards of concrete for a job. How many cubic meters is this?

17–111. 220 lb = __________ kg.

17–112. 53 kg = __________ lb.

17–113. A 100-pound sack of potatoes is equal to how many kilograms?

17–114. Number 1 spring wheat weighs 60 pounds per bushel (1.244 cubic feet). How many kilograms does the wheat weigh per cubic decimeter?

17–115. A liter of milk equals how many ounces?

17–116. A spark gap is 1.5 millimeters. How many thousandths of an inch is this?

17–117. A closet shelf is 1.83 meters long by 61 centimeters wide. How long and wide is the shelf to the nearest inch?

17–118. A garden is 150 feet by 30 feet. How many square meters is the garden?

17–119. A jet flies from New York to San Francisco, a distance of 2560 miles, in $4\frac{1}{2}$ hours. How many kilometers per hour does the jet fly?

17–120. Determine the cost of carpeting a room 8 meters by 4.5 meters at $14.50 per square yard.

17–121. How tall is a 6 foot 5 inch person in centimeters?

17–122. What is the volume in kiloliters of a pool 50 feet by 25 feet and 6 feet deep?

17–123. Father Andy Marthaler drove 600 kilometers on 93.5 liters of gasoline on a Canadian fishing trip. Determine his average miles per gallon.

17–124. Five quarts of antifreeze are added to a 22-quart cooling system. How many liters of antifreeze would be needed for a 100-quart cooling system?

17–125. Water expands 9% when becoming ice. How large a cube of ice in cubic centimeters would you get from 0.57 kilogram of water?

THINK TIME

In time, personal data will probably be known in metric measurement as well as U.S. measurement. Complete your personal data in the space provided in Figure 17–16.

Measure	*Type of Unit*	*Estimate*	*Actual*
Height	centimeters		
Weight	kilograms		
Chest/bust	centimeters		
Waist	centimeters		
Hips	centimeters		
Foot	centimeters		

Figure 17–16 Personal Data

Estimate your measurements; this will help you relate measurement to the metric system. Try estimating and measuring other common objects that are listed in Figure 17–17. Add them to the list.

Measure	*Type of Unit*	*Estimate*	*Actual*
Height of doorway	meters		
Weight of car	kilograms		
Distance to work	kilometers		
Amount of rainfall	centimeters		
Present temperature	Celsius		

Figure 17–17 Measure of Common Objects

PROCEDURES TO REMEMBER

1. To change from one unit to another in the metric system, move the decimal point the appropriate number of places.
   (a) To change a smaller unit to a larger unit, move the decimal point the appropriate number of places to the left.
   (b) To change a larger unit to a smaller unit, move the decimal point the appropriate number of places to the right.
2. To change from Celsius to Fahrenheit, multiply the Celsius degrees by 1.8, and add 32 degrees to the product; the result is degrees Fahrenheit: $F = 1.8C + 32°$.
3. To change from Fahrenheit to Celsius, subtract 32 degrees from the given Fahrenheit degrees, and multiply the difference by $^5/_9$; the result equals degrees Celsius: $C = {}^5/_9\,(F - 32°)$.
4. To convert from U.S. to metric or metric to U.S. select the appropriate conversion fact and multiply by the conversion factor to obtain the new value.

CHAPTER SUMMARY

1. The seven basic units of measure in the SI metric system are meter (length), kilogram (mass or weight), kelvin (temperature), second (time), ampere (electricity), candela (luminous intensity), and mole (amount of substance).
2. Two common metric units that are not part of the SI metric system are liter (capacity) and Celsius (temperature).
3. The interrelationship between volume, capacity, and weight in the metric system is as follows:

Volume		*Capacity*		*Weight*
1 cm^3 of water	=	1 mL of water	=	1 g of water
or		or		or
1 dm^3 of water	=	1 L of water	=	1 kg of water

Figure 17–18

4. Greek prefixes identify units *larger* than the base unit: *kilo* = 1000, *hecto* = 100, and *deka* = 10.
5. Latin prefixes identify units *smaller* than the base unit: *deci* = 0.1, *centi* = 0.01, and *milli* = 0.001.
6. Metric units, letter units, power of 10, and decimal units in the metric system are shown in Figure 17–2.
7. *Mass* is the quantity of material contained in a given body.
8. *Weight* is the measure of the force of gravity upon a given body.

CHAPTER TEST

T–17–1. 33 km = __________ m.

T–17–2. 1748 mL = __________ L.

T–17–3. 5280 m = __________ km.

T–17–4. 31.5 cm = __________ m.

T–17–5. 84 sq in. = __________ cm^2.

T–17–6. 0.527 m^2 = __________ sq in.

T–17–7. 500 mL = __________ L.

T–17–8. 31.5 gal = __________ L.

T–17–9. 21.5 L = __________ gal.

T–17–10. 1500 g of water = __________ mL.

T–17–11. 44.2 dag = __________ kg.

T–17–12. 53,000 kg = __________ metric tons.

T–17–13. 5000 mg = __________ g.

T–17–14. 35.3 dg = __________ g.

T–17–15. 26,400 ft = __________ km.

T–17–16. 100 ft = __________ m.

T–17–17. 90°C = __________ °F.

T–17–18. 95°F = __________ °C.

T–17–19. A house contains 4000 square feet of floor space. How many square meters is this?

T–17–20. A 12-millimeter wrench has an opening of how many thousandths of an inch?

T–17–21. How many inches in diameter is a 155-millimeter howitzer?

T–17–22. A mixture of 8 grams of salt and 8 grams of baking soda is poured into 225 milliliters of water. What is the total weight of the solution?

T–17–23. A radio station broadcasts at 100 megahertz. How many kilohertz is this?

T–17–24. At a cost of $12.50 per square meter, how much would it cost to blacktop a 50-foot by 16-foot driveway?

T–17–25. Two liters of concentrated liquid fertilizer are added to water to make a total of 10 gallons of plant food. How many milliliters of fertilizer are added per gallon?

FUNDAMENTALS OF TRIGONOMETRY

OBJECTIVES

After studying this chapter, you will be able to:

18.1. Understand basic definitions and properties of trigonometry.
18.2. Discuss right and similar triangles and their properties.
18.3. Understand definitions of trigonometric ratios and their properties.
18.4. Determine the values of trigonometric functions.

SELF-TEST

This test will measure your understanding of basic trigonometry. Successful completion will indicate that you are ready to do more complex trigonometry problems in Chapters 19 and 20.

S–18–1. How many degrees are in $\angle A$ in Figure 18–1?

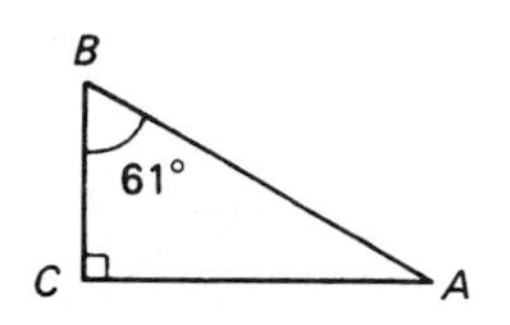

Figure 18–1

S–18–2. How many degrees are in $\angle E$ in triangle DEF in Figure 18–2?

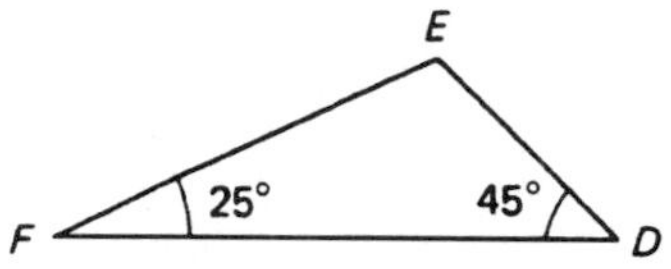

Figure 18–2

S–18–3. Letter the sides of triangle GHI in Figure 18–3.

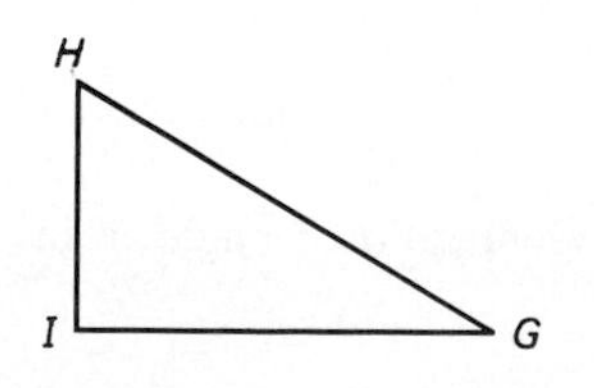

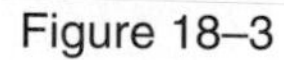
Figure 18–3

S–18–4. Letter the sides of triangle MLK in Figure 18–4.

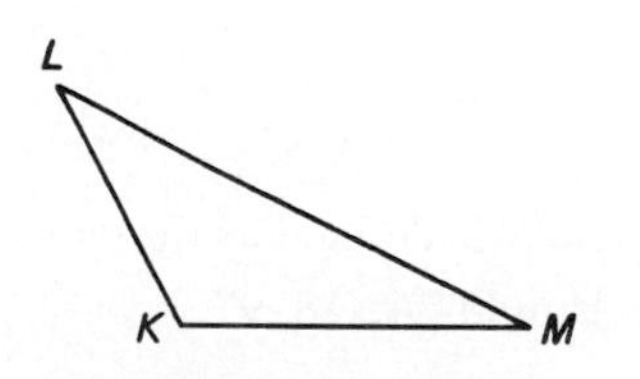

Figure 18–4

S–18–5. Letter the angles of the triangle in Figure 18–5.

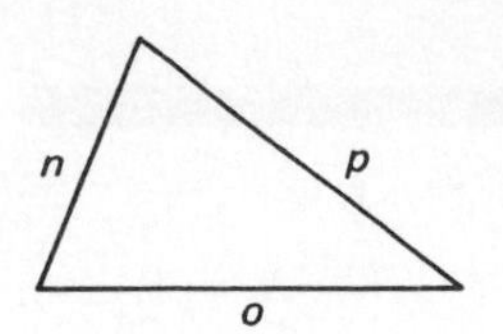

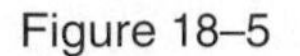
Figure 18–5

S–18–6. Given similar triangles *QRS* and *TUV*, find *q* in Figure 18–6.

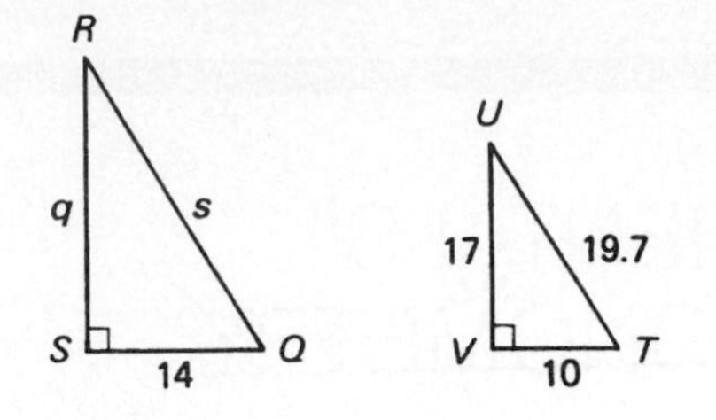

Figure 18–6

S–18–7. Find *s* in Figure 18–6.

S–18–8. Given similar triangles *ABC* and *DEF* in Figure 18–7, find ∠*A*.

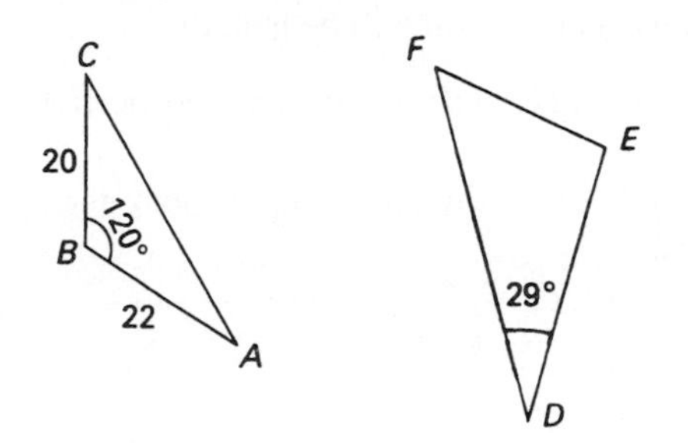

Figure 18–7

S–18–9. Find the size of ∠*F* in Figure 18–7.

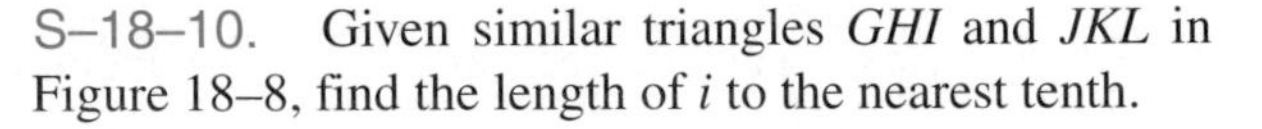
S–18–10. Given similar triangles *GHI* and *JKL* in Figure 18–8, find the length of *i* to the nearest tenth.

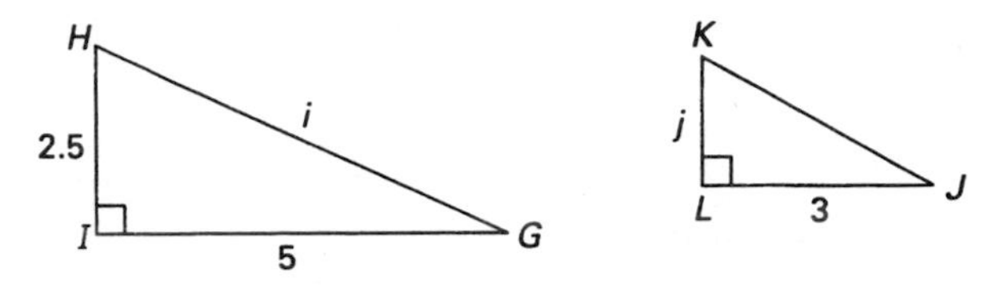

Figure 18–8

S–18–11. Find the length of *j* to the nearest tenth in Figure 18–8, given the similar triangles *GHI* and *JKL*.

S–18–12. Find the length of *l* to the nearest tenth in Figure 18–8, given the similar triangles *GHI* and *JKL*.

S–18–13. The sine of angle *N* is defined as the ratio $\frac{n}{o}$. What is the sine of angle *M* in Figure 18–9?

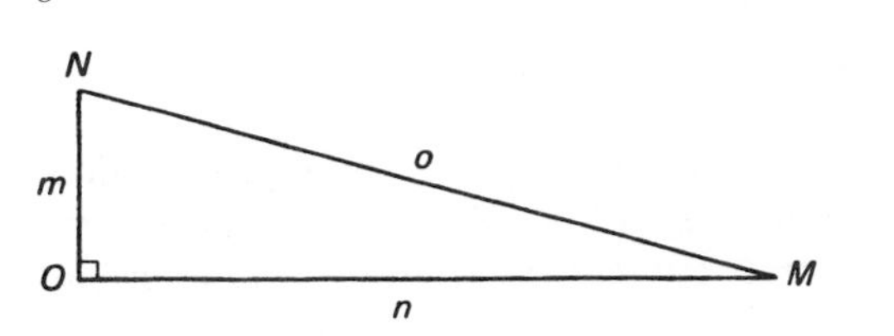

Figure 18–9

S–18–14. What is the tangent ratio of angle *N* in Figure 18–9?

S–18–15. What is the cosine ratio of angle M in Figure 18–9?

S–18–16. What trigonometric function equals the ratio $\frac{q}{r}$ in Figure 18–10?

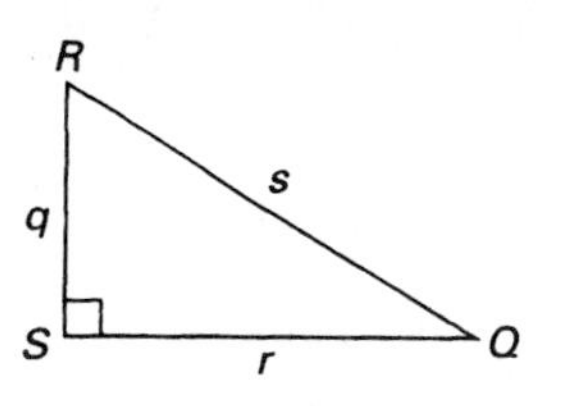

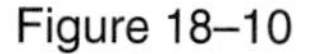
Figure 18–10

S–18–17. In Figure 18–10, what trigonometric function equals the ratio $\frac{r}{q}$?

S–18–18. In Figure 18–10, what two trigonometric functions equal the ratio $\frac{r}{s}$?

S–18–19. If the trigonometric function of the sine of angle A is 0.5000, what is the size of angle A?

S–18–20. If the trigonometric function of the tangent of angle R is 0.577, what is the size of angle R?

18.1 BASIC DEFINITIONS AND PROPERTIES OF TRIGONOMETRY

Trigonometry is a combination of arithmetic, algebra, and geometry that is used by tradespeople and technicians to solve applied problems. The word *trigonometry* is derived from the Greek words *trigonon* and *metria,* which mean "triangle measure." However, trigonometry has expanded and includes the solution of problems dealing with squares, circles, rectangles, and other geometric figures. Trigonometry, or "trig," as it is commonly called, is essential for solving certain problems—solutions that could not be found by any other branch of mathematics.

All triangles have six parts: three sides and three angles. Finding the six parts of a triangle is called *solving the triangle.* A *right triangle* is a triangle that has a right or 90° angle.

Figure 18–11 is a right triangle. Study the names of

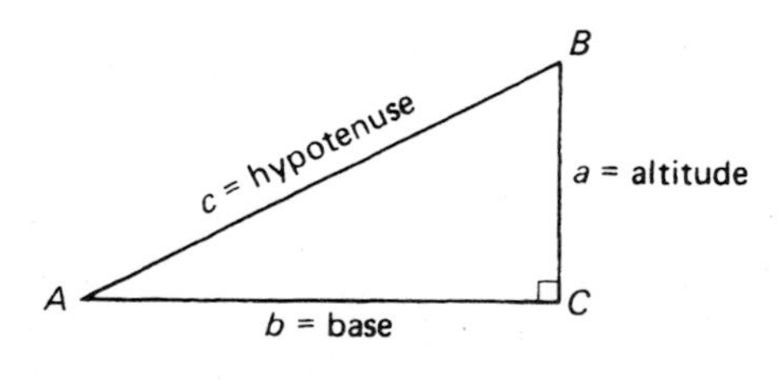

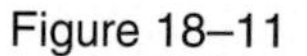
Figure 18–11

the parts of the triangle. Angle C is a right angle. The letter C is generally assigned to the right angle. The symbol ⊾ is used to identify a right angle. Angles are identified by capital letters. Each side is identified by a lowercase letter, the same letter as that given to the angle opposite the side. Thus side a is opposite angle A, side b is opposite angle B, and side c is opposite angle C.

Each of the sides has a name. Side a is the altitude, side b is the base, and side c is the hypotenuse. However, the positioning and the shape of the right triangle may not always identify a base or altitude. The side opposite the right angle is always called the *hypotenuse.*

The problems and solutions that follow will illustrate other properties of right triangles.

Example 18–1: Find the Sum of the Angles of a Right Triangle

PROBLEM: Find the sum in degrees of the angles of the right triangle shown in Figure 18–12.

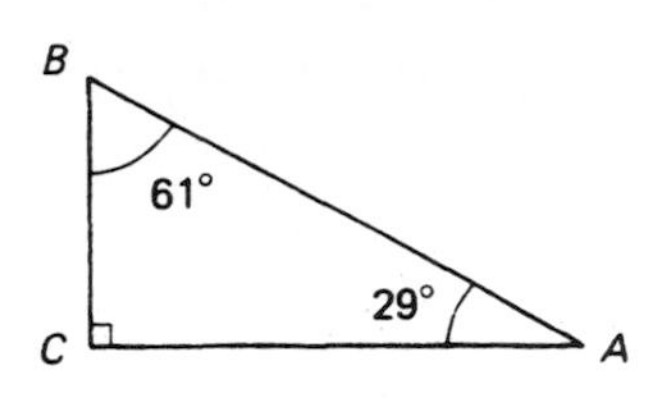

Figure 18–12

SOLUTION: The sum, in degrees, of the angles of any triangle will always be 180°. Measure the two unknown angles with a protractor given that $\angle C$ equals 90°. If only one angle is unknown, subtract the sum of the two known angles from 180°. Then check by adding the degrees of the angles to equal 180. Thus $90° + 61° + 29° = 180°$.

Another way to prove the total degrees of a triangle is 180 is to tear the angles from a paper triangle and arrange them together to form a straight angle or 180°. See Figure 18–13. Thus the sum of the interior angles of any triangle is 180°.

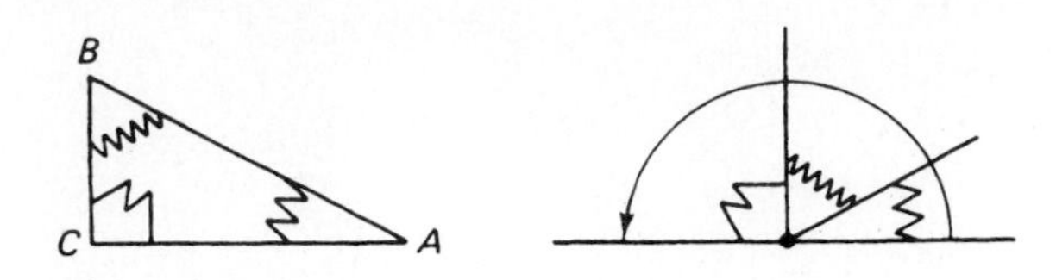

Figure 18–13

Example 18–2: Find the Sum of the Angle of any Triangle

PROBLEM: Find the total degrees of the angles of *any* triangle as shown by Figure 18–14.

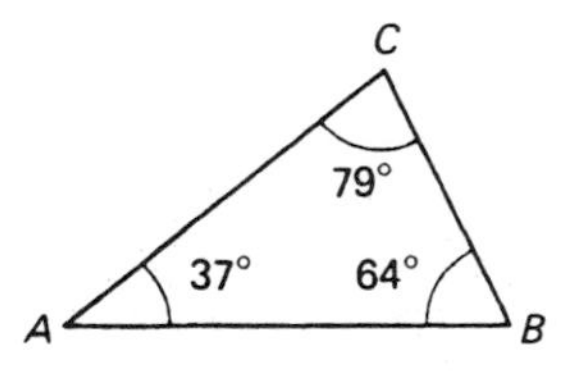

Figure 18–14

SOLUTION: Measure two unknown angles with a protractor if necessary. If one angle is known, measure only one angle. Second, subtract the sum of the degrees of the two known angles from 180°. Third, check by adding the degrees of the angles to total 180°. Thus $37° + 64° + 79° = 180°$.

EXERCISE 18–1 DEFINITIONS AND PROPERTIES OF TRIANGLES

Name the triangles and their parts as indicated in the following problems.

18–1. What type of triangle is ABC in Figure 18–15?

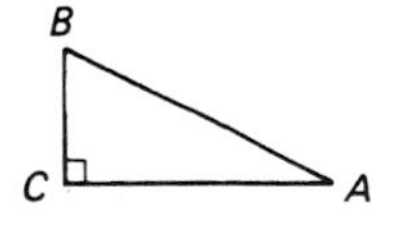

Figure 18–15

18–2. What type of angle is $\angle C$?

18–3. Letter the sides of the triangle in Figure 18–16.

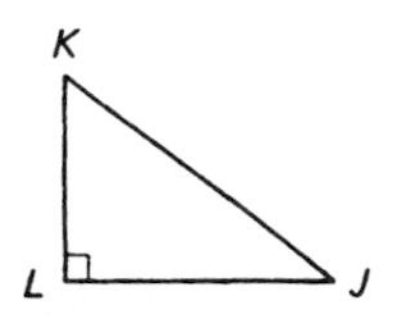

Figure 18–16

18–4. Letter the sides of the triangle in Figure 18–17.

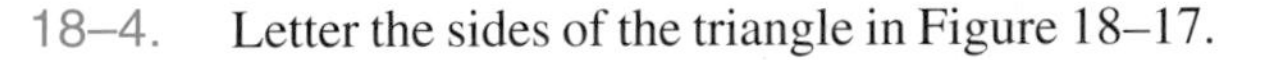

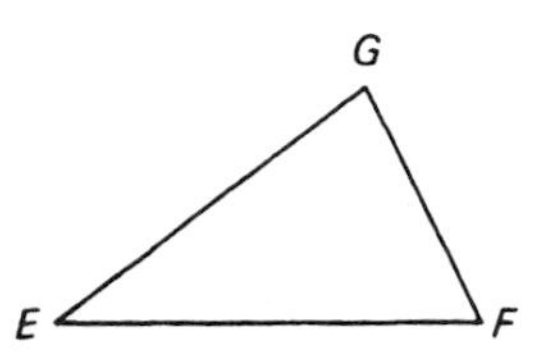

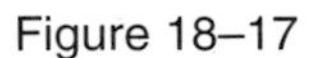

Figure 18–17

18–5. Letter the angles of the triangle in Figure 18–18.

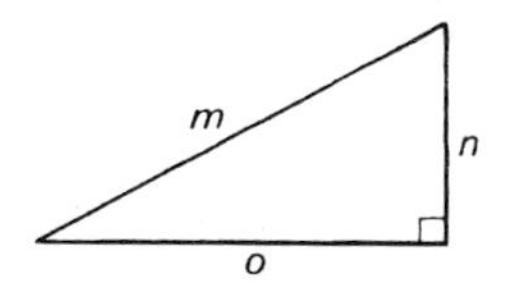

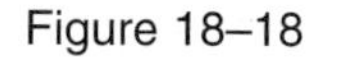
Figure 18–18

18–6. Label the sides of the triangle in Figure 18–19.

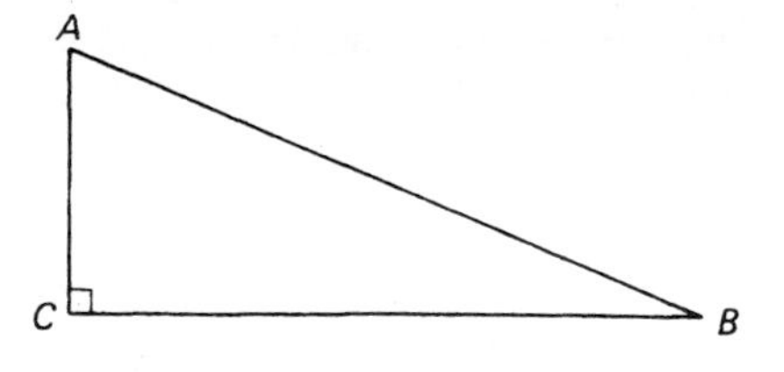

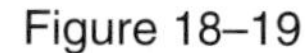
Figure 18–19

18–7. Label the sides of the triangle in Figure 18–20.

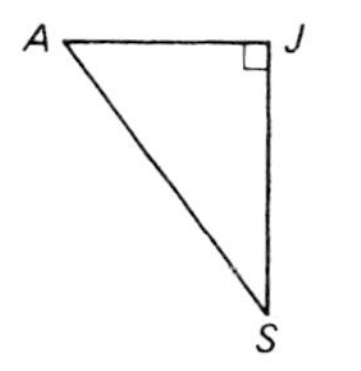

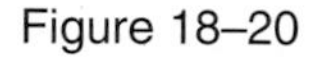
Figure 18–20

18–8. Determine the size of the missing angle in Figure 18–21.

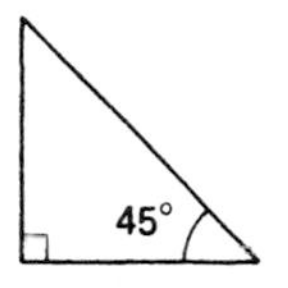

Figure 18–21

18–9. Determine the size of the missing angle in Figure 18–22.

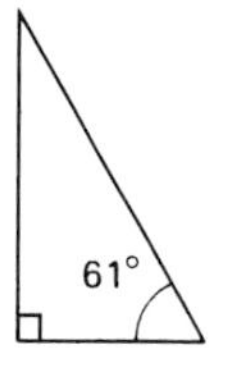

Figure 18–22

18–10. Determine the size of the missing angle in Figure 18–23.

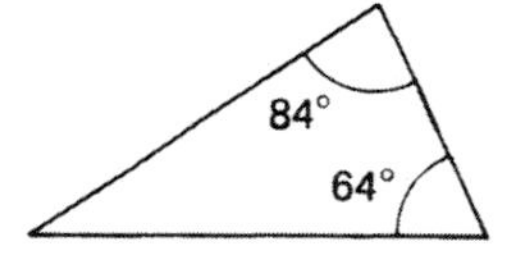

Figure 18–23

18.2 RIGHT AND SIMILAR TRIANGLES AND THEIR PROPERTIES

Triangles are *similar* if their corresponding angles are equal. The triangles in Figure 18–24 are similar right triangles. The equality of angles can be shown by comparing the ratios of known sides. Study the triangles and the ratios given below.

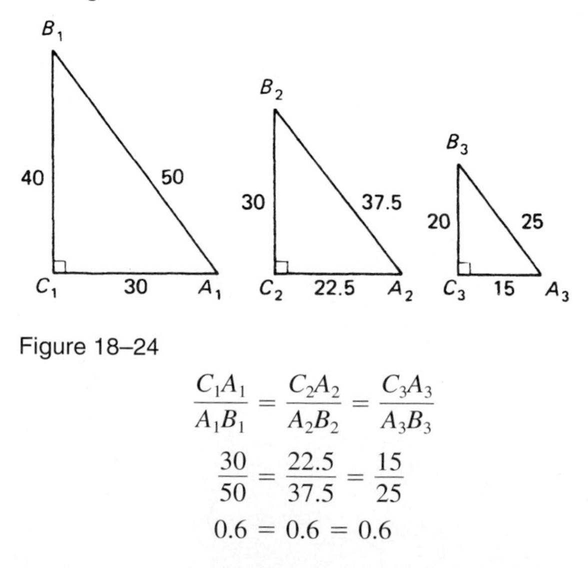

Figure 18–24

$$\frac{C_1A_1}{A_1B_1} = \frac{C_2A_2}{A_2B_2} = \frac{C_3A_3}{A_3B_3}$$

$$\frac{30}{50} = \frac{22.5}{37.5} = \frac{15}{25}$$

$$0.6 = 0.6 = 0.6$$

Equal ratios of the lengths of the sides indicate equality of angles. Study the triangles shown in Figure 18–25, and the following ratios, which prove that the corresponding angles are equal.

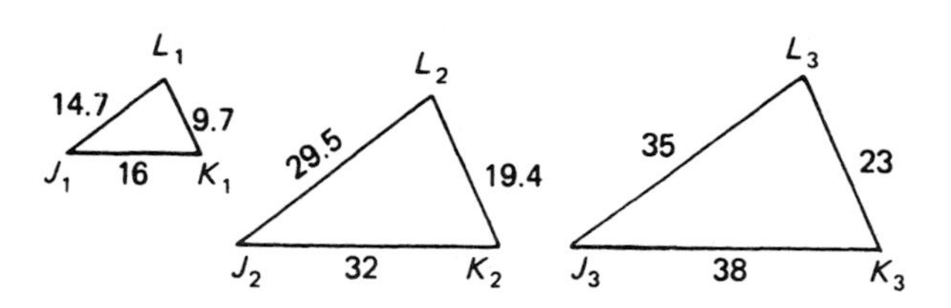

Figure 18–25

$$\frac{J_1L_1}{J_1K_1} = \frac{J_2L_2}{L_2K_2} = \frac{J_3L_3}{J_3K_3}$$

$$\frac{14.7}{16} = \frac{29.5}{32} = \frac{35}{38}$$

$$0.92 = 0.92 = 0.92$$

The fact that the ratios of the sides of similar triangles are equal is a basic principle of trigonometry. This principle is the foundation of finding solutions to trigonometry problems.

Example 18–3: Solve Similar Triangles

PROBLEM: Given two similar triangles (see Figure 18–26), determine the lengths of the remaining sides.

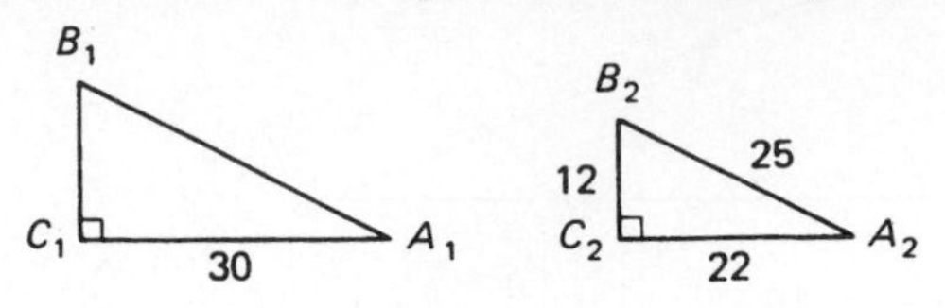

Figure 18–26

SOLUTION: Corresponding sides of similar triangles have similar ratios, thus ratios can be set up and unknowns can be found using simple algebra. Set up the ratios.

$$\frac{C_1A_1}{C_2A_2} = \frac{A_1B_1}{A_2B_2} = \frac{B_1C_1}{B_2C_2}$$

Substitute known values into the proper proportion:

$$C_1A_1 = 30 \qquad C_2A_2 = 22$$
$$A_2B_2 = 25 \qquad B_2C_2 = 12$$

$$\frac{C_1A_1}{C_2A_2} = \frac{A_1B_1}{A_1B_2}$$

$$\frac{30}{22} = \frac{A_1B_1}{25}$$

Solve for the unknown A_1B_1.

$$\frac{30 \times 25}{22} = A_1B_1$$
$$34.1 = A_1B_1$$

Then substitute known values.

$$C_1A_1 = 30 \qquad C_2A_2 = 22 \qquad B_2C_2 = 12$$

$$\frac{C_1A_1}{C_2A_2} = \frac{B_1C_1}{B_1C_2}$$
$$\frac{30}{22} = \frac{B_1C_1}{12}$$

Solve for the unknown B_1C_1.

$$\frac{30 \times 12}{22} = B_1C_1$$
$$B_1C_1 = 16.\overline{36}$$
$$B_1C_1 = 16.4$$

EXERCISE 18–2 PROPERTIES OF SIMILAR AND RIGHT TRIANGLES

Solve the following similar triangles.

18–11. Given the similar triangles $A_1B_1C_1$ and $A_2B_2C_2$ in Figure 18–27, solve for A_2B_2.

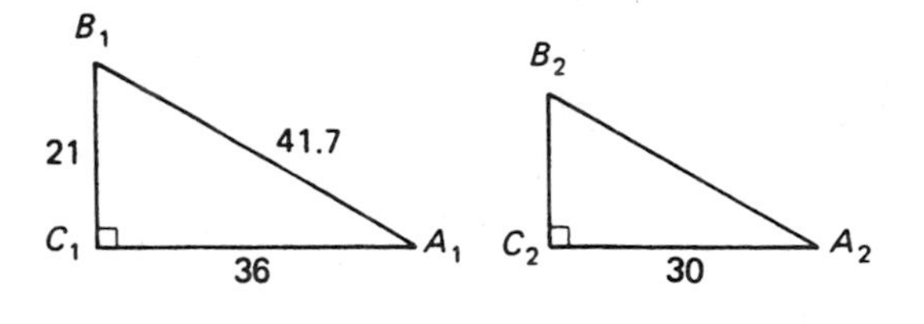

Figure 18–27

18–12. Solve for B_2C_2 in Figure 18–27.

18–13. Given the similar triangles $J_1K_1L_1$, and $J_2K_2L_2$ in Figure 18–28, solve for J_1K_1.

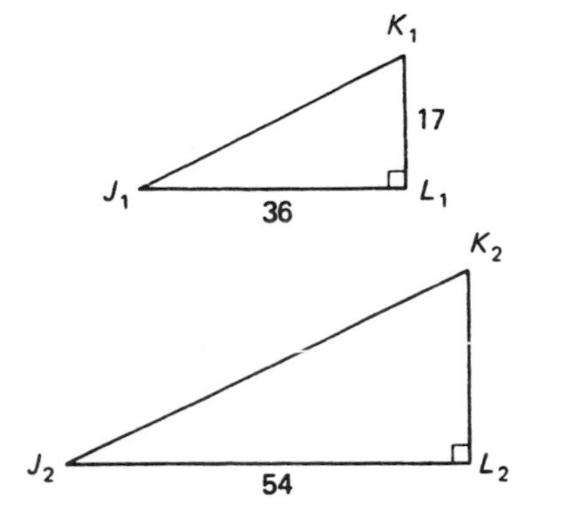

Figure 18–28

18–14. Solve for L_2K_2 in Figure 18–28, given the similar triangles JKL and $J_2K_2L_2$.

18–15. Solve for K_2J_2 in Figure 18–28, given the similar triangles *JKL* and $J_2K_2L_2$.

18–16. Given the similar triangles *DEF* and *XYZ* in Figure 18–29, solve for *y*.

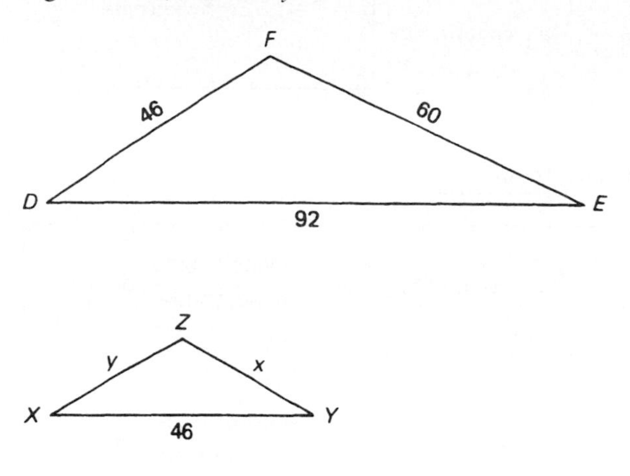

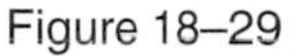
Figure 18–29

18–17. Solve for *x* in Figure 18–29, given the similar triangles *DEF* and *XYZ.*

18–18. Given the similar triangles *MNO* and *EFG* in Figure 18–30, solve for *e*.

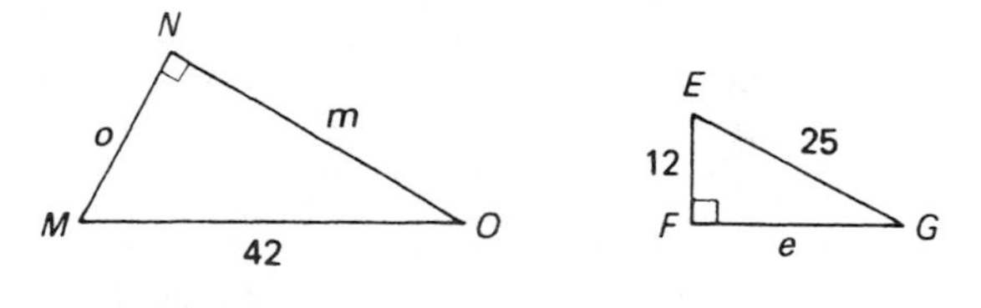

Figure 18–30

18–19. Solve for *o* in Figure 18–30, given the similar triangles *MNO* and *EFG.*

18–20. Solve for *m* in Figure 18–30, given the similar triangles *MNO* and *EFG.*

18.3 DEFINITIONS OF TRIGONOMETRIC RATIOS AND THEIR PROPERTIES

The sides of right triangles are named in relation to their angles. In triangle *ABC,* Figure 18–31, angle *A* indicates that side *b* is the adjacent side, side *a* is the opposite side, and *c* is the hypotenuse.

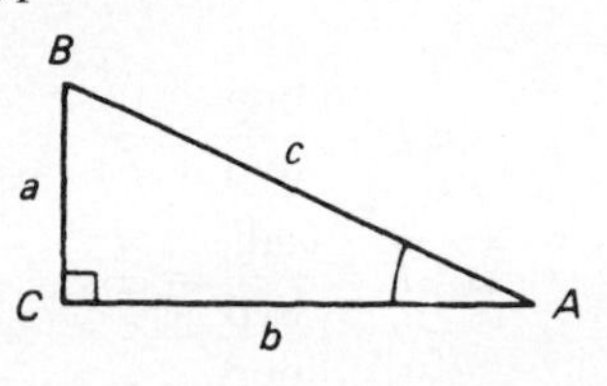

Figure 18–31

Regarding angle *B,* side *b* is the opposite side, side *a* is the adjacent side, and side *c* is the hypotenuse. A rule to follow for naming the sides of a right triangle concerning particular angles is: (1) the side opposite the right angle is called the *hypotenuse*; (2) the side that is a part of the angle but is not the hypotenuse is called the *adjacent side*; and (3) the side opposite or farthest from the angle is called the *opposite side.*

The ratios of the sides of right triangles are identified by specific names. These ratios and their identification are necessary in solving trig problems. In triangle *ABC,* angle *A* determines the names of the sides of the triangle. See Figure 18–32. The names of the trig ratios concerning angle *A* are as follows:

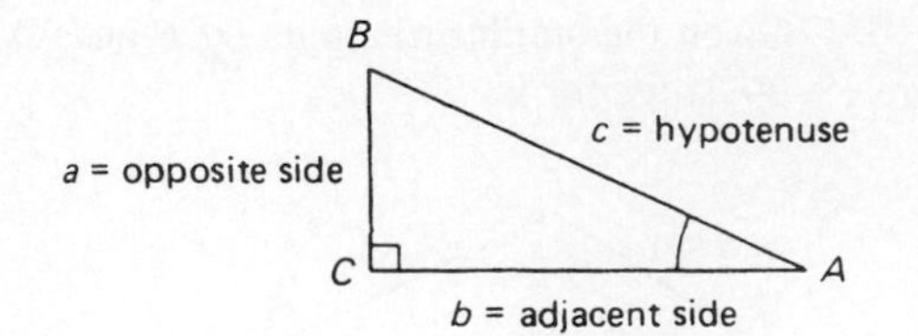

Figure 18–32

$$\text{sine of angle } A = \frac{\text{opposite side}}{\text{hypotenuse}} \quad \text{or}$$

$$\sin A = \frac{\text{opp}}{\text{hyp}} = \frac{a}{c}$$

$$\text{cosine of angle } A = \frac{\text{adjacent side}}{\text{hypotenuse}} \quad \text{or}$$

$$\cos A = \frac{\text{adj}}{\text{hyp}} = \frac{b}{c}$$

$$\text{tangent of angle } A = \frac{\text{opposite side}}{\text{adjacent}} \quad \text{or}$$

$$\tan A = \frac{\text{opp}}{\text{adj}} = \frac{a}{b}$$

There are three additional trigonometric functions that are reciprocals of the sin, cos, and tan functions. They are defined as follows in relation to angle A in Figure 18–32:

$$\text{cotangent of angle } A = \cot A = \frac{\text{adj}}{\text{opp}}$$

$$\cot A = \frac{b}{a} \quad \text{which is the reciprocal of}$$

$$\tan A = \frac{a}{b}$$

$$\text{secant of angle } A = \sec A = \frac{\text{hyp}}{\text{adj}}$$

$$\sec A = \frac{c}{b} \quad \text{which is the reciprocal of}$$

$$\cos A = \frac{b}{c}$$

$$\text{cosecant of angle } A = \csc A = \frac{\text{hyp}}{\text{opp}}$$

$$\csc A = \frac{c}{a} \quad \text{which is the reciprocal of}$$

$$\sin A = \frac{a}{c}$$

It is necessary to know the trig ratios to solve trig problems. You should know the trig ratios as well as the multiplication facts. The ratios of the sides of right triangles are also known as the *trigonometric functions*. Since the angles will not always be designated as A, B, and C, it is important to identify the trigonometric functions with the names of the sides and not with given letters. The two examples illustrate how to define the three basic trigonometric functions. Remember when trig ratios are defined, they are given as the ratio of the sides of the given right triangle.

Example 18–4: Identify the Trig Ratios

PROBLEM: Identify the three basic trigonometric functions of angle D and angle E in triangle DEF in Figure 18–33.

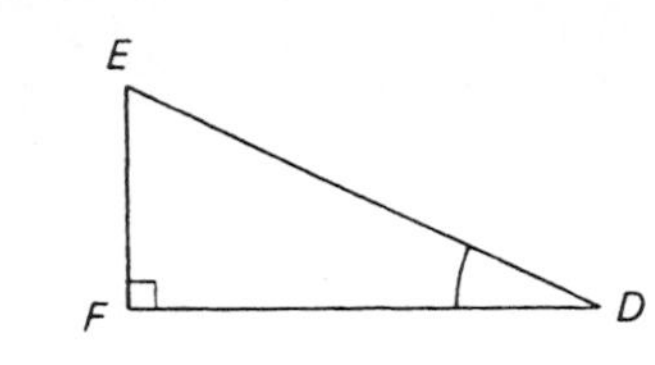

Figure 18–33

SOLUTION: Name the sides in relation to angle D in Figure 18–34. Identify the trigonometric functions.

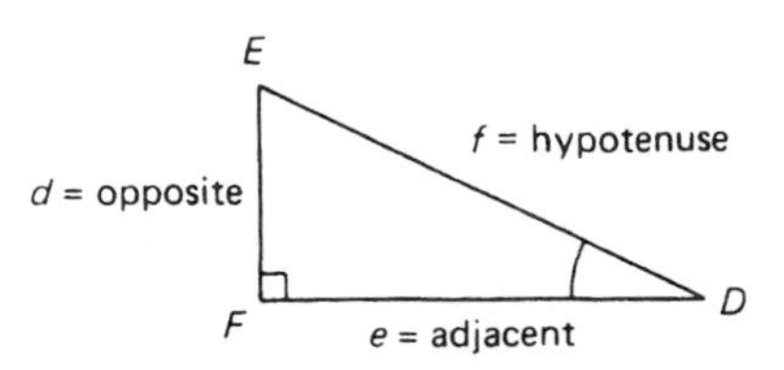

Figure 18–34

$$\sin D = \frac{\text{opp}}{\text{hyp}} = \frac{d}{f}$$

$$\cos D = \frac{\text{adj}}{\text{hyp}} = \frac{e}{f}$$

$$\tan D = \frac{\text{opp}}{\text{adj}} = \frac{d}{e}$$

Then identify the sides in relation to angle E as shown in Figure 18–35. Finally, identify the trigonometric functions.

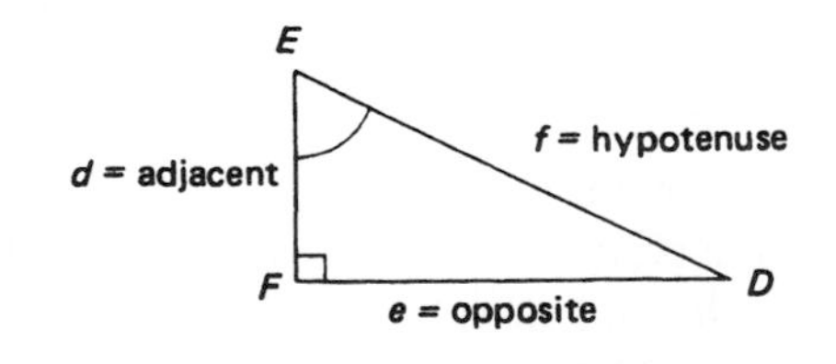

Figure 18–35

$$\sin E = \frac{\text{opp}}{\text{hyp}} = \frac{e}{f}$$

$$\cos E = \frac{\text{adj}}{\text{hyp}} = \frac{d}{f}$$

$$\tan E = \frac{\text{opp}}{\text{adj}} = \frac{e}{d}$$

Example 18–5: Identify the Trig Ratios

PROBLEM: Identify the basic trigonometric functions of angle J and angle K in triangle JKL in Figure 18–36.

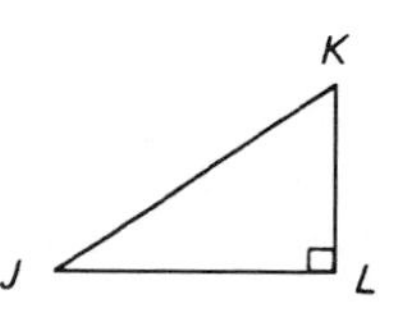

Figure 18–36

SOLUTION: Name the sides in relation to angle J as shown in Figure 18–37. Identify the trigonometric functions.

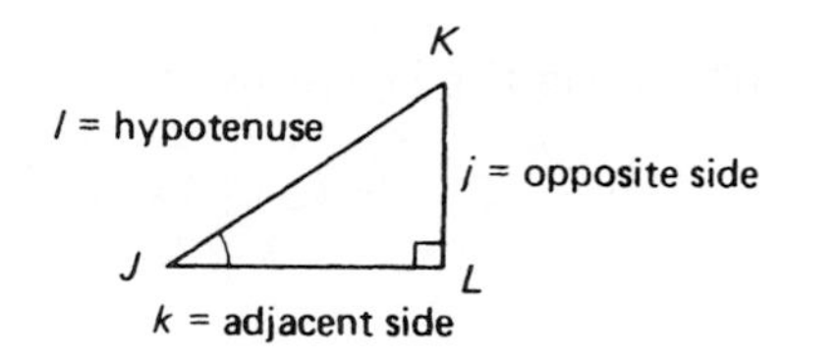

Figure 18–37

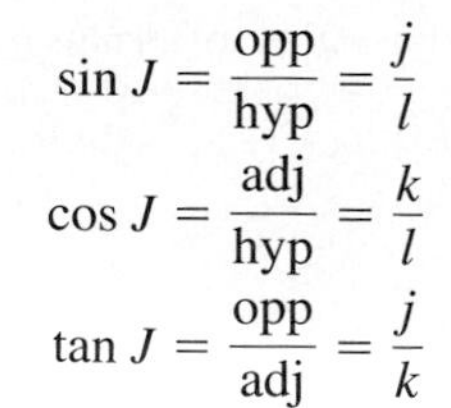

$$\sin J = \frac{\text{opp}}{\text{hyp}} = \frac{j}{l}$$
$$\cos J = \frac{\text{adj}}{\text{hyp}} = \frac{k}{l}$$
$$\tan J = \frac{\text{opp}}{\text{adj}} = \frac{j}{k}$$

Then identify the sides in relation to angle K as shown in Figure 18–38, and identify the trigonometric functions.

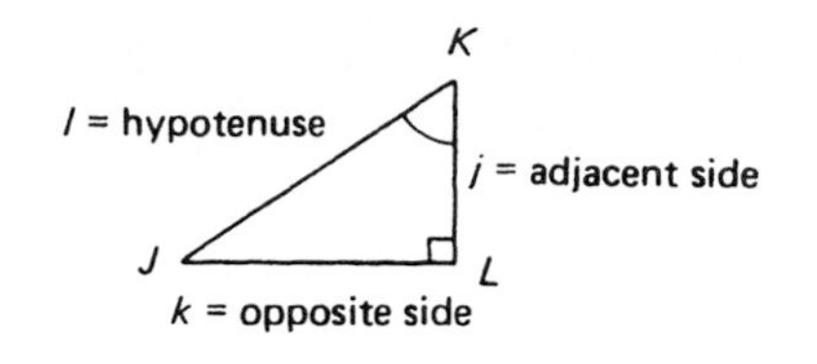

Figure 18–38

$$\sin K = \frac{\text{opp}}{\text{hyp}} = \frac{k}{l}$$
$$\cos K = \frac{\text{adj}}{\text{hyp}} = \frac{j}{l}$$
$$\tan K = \frac{\text{opp}}{\text{adj}} = \frac{k}{j}$$

EXERCISE 18–3 TRIGONOMETRY RATIOS

Identify the trigonometric functions as indicated in the following problems.

18–21. Given the right triangle HIJ in Figure 18–39, what is the sine of $\angle H$?

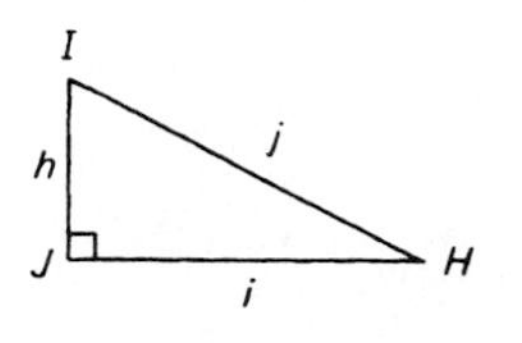

Figure 18–39

18–22. What is the tangent of $\angle I$, Figure 18–39?

18–23. What is the cosine of $\angle H$, Figure 18–39?

18–24. What is the sine of $\angle I$, Figure 18–39?

18–25. What is the cosine of $\angle I$, Figure 18–39?

18–26. What is the tangent of $\angle H$, Figure 18–39?

18–27. Letter the sides of triangle *XYZ* shown in Figure 18–40.

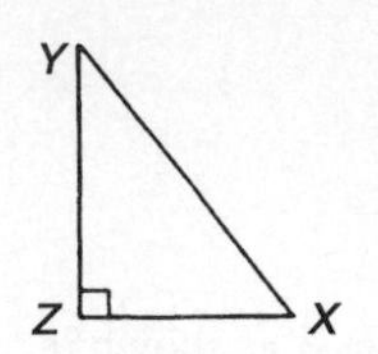

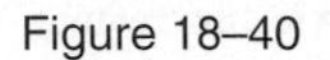
Figure 18–40

18–28. What is sin *X* in Figure 18–40?

18–29. What is tan *X* in Figure 18–40?

18–30. What is cos *Y* in Figure 18–40?

18–31. What is tan *Y* in Figure 18–40?

18–32. What is sin *Y* in Figure 18–40?

18–33. What is cos *X* in Figure 18–40?

18–34. Letter the angles of the triangle in Figure 18–41.

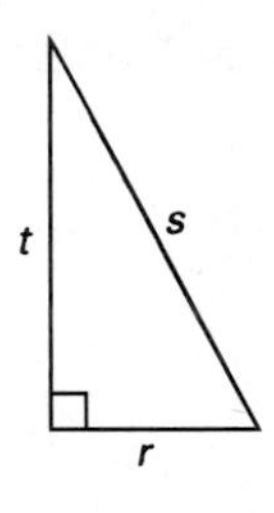

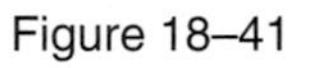
Figure 18–41

18–35. What is sin *R* in Figure 18–41?

18–36. What is cos *T* in Figure 18–41?

18–37. What is tan *T* in Figure 18–41?

18–38. Complete the lettering of the triangle in Figure 18–42.

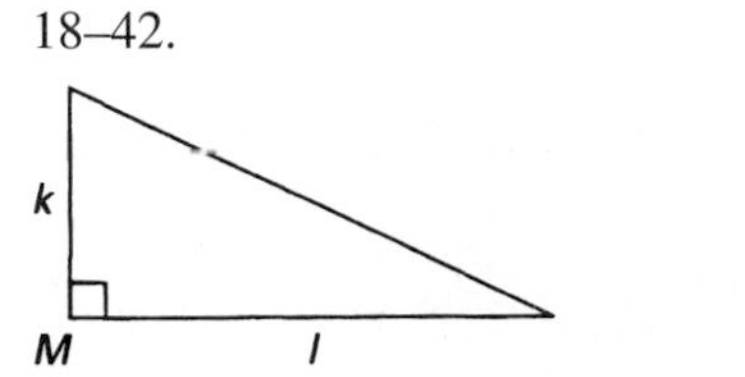

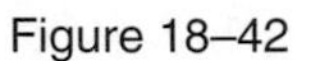
Figure 18–42

18–39. What is $\sin K$ in Figure 18–42?

18–40. What is $\tan K$ in Figure 18–42?

18–41. Given the ratio $\frac{k}{l}$ of $\angle L$, what are the trigonometric functions in Figure 18–42?

18–42. Given the ratio $\frac{k}{m}$ of $\angle K$ in Figure 18–42, what are the trigonometric functions?

18–43. Given the ratio $\frac{l}{m}$ of $\angle L$ in Figure 18–42, what are the trigonometric functions?

18–44. Complete the lettering of the triangle in Figure 18–43.

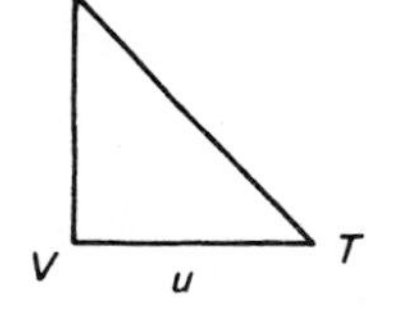

Figure 18–43

18–45. Given the ratio $\frac{t}{u}$ of $\angle T$, Figure 18–43, what are the trigonometric functions?

18–46. Given the ratio $\frac{u}{v}$ of $\angle T$, Figure 18–43, what are the trigonometric functions?

18–47. Given the ratio $\frac{u}{t}$ of $\angle T$, Figure 18–43, what are the trigonometric functions?

18.4 DETERMINING THE VALUES OF TRIGONOMETRIC FUNCTIONS

A trigonometric function of an angle has a constant value. In the examples that follow, three procedures for determining the values of trigonometric functions of 30°, 45°, and 60° angles will be illustrated.

Example 18–6: Determine the Value of the Trig Functions of a 30° Angle

PROBLEM: Determine the values of the three basic trigonometric functions (sin, cos, and tan) of a 30° angle.

SOLUTION: Draw a 30°–60° right triangle as

shown in Figure 18–44 and letter as *ABC*. Label the

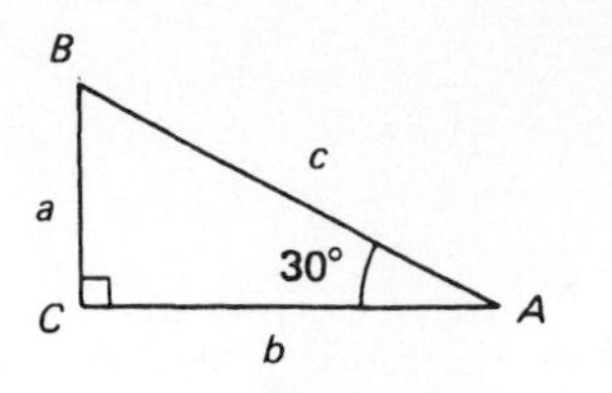

Figure 18–44

sides with their lengths as shown in Figure 18–45. The

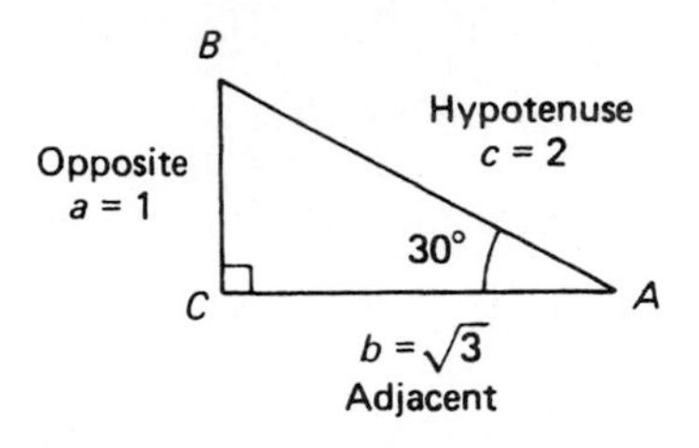

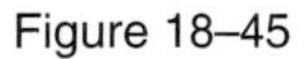
Figure 18–45

length of the hypotenuse, *c*, is given as 2 units. The length of side *a* opposite the 30° angle is one half of the length of the hypotenuse. Thus the length of side *a* is 1. In a right triangle, the square of the hypotenuse is equal to the sum of the squares of the other two sides. This is known as the *Pythagorean theorem.* Use the Pythagorean theorem, $a^2 + b^2 = c^2$, accordingly to determine the value of side *b*.

$$\begin{aligned} b &= \sqrt{c^2 - a^2} \\ &= \sqrt{(2)^2 - (1)^2} \\ &= \sqrt{4 - 1} \\ &= \sqrt{3} \end{aligned}$$

Label the length of *b* as shown in Figure 18–45. Determine the sine of 30°.

$$\sin 30^\circ = \frac{\text{opp}}{\text{hyp}} = \frac{1}{2} = 0.500$$

Then determine the cosine of 30°.

$$\cos 30^\circ = \frac{\text{adj}}{\text{hyp}} = \frac{\sqrt{3}}{2} = \frac{1.732}{2} = 0.866$$

Determine the tangent of 30°.

$$\begin{aligned} \tan 30^\circ &= \frac{\text{opp}}{\text{adj}} = \frac{1}{\sqrt{3}} = \frac{1}{\sqrt{3}}\left(\frac{\sqrt{3}}{\sqrt{3}}\right) \\ &= \frac{\sqrt{3}}{3} = \frac{1.732}{3} = 0.577 \end{aligned}$$

Thus the values, to the nearest thousandth, of the basic trigonometric functions of a 30° angle are

$$\begin{aligned} \sin 30^\circ &= 0.500 \\ \cos 30^\circ &= 0.866 \\ \tan 30^\circ &= 0.577 \end{aligned}$$

Example 18–7: Determine the Value of the Trig Functions of a 45° Angle

PROBLEM: Determine the values of the three basic trigonometric functions (sin, cos, and tan) of a 45° angle.

SOLUTION: Draw a 45°–45° right triangle as shown by triangle *DEF* in Figure 18–46.

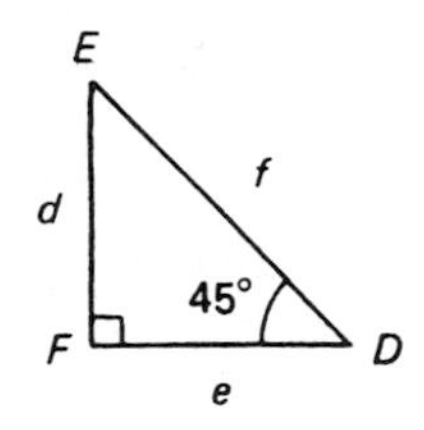

Figure 18–46

Label the length of the sides as indicated in triangle *DEF* in Figure 18–47.

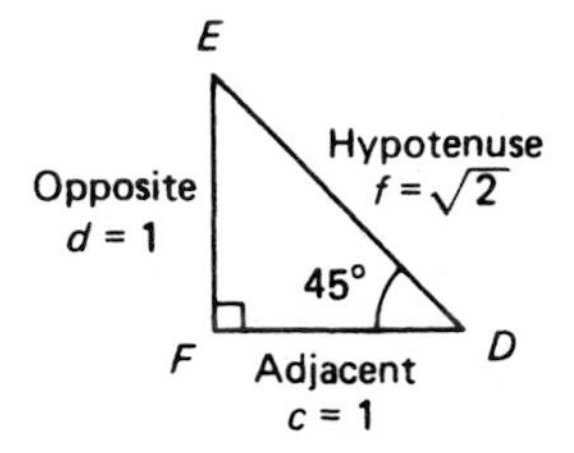

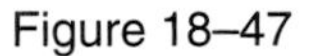
Figure 18–47

The lengths of the adjacent side *e* and the opposite side *d* are given as 1. The lengths of the two sides of a 45° right triangle are equal. Use the Pythagorean theorem, $f^2 = d^2 + e^2$, to determine the value of side *f*.

$$\begin{aligned} f &= \sqrt{d^2 + e^2} \\ &= \sqrt{(1)^2 + (1)^2} \\ &= \sqrt{1 + 1} \\ &= \sqrt{2} \end{aligned}$$

Determine the sine of 45°.

$$\begin{aligned} \sin 45^\circ &= \frac{\text{opp}}{\text{hyp}} = \frac{1}{\sqrt{2}} = \frac{1}{\sqrt{2}}\left(\frac{\sqrt{2}}{\sqrt{2}}\right) \\ &= \frac{\sqrt{2}}{2} = \frac{1.414}{2} = 0.707 \end{aligned}$$

Determine the cosine of 45°.

$$\tan 45^\circ = \frac{\text{opp}}{\text{adj}} = \frac{1}{\sqrt{2}} = \frac{1}{\sqrt{2}}\left(\frac{\sqrt{2}}{\sqrt{2}}\right)$$

$$= \frac{\sqrt{2}}{2} = \frac{1.414}{2} = 0.707$$

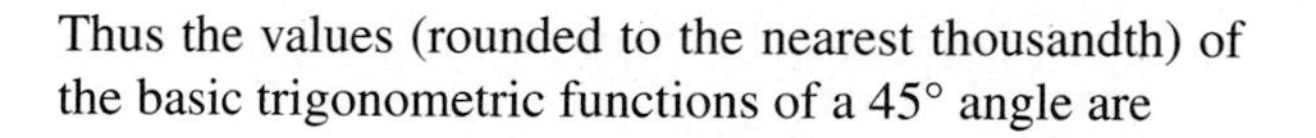

Thus the values (rounded to the nearest thousandth) of the basic trigonometric functions of a 45° angle are

$$\sin 45^\circ = 0.707$$
$$\cos 45^\circ = 0.707$$
$$\tan 45^\circ = 1.000$$

EXERCISE 18–4 FIND THE VALUE OF TRIG RATIOS

Determine the values of the trig functions as indicated in the following problems.

18–48. Given the triangle *ABC*, Figure 18–48, find *b*.

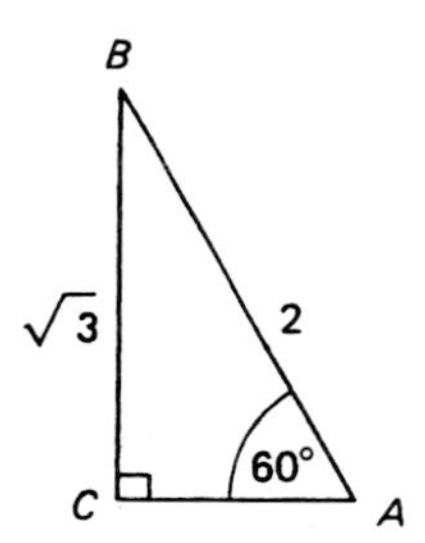

Figure 18–48

18–49. Find $\angle B$, Figure 18–48.

18–50. Find the sine of 60° in Figure 18–48.

18–51. Find the cosine of 60° in Figure 18–48.

18–52. Find the tangent of 60° in Figure 18–48.

18–53. Given the right triangle *JKL*, Figure 18–49, determine the value of *l*.

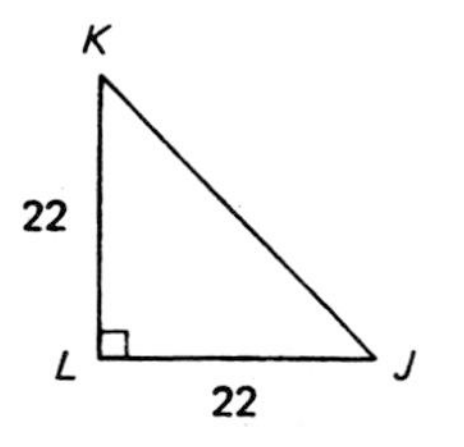

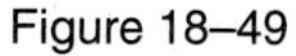

Figure 18–49

18–54. Determine the tangent of $\angle J$, in Figure 18–49.

18–55. Determine the sine of $\angle K$ in Figure 18–49.

18–56. Determine the cosine of $\angle J$ in Figure 18–49.

18–57. How many degrees are there in angle J in Figure 18–49?

18–58. Angle K equals how many degrees in Figure 18–49?

18–59. Given right triangle PQR, Figure 18–50, determine the value of side p.

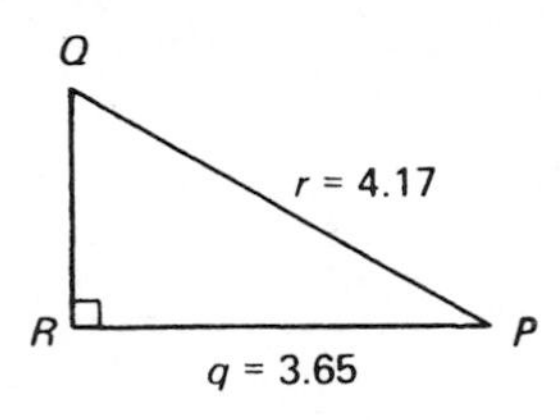

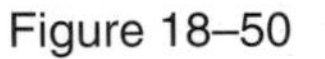

Figure 18–50

18–60. Determine the value of the tangent of $\angle P$ in Figure 18–50.

18–61. Determine the value of sine of $\angle Q$ in Figure 18–50.

18–62. Determine the value of the tangent of $\angle Q$ in Figure 18–50.

18–63. Determine the value of the cosine of $\angle Q$ in Figure 18–50.

18–64. How many degrees are there in $\angle Q$ in Figure 18–50?

18–65. How many degrees are there in $\angle P$ in Figure 18–50?

THINK TIME

Many students find it helpful to have a word to help them remember several things. Three acronyms, *soph, cash,* and *topa,* may help you remember the trig functions. The meanings of the acronyms are as follows:

soph (*s*ine = *o*pposite over *h*ypotenuse)
cash (*c*osine = *a*djacent over *h*ypotenuse)
topa (*t*angent = *o*pposite over *a*djacent)

PROCEDURES TO REMEMBER

1. To find the unknown sides of similar triangles:
   (a) Set up ratios between similar sides.
   (b) Substitute the given values for the sides.
   (c) Solve for the unknown sides.
2. The formulas for basic trigonometric functions are:

$$\sin A = \frac{\text{opposite side}}{\text{hypotenuse}} = \frac{\text{opp}}{\text{hyp}}$$

$$\cos A = \frac{\text{adjacent side}}{\text{hypotenuse}} = \frac{\text{adj}}{\text{hyp}}$$

$$\tan A = \frac{\text{opposite side}}{\text{adjacent side}} = \frac{\text{opp}}{\text{adj}}$$

$$\cot A = \frac{\text{adjacent side}}{\text{opposite side}} = \frac{\text{adj}}{\text{opp}}$$

$$\sec A = \frac{\text{hypotenuse}}{\text{adjacent side}} = \frac{\text{hyp}}{\text{adj}}$$

$$\csc A = \frac{\text{hypotenuse}}{\text{opposite side}} = \frac{\text{hyp}}{\text{opp}}$$

CHAPTER SUMMARY

The following definitions are important when solving problems with trigonometric functions.

1. All triangles have six parts: three sides and three angles.
2. Finding the six parts of a triangle is called solving the triangle.
3. A right triangle is a triangle that has a right (90°) angle.
4. Capital letters are used to identify angles.
5. Lowercase letters are used to identify sides.
6. In a right triangle, the side opposite the right angle is the hypotenuse, the side opposite or farthest from a given angle is the opposite side, and the side next to a given angle is the adjacent side.
7. The sum of the degrees of the interior angles of a triangle is 180°.
8. Triangles are similar if their corresponding angles are equal.
9. Identical ratios of the lengths of the sides of triangles indicate that the angles are equal.
10. The ratios of the sides of right triangles are called the trigonometric functions.
11. The value of the side opposite a 30° angle of a right triangle is one half the hypotenuse.
12. In a right triangle, the square of the hypotenuse is equal to the sum of the squares of the other two sides.

CHAPTER TEST

T–18–1. How many degrees are in angle *B*, triangle *ABC*, in Figure 18–51?

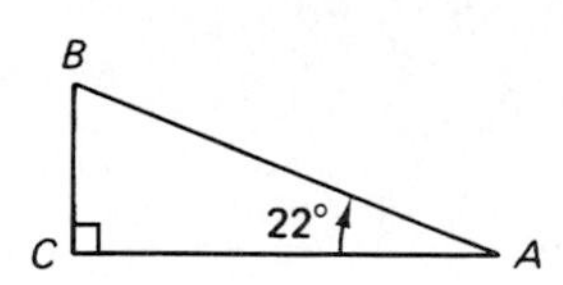

Figure 18–51

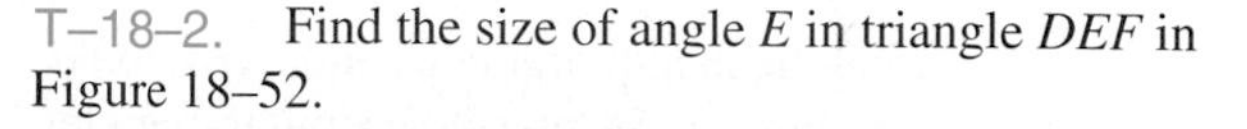

T–18–2. Find the size of angle *E* in triangle *DEF* in Figure 18–52.

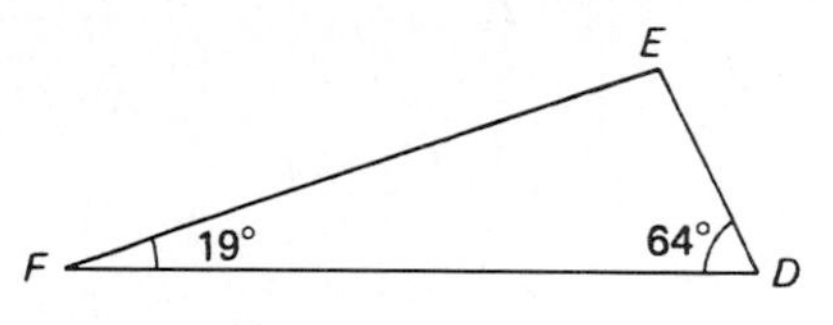

Figure 18–52

T–18–3. Letter the sides of triangle *GHI* in Figure 18–53.

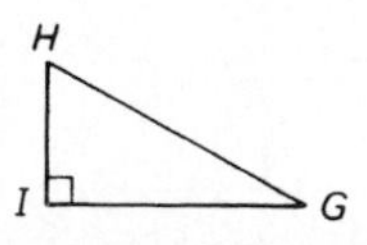

Figure 18–53

T–18–4. Letter the sides of triangle *XYZ* in Figure 18–54.

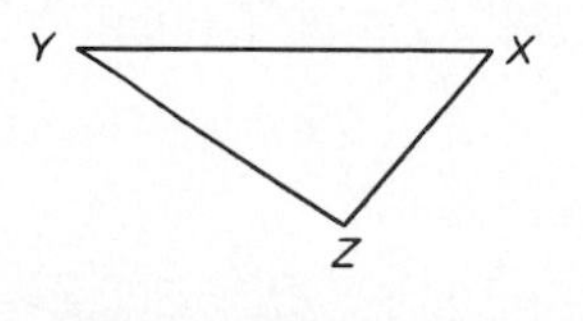

Figure 18–54

T–18–5. Letter the angles of the triangle in Figure 18–55.

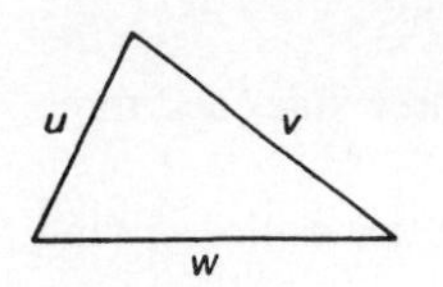

Figure 18–55

T–18–6. Given similar triangles *JKL* and *MNO*, find the length of *m* in triangle *MNO* in Figure 18–56.

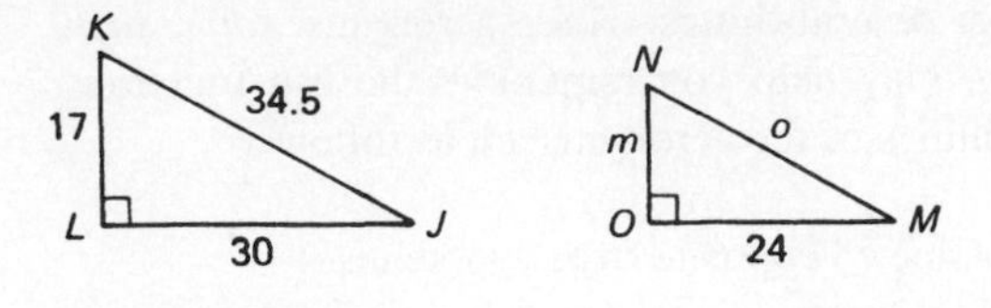

Figure 18–56

T–18–7. Find the length of *o* in triangle *MNO* in Figure 18–56, given the similar triangles *PQR* and *STU*.

T–18–8. Given similar triangles *PQR* and *STU*, find the length of *r* in triangle *PQR* in Figure 18–57.

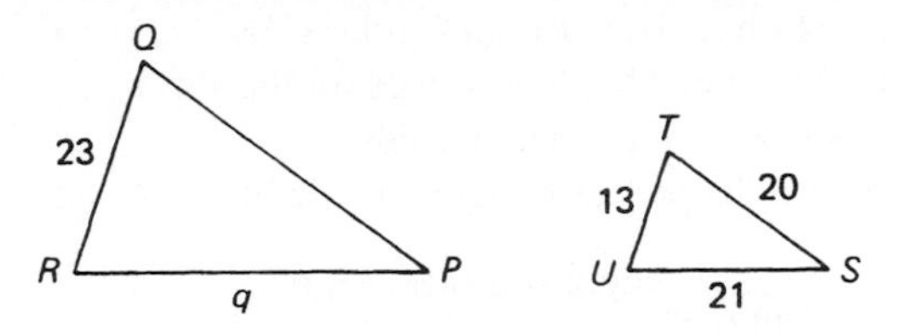

Figure 18–57

T–18–9. Find the length of *q* in triangle *PQR* in Figure 18–57, given the similar triangles *PQR* and *STU*.

T–18–10. Given similar triangles *XYZ* and *ABC*, determine the length of *z* in triangle *XYZ* to the nearest tenth in Figure 18–58.

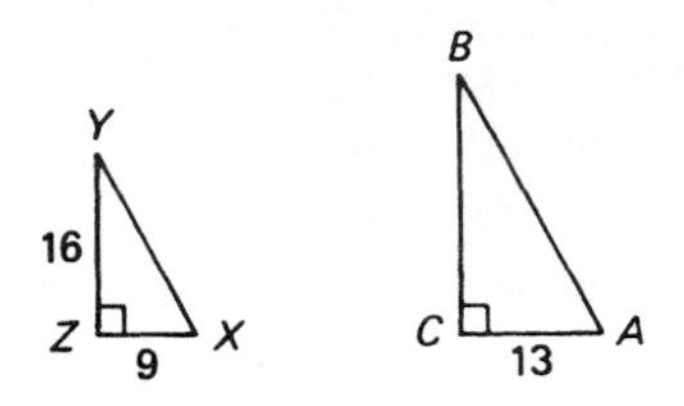

Figure 18–58

T–18–11. Find the length of *a* in triangle *ABC* to the nearest tenth in Figure 18–58, given the similar triangles *XYZ* and *ABC*.

T–18–12. Find the length of *c* in triangle *ABC* to the nearest tenth in Figure 18–58, given the similar triangles *XYZ* and *ABC*.

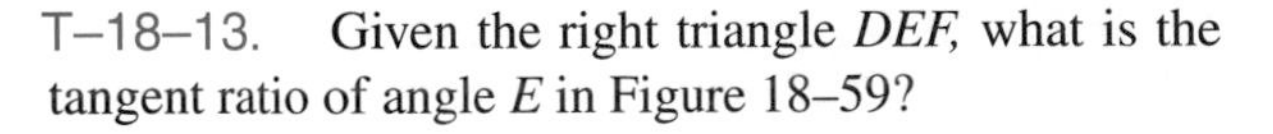

T–18–13. Given the right triangle *DEF*, what is the tangent ratio of angle *E* in Figure 18–59?

T–18–14. What is the cosine ratio of angle *D* in Figure 18–59?

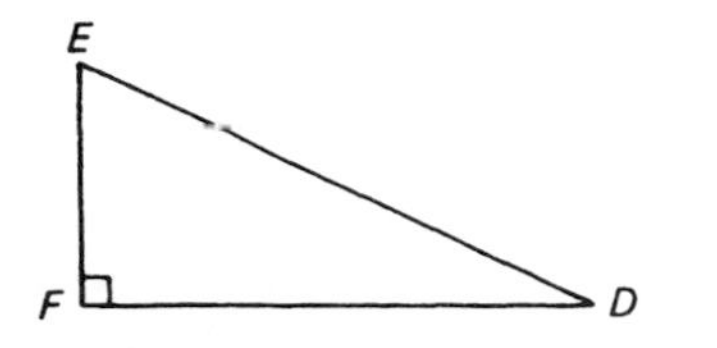

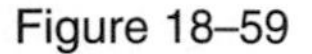

Figure 18–59

T–18–15. What is the sine ratio of angle E in Figure 18–59?

T–18–16. In Figure 18–60, given the right triangle GHI, what trigonometric functions equal the ratio $\frac{h}{g}$?

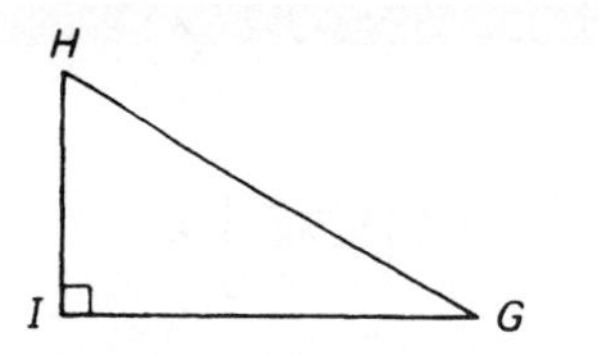

Figure 18–60

T–18–17. In Figure 18–60, what trigonometric functions equal the ratio $\frac{h}{i}$?

T–18–18. In Figure 18–60, what trigonometric functions equal ratio $\frac{g}{i}$?

T–18–19. If the tangent of angle D is 1.000, what is the size of angle D?

T–18–20. If the sine of angle K is 0.866, what is the size of angle K?

19

SOLUTION OF RIGHT TRIANGLES

OBJECTIVES

After studying this chapter, you will be able to:

19.1. Find the values of trig functions.
19.2. Find the angle given the trig functions.
19.3. Solve right triangles.
19.4. Understand the application of right triangles.

SELF-TEST

This test will measure your ability to use trigonometric functions to solve right triangles. You will need a scientific calculator to solve these problems. Satisfactory completion will indicate your mastery of these skills and your readiness to solve oblique triangles.

S–19–1. Find the sine of 43°.

S–19–2. Find the cosine of 5°40′.

S–19–3. Find the tangent of 48°20′.

S–19–4. Find the angle whose sine is 0.2186.

S–19–5. Find the angle whose tangent is 4.1653.

S–19–6. Find the sine of 11°37′.

S–19–7. Find the angle whose cosine is 0.5476.

S–19–8. Determine the degrees in angle B of triangle ABC in Figure 19–1.

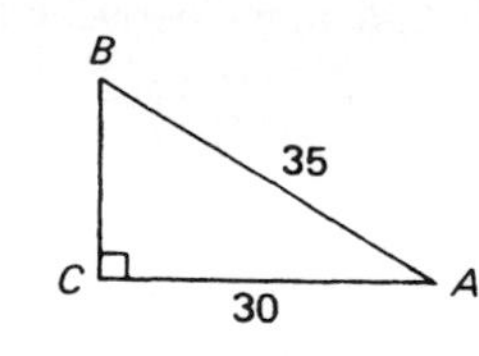

Figure 19–1

S–19–9. Given triangle PQR in Figure 19–2, determine the length of q.

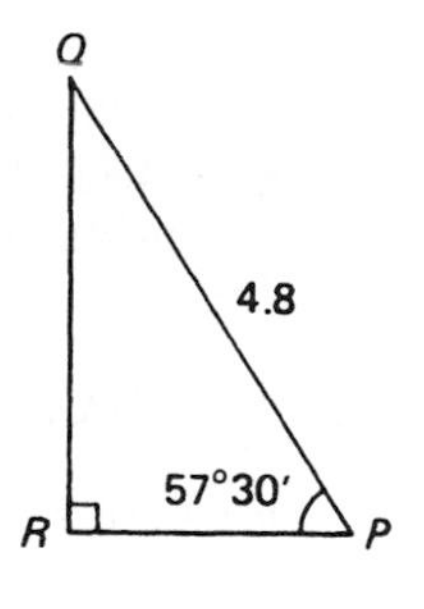

Figure 19–2

S–19–10. Find the length of p in triangle PQR in Figure 19–2.

S–19–11. Given triangle DEF, determine the size of angle D in Figure 19–3.

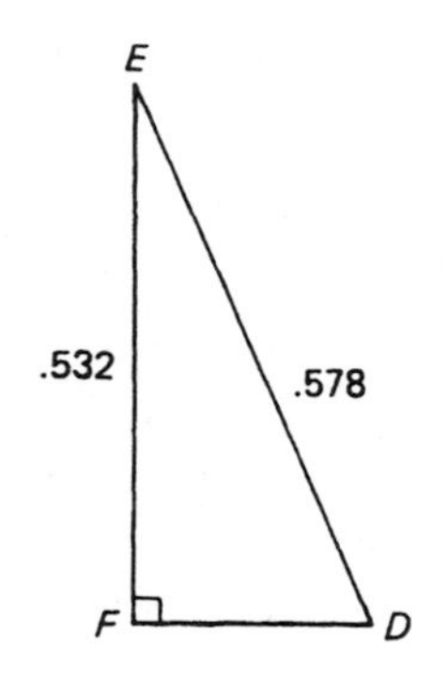

Figure 19–3

S–19–12. Determine the length of e in triangle DEF in Figure 19–3.

S–19–13. An airplane flies 1250 miles northeast. How far east is the plane from its original position?

S–19–14. Determine the taper angle of a crankshaft having a taper of 0.75 inch per foot.

S–19–15. Four bolt holes are to be drilled on the circumference of a 6-inch diameter hub of a wheel. Determine the straight-line distance between the holes.

19.1 FIND THE VALUES OF TRIG FUNCTIONS

It is essential to know trigonometric functions or ratios to solve triangles. Trig functions can be found in trig tables; however, using a scientific calculator is much more efficient. Some trig functions are as easily found; for example, angles of 30°, 45°, and 60° were determined in Chapter 18.

When working with trigonometry, occasionally angles will be given in degrees and minutes. Minutes are converted to a decimal part of a degree. For example, 38° 42′ is read 38 degrees and 42 minutes or, converted to a decimal, $38^{42}/_{60}{}^{\circ} = 38.7^{\circ}$. To convert 38.7° to degrees and minutes, do the following:

$$\begin{aligned} 38.7^{\circ} &= 38^{\circ} + 0.7^{\circ} \times \frac{60'}{1^{\circ}} \\ &= 38^{\circ} + 0.7^{\circ} \times 60' \\ &= 38^{\circ} + 42' \\ &= 38^{\circ}42' \end{aligned}$$

The use of a scientific calculator is easier and more efficient than any table of trigonometric functions. Scientific calculators give instantaneous and accurate values of trig functions. Scientific calculators have keys for sin, cos, and tan. The other trig functions—cot, sec, and csc—can be obtained by using the reciprocal key.

The decimal degree notation is the standard notation on most electronic calculators. Some scientific calculators have a key that automatically converts angles given in degrees, minutes, and seconds into decimal degrees; the inverse key will reverse this process. Consult your calculator manual to be certain of the proper procedure.

The following examples will show how to use a scientific calculator to find trig functions. Many calculators operate in both degrees and radians. Be certain to determine which units you are using before using your calculator.

Example 19–1: Find the Sine Value Given the Angle

PROBLEM: Use a scientific calculator to find sin 37.4°.

SOLUTION: Enter 37.4 and press [sin] to obtain 0.6073758; round to 0.6074. Thus sin 37.4° = 0.6074.

Example 19–2: Find the Tangent Value Given the Angle

PROBLEM: Use a scientific calculator to find tan 76° 42′.

SOLUTION: Convert 42′ into a decimal part of a degree by dividing 60 into 42, since 1° = 60′. Thus

$$76^{\circ}42' = 76^{\circ} + \left(\frac{42}{60}\right)^{\circ} = 76.7^{\circ}$$

Then enter 76.7° and press [tan] to obtain 4.2302977; round to 4.2303. Therefore, tan 76°42′ = tan 76.7° = 4.2303.

EXERCISE 19–1 FIND TRIG FUNCTIONS WITH A CALCULATOR

Use a scientific calculator to find the functions of the angles indicated.

19–1. Find the tangent value of 31°.

19–2. Find the sine value of 21°.

19–3. Find the cosine value of 30°30′.

19–4. Find the cosine value of 86°30′.

19–5. Find the tangent value of 63°50′.

19–6. Find the cosine value of 37°20′.

19–7. Find tan 17°40′.

19–8. Find sin 38°50′.

19–9. Find tan 78°10′.

19–10. Find sin 45°20′.

19.2 FIND AN ANGLE GIVEN THE TRIG FUNCTION

Using the skills learned, but in reverse order, you can determine the degrees of the angle of a given trigonometric function. The examples will show how to locate the degrees of angles of trigonometric functions given the trigonometric function.

Example 19–3: Find the Angle Given the Sine Value

PROBLEM: Find the angle whose sine is 0.3907, to the nearest tenth of a degree.

SOLUTION: Enter 0.3907, press [inv], and then press [sin] to obtain 23. Thus

$$\sin A = 0.3907$$
$$A = 23°$$

Example 19–4: Find the Angle Given the Tangent Value

PROBLEM: Find the angle whose tangent is 2.2113 to the nearest tenth of a degree.

SOLUTION: Enter 2.2113, press [inv], then press [tan] to obtain 65.666439; round to 65.7. Thus

$$\tan A = 2.2113$$
$$\angle A = 65.7$$

Example 19–5: Find the Angle Given the Cosine Value

PROBLEM: Find the angle whose cosine is 0.7880, to the nearest tenth of a degree.

SOLUTION: Enter 0.7880, press [inv], and then press [cos] to obtain 38. Thus

$$\cos A = 0.7880$$
$$A = 38°$$

Example 19–6: Find the Angle Given the Tangent Value

PROBLEM: Find the angle K whose tangent is 0.2600, to the nearest degree, minute, and second.

SOLUTION: Enter 0.2600, press [inv], then press [tan] to obtain 14.574216; round to 14.6°. To convert 14.574216 to minutes and seconds do as follows:

Note $1° = 60'$, thus

$$14.574216° = 14° + \left(0.574216° \times \frac{60'}{1°}\right)$$
$$= 14°34.452966'$$

Then

$$= 14°34' + \left(0.452966' \times \frac{60''}{1'}\right) \quad (1' = 60'')$$
$$= 14°34'27''$$

Thus $\tan K = 0.2600$

$$K = 14°34'27''$$

Some calculators have an automatic conversion key to convert from degrees to radians and radians to degrees. Be sure to determine the correct unit before obtaining the trig function.

In some scientific and technical fields, angles are measured in a unit called radians. By definition:

$$1 \text{ radian} = \frac{360°}{2\pi} \text{ degrees or } 57.296° \text{ or } 57.3°$$

One radian is the angle at the center of a circle that corresponds to an arc exactly 1 radian in length. Figure 19–4 shows the radian degree comparison.

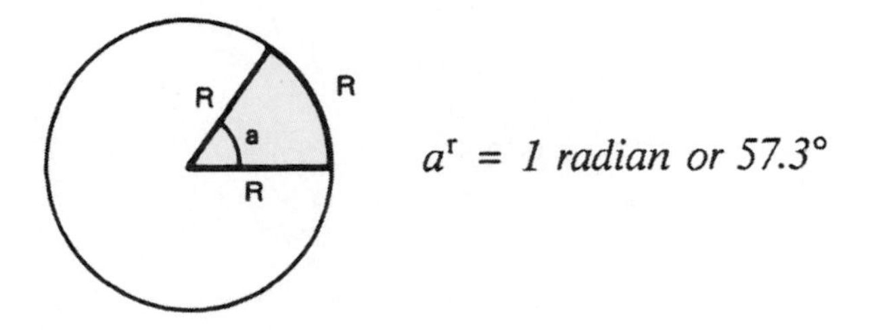

Figure 19–4

The following ratios demonstrate the radian degree conversion.

$$\frac{\text{angle } a \text{ in radians}}{\text{central angle of the circle}} = \frac{\text{arc length}}{\text{circumference of circle}}$$

$$\frac{a^r}{360°} = \frac{R}{2\pi R}$$

$$\left(\frac{360°}{1}\right)\frac{a^r}{360°} = \frac{R}{2\pi R}\left(\frac{360°}{1}\right)$$

$$a^r = \frac{360°}{2\pi}$$

$$= 57.296°$$

$$= 57.3°$$

Note: A small r above and to the right of the angle indicates radian measure; this symbol may be used similar to the degree symbol °.

$$1^r = 57.3°$$

You should complete problems 19–11 to 19–20 using a scientific calculator to practice finding angles and functions. For the remainder of the book, a scientific calculator will be used to solve problems.

EXERCISE 19–2 FIND ANGLES WITH A CALCULATOR

Use an electronic calculator to find the angles given the trigonometric functions.

19–11. Find angle A given $\sin A = 0.1219$.

19–12. Find angle X given $\tan X = 1.2954$.

19–13. Determine angle R when $\cos R = 0.4566$.

19–14. How large is angle W when $\tan W = 20.206$?

19–15. Given $\sin J = 0.6225$. Find angle J.

19–16. $\cos P = 0.9528$; $\angle P =$

19–17. $\tan K = 2.9319$; $\angle K =$

19–18. $\sin M = 0.0640$; $\angle M =$

19–19. $\cos N = 0.4488$; $\angle N =$

19–20. $\sin Y = 0.9644$; $\angle Y =$

19.3 SOLVING RIGHT TRIANGLES

Using the skills you have developed using trigonometric functions and angles, you will be able to solve right triangles. To solve a triangle means to find the unknown sides and angles when some of the sides and angles are given. The examples show ways to solve for unknown angles and sides.

Example 19–7: Given a Right Triangle, the Hypotenuse, and an Angle, Find a Side

PROBLEM: Given right triangle ABC shown in Figure 19–5, find side a.

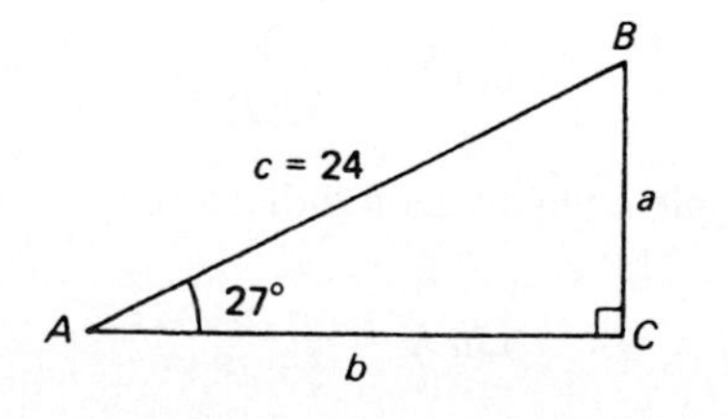

Figure 19–5

SOLUTION: Set up a trig ratio using the given values. This ratio would involve the opposite side a and the hypotenuse c and the given angle A. Use the trig function:

$$\sin A = \frac{\text{opp}}{\text{hyp}}$$

Thus

$$\sin 27° = \frac{a}{24}$$

Find sin 27° using a scientific calculator. Then solve for a. Remember that the "loop" terms equal side a.

$$0.4540 = \frac{a}{24}$$

$$24(0.4540) = a$$

$$10.896 = a$$

Example 19–8: Find the Side of a Right Triangle Given an Angle and a Side

PROBLEM: Given right triangle ABC shown in Figure 19–6, determine the length of side b.

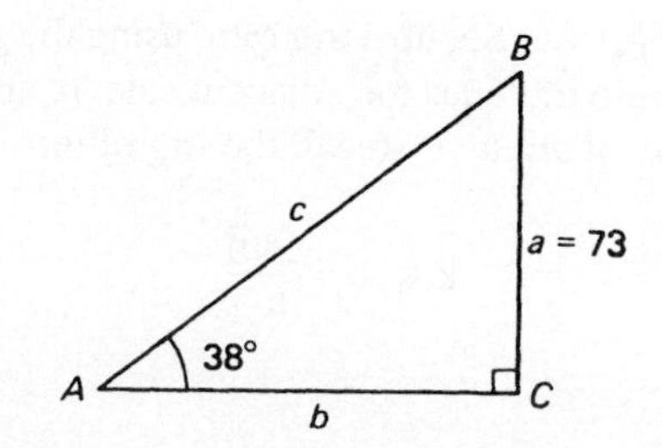

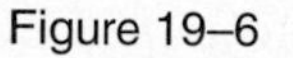

Figure 19–6

SOLUTION: Set up a ratio using the given values. This ratio would involve the opposite side, a, and the adjacent side, b, of angle A. Remember the ratio

$$\tan A = \frac{\text{opp}}{\text{adj}}$$

Thus

$$\tan 38^\circ = \frac{73}{b}$$

Find tan 38° using a scientific calculator. Then solve by multiplying both sides by b.

$$0.7813 = \frac{73}{b}$$

$$b(0.7813) = 73$$

Divide both sides by 0.7813.

$$\frac{b(0.7813)}{0.7813} = \frac{73}{0.7813}$$

$$b = 93.434$$

$$= 93.4$$

Round to the nearest tenth. Remember, complete accuracy cannot be obtained through calculation.

Example 19–9: Find the Hypotenuse of a Right Triangle Given an Angle and a Side

PROBLEM: Given right triangle ABC shown in Figure 19–7, determine the length of side c.

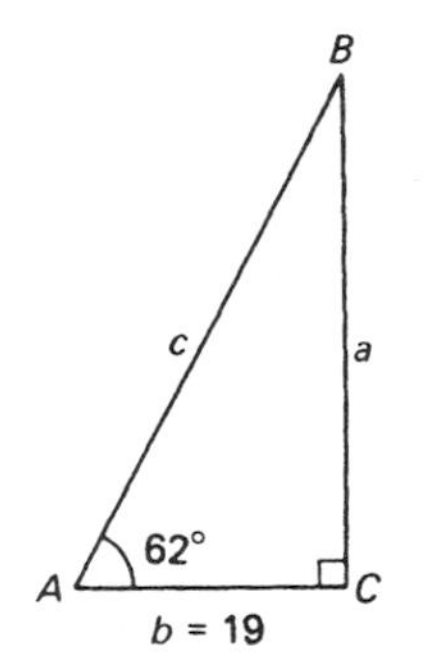

Figure 19–7

SOLUTION: Set up a trig ratio using the given values. This ratio includes the adjacent side, b, and the hypotenuse, c, of angle A. Recall the trig ratio:

$$\cos A = \frac{\text{adj}}{\text{hyp}}$$

Thus

$$\cos 62^\circ = \frac{19}{c}$$

Find cos 62° by using a scientific calculator to complete the ratio. Solve by multiplying both sides of the equation by c.

$$0.4695 = \frac{19}{c}$$

$$c(0.4695) = 19$$

Then divide both sides by 0.4695.

$$\frac{c(0.4695)}{0.4695} = \frac{19}{0.4695}$$

$$c = 40.2556$$

$$= 40.3$$

Round to the nearest tenth.

Example 19–10: Find the Angle of a Right Triangle Given Two Sides

PROBLEM: Given the triangle JKL shown in Figure 19–8, find angle K.

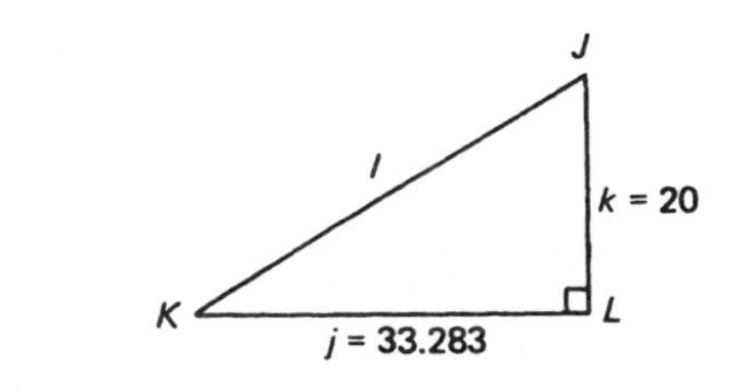

Figure 19–8

SOLUTION: Set up a trig ratio using the given values. This ratio would include the opposite side, $k = 20$, and the adjacent side, $j = 33.283$, of angle D. Recall that

$$\tan K = \frac{\text{opp}}{\text{adj}}$$

Thus

$$\tan K = \frac{20}{33.283}$$

Then complete the division indicated.

$$\tan K = \frac{20}{33.283}$$

$$= 0.6009$$

Use a scientific calculator to find the angle K, whose tangent is 0.6009. Thus angle $K = 31^\circ$.

Example 19–11: Solve a Right Triangle Given Two Sides

PROBLEM: Solve the right triangle ABC shown in Figure 19–9.

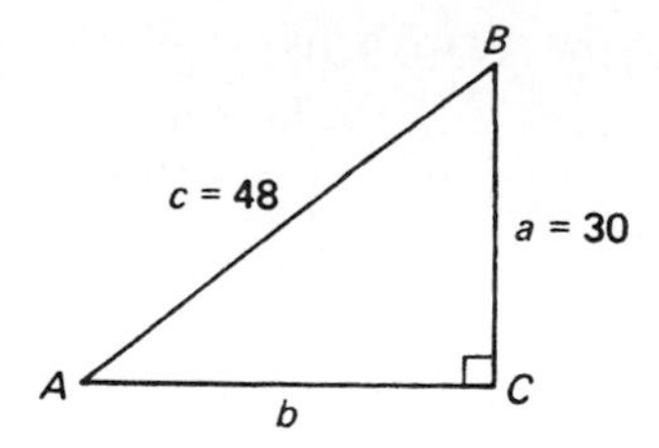

Figure 19–9

SOLUTION: The phrase "solve the right triangle" means to find all of the unknown parts of the right triangle. Therefore, angles A and B and side b must be found. First, find angle A. Set up a trig ratio using the given values: the opposite sides, $a = 30$, and the hypotenuse, $c = 48$. Remember the trig ratio:

$$\sin A = \frac{\text{opp}}{\text{hyp}}$$

Thus

$$\begin{aligned} \sin A &= \frac{30}{48} \\ &= 0.625 \end{aligned}$$

Use a scientific calculator to find angle A, whose sin is 0.6250. Thus $A = 38.682° = 38.7°$, or $38°40'$. The sum of angles A and B equals 90°, so

$$\begin{aligned} \angle A + \angle B &= 90° \\ 38.7° + \angle B &= 90° \\ \angle B &= 90° - 38.7° \\ &= 53.3° \end{aligned}$$

Then find side b. Set up a trip ratio using the given values: angle A, 38.7°, and the hypotenuse, 48. Recall the trig function:

$$\cos A = \frac{\text{adj}}{\text{hyp}}$$

Thus

$$\cos 38.7° = \frac{a}{48}$$

Use a scientific calculator to find $\cos 38.7° = 0.7804$. Substitute this into the trig ratio.

$$\cos 38.7° = \frac{b}{48} \text{ or } 0.7808 = \frac{b}{48}$$

Solve by multiplying both sides by 48. Thus $b = 37.5$.

$$\begin{aligned} 48(0.7808) &= \frac{b}{48}(48) \\ 37.4784 &= b \\ b &= 37.5 \end{aligned}$$

Therefore, we have solved triangle ABC by finding all the unknown parts: $\angle A = 38.7°$, $\angle B = 53.3°$, and $b = 37.5$.

Example 19–12: Solve a Right Triangle Given Angle and the Hypotenuse

PROBLEM: Given the following parts of a right triangle, PQR $\angle P = 23°$, $\angle R = 90°$, and side $r = 55$, solve the right triangle.

SOLUTION: First, sketch triangle PQR as shown in Figure 19–10. Find angle Q. The sum of angles P and Q equals 90°, so

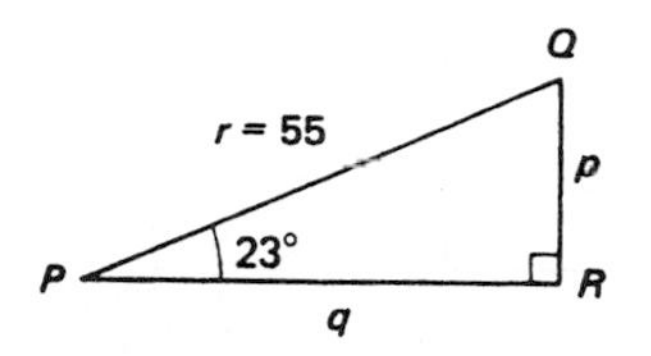

Figure 19–10

$$\begin{aligned} \angle P + \angle Q &= 90° \\ 23° + \angle Q &= 90° \\ \angle Q &= 90° - 23° \\ &= 67° \end{aligned}$$

Then find side p. Set up a trig ratio using the given values: angle $p = 23°$, and the hypotenuse $r = 55$.

$$\begin{aligned} \sin A &= \frac{\text{opp}}{\text{hyp}} \\ \sin 23° &= \frac{p}{55} \end{aligned}$$

Use a calculator to find $\sin 23° = 0.3907$. Substitute the value into the ratio.

$$\begin{aligned} \sin 23° &= \frac{p}{55} \\ 0.3907 &= \frac{p}{55} \end{aligned}$$

Solve by multiplying both sides by 55.

$$\begin{aligned} 55(0.3907) &= \frac{p}{55}(55) \\ 21.4885 &= p \\ p &= 21.5 \end{aligned}$$

Then find side q. Set up a trig ratio including the given values: angle $P = 23°$, and the hypotenuse $c = 55$.

$$\begin{aligned} \cos P &= \frac{\text{adj}}{\text{hyp}} \\ \cos 23° &= \frac{q}{55} \end{aligned}$$

Use a scientific calculator to find cos 23° = 0.9205. Substitute this value into the trig ratio and solve.

$$\cos 23° = \frac{q}{55} \quad \text{or} \quad 0.9205 = \frac{q}{55}$$

$$(55)0.9205 = \frac{q}{55}(55)$$

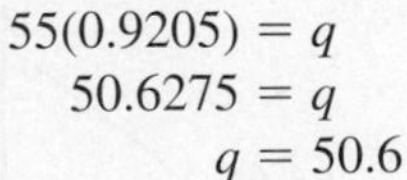

$$55(0.9205) = q$$
$$50.6275 = q$$
$$q = 50.6$$

Therefore, we have solved triangle *PQR* by finding all the unknown parts: $\angle Q = 67°$, $p = 21.5$, and $q = 50.6$.

EXERCISE 19–3 SOLVE TRIANGLES USING TRIG FUCNTIONS

Find the value of the unknown sides and angles as indicated.

19–21. Given the triangle *ABC*, Figure 19–11, find *c*.

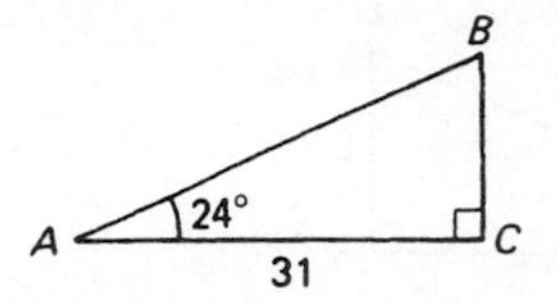

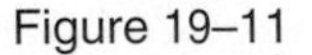
Figure 19–11

19–22. Given the triangle *DEF*, Figure 19–12, find *f*.

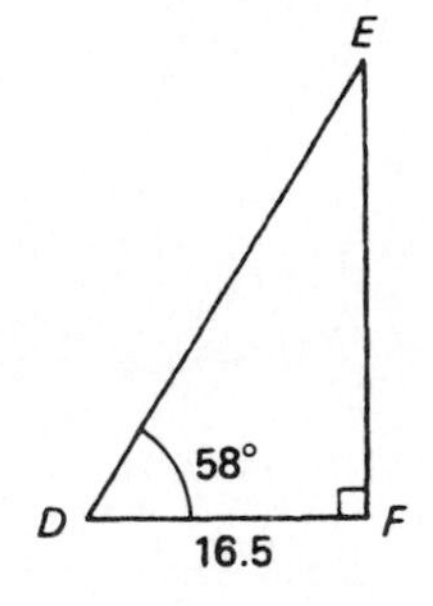

Figure 19–12

19–23. Find $\angle G$ in triangle *GHI*, Figure 19–13.

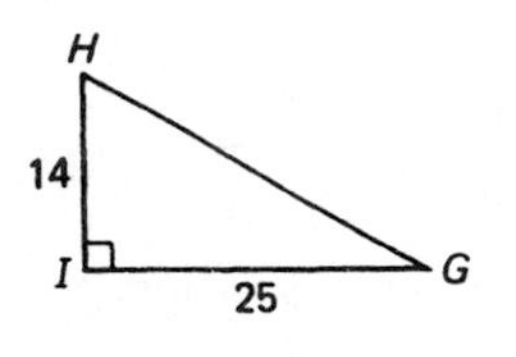

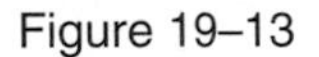
Figure 19–13

19–24. Find the length of side *l* in triangle *JKL*, Figure 19–14.

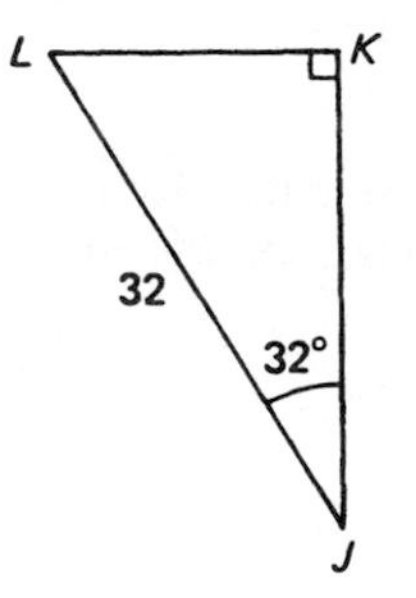

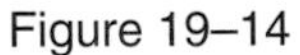
Figure 19–14

19–25. Find the length of side *n* in triangle *MNO*, Figure 19–15.

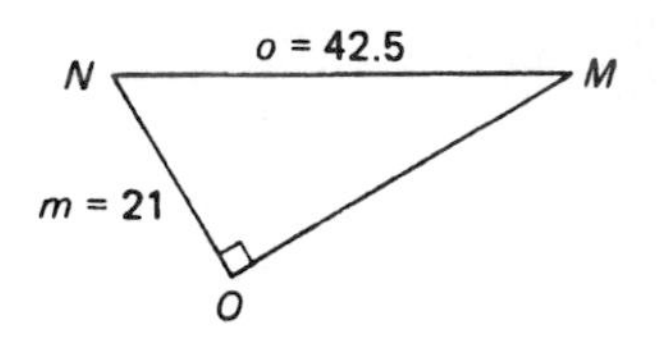

Figure 19–15

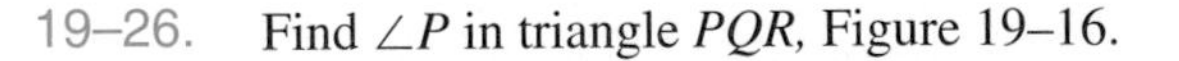
19–26. Find $\angle P$ in triangle *PQR*, Figure 19–16.

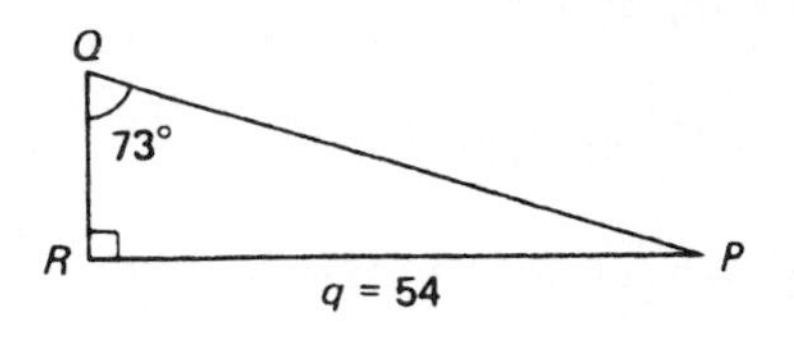

Figure 19–16

19–27. Find side r in triangle PQR, Figure 19–16.

19–28. Find side p in triangle PQR, Figure 19–16.

19–29. Find the degrees in $\angle S$ in triangle STU, Figure 19–17.

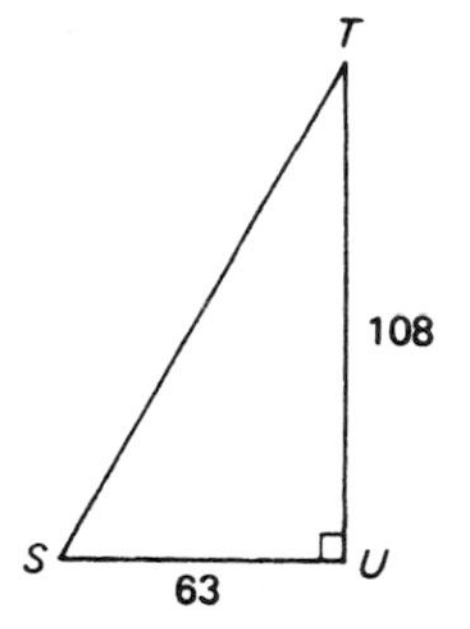

Figure 19–17

19–30. Determine $\angle T$ in triangle STU, Figure 19–17.

19–31. Find side u in triangle STU, Figure 19–17.

19–32. Find $\angle X$ in triangle XYZ, Figure 19–18.

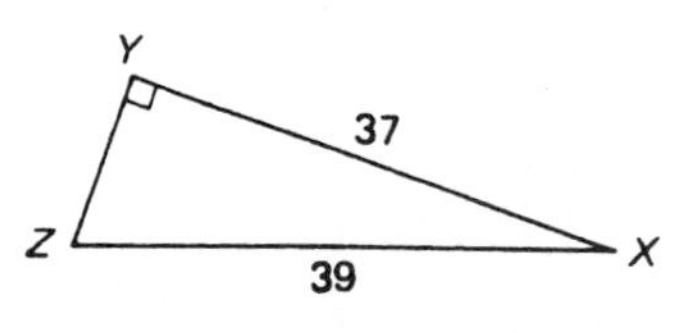

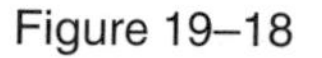

Figure 19–18

19–33. Find $\angle Y$ in triangle XYZ, Figure 19–18.

19–34. Find side x in triangle XYZ, Figure 19–18.

19–35. Find the degrees in $\angle B$, triangle ABC, Figure 19–19.

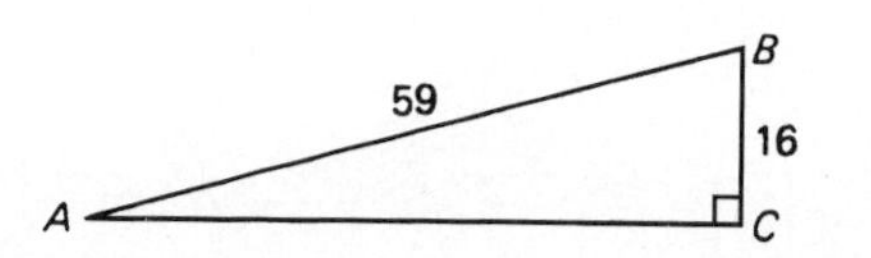

Figure 19–19

19–36. Find $\angle A$, triangle ABC, Figure 19–19.

19–37. Find side b, triangle ABC, Figure 19–19.

19–38. Given the right triangle, find $\angle D$, triangle DEF, Figure 19–20.

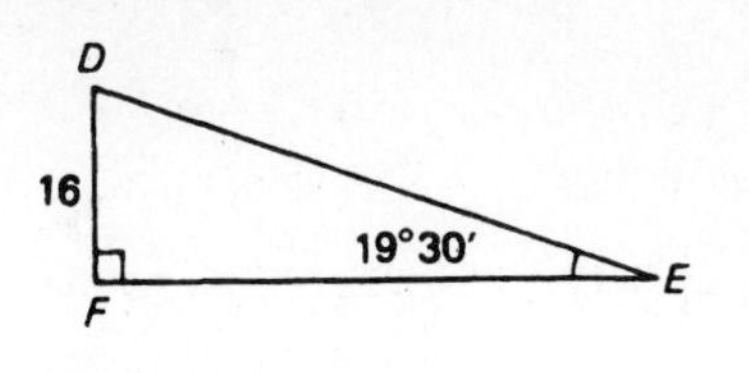

Figure 19–20

19–39. Find side d, triangle DEF, Figure 19–20.

19–40. Find side f, triangle DEF, Figure 19–20.

19–41. Sketch triangle GHI, with $\angle I = 90°$, $H = 21°50'$, and side $g = 175$.

19–42. Find side h, given the sketch of triangle GHI in problem 19–41.

19–43. Find $\angle G$ of triangle GHI in problem 19–41.

19–44. Find side i of triangle GHI in problem 19–41.

19–45. Find side j in triangle JKL when side $l = 12.48$, $\angle L = 90°$, and $\angle J = 83°45'$.

19.4 APPLICATIONS OF RIGHT TRIANGLES

The ability to solve right triangles is essential when solving applied problems. In fact, some applied problems can be solved only by using trigonometry. In many problems, the use of trigonometry can help you solve problems more easily and efficiently.

The examples that follow show applications of the solution of right triangles to applied problems. Your imagination will help you find other uses for your ability to solve right triangles.

Example 19–13: Trig Application Concerning Building Construction

PROBLEM: A contractor builds an A-frame cabin 24 feet high with a peak angle of 55°. Find the length of the rafters.

SOLUTION: Draw a sketch of the cabin and indicate the given values as shown in Figure 19–21.

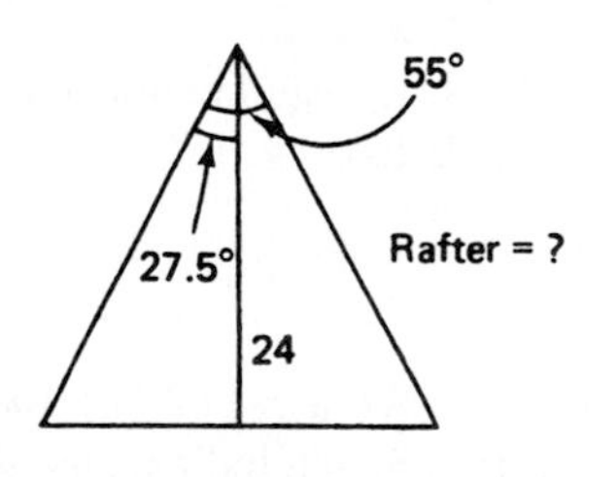

Figure 19–21

Set up a trig ratio using the given values, 27.5°, and the height, 24 feet.

$$\cos A = \frac{\text{adj}}{\text{hyp}} \quad \text{or} \quad \cos 27.5° = \frac{24 \text{ ft}}{\text{rafter length}}$$

Find the value of cos 27.5°, and solve for the rafter length.

$$\begin{aligned} \cos 27.5° &= \frac{24 \text{ ft}}{\text{rafter length}} \\ 0.8870 &= \frac{24 \text{ ft}}{\text{rafter length}} \\ \text{Rafter length}(0.8870) &= 24 \text{ ft} \\ \text{Rafter length} &= \frac{24 \text{ ft}}{0.8870} \\ &= 27.057 \text{ ft} \\ &= 27.1 \text{ ft} \end{aligned}$$

Example 19–14: Trig Application Concerning Navigation

PROBLEM: The captain of an iron-ore freighter spots a lighthouse at an angle of 2° 30′ from his position. The map indicates the light is 158 feet above the surface of the water. How far is the ship from the lighthouse?

SOLUTION: Draw a sketch and label the given quantities as shown in Figure 19–22. Set up a trig ratio

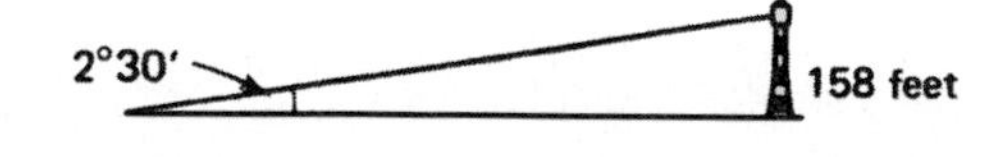

Figure 19–22

using the given values: 2°30′ and the height of the light, 158 feet. Recall that

$$\tan A = \frac{\text{opp}}{\text{adj}}$$

Thus

$$\tan 2°30' = \frac{\text{opp}(158 \text{ ft})}{\text{adj(distance from light)}}$$

Find the value of tan 2°30′ or 2.5° $\left(2\frac{30}{60}°\right)$ and solve for the distance.

$$\begin{aligned} \tan 2°30' &= \frac{158 \text{ ft}}{\text{distance}} \\ 0.0437 &= \frac{158 \text{ ft}}{\text{distance}} \\ \text{distance} &= \frac{158 \text{ ft}}{0.0437} \\ &= 3615.5606 \text{ ft} \\ &= 3615 \text{ ft} \end{aligned}$$

Example 19–15: Trig Application Concerning Properties of Right Triangles

PROBLEM: Find the missing dimension x in Figure 19–23.

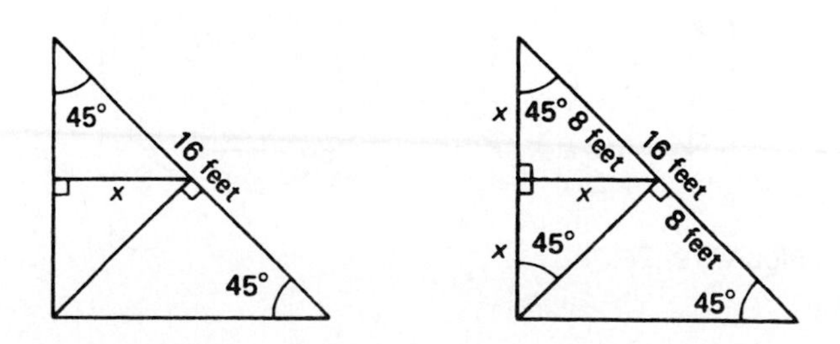

Figure 19–23

SOLUTION: Study Figure 19–23, recalling the properties of a 45° right triangle, and indicate the facts on the sketch. Note that all values of x are equal. Set up a trig ratio using the given values: 45° and the length of one side, 8 feet.

$$\sin A = \frac{\text{opp}}{\text{hyp}}$$

Thus

$$\sin 45° = \frac{\text{opp}(x)}{\text{hyp}} = \frac{x}{8 \text{ ft}}$$

Find the sin 45° and solve for x.

$$\sin 45° = \frac{x}{8 \text{ ft}}$$

$$0.7071 = \frac{x}{8 \text{ ft}}$$

Multiply both sides by 8 feet.

$$\begin{aligned} x &= (8 \text{ ft})(\sin 45°) \\ &= (8 \text{ ft})(0.7071) \\ &= 5.6568 \text{ ft} \end{aligned}$$

Convert 0.6568 feet to inches.

$$0.6568 \times 12 \text{ in.} = 7.8816 \text{ in.}$$

The inches could be rounded to the nearest eighth of an inch, which would be $7^7/_8$ inches. Thus the length of the missing dimension is 5 feet $7^7/_8$ inches.

Example 19–16: Trig Application Concerning Circles

PROBLEM: Find the depth of cut required to mill a square on the end of a $1^1/_2$-inch shaft, shown in Figure 19–24.

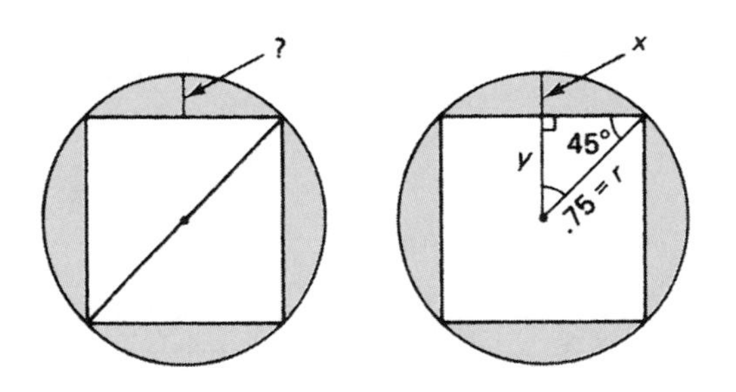

Figure 19–24

SOLUTION: Study Figure 19–24, and enter the given values as shown on the second sketch. Recall the facts of a 45° right triangle. Letter the missing dimension x. Use the letter y as a leg of the right triangle as shown. Thus $x + y = 0.75$ inch. Find the length of y. Set up a trig ratio involving the given values: 45° and the hypotenuse, 0.75 inch.

$$\sin A = \frac{\text{opp}}{\text{hyp}}$$

$$\sin 45° = \frac{y}{0.75}$$

Find the sin of 45° and solve for y.

$$0.7071 = \frac{y}{0.75 \text{ in.}}$$

Multiply both sides of the equation by 0.75 inch.

$$\begin{aligned} y &= (0.75 \text{ in.})(0.7071) \\ &= 0.530325 \text{ in.} \\ &= 0.53 \text{ in.} \end{aligned}$$

Then solve for x.

$$\begin{aligned} x + y &= 0.75 \text{ in.} \\ x + 0.53 \text{ in.} &= 0.75 \text{ in.} \\ x &= 0.75 \text{ in.} - 0.53 \text{ in.} \\ x &= 0.22 \text{ in.} \end{aligned}$$

Example 19–17: Trig Application Concerning Missiles

PROBLEM: If a missile is launched at 57° at 220 miles per hour, how fast is it traveling horizontally?

SOLUTION: Draw a sketch of the problem as shown in Figure 19–25, and indicate the given values,

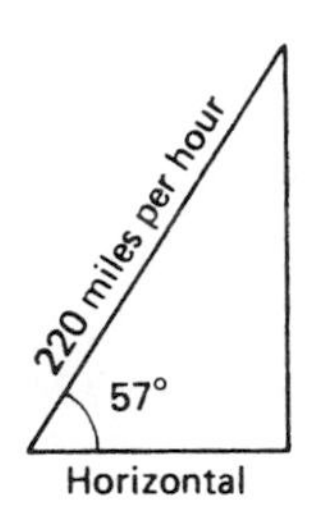

Figure 19–25

220 miles per hour and 57°. Then find the horizontal distance. Recall the trig function that involves the adjacent side and hypotenuse of the given angle.

$$\cos A = \frac{\text{adj}}{\text{hyp}}$$

Substitute the given values and solve for the horizontal speed.

$$\cos 57° = \frac{\text{horizontal}}{220 \text{ mph}}$$

$$\begin{aligned} \text{Horizontal} &= (220 \text{ mph})(\cos 57°) \\ \text{Horizontal} &= (220 \text{ mph})(0.5446) \\ \text{Horizontal} &= 119.812 \text{ mph} \\ \text{Horizontal} &= 119.8 \text{ mph} \end{aligned}$$

EXERCISE 19–4 SOLVE APPLIED PROBLEMS USING TRIG FUNCTIONS

The following problems will help develop your skills solving applied problems using trigonometry. Draw sketches to better understand the problem you are solving. Refer to the examples if necessary for assistance.

19–46. A triangle is the most rigid of all geometric forms. Determine the length of a diagonal for a cattle gate 10 feet long and 4 feet high.

19–47. A rocket is launched at an angle of 82°30′ with a speed of 2800 kilometers per hour. What is its vertical speed in kilometers per hour?

19–48. A roof is to rise 20 feet in a horizontal distance of 32 feet. What is the angle of slope of the roof with the horizon?

19–49. What length of ladder is needed to paint the top of the gable of a house if the gable is 18 feet 6 inches high and the base of the ladder is 10 feet from the house?

19–50. Find the missing dimension, x, in Figure 19–26.

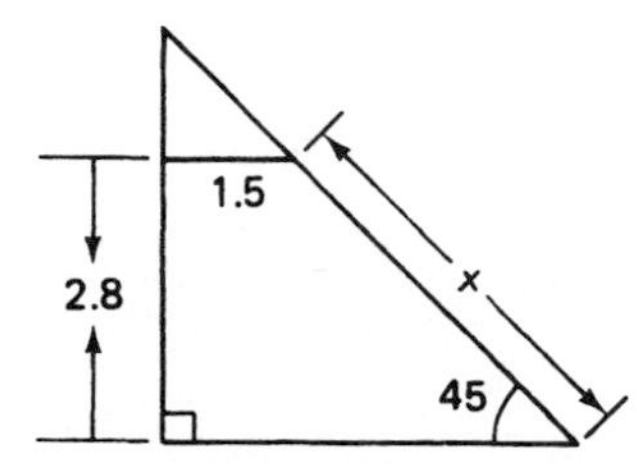

Figure 19–26

19–51. A utility pole is supported by a guy wire 35 feet from the ground. If the wire is inclined at an angle of 49°30′ with the ground, how many feet of wire will be needed?

19–52. Point B is 80 air miles north of point A. Point C is 125 air miles east of point B. What is the shortest air mileage between points A and C? (*Note:* Directions are at 90° angles to each other.)

19–53. Find the length, in feet, of the air ventilation shaft of the tunnel shown in Figure 19–27.

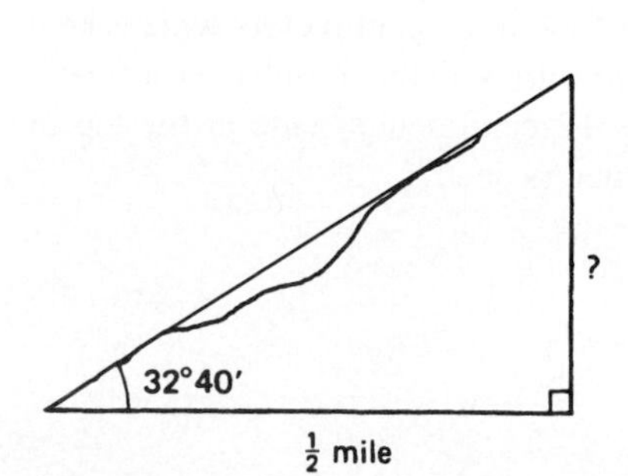

Figure 19–27

19–54. A land surveyor measures 50 feet along a river bank. Then with a transit point directly across the river from the original point, he reads an angle of 76°. How wide is the river?

19–55. Find the missing dimension of a flat-head screw shown in Figure 19–28.

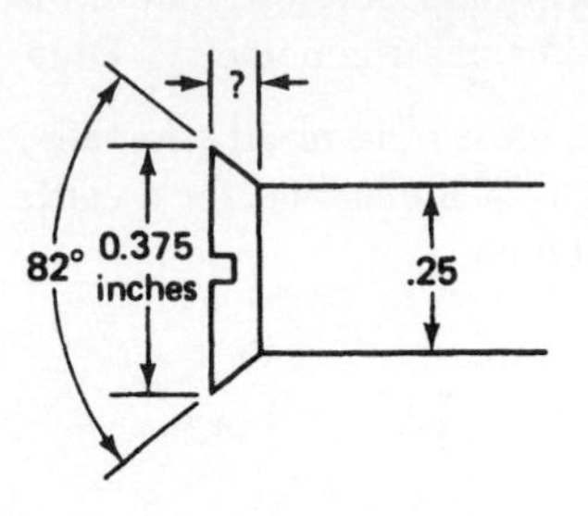

Figure 19–28

19–56. Find the taper angle if there is a $^1/_2$-inch taper in $2^1/_2$ inches.

19–57. The Washington Monument is 555 feet high. What is the angle of elevation to the top from a distance of 1 mile?

19–58. A freeway shown in Figure 19–29 is constructed through a city at an angle of 68° to the local streets. If the streets are 380 feet apart, what length will the freeway be between intersections of north/south streets?

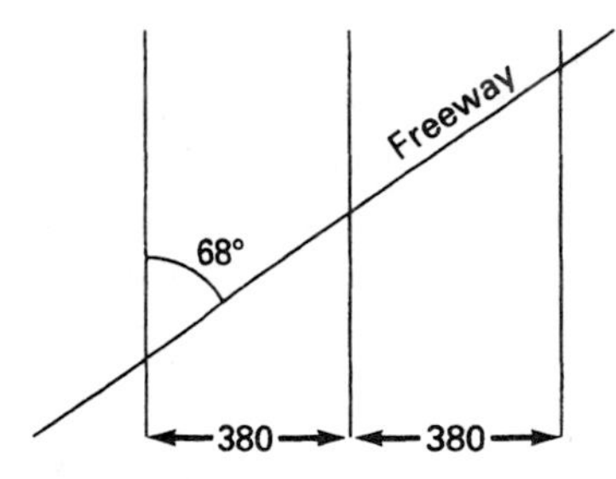

Figure 19–29

19–59. Find the volume of the right prism shown in Figure 19–30.

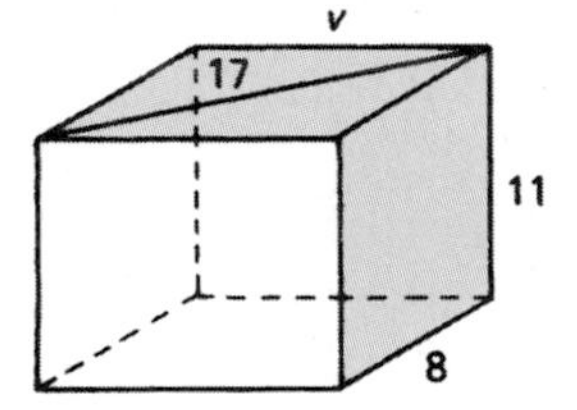

Figure 19–30

19–60. A sheet of metal 25 centimeters wide is bent to form a V-shaped trough. Will the trough have a greater capacity when it is 12 centimeters wide at the top or when it is 10 centimeters deep?

19–61. A helicopter flying at an altitude of 1 mile observes a heliport landing area at an angle of depression of 24°20′. How far, in feet, is the helicopter from the landing area?

19–62. A pentagonal flower garden is constructed inside a 16-foot-diameter circle. How much decorative fence must be ordered to fence the garden?

19–63. If the slope of the land is 4° 10′, how far will the water be backed up behind a 45-foot dam?

19–64. The end of an 8-centimeter diameter shaft is milled as a triangle shown in Figure 19–31. Determine the radius of the smaller circle.

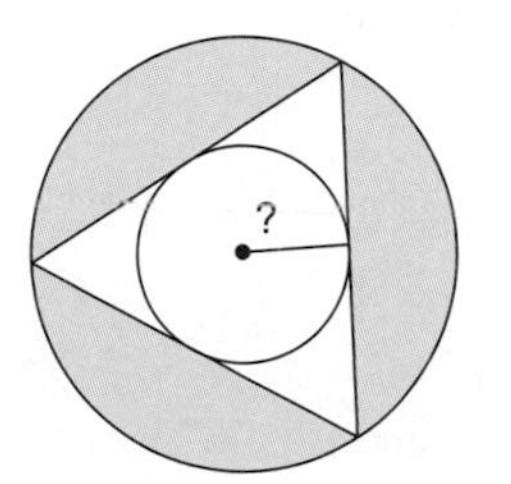

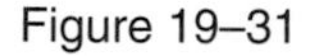
Figure 19–31

19–65. The 30° arc of a circular curve of a mountain road has a chord of 75 feet. Find the radius of the curve.

THINK TIME

The procedures used to solve a right triangle can be used to solve everyday problems such as: How high is a building or tree? How far has one sailed, flown, or snowmobiled in a particular direction? The use of these procedures to solve problems is limited only to the creative thinking of the person. For example, designers of golf courses determine distances aross natural barriers. Find the shortest distance, from the tee to the hole, in Figure 19–32.

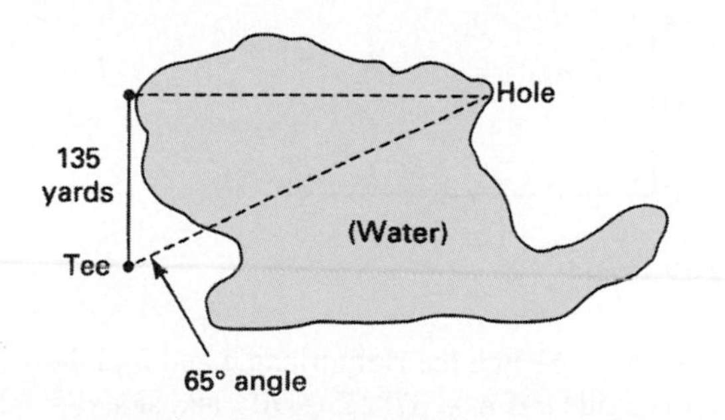

Figure 19–32

PROCEDURES TO REMEMBER

1. To find the trigonometric functions of angles:
   (a) Use a scientific calculator.
   (b) Locate the degree of the given angle.
   (c) Read the appropriate trigonometric function of the angle.
2. To determine the angle of a trigonometric function:
   (a) Use a scientific calculator.
   (b) Locate the given trigonometric value.
   (c) Read the appropriate angle for the given trigonometric function.
3. To find trig functions using a calculator, enter the angle and press the key for the desired function.
4. To find an angle given a function using a calculator: enter the function, press the inverse or second function key, and then press the given function key.
5. To find the side of a right triangle:
   (a) Set up a trig ratio using the given values.
   (b) Solve the ratio for the unknown side.
   (c) Round accordingly.
6. To find an angle of a right triangle:
   (a) Set up a trig ratio using the given sides to find the value of the trig function.
   (b) Solve the ratio for the trig function.
   (c) Use a table or calculator to find the angle.

CHAPTER SUMMARY

The following definitions and procedures are essential to solve right triangles.

1. *Interpolation* is the process of finding the value of a trig function or angle using known trig functions or angles or a table of trigonometric functions. Determination of the interpolation is

not necessary; when using a calculator, it is done automatically.

2. The determination of the *arc function* is based on the value of a trigonometric function and the angle it represents.
3. When the tangent value is greater than 1, the angle is greater than 45°.
4. Sine values *increase* as the size of the angle increases.
5. Cosine values *decrease* as the size of the angle increases.
6. Tangent values *increase* as the size of the angle increases.
7. To solve a right triangle means to find all the unknown parts of the right triangle.

CHAPTER TEST

T–19–1. Find the tangent of 76°.

T–19–2. Find the cosine of 32°40′.

T–19–3. Sin 81°20′ = ______.

T–19–4. Find the angle whose cosine is 0.4823.

T–19–5. Cos A = 0.7679; A = ______.

T–19–6. Find the cosine of 29°33′.

T–19–7. Tan A = 3.6796; A = ______.

T–19–8. Find the length of side b in triangle ABC, Figure 19–33.

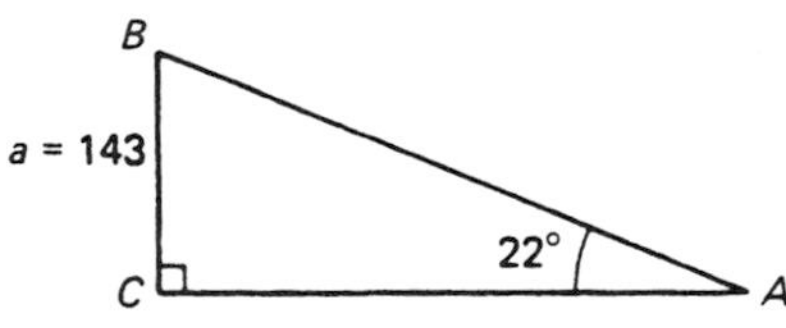

Figure 19–33

T–19–9. Given triangle JKL, Figure 19–34, find angle K.

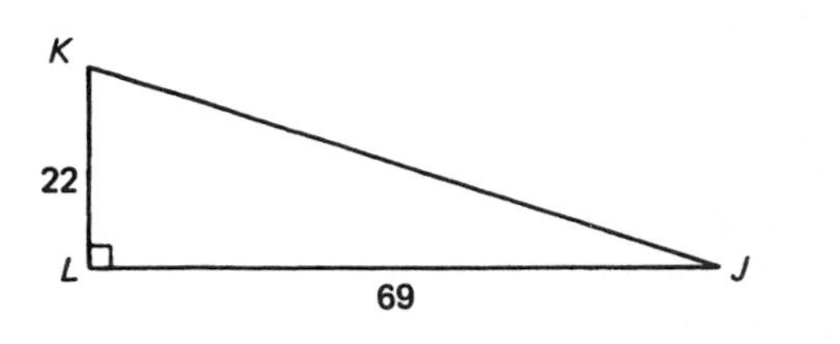

Figure 19–34

T–19–10. Sketch the right triangle and find the length of side a if A = 37°, C = 90°, and side c = 303 centimeters.

T–19–11. Find side s in triangle RST, Figure 19–35.

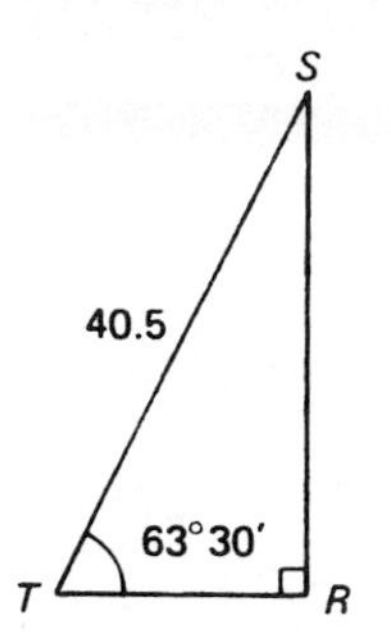

Figure 19–35

T–19–12. Find the length of side t in triangle RST, Figure 19–35.

T–19–13. A 24-foot ladder leans against a wall. Find the angle the ladder makes with the wall if the ladder is $9\frac{1}{2}$ feet from the base of the wall.

T–19–14. From point A, a sailboat sails $1\frac{1}{2}$ miles due west to point B and then 4 miles south to point C. What is the distance from point A to point C?

T–19–15. What is the largest square that can be milled from the end of a $1\frac{1}{2}$-inch shaft?

20

SOLUTION OF OBLIQUE TRIANGLES

OBJECTIVES

After studying this chapter, you will be able to:

20.1. Understand the properties of oblique triangles and solve using the Law of Sines.
20.2. Solve Oblique triangles using the Law of Cosines.
20.3. Understand applications of the Law of Sines and the Law of Cosines.
20.4. Solve applied problems involving oblique triangles.

SELF-TEST

This test will evaluate your ability to solve oblique triangles. If you solve these oblique triangle problems, your trigonometry skills in the solution of oblique triangles should be adequate. Be sure to show your work so that a possible area of concern can be identified.

S–20–1. Find the cos 119°.

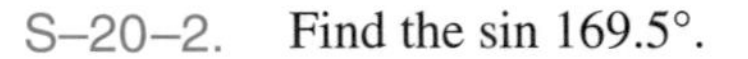

S–20–2. Find the sin 169.5°.

S–20–3. Find side a in the oblique triangle ABC in Figure 20–1.

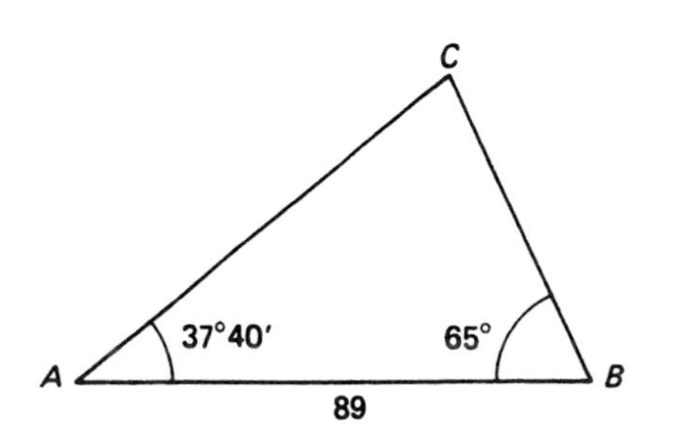

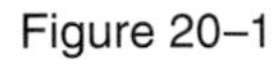

Figure 20–1

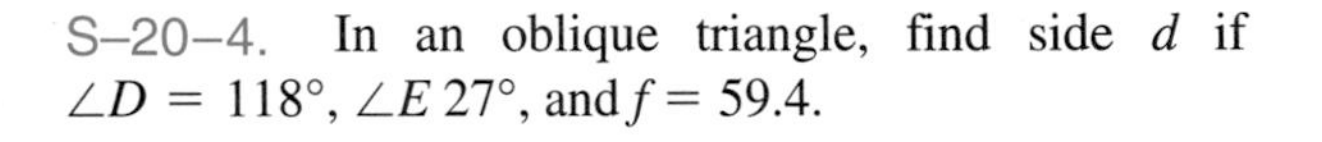

S–20–4. In an oblique triangle, find side d if $\angle D = 118°$, $\angle E\ 27°$, and $f = 59.4$.

S–20–5. Find angle I of oblique triangle GHI in Figure 20–2.

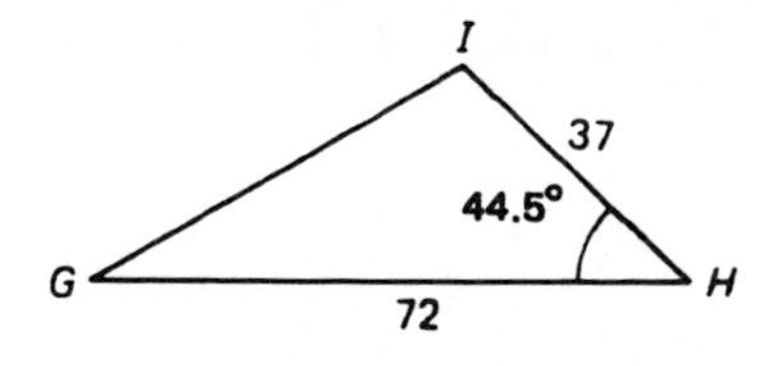

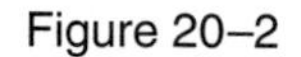
Figure 20–2

S–20–6. Find side k of triangle JKL if side $j = 426$, side $l = 304$, and $\angle K = 32°40'$.

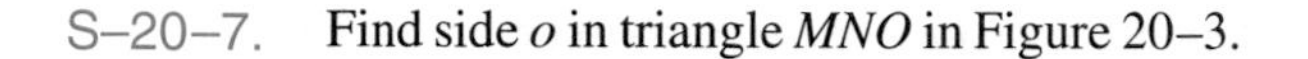
S–20–7. Find side o in triangle MNO in Figure 20–3.

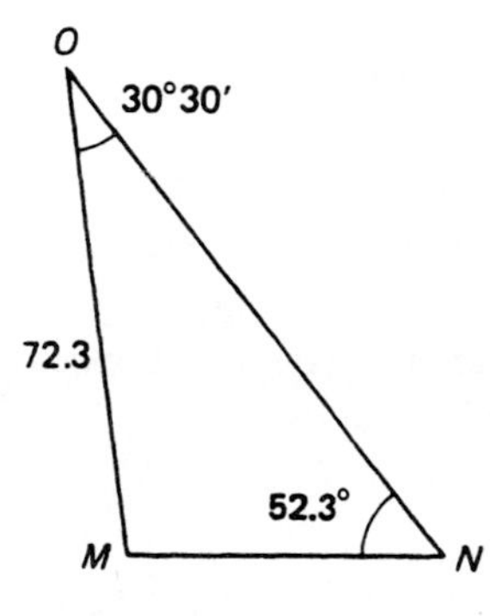

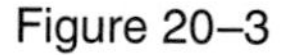
Figure 20–3

S–20–8. Find $\angle M$ in triangle MNO in problem S–20–7.

S–20–9. From problem S–20–7, find side m in triangle MNO.

S–20–10. Find the shortest distance across the ravine from point A to point B in Figure 20–4. Use the sketch as a guide.

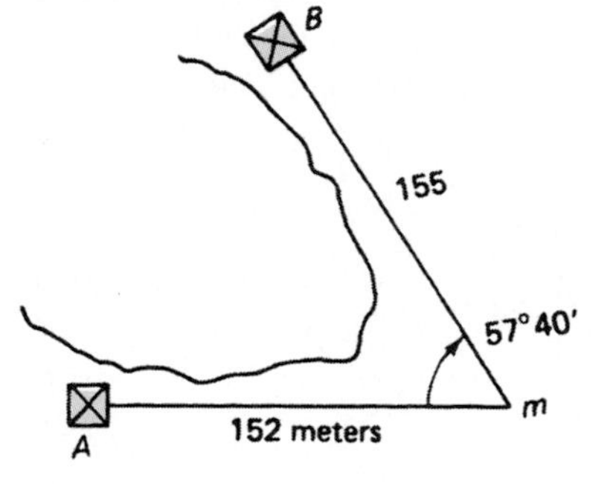

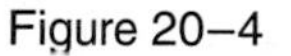
Figure 20–4

20.1 PROPERTIES OF OBLIQUE TRIANGLES AND THE LAW OF SINES

Recall that oblique triangles do not contain a right angle. The solution of oblique triangles is more difficult than the solution of right triangles. The Laws of Sines and Cosines are used to solve oblique triangles. Some oblique triangles can be divided into two or more right triangles for an easy solution. The sum of the interior angles of all triangles is 180°. There are six parts of all triangles: three sides and three angles.

Two facts concerning sines and cosines of angles greater than 90° are important before the Laws of Sines and cosines are considered. If an angle is greater than 90° and less than 180°, the sine is equivalent to the sine of 180° minus the given angle. Thus if $\angle A > 90°$, then

$$\sin A = \sin(180° - \angle A)$$

So if $\angle A = 127°$,

$$\sin 127° = \sin(180° - 127°)$$
$$= \sin 53°$$

A scientific calculator will give the correct sine value regardless of the size of the angle.

If an angle is greater than 90° and less than 180°, the cosine is equal to a negative cosine value of 180° minus the given angle. Thus if $\angle A > 90°$, then

$$\cos A = -\cos(180° - \angle A)$$

So if $\angle A = 164°$,

$$\cos 164° = -\cos(180° - 64°)$$
$$\cos 164° = -\cos 16°$$

Again, a scientific calculator will give the correct cosine value regardless of the size of the angle.

Remember, the cosine of an angle greater than 90° is always negative. Further proof of these concepts may be obtained in more advanced trigonometry texts.

To solve oblique triangles using the laws of sines one of two combinations must be given: two angles and one side or two sides and the angle opposite one of the sides. Thus at least three of the six parts of the triangle must be given. The parts that are given will determine which law will be used to solve the oblique triangle.

Although it is not essential to know how the Law of Sines is developed, a simple study of the process will help you understand the concept. Study the illustration and the procedure given using triangle ABC in Figure 20–5.

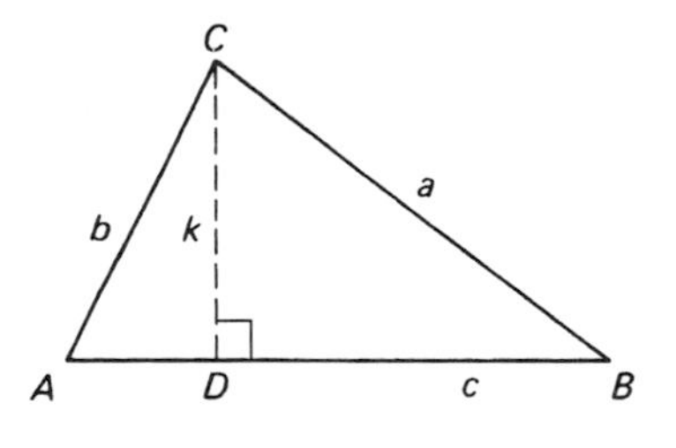

Figure 20–5

Draw a perpendicular from point C to point D on side c to form right triangles, ACD and BCD. State the sine ratios of $\angle A$ and $\angle B$.

$$\sin A = \frac{\text{opposite}}{\text{hypotenuse}} = \frac{k}{b}$$

$$\sin B = \frac{\text{opposite}}{\text{hypotenuse}} = \frac{k}{a}$$

Solve each ratio for k.

$$\sin A = \frac{k}{b} = b \sin A = k$$

$$\sin B = \frac{k}{a} = a \sin B = k$$

Since $k = k$,

$$b \sin A = a \sin B$$

Divide both sides of the equation by $\sin A \sin B$.

$$\frac{b \sin A}{\sin A \sin B} = \frac{a \sin B}{\sin A \sin B}$$

$$\frac{b}{\sin B} = \frac{a}{\sin A}$$

In a similar manner, a perpendicular could be dropped from point A which would result in ratios

$$\frac{b}{\sin B} = \frac{c}{\sin C}$$

Thus we have developed the law of sines:

$$\frac{a}{\sin A} = \frac{b}{\sin B} = \frac{c}{\sin C}$$

The Law of Sines states that the ratio of a side of a triangle to the sine of its angle is equal to the ratios of the other sides and their ratios.

The Law of Sines could also be stated in inverse form:

$$\frac{\sin A}{a} = \frac{\sin B}{b} = \frac{\sin C}{c}$$

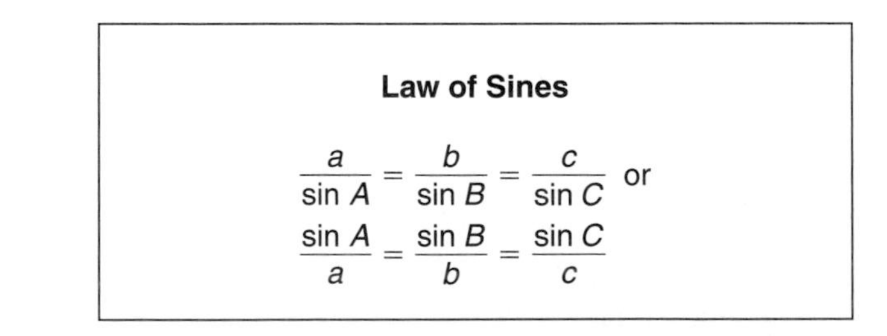

Law of Sines

$$\frac{a}{\sin A} = \frac{b}{\sin B} = \frac{c}{\sin C} \quad \text{or}$$

$$\frac{\sin A}{a} = \frac{\sin B}{b} = \frac{\sin C}{c}$$

To solve triangles using the law of sines, two or more ratios are used, and the given quantities are substituted in the ratios. The triangles may be lettered differently and other letters may be used. The following examples show how the Law of Sines may be used to solve oblique triangles.

Example 20–1: Solve an Oblique Triangle Given Two Angles and a Side

PROBLEM: Find side a in triangle ABC in Figure 20–6.

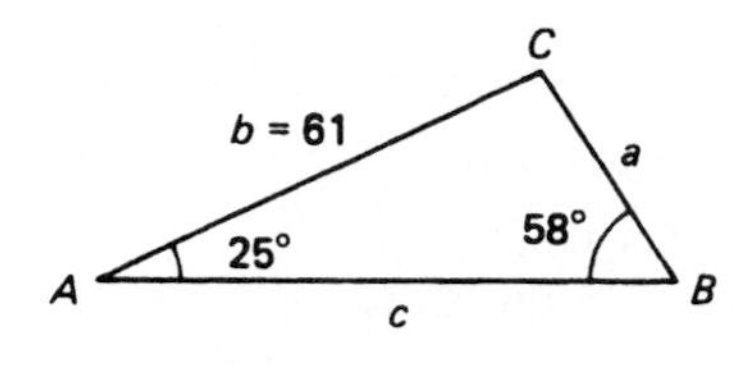

Figure 20–6

SOLUTION: Note the given quantities, $\angle A = 25°$, $\angle B = 58°$, and side $b = 61$. Two angles and the included side are given; therefore, the Law of Sines should be used:

$$\frac{a}{\sin A} = \frac{b}{\sin B} = \frac{c}{\sin C}$$

Select two ratios to find side a. The ratios are

$$\frac{a}{\sin A} = \frac{b}{\sin B}$$

Substitute the given values and solve.

$$\frac{a}{\sin 25°} = \frac{61}{\sin 58°}$$

$$\frac{a}{0.4226} = \frac{61}{0.8480}$$

Multiply both sides by 0.4226 and solve; thus $a = 30.4$.

$$(0.4226)\frac{a}{0.4226} = \frac{61}{0.8480}(0.4226)$$
$$a = 30.99 \text{ or } 30.4$$

Example 20–2: Solve an Oblique Triangle Given Two Sides and an Angle

PROBLEM: Find the size of $\angle C$ given the oblique triangle ABC in Figure 20–7.

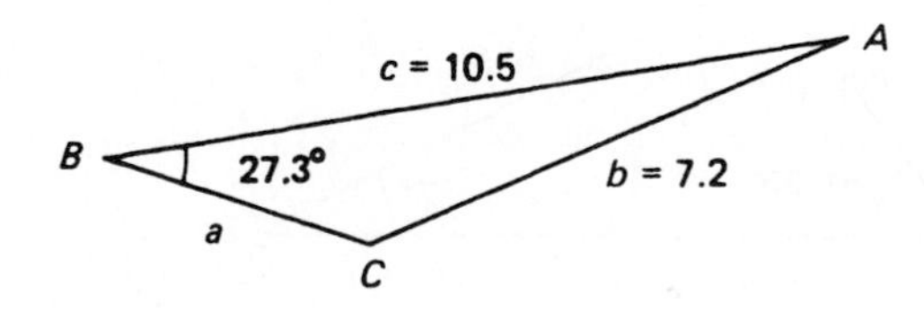

Figure 20–7

SOLUTION: Note the given quantities: $\angle B = 27.3°$, side $b = 7.2$, and side $c = 10.5$. Two angles and one side are given; therefore, the Law of Sines should be used. Set up two ratios to find the size of $\angle C$. The ratios are

$$\frac{b}{\sin B} = \frac{c}{\sin C}$$

Substitute the given values and solve for angle C.

$$\frac{b}{\sin B} = \frac{c}{\sin C}$$

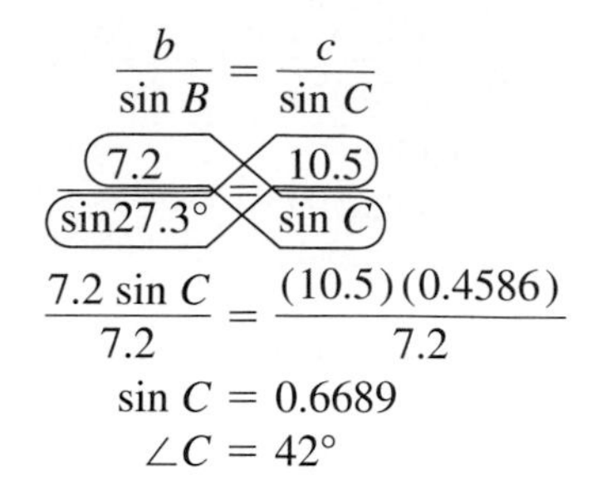

$$\frac{7.2 \sin C}{7.2} = \frac{(10.5)(0.4586)}{7.2}$$
$$\sin C = 0.6689$$
$$\angle C = 42°$$

From the sketch, $\angle C$ is greater than 90°. Using the definition $\sin A = \sin(180° - \angle A)$, substitute as follows:

$$\sin 42° = \sin(180° - 42°)$$
$$= \sin 138°$$

Thus $\angle C = 138°$

Example 20–2: Solve an Oblique Triangle Given Two Angles and a Side.

PROBLEM: Solve oblique triangle ABC shown in Figure 20–8. To solve an oblique triangle means to find all the sides and angles.

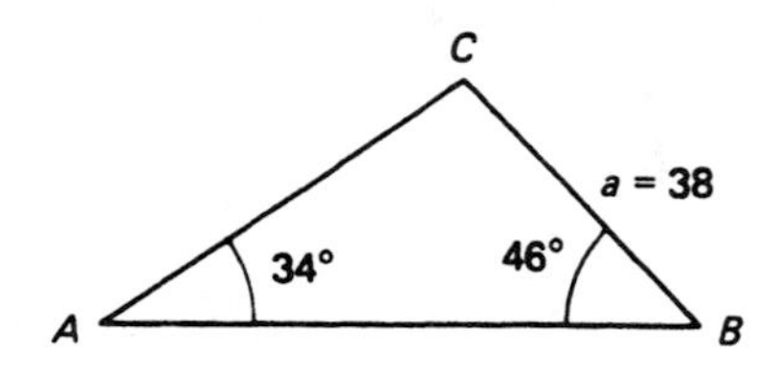

Figure 20–8

SOLUTION: Note that $\angle A = 34°$, $\angle B = 46°$, and side $a = 38$. $\angle C$ and sides b and c are to be found. We are given two sides and one angle; the Law of Sines should be used to solve the triangle. First find $\angle C$.

$$\angle A + \angle B + \angle C = 180°$$
$$24° + 46° + \angle C = 180°$$
$$\angle C = 180 - 34° - 46°$$
$$= 100°$$

Then set up two ratios to find side b. The ratios are

$$\frac{a}{\sin A} = \frac{b}{\sin B}$$

Substitute the given values and solve.

$$\frac{a}{\sin A} = \frac{b}{\sin B}$$

$$\frac{38}{\sin 34^\circ} = \frac{b}{\sin 46^\circ}$$

$$\frac{38}{0.5592} = \frac{b}{0.7193}$$

$$\frac{(38)(0.7193)}{0.5592} = b$$

$$48.8794 = b$$

$$48.9 = b$$

Next set up two ratios to find side c. The ratios are

$$\frac{a}{\sin A} = \frac{c}{\sin C}$$

EXERCISE 20–1 SOLVE TRIANGLES USING THE LAW OF SINES

Using the Law of Sines, find the size of angles and sides as indicated.

20–1. Find side b in triangle ABC, Figure 20–9.

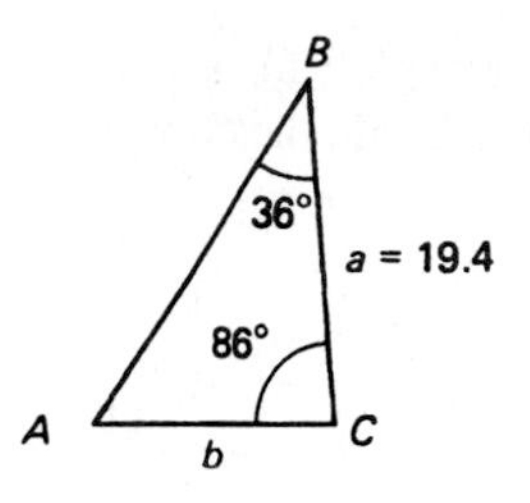

Figure 20–9

20–3. Find the size of $\angle B$ given triangle ABC in Figure 20–10.

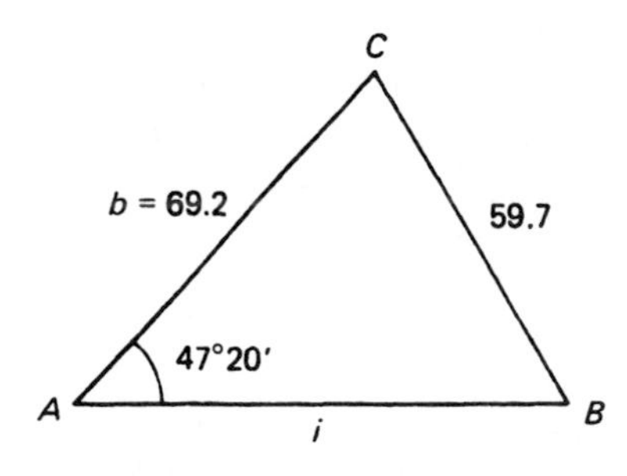

Figure 20–10

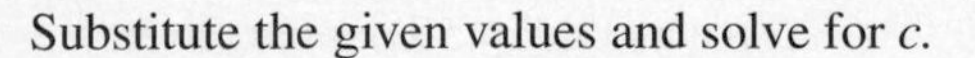
Substitute the given values and solve for c.

$$\frac{a}{\sin A} = \frac{c}{\sin C}$$

$$\frac{38}{\sin 34^\circ} = \frac{c}{\sin 100^\circ}$$

$$\frac{38}{0.5592} = \frac{c}{0.9848}$$

$$\frac{38 \times 0.9848}{0.5592} = c$$

$$66.9213 = c$$

$$c = 66.9$$

20–2. Given $\angle A = 24^\circ$, $\angle C = 65^\circ$, and side $b = 24.3$, determine the length of side c.

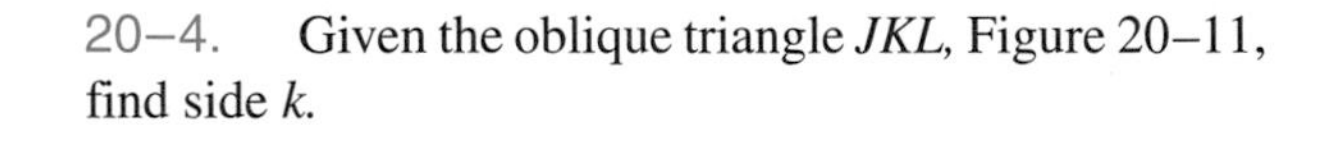
20–4. Given the oblique triangle JKL, Figure 20–11, find side k.

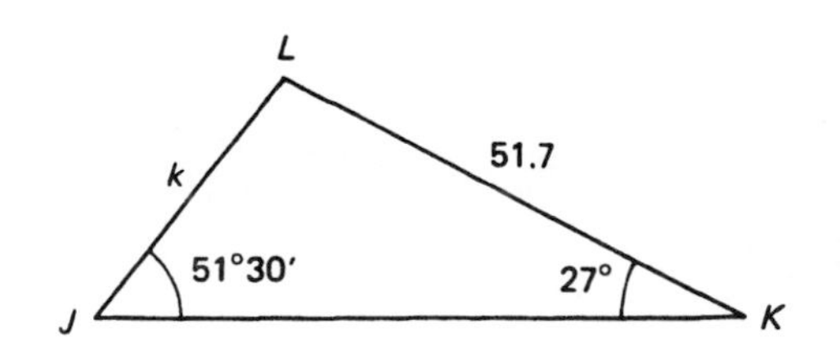

Figure 20–11

20–5. Given $\angle A = 98.1°$, $\angle B = 43.7°$, and side $b = 78.23$, find side a.

20–6. Find side c, given the triangle ABC in Figure 20–12.

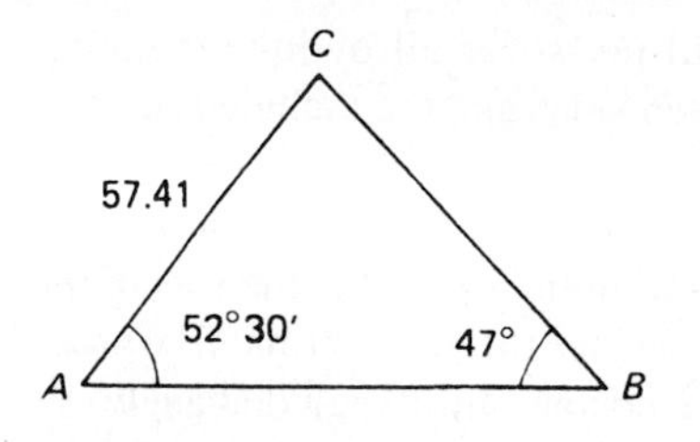

Figure 20–12

20–7. Given the triangle STU, find the size of angle U in Figure 20–13.

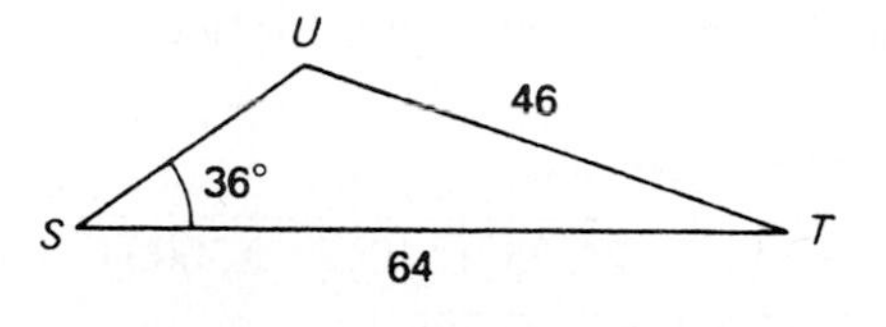

Figure 20–13

20–8. Find the size of $\angle T$ given triangle STU in Figure 20–13.

20–9. Find side t given triangle STU in Figure 20–13.

20–10. Given the triangle XYZ, find the size of $\angle Z$ in Figure 20–14.

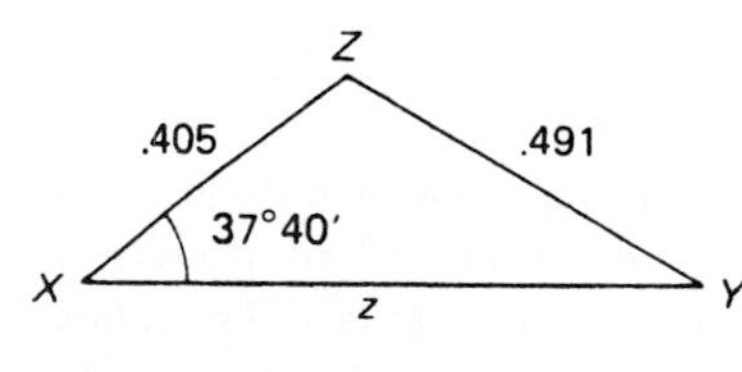

Figure 20–14

20–11. Find the size of $\angle Y$ given triangle XYZ in Figure 20–14.

20–12. Find z given triangle XYZ in Figure 20–14.

20–13. Solve the oblique triangle ABC given $\angle A = 63°$, $\angle B = 42°$, and side $c = 670$. (Remember to draw a sketch.)

20–14. From problem 20–13, find side a.

20–15. From problem 20–13, find side b.

20.2 SOLVE OBLIQUE TRIANGLES USING THE LAW OF COSINES

The Law of Sines will not solve all oblique triangles. Triangles with two given sides and the included angle or three given sides can be solved only by using the law of cosines. The Law of Cosines is stated as follows: The square of any side of a triangle is equal to the sum of the squares of the other two sides minus two times the product of the sides and the cosine of the included angle.

The Law of Cosines formulas are as follows:

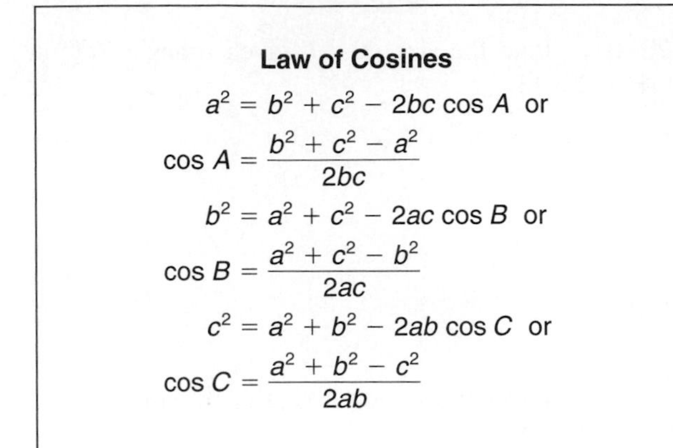

Law of Cosines

$$a^2 = b^2 + c^2 - 2bc \cos A \quad \text{or}$$

$$\cos A = \frac{b^2 + c^2 - a^2}{2bc}$$

$$b^2 = a^2 + c^2 - 2ac \cos B \quad \text{or}$$

$$\cos B = \frac{a^2 + c^2 - b^2}{2ac}$$

$$c^2 = a^2 + b^2 - 2ab \cos C \quad \text{or}$$

$$\cos C = \frac{a^2 + b^2 - c^2}{2ab}$$

The Law of Cosines is much more difficult to prove than the Law of Sines. The proof of the Law of Cosines is found in most advanced trigonometry books.

The Law of Cosines is more difficult to use than the Law of Sines; use it only when necessary. The following examples illustrate the Law of Cosines to solve oblique triangles.

Example 20–4: Find the Side of an Oblique Triangle Given Two Sides and the Included Angle

PROBLEM: Given oblique triangle ABC shown in Figure 20–15, find the length of side c.

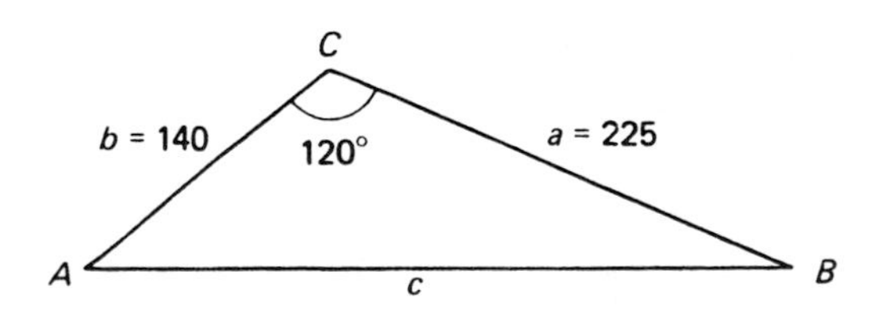

Figure 20–15

SOLUTION: Note the given values: side $a = 225$, side $b = 140$, and $\angle C = 120°$. Two sides and the included angle are given; thus the Law of Cosines should be used to solve the triangle. Select the Law of Cosines formula for side c.

$$c^2 = a^2 + b^2 - 2ab \cos C$$

or

$$\sqrt{c^2} = \sqrt{a^2 + b^2 - 2ab \cos C}$$

$$c = \sqrt{a^2 + b^2 - 2ab \cos C}$$

To find the cosine value of $\angle C$ (120°), remember that $\angle C$ is greater than 90°. Thus

$$\cos A = \cos(180° - \angle A)$$

$$\cos 120° = -\cos(180° - 120°)$$

$$\cos 120° = -\cos 60°$$

$$\cos 120° = -0.5000$$

A scientific calculator will indicate the cos 120° = −0.5000.

Substitute the known values and solve for side c.

$$c = \sqrt{a^2 + b^2 - 2ab \cos C}$$

$$= \sqrt{225^2 + 140^2 - 2(225)(140)(-0.5000)}$$

$$= \sqrt{50{,}625 + 19{,}600 + 31{,}500}$$

$$= \sqrt{101{,}725}$$

$$= 318.9$$

Example 20–5: Find an Angle of an Oblique Triangle Given the Three Sides

PROBLEM: Given the oblique triangle JKL shown in Figure 20–16, find the size of $\angle J$.

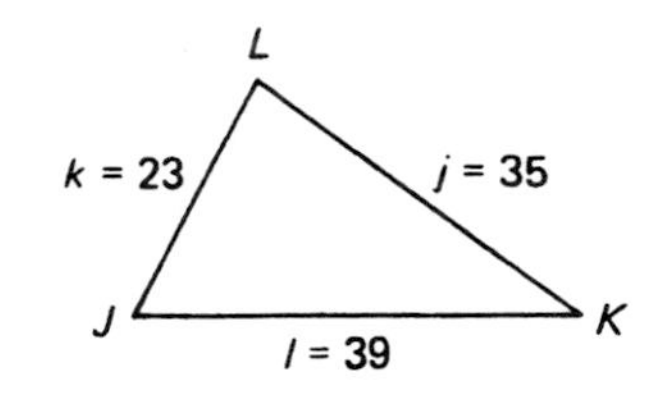

Figure 20–16

SOLUTION: Note the given values: side $j = 34$, side $k = 23$, and side $l = 39$. Three sides are given; thus the Law of Cosines should be used to solve the triangle. Select the Law of Cosines formula to solve for $\angle J$. Recall that the square of a side of any triangle is equal to the sum of the squares of the other two sides minus two times the product of those two sides and the cosine of the included angle. Hence solving the formula for angle J:

$$\cos J = \frac{k^2 + l^2 - j^2}{2kl}$$

Substitute known values and solve for $\cos J$.

$$\cos J = \frac{(23)^2 + (39)^2 - (35)^2}{2(23)(39)}$$

$$\cos J = \frac{529 + 1521 - 1225}{1794}$$

$$\cos J = \frac{825}{1794}$$

$$\cos J = 0.4599$$

$$\angle J = 62.6^\circ$$

Example 20–6: Solve an Oblique Triangle Given Two Sides and the Included Angle

PROBLEM: Solve oblique triangle ABC, Figure 20–17, $\angle A = 110^\circ$, $b = 24$, and $c = 15$.

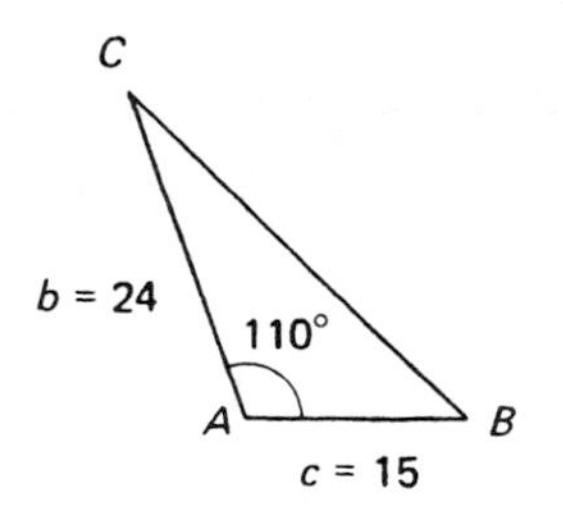

Figure 20–17

SOLUTION: Note the given values: $\angle A$, b and c; solve for $\angle B$, $\angle C$, and a. Two sides and the included angle are given; thus the Law of Cosines should be used to solve the triangle.

Select the Law of Cosines formula to solve for a.

$$a^2 = b^2 + c^2 - 2bc \cos A$$

or

$$\sqrt{a^2} = \sqrt{b^2 + c^2 - 2bc \cos A}$$

$$a = \sqrt{b^2 + c^2 - 2bc \cos A}$$

Find the cosine value of $\angle A$ (110°). Recall that $\angle A$ is greater than 90° and use the formula $\cos A = -\cos(180^\circ - \angle A)$:

$$\cos 110^\circ = -\cos(180^\circ - 110^\circ)$$

$$-\cos 70^\circ = -0.3420$$

$$\cos 110^\circ = -0.3420$$

Then substitute the given values and solve for side a.

$$\sqrt{a^2} = \sqrt{b^2 + c^2 - 2bc \cos A}$$

$$= \sqrt{(24)^2 + (15)^2 - 2(24)(15)(-0.3420)}$$

$$= \sqrt{576 + 225 + 246.24}$$

$$= \sqrt{1047.24}$$

$$= 32.36$$

Note that the multiplication of two negatives $-2(24)(15)$ and -0.3420 resulted in a positive 246.24. Select the Law of Sines to find $\angle B$. Use:

$$\frac{a}{\sin A} = \frac{b}{\sin B}$$

Use a scientific calculator to find the sine of 110° or the formula:

$$\sin 110^\circ = 0.9397$$

Substitute the values and solve for $\sin B$.

$$\frac{a}{\sin A} = \frac{b}{\sin B}$$

$$\frac{32.36}{0.9397} = \frac{24}{\sin B}$$

$$(\sin B)(32.36) = (24)(0.9397)$$

$$\frac{\sin B\,(32.36)}{32.36} = \frac{(24)(0.9397)}{32.36}$$

$$\sin B = 0.6969$$

Thus

$$\angle B = 44.2^\circ$$

Then find $\angle C$.

$$\angle A + \angle B + \angle C = 180^\circ$$

$$\angle C = 180 - 110^\circ - 44.2^\circ$$

$$= 25.8^\circ$$

EXERCISE 20–2 SOLVE TRIANGLES USING THE LAW OF SINES AND COSINES

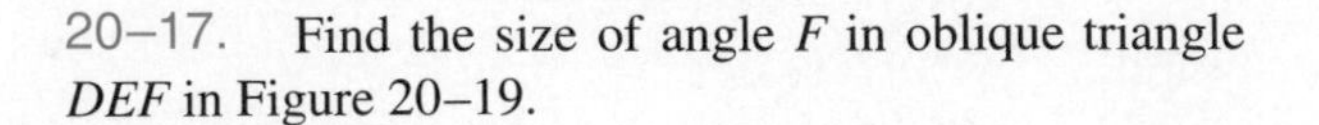

Use the Law of Sines and the Law of Cosines to find the angles and sides as indicated.

20–16. Find side b given triangle ABC in Figure 20–18.

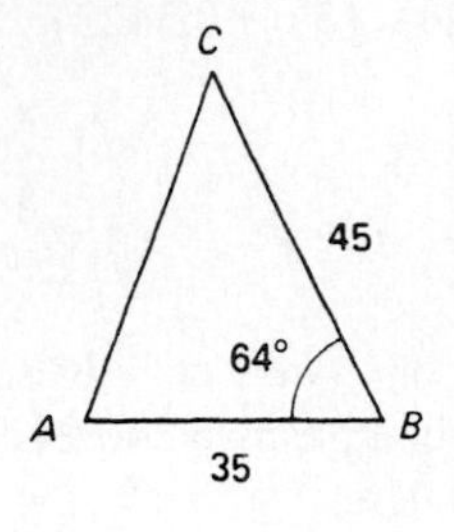

Figure 20–18

20–17. Find the size of angle F in oblique triangle DEF in Figure 20–19.

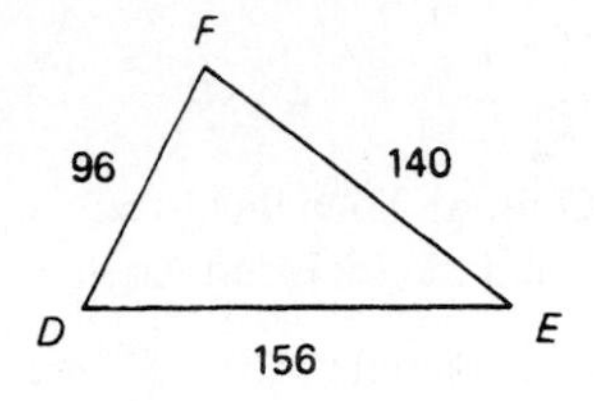

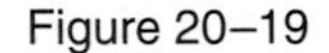

Figure 20–19

20–18. Given oblique triangle GHI, find side i in Figure 20–20.

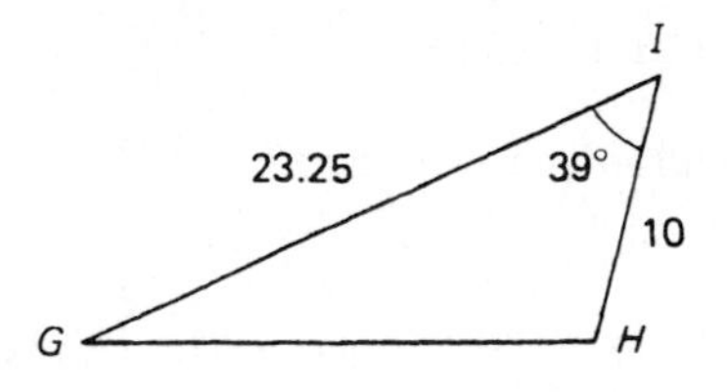

Figure 20–20

20–19. Find side k given oblique triangle JKL in Figure 20–21.

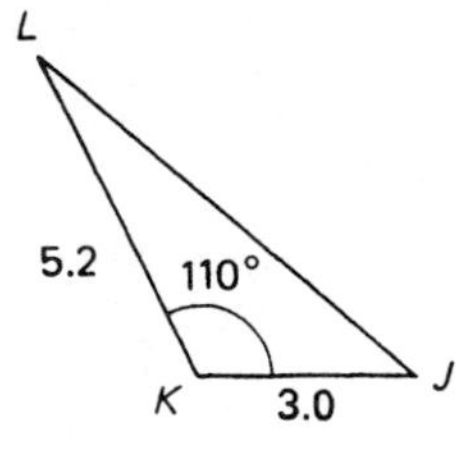

Figure 20–21

20–20. Find the size of angle R in oblique triangle PQR when side $p = 61$, side $q = 65$, and side $r = 76$.

20–21. Find side m in oblique triangle MNO when side $n = 48$, side $o = 28.6$, and $\angle M = 128°$.

20–22. Given triangle TUV, find angle U in Figure 20–22.

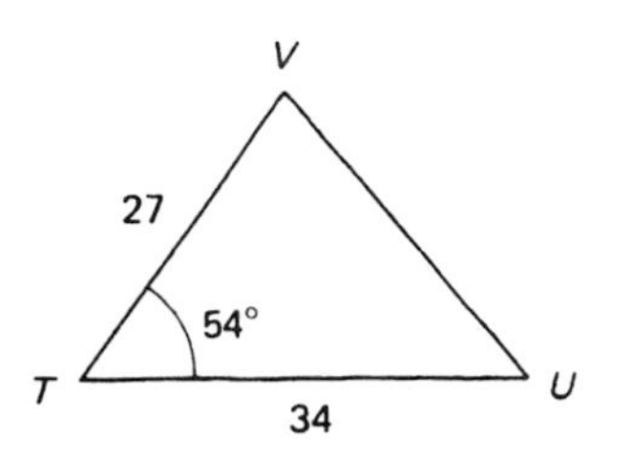

Figure 20–22

20–23. From problem 20–22, find angle V.

20–24. From problem 20–22, find side t.

20–25. Given the oblique triangle GHI in Figure 20–23, find angle H.

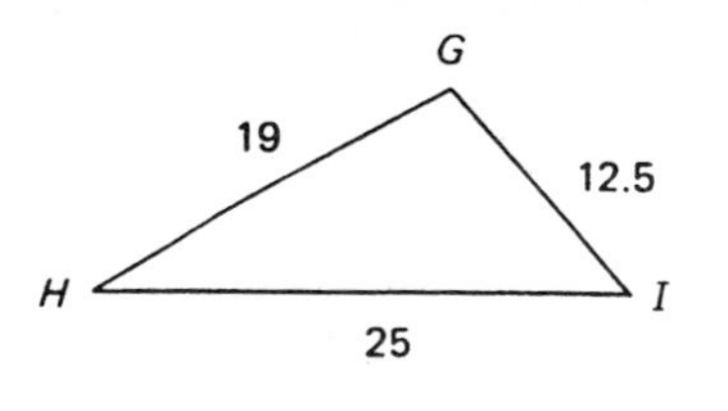

Figure 20–23

20–26. From problem 20–25, find $\angle I$.

20–27. From problem 20–25, find $\angle G$.

20–28. Solve the oblique triangle ABC given sides $a = 4.3$, $b = 8.4$, and $c = 5.2$, for $\angle A$. (Remember to draw a sketch.)

20–29. From problem 20–28, find $\angle B$.

20–30. From problem 20–28, find $\angle C$.

20.3 APPLICATIONS OF THE LAW OF SINES AND THE LAW OF COSINES

The Laws of Sines and Cosines have many uses in business and industry. Problems concerning navigation, alternating current, and vectors are just a few of the situations where the skills of solving oblique triangles are necessary. The examples that follow illustrate some of these applications.

Example 20–7: Oblique Triangle Application Given Two Sides and an Angle

PROBLEM: A light plane flew for 85 kilometers and then discovered a course error of 8°30′. After the course was corrected the destination was reached in 120 kilometers. What should have been the distance if the course error had not been made?

SOLUTION: Make a sketch, Figure 20–24, and

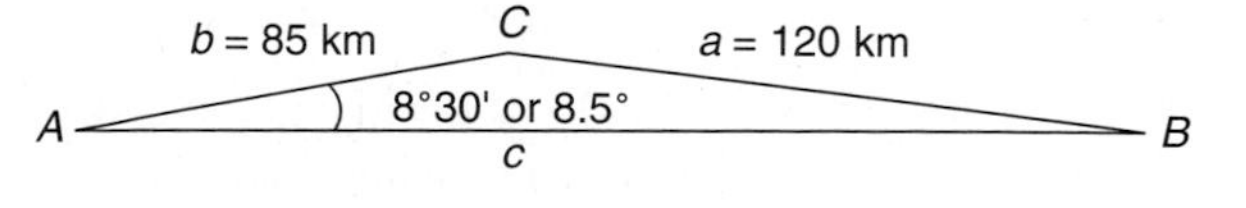

Figure 20–24

identify the given values. Select the Law of Sines formula to solve for $\angle B$ and solve.

$$\frac{a}{\sin A} = \frac{b}{\sin B}$$

$$\frac{120 \text{ km}}{\sin 8.5°} = \frac{85 \text{ km}}{\sin B}$$

$$(\sin B)(120 \text{ km}) = (85 \text{ km})(0.1478)$$

$$\frac{\sin B(120 \text{ km})}{120 \text{ km}} = \frac{(85 \text{ km})(0.1478)}{120 \text{ km}}$$

$$\sin B = 0.1047$$

$$\angle B = 6°$$

Then find $\angle C$.

$$\angle A + \angle B + \angle C = 180°$$

$$8.5° + 6° + \angle C = 180°$$

$$\angle C = 180 - 8.5° - 6°$$

$$= 165.5°$$

Select the Law of Sines ratio to solve for c.

$$\frac{a}{\sin A} = \frac{c}{\sin C}$$

$$\frac{120 \text{ km}}{\sin 8.5°} = \frac{c}{\sin 165.5°}$$

Recall: $\sin 165.5° = 0.2504$

Thus $\dfrac{120 \text{ km}}{0.1478} = \dfrac{c}{0.2504}$

$$\frac{(120 \text{ km})(0.2504)}{0.1478} = c$$

$$203.3 \text{ km} = c$$

The correct course distance should have been 203.3 kilometers.

Example 20–8: Oblique Triangle Application Given Two Sides and the Included Angle

PROBLEM: Given Figure 20–25 representing two concurrent forces acting on a pin connection on a shearing machine, find the resultant.

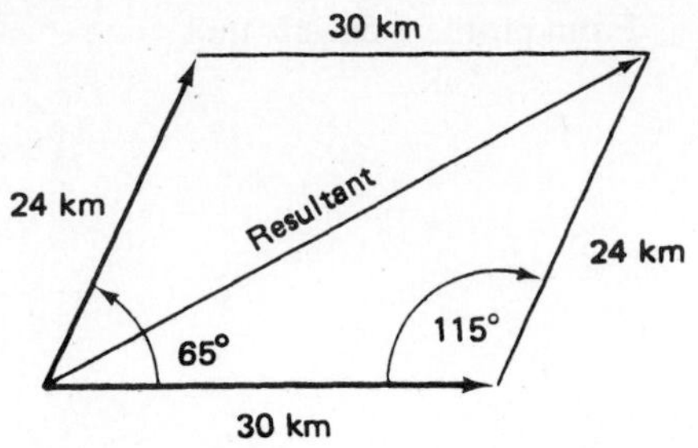

Figure 20–25

SOLUTION: Note the given values and the nature of the problem. Two forces acting concurrently form a parallelogram, and the resultant cuts the parallelogram into two oblique triangles. Select the Law of Cosines formula to find the resultant.

$$c^2 = a^2 + b^2 - 2ab \cos C$$

Resultant =

$$\sqrt{(\text{force } 1)^2 + (\text{force } 2)^2 - 2(\text{force } 1)(\text{force } 2)\cos 155°}$$

$$= \sqrt{(24)^2 + (30)^2 - 2(24)(30)(\cos 115°)}$$

Find the cosine of 115° and solve.

$$\cos 115° = -0.4226$$

$$\text{Resultant} = \sqrt{(24)^2 + (30)^2 - 2(24)(30)(-0.4226)}$$

$$= \sqrt{576 + 900 + 608.544}$$

$$= \sqrt{2084.544}$$

$$= 45.6568$$

$$= 45.7 \text{ km}$$

EXERCISE 20–3 SOLVE APPLIED PROBLEMS USING THE LAW OF SINES AND COSINES

Using the Laws of Sines and Cosines, solve the following applied problems. Make a sketch whenever possible.

20–31. From Figure 20–26, showing a triangular-shaped lot, find the length of the missing side.

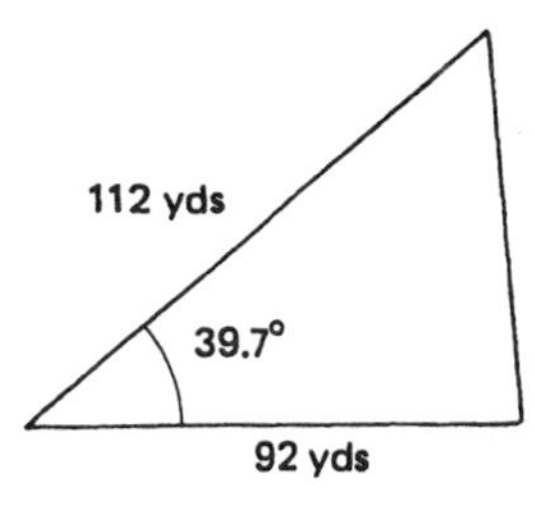

Figure 20–26

20–32. If 18-foot rafters are used for the building shown in Figure 20–27, find the width of the building.

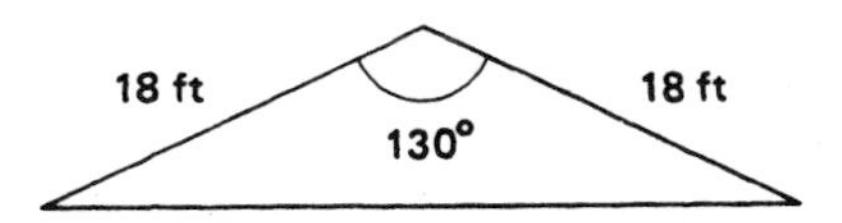

Figure 20–27

20–33. Find the angle of the rafters with the horizon in problem 20–32.

20–34. Find the resultant force in Figure 20–28.

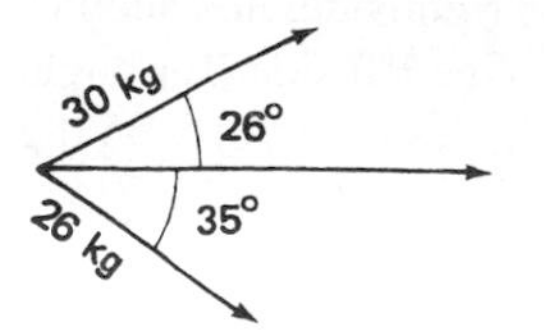

Figure 20–28

20–35. Find the resultant force in Figure 20–29.

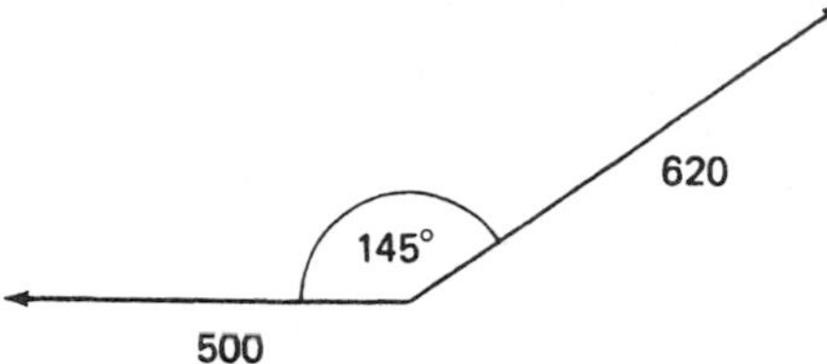

Figure 20–29

20–36. Determine the lengths of the guy wires on the utility pole in Figure 20–30.

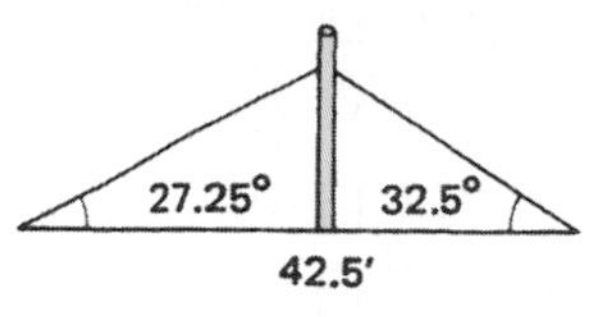

Figure 20–30

20–37. Find the length of the duck-hunting lake shown in Figure 20–31.

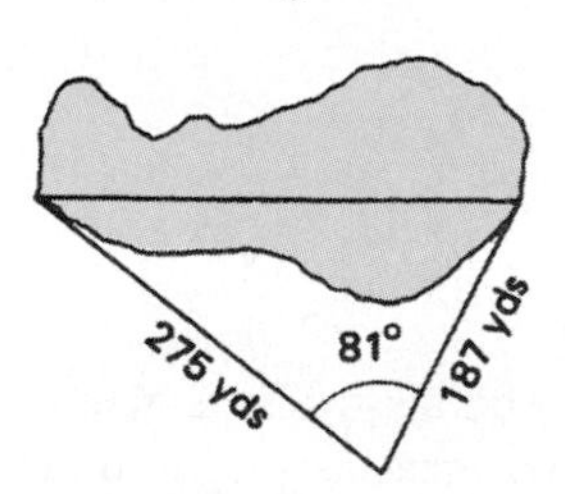

Figure 20–31

20–38. A surveyor measures the length of one side of a tract of land to be 520 rods. From each end of this side, he sights a point and finds the angles to be 74.2° and 65°. Find the length of the other sides.

20–39. An airplane flies west at 240 miles per hour for 4 hours. A north wind causes the plane to drift to the south at 6.3. Determine how far the plane must fly to make the correction as indicated in Figure 20–32 and reach the intended destination.

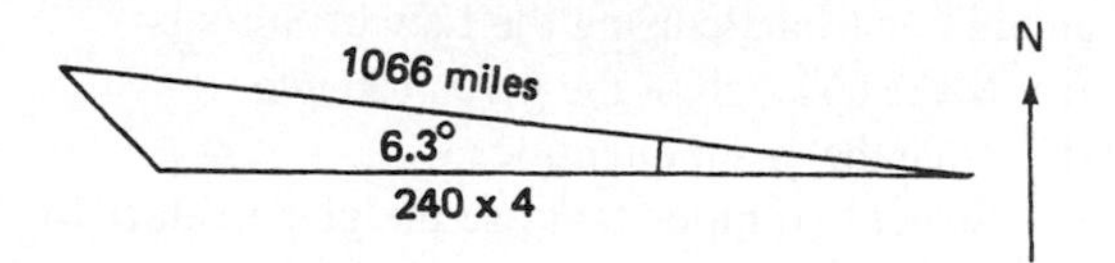

Figure 20–32

20–40. A helicopter is observed from two ground stations 3560 meters apart. If the angles of elevation are 54° and 28.3°, how high is the helicopter?

20–41. Find the length of an air tunnel to be constructed by the Montana Highway Department for a mountain highway shown in Figure 20–33.

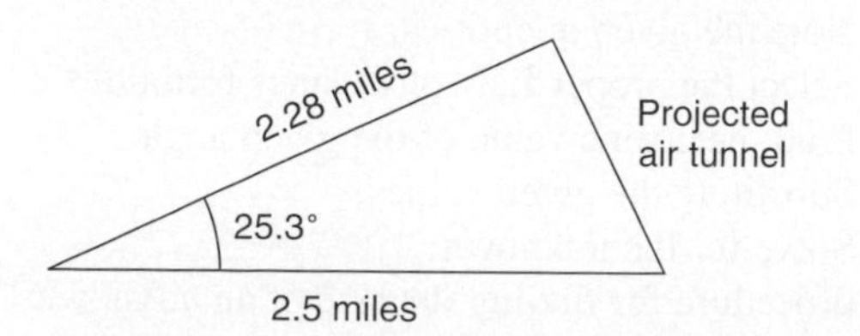

Figure 20–33

20–42. Two freeways intersect at an angle of 132°. If two cars start from the intersection and travel on the rays of the angle, how far apart will they be in 4 hours if one car is traveling at 65 miles per hour and the other car is traveling at 70 miles per hour?

20–43. A snowmobile is stuck in a snowbank. A force of 100 pounds is pulling on the right side, and another force of 185 pounds is pulling on the left side. The angle between the forces is 50. Find the resultant force on the snowmobile.

20–44. A hunter walks in straight lines for 281, 385, and 582 meters. Determine the smallest angle that he makes in his search for game.

20–45. Find the height of the building shown in Figure 20–34.

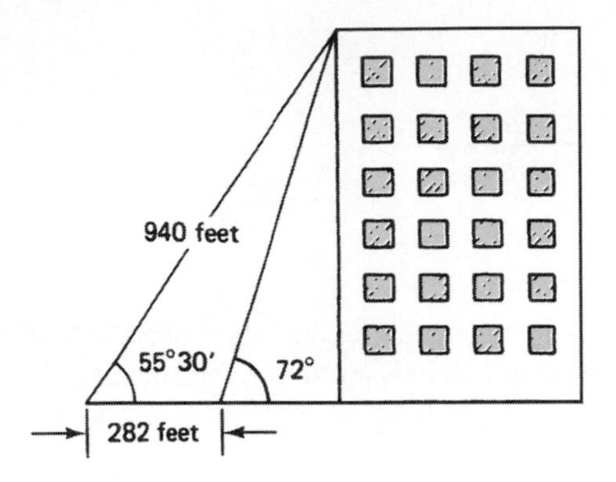

Figure 20–34

THINK TIME

There are many uses for the Laws of Sines and Cosines. Think of ways that you can solve oblique triangles to assist you in your occupation or at home. It may take some imagination to think of uses, but you will find that the Laws of Sines and Cosines will save you time and work.

The example shows how solving oblique triangles can be used to develop recreational and natural resource facilities. A park and recreation director decides to set up a marsh trail shown in Figure 20–35. Because of the low ground, a boardwalk is to be built. She is able to place the walk around the water areas as shown in the sketch. After she has completed the boardwalk around the marsh, she decides to place bridges across the water areas from point D to G and point B to G. Find the length of these bridges. Note that $\angle x = \angle y$.

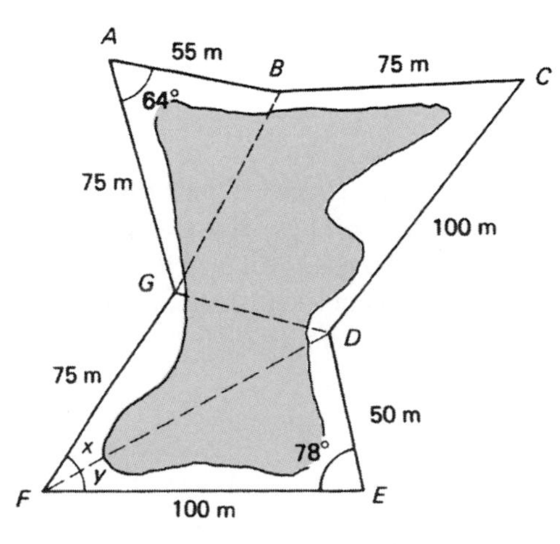

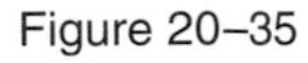
Figure 20–35

PROCEDURES TO REMEMBER

1. The procedure for finding the size of an angle of an oblique triangle using the Law of Sines is:
   (a) Make a sketch of the given triangle.
   (b) Note the given quantities.
   (c) Select two ratios that use the given values to find the side.
   (d) Substitute the given values.
   (e) Solve the ratios for the unknown.
2. The procedure for finding the side of an oblique triangle using the Law of Cosines is:
   (a) Make a sketch of the triangle to be solved.
   (b) Note the given quantities.
   (c) Select the proper Law of Cosines formula.
   (d) Find the cosine value of the given angle.
   (e) Substitute the given values.
   (f) Solve for the unknown.
3. The procedure for finding the size of an unknown angle of an oblique triangle, given three sides, using the Law of Cosines is:

(a) Make a sketch of the triangle.
(b) Note the given quantities.
(c) Select the proper Law of Cosines formula for the unknown angle.
(d) Substitute the given values and solve.

CHAPTER SUMMARY

1. The following definitions are necessary when solving oblique triangles using the Laws of Sines and Cosines.
   (a) Oblique triangles do not contain a right angle.
   (b) If an angle is greater than 90°, the sine of that angle can be determined by using a scientific calculator or the formula: $\angle A > 90°$; then $\sin A = \sin(180° - \angle A)$.
   (c) If an angle is greater than 90°, the cosine of that angle can be determined by using a scientific calculator or the formula: $\angle A > 90°$; then $\cos A = -\cos(180° - \angle A)$.
   (d) If the Laws of Sines and Cosines are to be used, one of the following conditions must exist:
      (1) Two angles and one side
      (2) Two sides and the angle opposite one of the sides
      (3) Two sides and the angle between the two sides
      (4) All three sides
   (e) The Law of Sines states that the ratio of a side to the sine of its angle is equal to the ratios of the other sides and the sines of their angles.
   (f) The Law of Cosines states that the square of any side of a triangle is equal to the sum of the squares of the other two sides minus two times the product of those sides and the cosine of the included angle.
2. The following formulas can be used to solve oblique triangles:
   (a) Law of Sines

$$\frac{a}{\sin A} = \frac{b}{\sin B} = \frac{c}{\sin C} \quad \text{or}$$

$$\frac{\sin A}{a} = \frac{\sin B}{b} = \frac{\sin C}{c}$$

   (b) Law of Cosines

$$a^2 = b^2 + c^2 - 2bc \cos A \quad \text{or}$$

$$\cos A = \frac{b^2 + c^2 - a^2}{2bc}$$

$$b^2 = a^2 + c^2 - 2ac \cos B \quad \text{or}$$

$$\cos B = \frac{a^2 + c^2 - b^2}{2ac}$$

$$c^2 = a^2 + b^2 - 2ab \cos C \quad \text{or}$$

$$\cos C = \frac{a^2 + b^2 - c^2}{2ab}$$

CHAPTER TEST

T–20–1. Find the sine of 159°.

T–20–2. Find the cosine of 99.45°.

T–20–3. Given the oblique triangle *ABC*, Figure 20–36, find the side *b*.

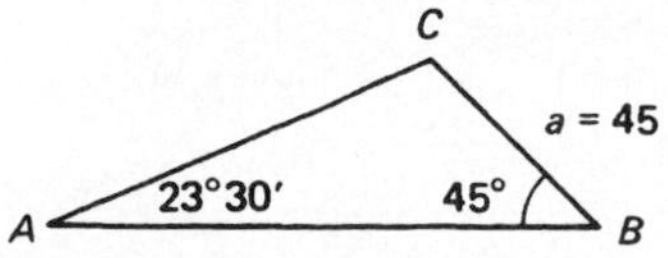

Figure 20–36

T–20–4. Given $\angle A = 65°$, side $a = 14.4$, and side $b = 12.2$, find $\angle B$.

T–20–5. Given the oblique triangle GHI, Figure 20–37, find $\angle I$.

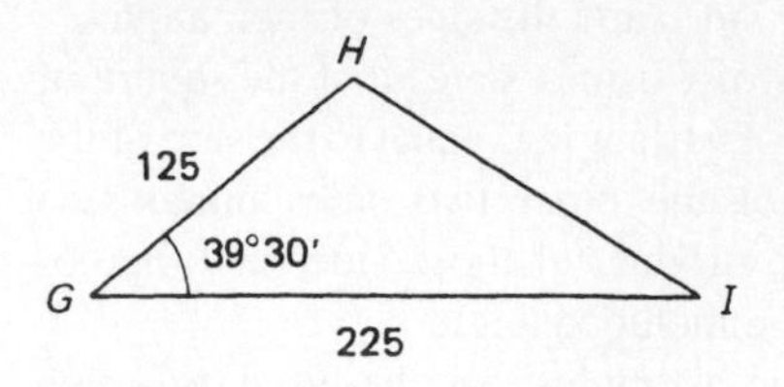

Figure 20–37

T–20–6. Given side $j = 65$, side $k = 40$, and angle $L = 37°20'$, sketch the triangle and find l.

T–20–7. Given triangle MNO in Figure 20–38, find $\angle O$.

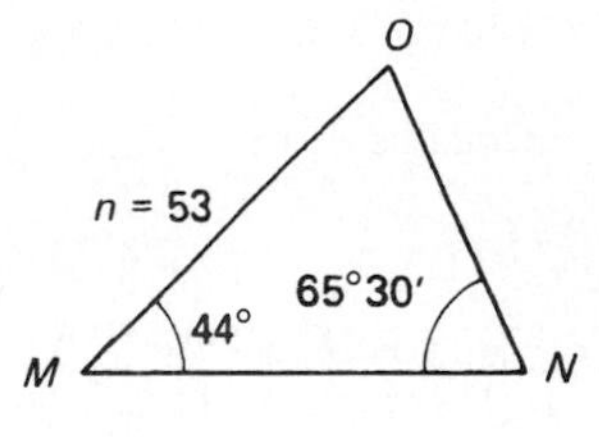

Figure 20–38

T–20–8. From problem T–20–7, find m.

T–20–9. From problem T–20–7, find o.

T–20–10. Two forces acting at right angles produce a resultant force of 1600 pounds. One of the forces has a magnitude of 960 pounds. Find the other force.

appendix

CONSTRUCTION OF SIMPLE GEOMETRIC FIGURES

OBJECTIVES

After studying this appendix, you will be able to:

1. Understand basic definitions and constructions of geometric figures.
2. Construct perpendicular, parallel, and tangent lines.
3. Construct triangles, squares, and hexagons.

SELF-TEST

This test will measure your ability to construct geometric forms. To complete this test, use a compass and straightedge. Construction of the figures will show your ability to construct geometric figures and your readiness to advance. Problems you are unable to solve will indicate areas for improvement. When necessary, make your constructions on a separate sheet of paper. Successful completion will indicate that you know the basic construction techniques of geometry.

S–1. Bisect $\overline{XY}$.

X Y

S–2. Bisect $\angle ABC$.

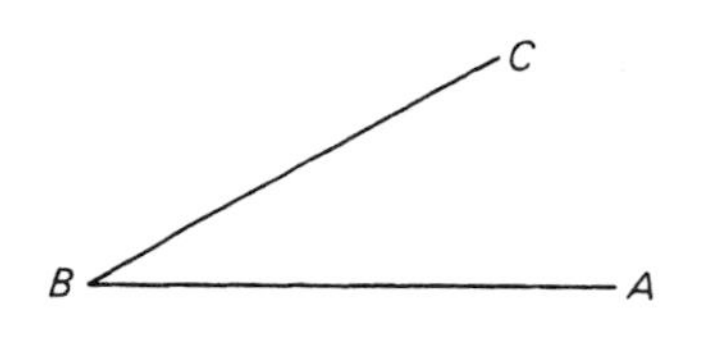

S–3. Copy $\angle FOB$.

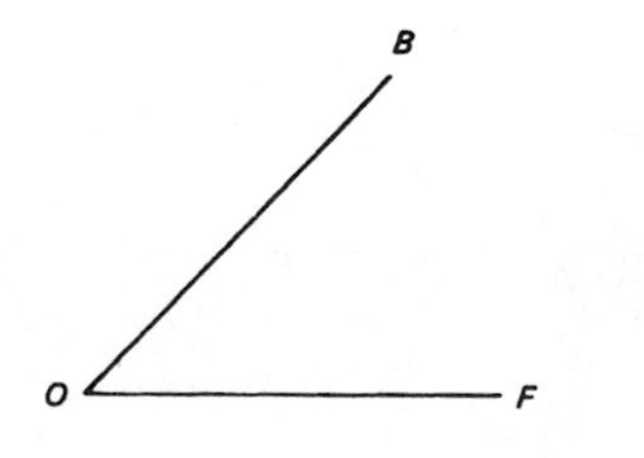

S–4. Divide $\overline{GJ}$ into three equal parts.

G J

S–5. Construct a perpendicular on $\overline{JY}$ at point L.

S–6. Construct a perpendicular to $\overline{KC}$ from point M.

S–7. Construct a line parallel to $\overline{ND}$.

S–8. Construct tangents to the circle from point P.

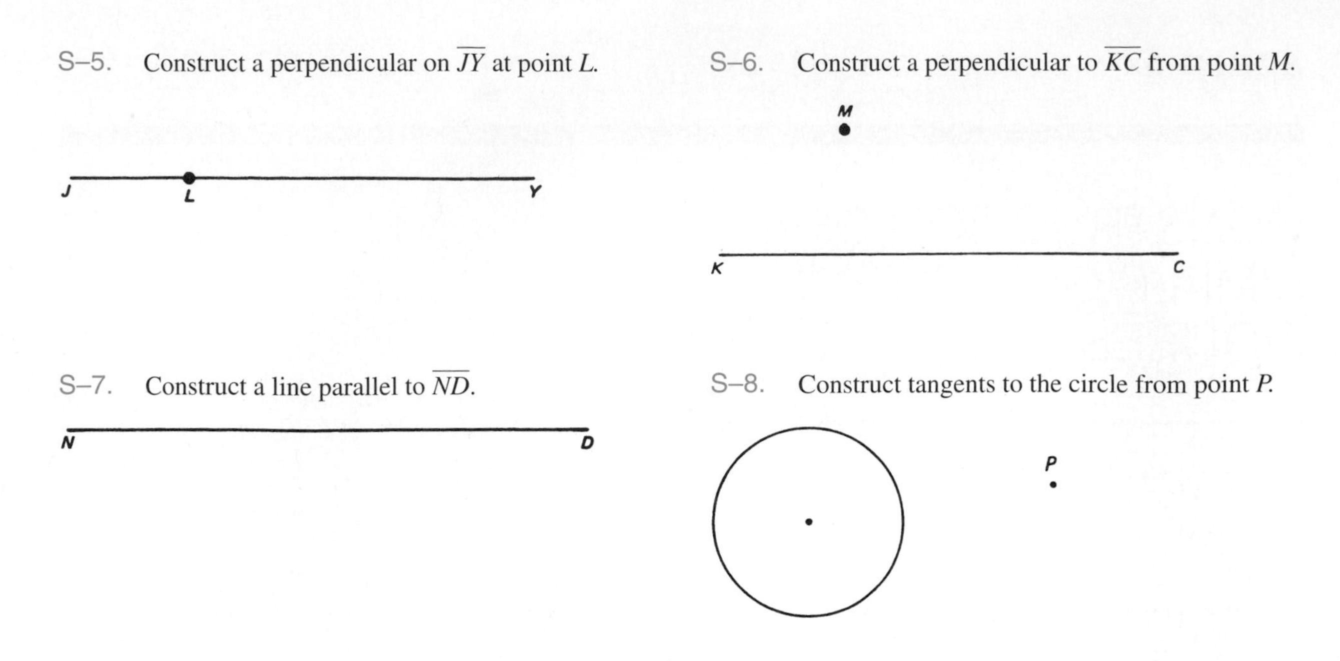

S–9. Construct a square with sides equal to $\overline{AB}$.

A

B

S–10. Construct a triangle from segments $\overline{AB}$, $\overline{CD}$, and $\overline{EF}$.

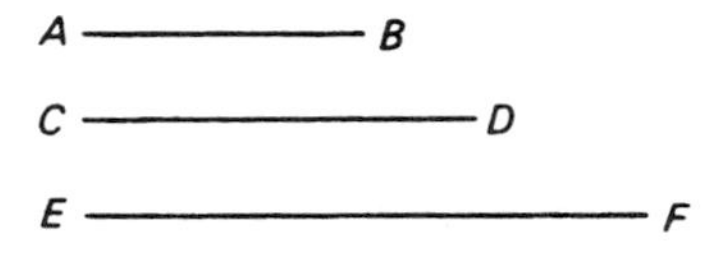

1. BASIC DEFINITIONS AND CONSTRUCTIONS

Tradespeople and technicians should be able to construct simple sketches of their work. Sketches help to clarify and save valuable time. The purpose of this chapter is to teach the skills necessary to construct basic geometric figures.

A *compass* is a tool used to draw circles or portions of circles. It is also used to copy lengths of line segments. One type of compass is shown.

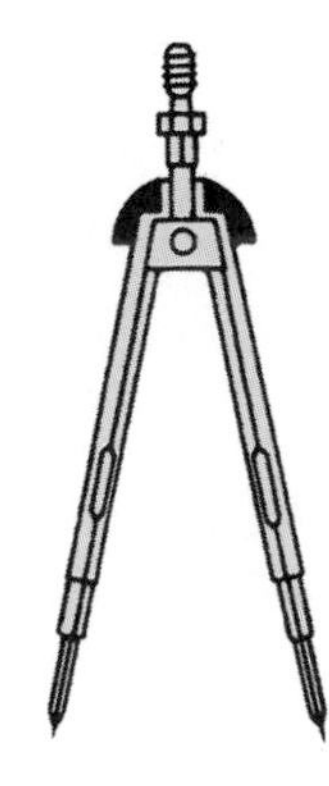

A *straightedge* is used to draw lines. A ruler is generally used as a straightedge, but anything may be used if it has a straight edge. The opening between two points of the compass is called the *radius*. The letters O and P are frequently used as point of identification in geometric construction.

Constructions are best learned through practice. The first figure constructed is a circle. Open the compass so the metal tip and lead tip are an inch or two apart. Place the metal tip on point P at the right. Draw or "swing" a circle with the compass. Point P is the center of the circle. Practice constructing circles similar to those shown below.

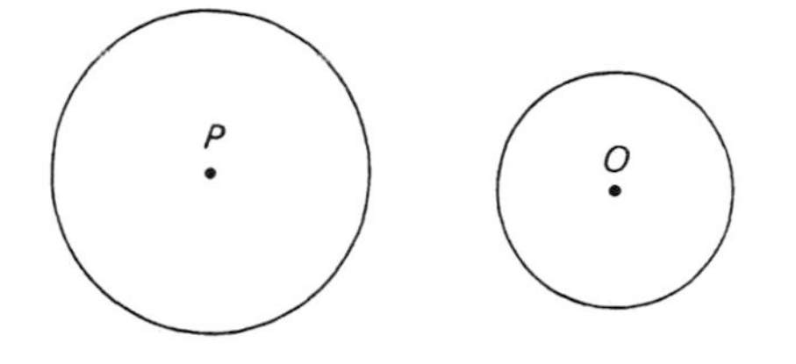

Use your straightedge to draw a line from the center O to point A on the circle. This line, OA, is the radius of the circle. Draw-in radius $\overline{OA}$ and radius $\overline{OB}$. They are called *radii* of the circle. (*Radii* is the plural of *radius*.) Note that $\overline{OB}$ and $\overline{OC}$ form a straight line that passes through the center point O. Thus BOC is the *diameter* of the circle. The curved line AC in the figure is called an *arc*. An *arc* is a portion of the circumference of a circle. The symbol of the circumference for an arc is $\frown$, which is placed over appropriate letters: $\overset{\frown}{AC}$. A circle may have several arcs. $\overset{\frown}{AB}$ and $\overset{\frown}{BC}$ are also arcs of circle O.

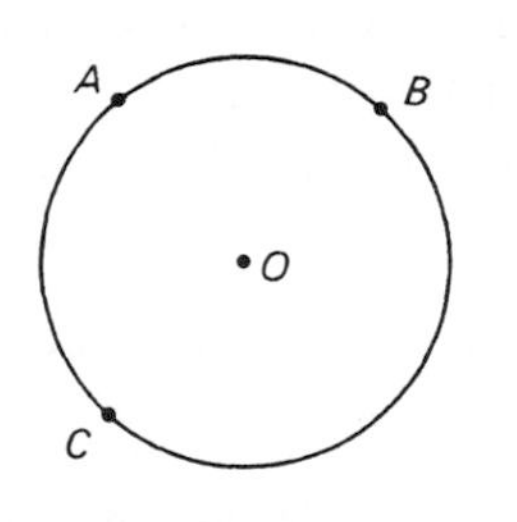

Given the radius $\overline{LM}$, construct a circle using M as the center point.

Place two additional letters, J and G, on the circumference of the circle. Identify the three arcs of circle M.

A compass and a straightedge are used to copy a line segment. Copy line segment $\overline{JG}$, illustrated below, by drawing a "working line" with a straightedge that is longer than the given line segment.

Open the compass to the length of line segment $\overline{JG}$. Place the metal point on J and the lead point on G. Then place the metal point on J of the working line and "arc" the working line. At the point where the arc crosses the working line, identify point G.

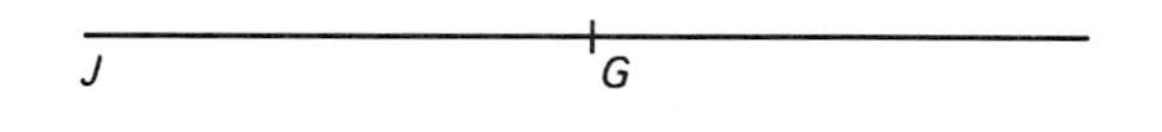

Copy the lines shown below on the right side of the page.

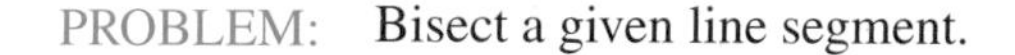

Example 1: Bisect a Line

PROBLEM: Bisect a given line segment.

K C

SOLUTION: To bisect a line means to divide it into two equal parts. First, open the compass to a position greater than half of the length of the line segment. Place the metal tip at point K and swing an arc as shown.

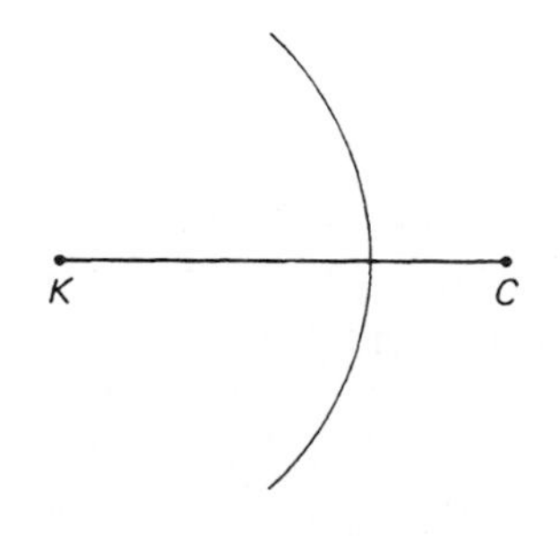

Second, keep the compass at the same opening and swing an arc from point C. Label the points where the arcs intersect as X and Y.

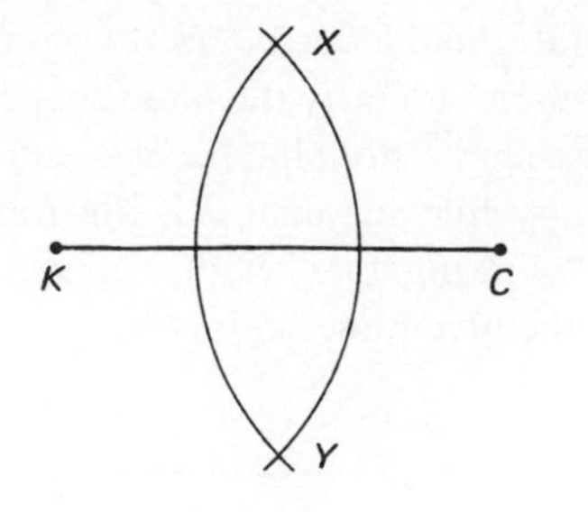

Third, sketch a line from the point of intersection X to the point of intersection Y. Where line XY intersects line $\overline{KC}$, label the point Z. Thus point Z is the midpoint of line KC.

Example 2: Bisect an Angle

PROBLEM: Bisect a given angle.

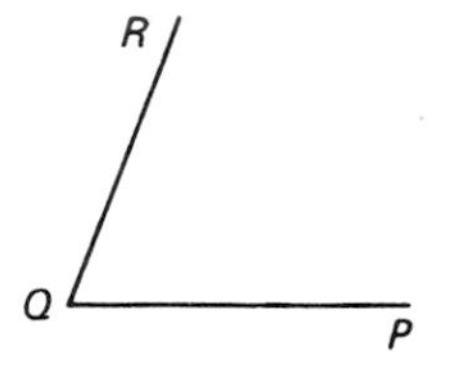

SOLUTION: To bisect an angle means to divide it into two equal angles. First, place the metal tip of your compass at the vertex of the angle (point Q) and swing an arc through the rays of the angle making point A and B. The compass opening should be $\frac{1}{2}$ to $\frac{3}{4}$ the length of $\overrightarrow{PO}$. Label points A and B.

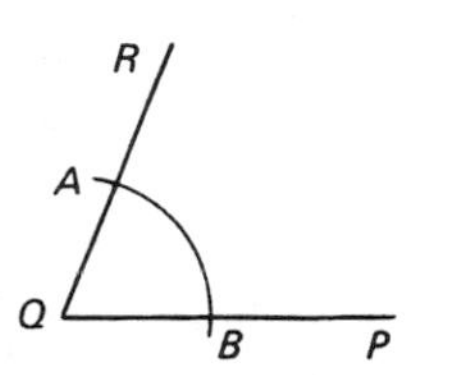

Second, swing arcs from A and B and label their point of intersection as point D.

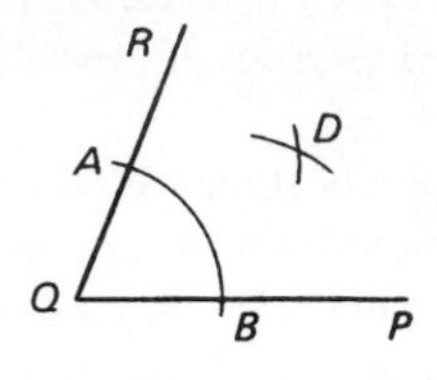

Third, draw line QD. Line QD bisects angle PQR to form two equal angles.

Thus angle PQD = angle RQD or $\angle PQD = \angle RQD$

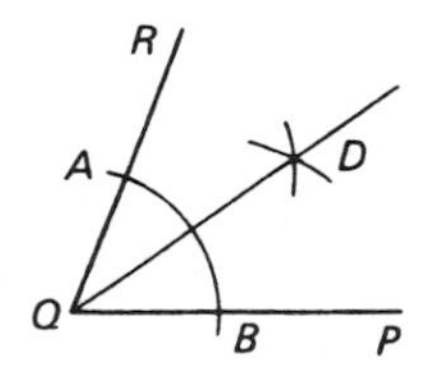

Example 3: Copy an Angle

PROBLEM: Copy the given angle.

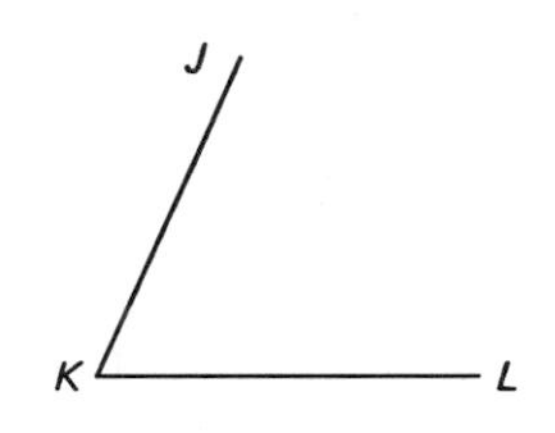

SOLUTION: To copy an angle means to produce another angle that is equal to the given angle. First, draw a working line and label it QR. This line should be somewhat longer than line KL above.

Second, swing an arc from point K above to form points N and M.

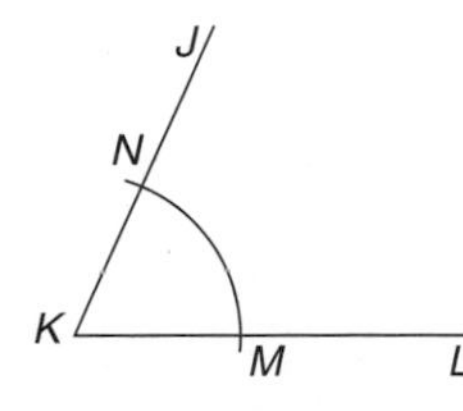

Third, swing a similar arc on the working line from point *Q*. Identify the point of intersection as point *X*.

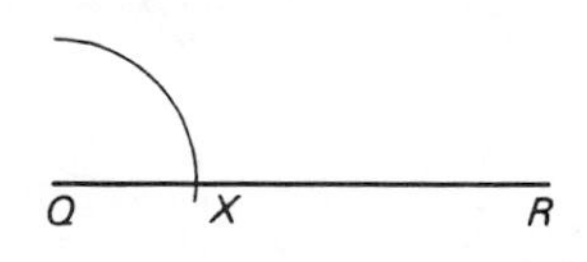

Fourth, set the compass so that the metal tip is on point *M* and the lead tip is on point *N*. Fifth, place the metal tip on point *X* above and swing an arc so that it intersects the other arc. Label this point of intersection as point *Y*.

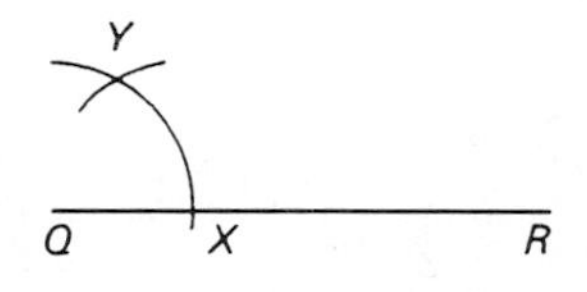

Sixth, draw a line from point *Q* through point *Y* to form line *QP*. Thus angle *LKJ* = angle *RQP*,

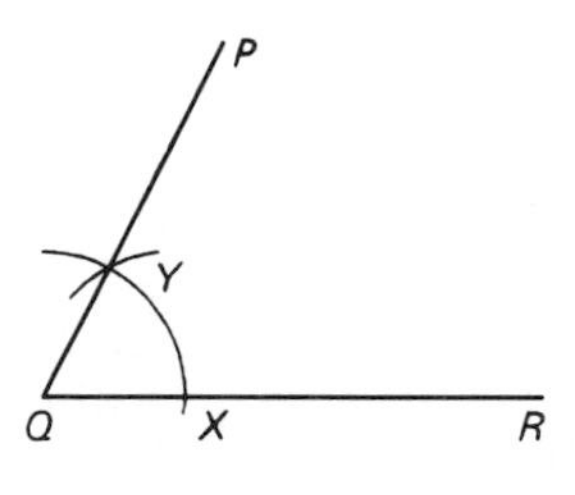

Example 4: Divide a Line Segment

PROBLEM: Divide a line into three equal parts.

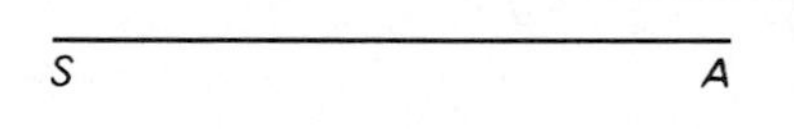

SOLUTION: To divide the given line $\overline{SA}$ into three equal parts, first draw a working line about 20 to 50° above line *SA* to form working angle *MSA*.

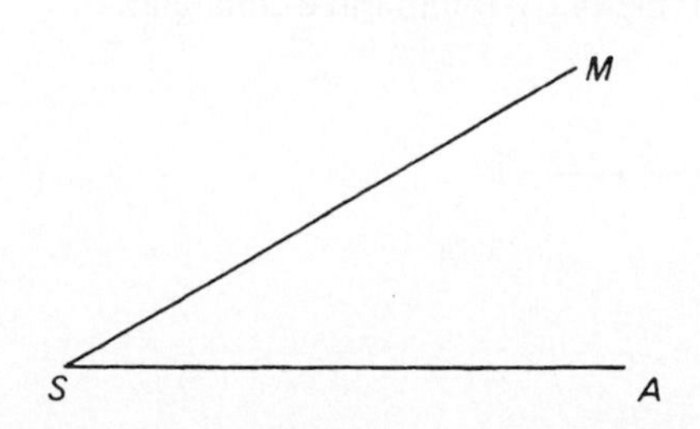

Second, divide the line *SM* into three equal parts by opening the compass to about $^1/_3$ of *SM* and swing an arc to *E*. From point *E*, swing the same arc to *F*. From *F*, swing the same arc to *G*. Draw a line from point *G* to point *A*.

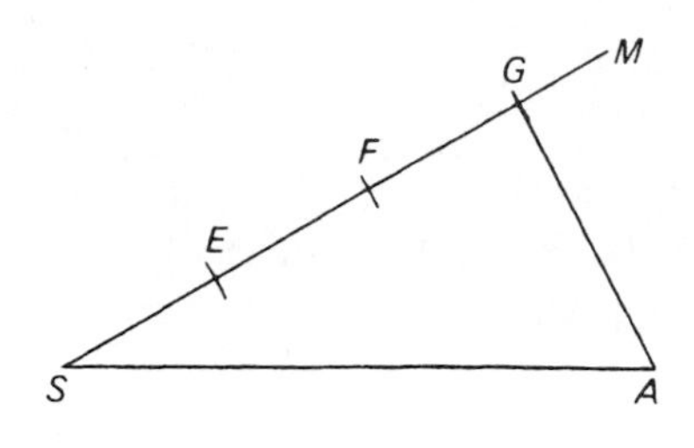

Third, using point *F* as a center, copy angle *SGA* and draw line *FB*.

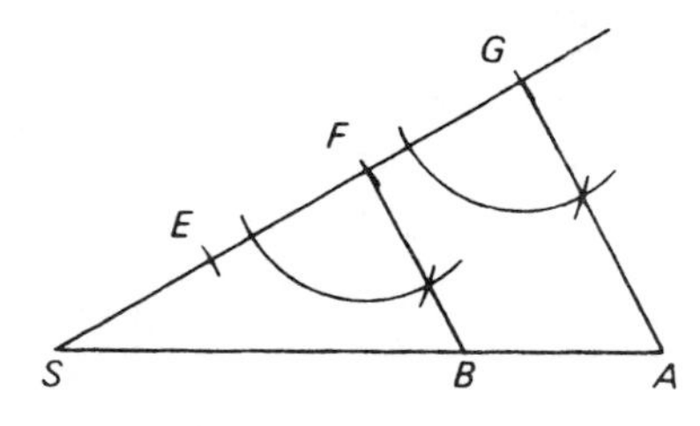

Fourth, using point *E* as a center, copy angle *SGA* and draw line *EK*.

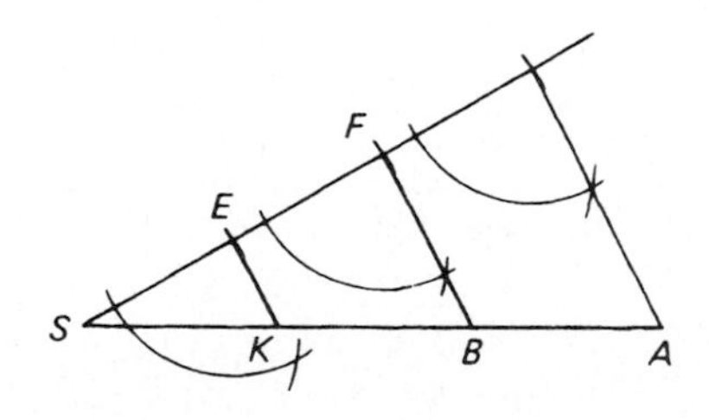

Thus $\overline{SK} = \overline{KB} = \overline{BA}$.

The same process may be used to divide any straight line into any number of equal parts.

EXERCISE 1 BASIC CONSTRUCTIONS

Do the following constructions as indicated.

1. Bisect line $\overline{AB}$.
2. Bisect line $\overline{CD}$.
3. Bisect line $\overline{EF}$.
4. Bisect angle *ABC*.
5. Bisect angle *EFG*.
6. Bisect angle *HIJ*.
7. Copy angle *KLM*.
8. Copy angle *NOP*.
9. Copy angle *QRS*.
10. Divide line $\overline{AB}$ into three equal parts.

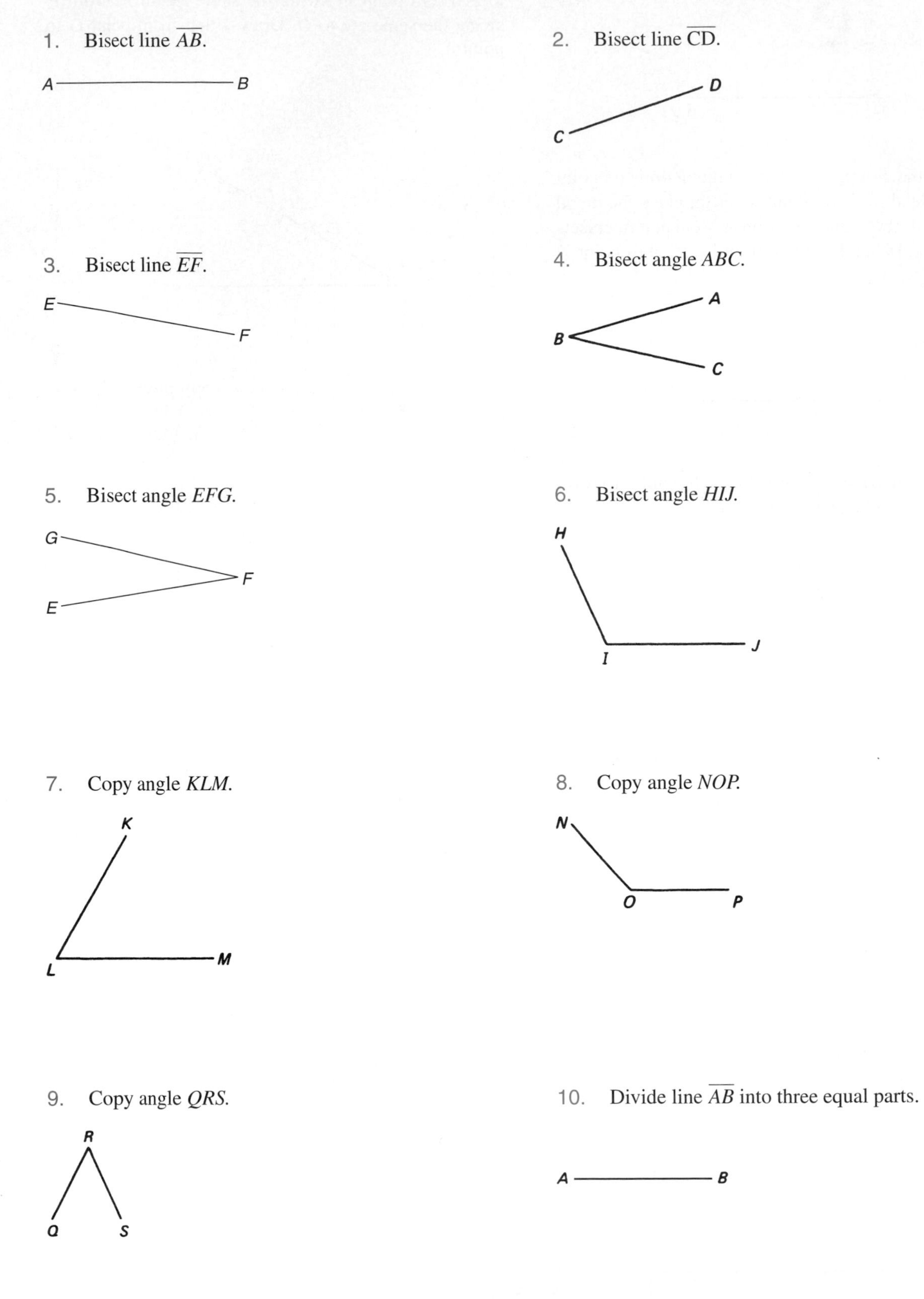

11. Divide line $\overline{CD}$ into four equal parts.

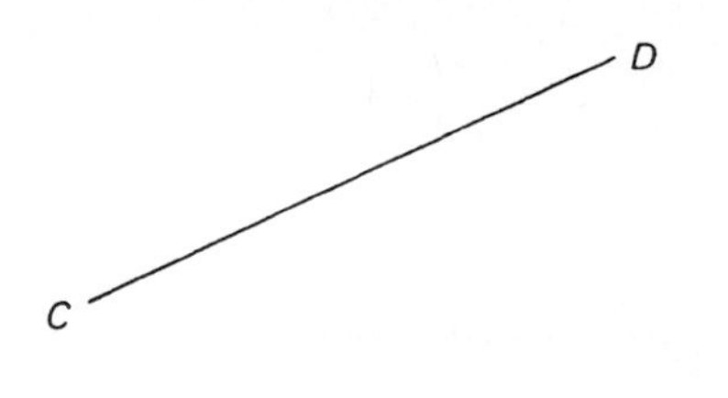

12. Divide line $\overline{EF}$ into five equal parts.

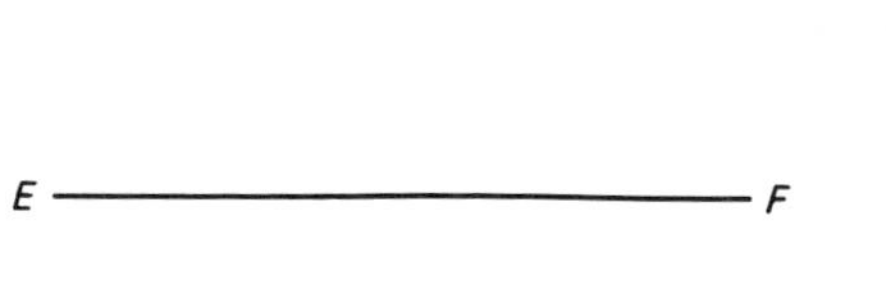

2. CONSTRUCTING PERPENDICULARS, PARALLELS, AND TANGENTS

Walls of a building are constucted "straight up from the floor" so they form 90° angles. Thus the walls are perpendicular to the floor. Therefore, when one line is *perpendicular* to another, it forms 90° angles. When two lines run side by side and never intersect, they are called *parallel* lines. When a line touches a circle at only one point, the line is *tangent* to the circle. In the next series of examples, you will learn how to construct perpendicular and parallel lines and tangents.

Example 5: Construct a Perpendicular

PROBLEM: Construct a perpendicular line from a point on a line.

SOLUTION: First, place the compass point on point *P*, swing arcs on each side of point *P*, and label them *X* and *Y*.

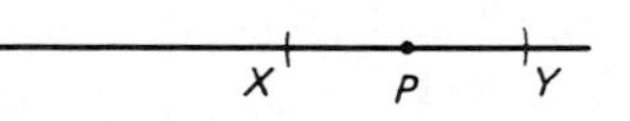

Second, open the compass and swing arcs above and below the line from points *X* and *Y* to form points *A* and *B*.

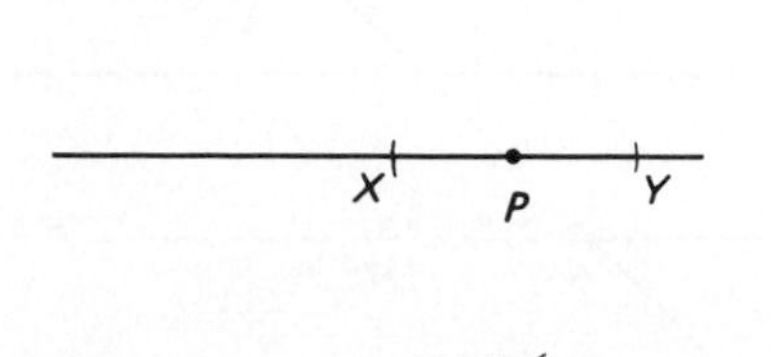

Third, draw *AP* using point *B* as a check point. $\overline{AP}$ will be perpendicular to the line at point *P*. Thus angles *YPA* and *XPA* are 90°, and *PA* is ⊥ to *XY*. Note ⊥ is the symbol for perpendicular.

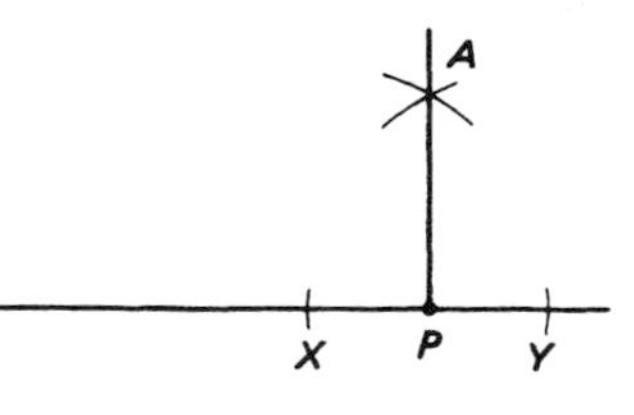

Example 6: Construct a Perpendicular from a Point

PROBLEM: Construct a perpendicular to line $\overline{AB}$ from point *P*.

SOLUTION: First, open the compass slightly more than the distance from point *P* to $\overline{AB}$ and arc the line in two places, *J* and *K*.

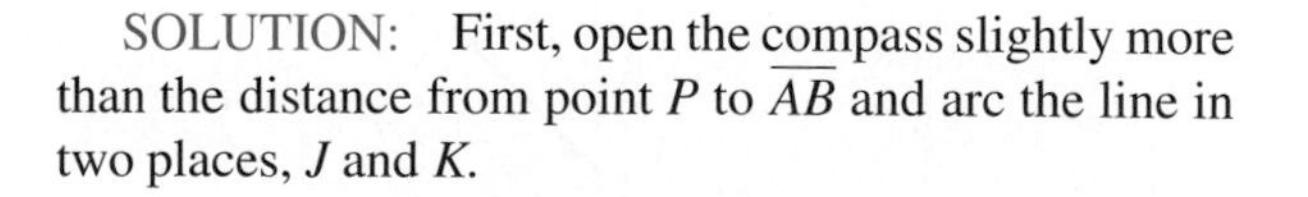

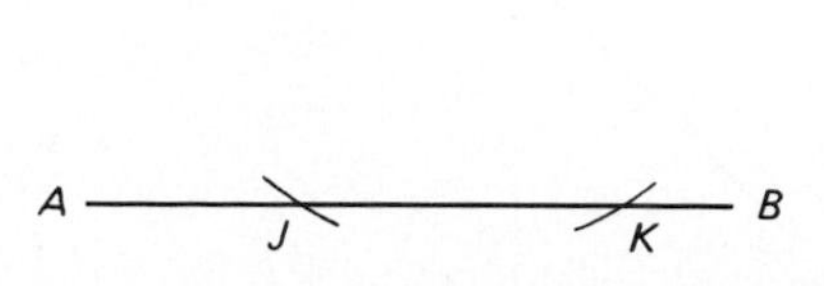

Second, using J and K as centers, swing arcs below $\overline{AB}$ and label the new point L.

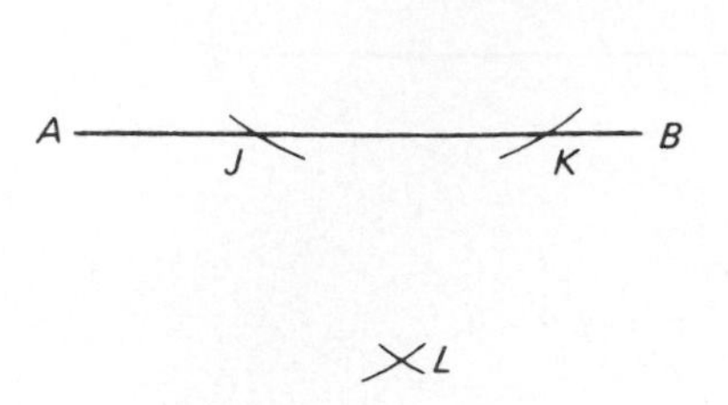

Third, draw the line segment from point P to $\overline{AB}$ using point L as the guide point to determine the line. Label the point of intersection as point Q. Thus PQ is perpendicular to $\overline{AB}$.

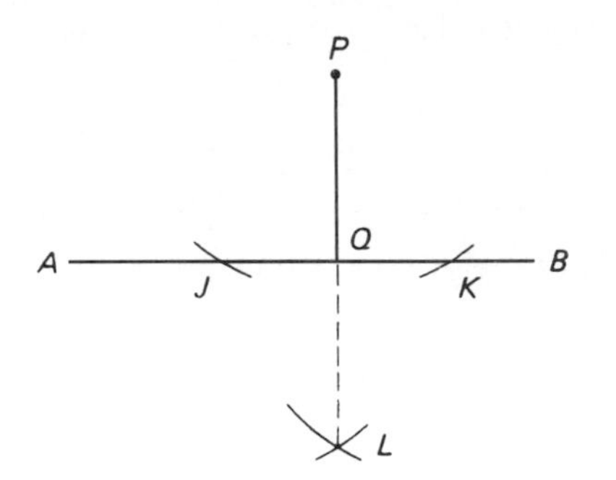

Example 7: Construct a Parallel Line

PROBLEM: Construct a line parallel to a given line.

X ——————————————— Y

SOLUTION: First, draw working line $\overline{JL}$, which intersects line $\overline{XY}$ at about a 45° angle, as shown.

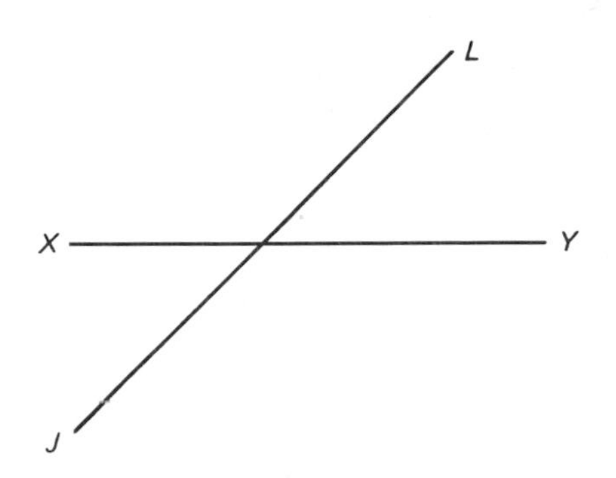

Second, swing an arc above $\overline{XY}$ at the intersection of $\overline{XY}$ and $\overline{JL}$.

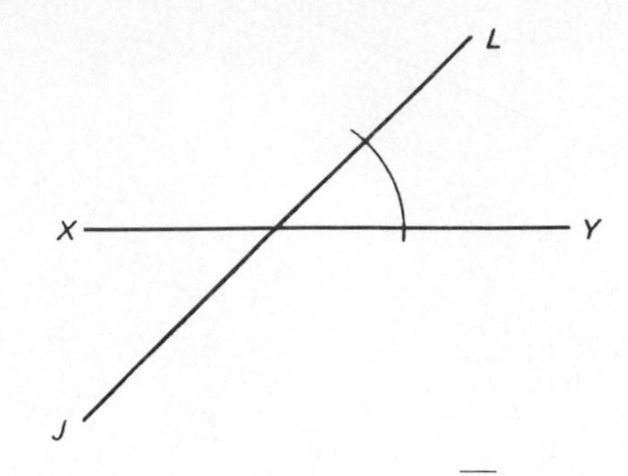

Third, swing an arc from a point on $\overline{JL}$ which might be identified as point K.

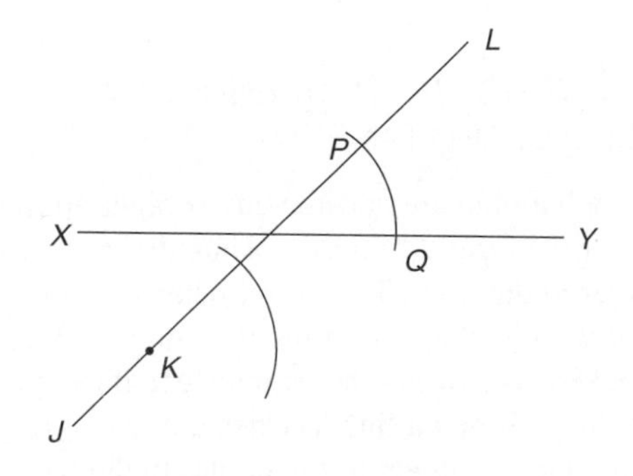

Fourth, open the compass a distance equal to PQ and swing an arc from the point where the previous arc intersected $\overline{JL}$ to intersect the previous arc at point W.

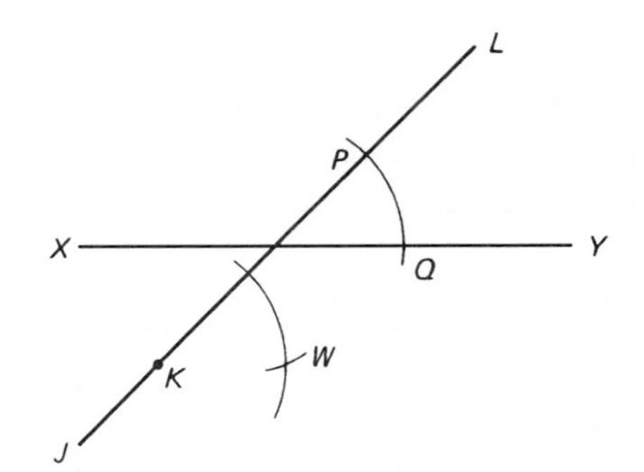

Fifth, draw line $\overline{VT}$ through points K and W. Thus $\overline{XY}$ is parallel to $\overline{VT}$, or $\overline{XY} \parallel \overline{VT}$.

Note $\parallel$ is the symbol for parallel lines.

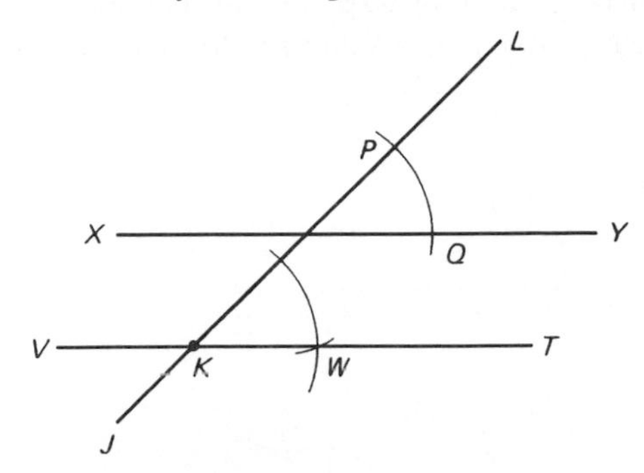

Example 8: Construct a Tangent

PROBLEM: Construct a tangent to a circle from a given point.

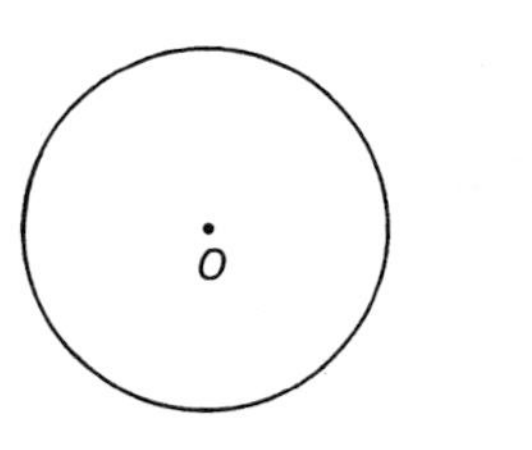

SOLUTION: First, draw a line from point A to the center of the circle, point O.

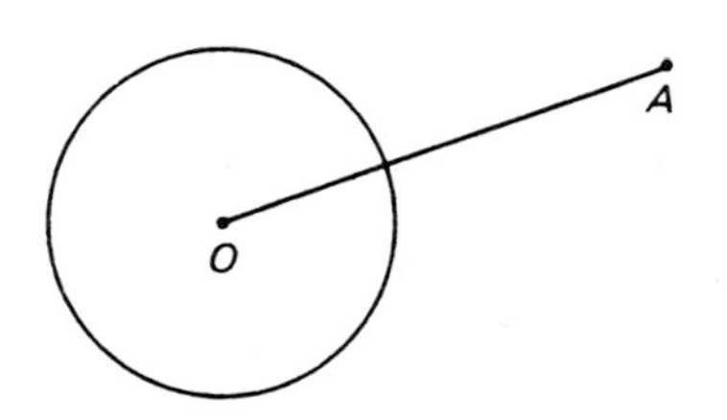

Second, construct a line that bisects $\overline{AO}$ as shown. Label the point of bisection as point K.

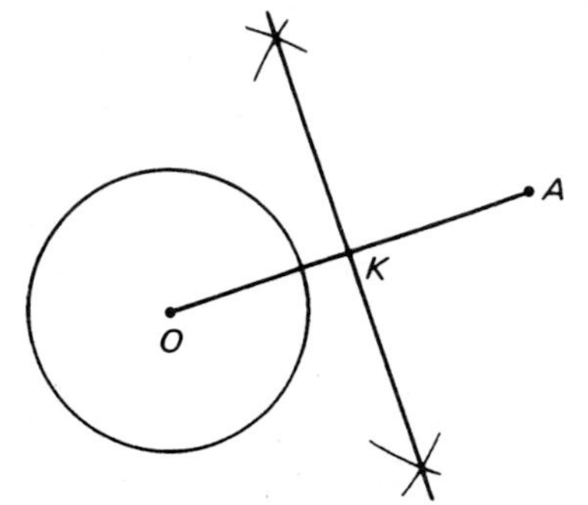

Third, swing an arc from point K to intersect point O and to create points E and D on the circle.

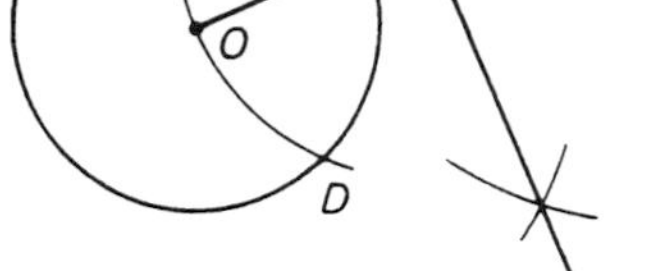

Fourth, draw rays AE *and* AD from point A to intersect at points E and D, respectively. Thus $\overrightarrow{AD}$ and AE are tangent to circle O.

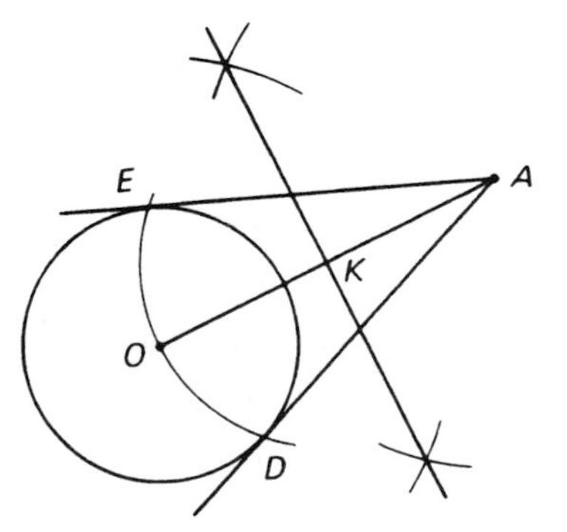

EXERCISE 2 CONSTRUCTING PERPENDICULAR, PARALLELS, AND TANGENT LINES

Construct the following perpendiculars, parallels, and tangents as indicated.

13. Construct a perpendicular to the line at point X.

X

14. Construct a perpendicular to point P.

P

15. Construct a perpendicular to point K.

16. Construct a perpendicular to the line from point L.

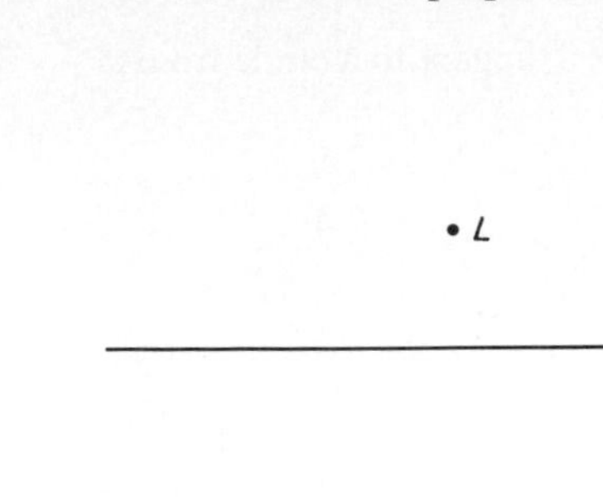

17. Construct a perpendicular to the line from point J.

18. Construct a perpendicular to the line from point K.

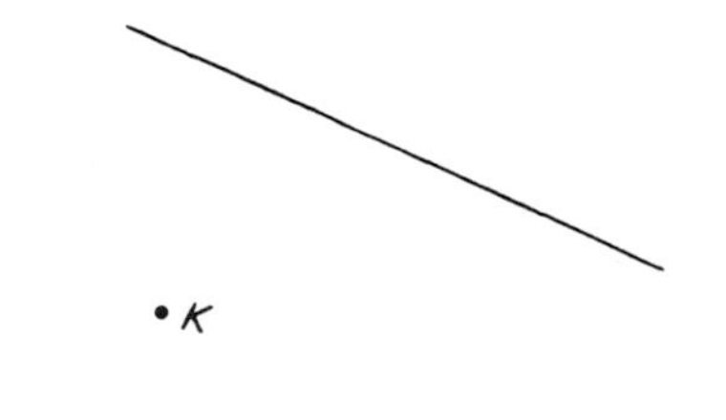

19. Construct a line parallel to the given line.

20. Construct a line parallel to the given line.

21. Construct a line parallel to the given line.

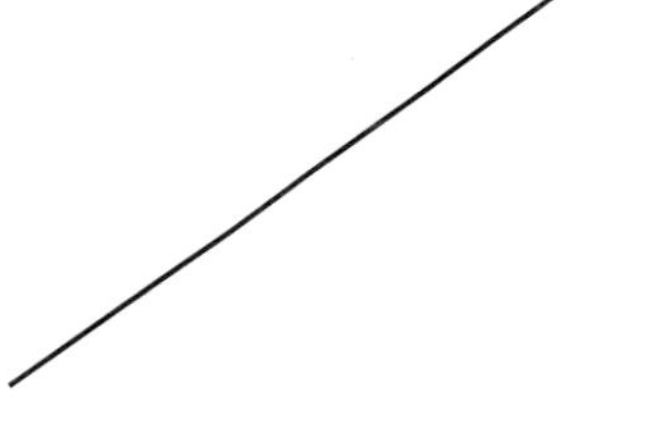

22. Construct tangents to circle L from point M.

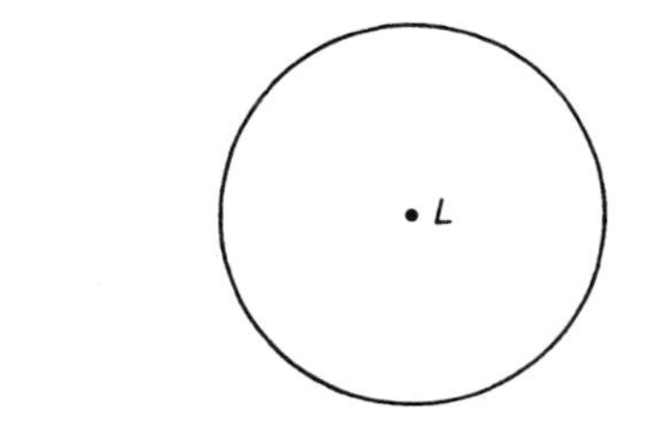

23. Construct tangents to circle X from point Y.

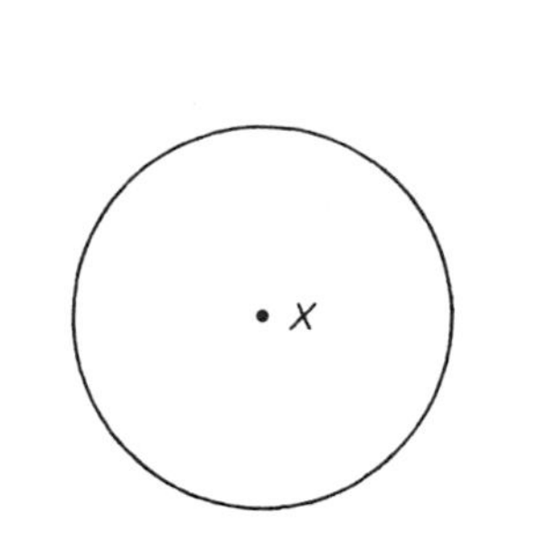

24. Construct tangents to circle S from point T.

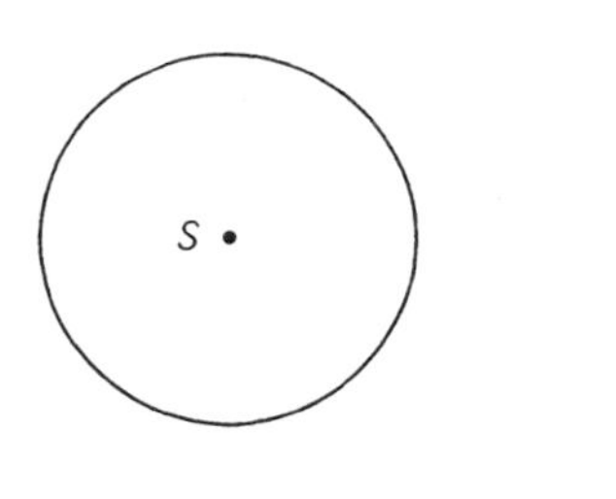

3. CONSTRUCT TRIANGLES, SQUARES, AND HEXAGONS

It is frequently helpful to construct simple geometric figures to understand complicated industrial problems. In this section, we construct triangles, squares, and hexagons.

Example 9: Construct a Triangle

PROBLEM: Construct a triangle from three given line segments.

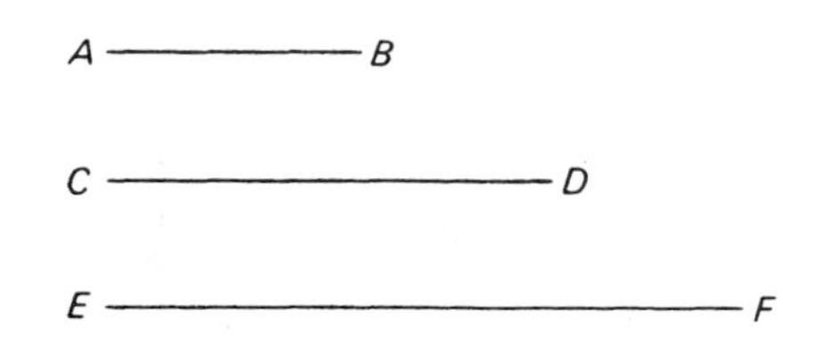

SOLUTION: First, draw a working line and label it.

X ——— Y

Second, measure line $\overline{EF}$ with a compass and enter the points on the working line XY.

Third, measure line $\overline{AB}$ with the compass and swing an arc from point E.

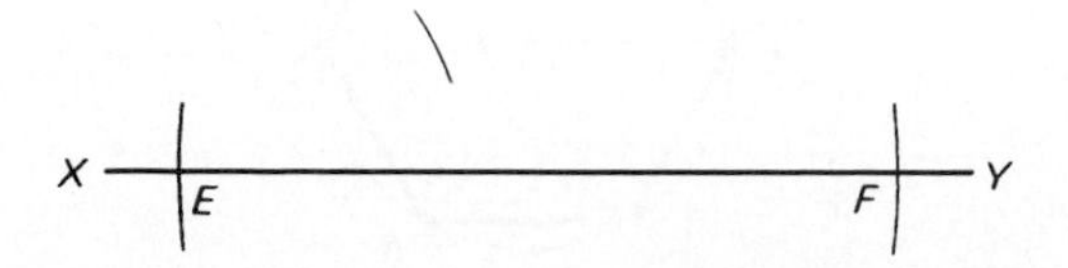

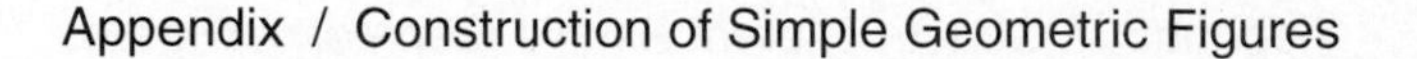

Fourth, measure line $\overline{CD}$ with the compass and swing an arc from point F so that it intersects the arc from point E. Designate the point of intersection as point G.

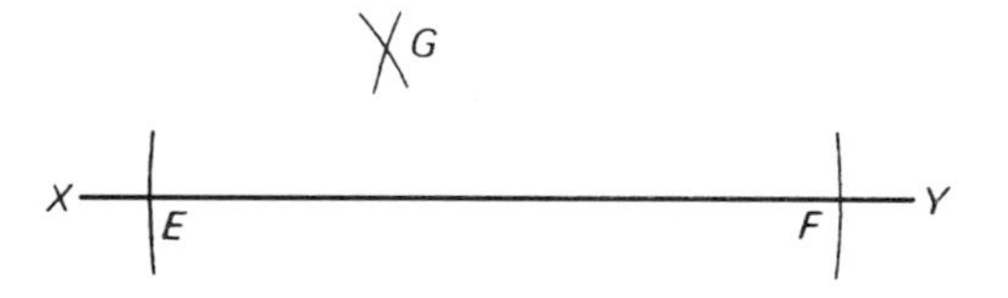

Fifth, draw in line segments $\overline{EG}$ and $\overline{FG}$.

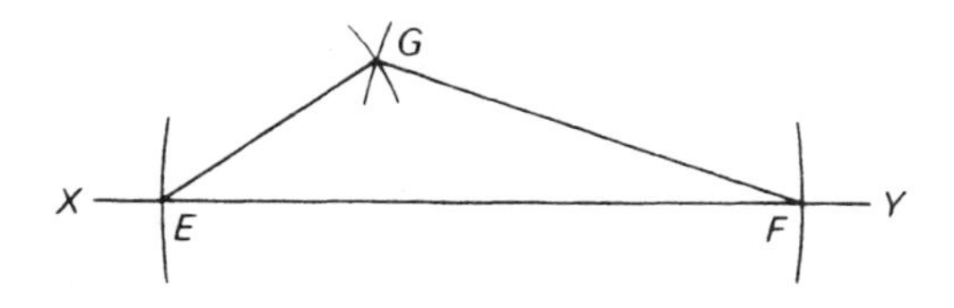

Sixth, measure with a compass to ensure that

$$\overline{AB} = \overline{EG}$$
$$\overline{CD} = \overline{FG}$$
$$\overline{EF} = \overline{EF}$$

Example 10: Construct a Square

PROBLEM: Construct a square given the length of one side.

SOLUTION: First, draw a working line.

Second, measure line $\overline{KJ}$ and enter the points on the working line.

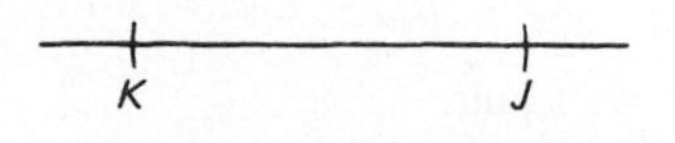

Third, construct perpendiculars at both points K and J.

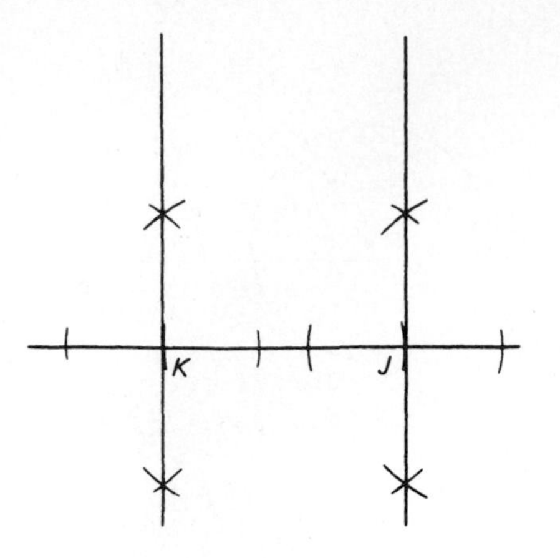

Fourth, measure the length of line segment $\overline{KJ}$ and strike an arc on the perpendiculars from points K and J. Enter points L and M.

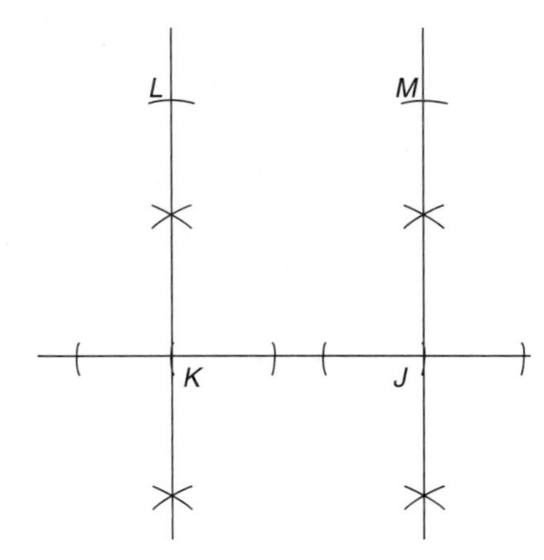

Fifth, draw in line segments $\overline{KL}$, $\overline{LM}$, and $\overline{JM}$.

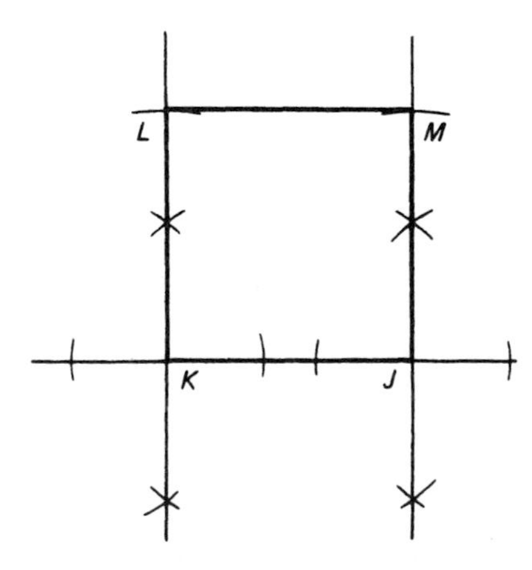

Sixth, measure with a compass to ensure that

$$\overline{KJ} = \overline{KL} = \overline{LM} = \overline{JM}$$

Thus $JKLM$ is a square.

Example 11: Construct a Hexagon

PROBLEM: Construct a regular hexagon, given the length of one side.

L——M

SOLUTION: All sides of a regular hexagon are equal and the sides are equal to the radius of the circle. First, draw a circle using line segment $\overline{LM}$ as the radius of the circle.

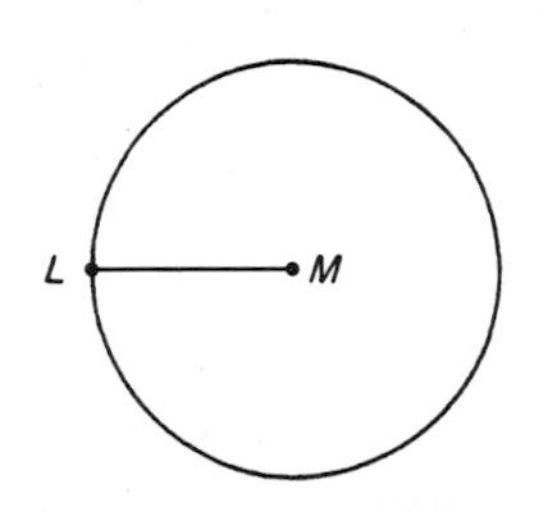

Second, using the same radius, swing an arc from point L to point A.

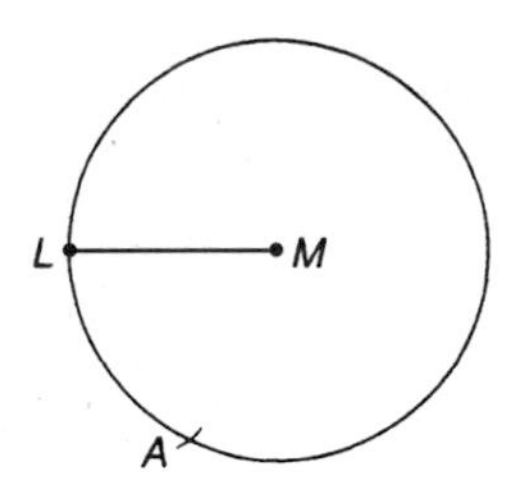

Third, using the same radius, swing an arc from point A to point B, from point B to point C to point D, and from point D to point E.

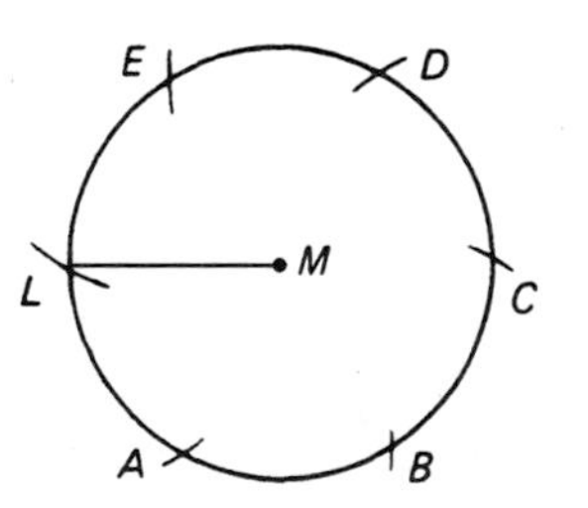

Fourth, draw in line segments $\overline{LA}$, $\overline{AB}$, $\overline{BC}$, $\overline{CD}$, $\overline{DE}$, and $\overline{EL}$.

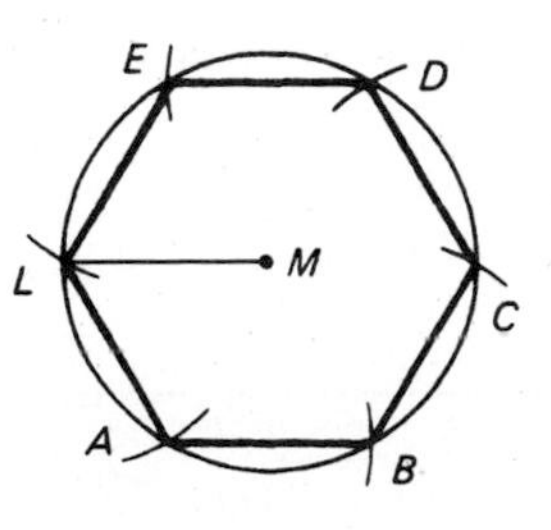

Example 12: Construct a Given Angle

PROBLEM: Construct a 60° angle.

SOLUTION: Recall that all angles of an equilateral triangle are equal to 60°. First, draw line segment CK on working line MO.

Second, from points C and K on working line MO swing arcs of equal lengths. Label the point of intersection as point H.

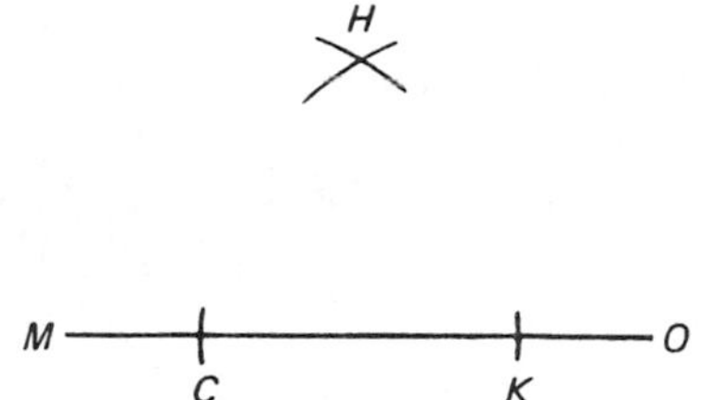

Third, draw line segments $\overline{CH}$ and $\overline{KH}$.

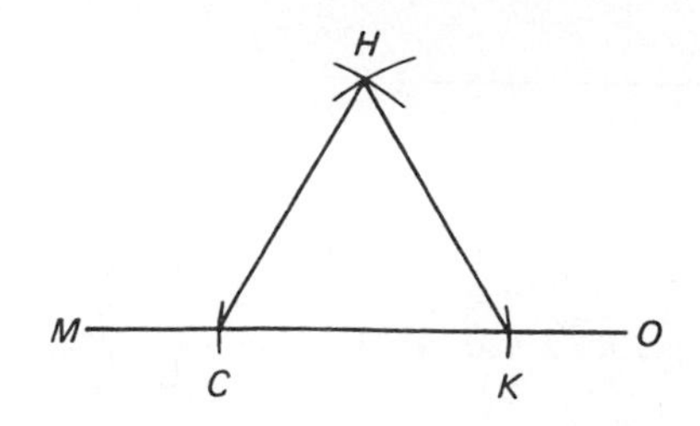

Fourth, since all sides are equal length, each interior angle is 60° and each exterior angle is 120°.

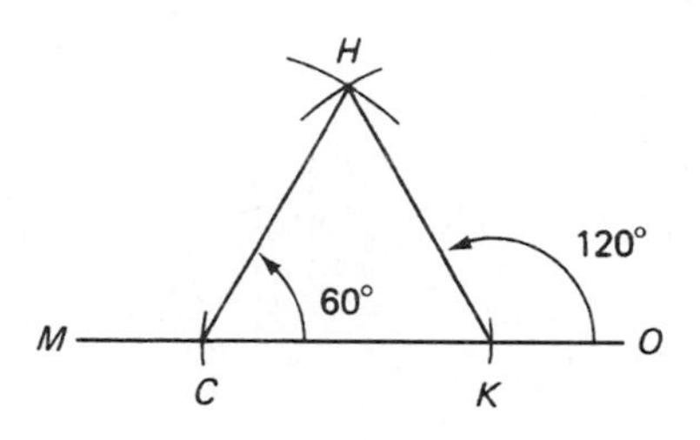

EXERCISE 3

Construct the following triangles, squares, rectangles, and angles as indicated.

25. Construct a triangle given the line segments.

A ——— B

C ——— D

E ——— F

26. Construct a triangle given the line segments.

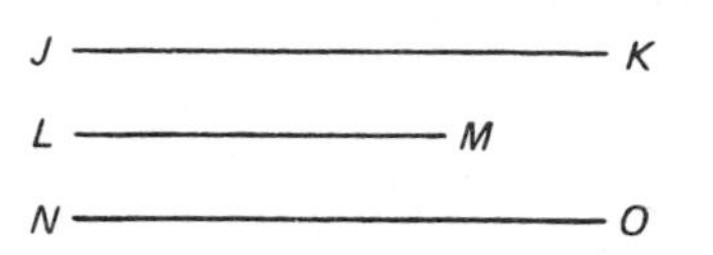

27. Construct a triangle given the line segments.

P ——— Q

R ——— S

T ——— U

28. Construct a square given line segment $\overline{AB}$.

A ——— B

29. Construct a rectangle given the line segments CD and EF.

C ———————— D

E ———————— F

30. Construct a rectangle given the line segments GH and IJ.

G ———————— H

I ———————— J

31. Construct a 120° angle.

32. Construct a 30° angle.

33. Construct a 45° angle.

34. Construct a 135° angle.

35. Construct a regular hexagon given the line segment $\overline{MN}$.

M ———————— N

36. Construct a parallelogram given the line segments OP and QR, with angles of 60° and 120°.

O ———————— P

Q ———————— R

THINK TIME

Corporations often create special symbols, called *logos*, to represent their companies in the minds of consumers. Study the three logos that appear below, and then design some of your own using the compass and straightedge.

PROCEDURES TO REMEMBER

1. To copy a line segment:
   (a) Draw a working line with a straightedge.
   (b) Measure the given line segment with a compass.
   (c) Indicate a point on the working line and swing an arc with the compass.
   (d) The new line segment extends from the point on the working into to the point where the arc crossed the line.

2. To bisect a line segment:
   (a) Open the compass to a position greater than half the length of the line segment.
   (b) Swing an arc from each endpoint.
   (c) Connect the points of intersection above and below the given line segment with a straightedge.
   (d) This perpendicular line will bisect the given line segment.

3. To bisect an angle:
   (a) Swing an arc from the vertex of the angle so that it intersects both rays of the angle.
   (b) Swing an arc from each of the intersection points on the rays.
   (c) From the point of intersection of these arcs, draw a line to the vertex using a straightedge.
   (d) This line bisects the angle, thus creating two equal angles.

4. To copy an angle:
   (a) Draw a working line with a straightedge.
   (b) Indicate a point on the working line to serve as a vertex.
   (c) Swing the compass on the given angle and on the working line.
   (d) Set the compass on the point of intersection of one of the two rays of the given angle, and open it so that it touches the point of intersection on both rays.
   (e) Set the compass on the point of intersection on the working line. Swing the same arc as measured in step (d).
   (f) Draw a line from the vertex of the working line through the intersecting arcs above the working line.
   (g) This line will be the second ray of the copied angle.

5. To divide a line into a given number of equal segments:
   (a) Draw a working line 20° to 50° above the given line segment so that the two lines form an angle with a common vertex.
   (b) Divide the working line into the given number of segments by swinging preset arcs from the vertex to the first intersection of the working line; then from the first intersection, swing the second preset arc; from the second intersection, swing the third preset arc; and so on.
   (c) Draw a line from the last intersection to the endpoint of the given line segment.
   (d) Copy each angle and draw lines to intersect the given line segment.
   (e) Each intersection will create equal line segments.

6. To construct a perpendicular line from a point on a line:
   (a) Place the compass on the given point on the line segment and swing an arc to cross the line segment at two different points.
   (b) From these two intersections, swing two arcs from each intersection one below and one above the given line segment.
   (c) Draw a straight line from these points of intersection to intersect the given line in a perpendicular fashion.

7. To construct a perpendicular from a given point:
   (a) Open the compass and swing an arc from the given point so that two intersections are made on the given segment.
   (b) From these points of intersection, swing two additional arcs so that they intersect on the opposite side of the line from the given point.
   (c) Connect the given point with the point of intersection of two arcs. This line will be perpendicular to the given line segment.

8. To construct a line parallel to a given line:
   (a) Construct a working diagonal intersecting the given line segment at about 45°.
   (b) Swing an arc from the point of intersection of the given line and the working line.
   (c) Swing an arc from another point on the working line to intersect the first one.
   (d) Draw a line from the point on the working line through the point of intersect of the two arcs.
   (e) The line will be parallel to the given line.

9. To construct a tangent to a circle from a given point:
   (a) Draw a line from the given point to the center of the circle.
   (b) Construct a bisector of the line drawn from the given point to the center of the circle.
   (c) Swing an arc from the midpoint of the line to intersect the center of the circle and the circumference of the circle at two points.
   (d) Draw lines from the given point to the two points on the circumference of the circle.
   (e) These two lines are tangents to the circle.

10. To construct a triangle from three given lines:
    (a) Draw a working line. Measure one of the given line segments with a compass, and enter this on the working line.
    (b) Measure a second line segment and swing an arc from one endpoint of the given line segment on the working line.
    (c) Measure the third line segment and swing an arc from the other endpoint of the given line segment so that the arc intersects the arc swing in step (b).
    (d) Draw lines from the point of intersection to each end of the given line segment.

11. To construct a square, given one side:
    (a) Draw a working line, measure the given side, and enter this on the working line.
    (b) Construct perpendiculars to the endpoints of the given side on the working line.

(c) Measure the length of the given side. From each of the endpoints of the given side, swing an arc on the perpendicular lines.
(d) Draw lines between the perpendiculars to form a square.

12. To construct a regular hexagon, given the length of one side:
 (a) Draw a circle using the line segment as the radius of the circle.
 (b) Using the same radius, swing an arc from the point where the radius touches the circle to cut the edge of the circle on both sides of the point.
 (c) From the point where the previous arc intersected the circle, swing an additional arc. Continue to do this until the edge of the circle has been cut by arcs into six equal lengths.
 (d) Draw line segments to connect adjacent intersections.

13. To construct a 60° angle:
 (a) Draw a line segment.
 (b) From one end of the line segment, swing an arc of equal length to the line segment.
 (c) From the other end of the line segment, swing an arc of equal length to the line segment.
 (d) At the point of intersection, draw line segments to the endpoints of the original line segments.
 (e) The inside angles of an equilateral triangle are 60° angles.

CHAPTER SUMMARY

1. The following definitions are necessary for the construction of simple geometric forms.
 (a) A *compass* is a tool used to draw circles or portions of circles and to measure line segments.
 (b) A *straightedge* is used to draw straight lines.
 (c) A *radius* is the opening between two points on a compass.
 (d) *Radii* is the plural of radius.
 (e) An *arc* is a portion of the circumference of a circle.
 (f) *Perpendicular lines* form angles of 90°.
 (g) Two lines are *parallel* when they run side by side and never intersect.
 (h) A *tangent* line touches a circle at only one point.

2. Procedures to remember, included in this appendix, are:
 (a) Copy a line segment.
 (b) Bisect a line segment.
 (c) Bisect an angle.
 (d) Copy an angle.
 (e) Divide a line segment into a given number of equal segments.
 (f) Construct a perpendicular line from a point on a line.
 (g) Construct a perpendicular from a given point.
 (h) Construct a line parallel to a given line.
 (i) Construct a tangent to a circle from a given point.
 (j) Construct a triangle from three given line segments.
 (k) Construct a square given one side.
 (l) Construct a regular hexagon given the length of one side.
 (m) Construct a 60° angle.

CHAPTER TEST

T–1. Copy line segment $\overline{ST}$.

S T

T–2. Copy angle *XYZ*.

X Y Z

T–3. Bisect $\overline{PQ}$.

P Q

T–4. Bisect angle *KLM*.

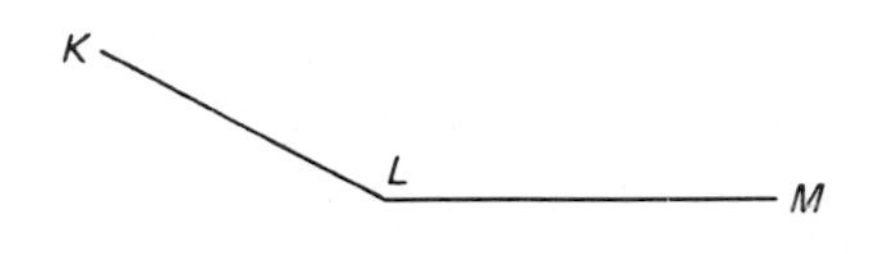

T–5. Divide $\overline{FG}$ into four equal parts.

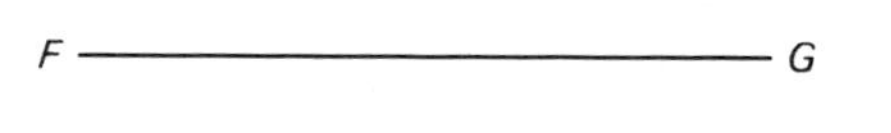

T–6. Construct a perpendicular on $\overline{AB}$ at X.

A X B

T–7. Construct a line parallel to $\overline{DE}$.

D E

T–8. Construct a perpendicular to $\overline{AB}$ from point Z.

A B

Z

T–9. Using $\overline{JK}$, construct a hexagon with each side equal in length to $\overline{JK}$.

J K

T–10. Construct a 60° angle.

T–11. Construct a tangent to circle O from point Y.

Y

T–12. Construct a triangle from the segments shown below.

T–13. Construct a square given side $\overline{LM}$.

L M

T–14. Construct a perpendicular to $\overline{WX}$.

W X

T–15. Divide $\overline{QR}$ into six equal parts.

Q R

GLOSSARY

Absolute value The distance between a number and zero on the number line.

Addition method A method for solving a system of linear equations that involves adding the equations.

Adjacent angles Angles that have a common vertex, share a common side, and have no interior points in common.

Acute angle An angle measuring between 0 degree and 90 degrees.

Adjacent side The side adjacent to an angle in a triangle.

Algebraic expression A mathematical expression that contain one or more variables.

Accuracy The number of significant digits a number contains.

Algebraic fraction A fraction in which the numerator and denominator both represent algebraic expressions.

Alternate exterior angles Angles formed by the intersection of a transversal with parallel lines, outside the parallel lines and on alternating sides of the transversal.

Alternate interior angles Angles formed by the intersection of a transversal with parallel lines, inside the parallel lines and on alternating side of the transversal.

Altitude A line drawn from any vertex of a figure to the opposite side.

Angle of depression An angle formed between the line of sight and the horizontal plane below the line of sight.

Angle of elevation The angle formed between the line of sight and the horizontal plane above the line of sight.

Angle The amount of rotation required to move a ray (a half-line with an endpoint) from one position to another.

Approximate number A number obtained from a measurement.

Arc A portion of a circumference.

Area The measure, in square units, of the interior region of a 2-dimensional figure or the surface of a 3-dimensional figure.

Arithmetic mean The sum of the data values divided by the total number of such values.

Ascending order The process of arranging data from smallest to largest.

Average The sum of the data values divided by the total number of such values.

Axes The plural of axis.

Axis The perpendicular number lines in a rectangular coordinate system.

Bar graph A graph that compares quantities using bars to represent data.

Base The side of a polygon on which the polygon rests.

Base of a percent The number for which the percent is found.

Bisect To cut or divide into equal parts.

Bisecting Finding the midpoint of a line segment, or finding the ray that divides an angle into two congruent angles.

Binomial A polynomial with two terms.

Borrow Regroup from one place value to a lower place value in order to subtract.

Cancel To remove equal factors from both sides of an equation or from the numerator and denominator of a fraction.

Capacity The maximum amount that can be contained by a solid—often refers to measurement of a liquid.

Cartesian coordinate system *See* rectangular coordinate system.

Celsius The metric system scale for measuring temperature. Zero degrees is the freezing point of water and 100 degrees is the boiling point of water at sea level.

Center (1) A point that is the same distance from all points on the circumference of a circle or the surface of a sphere. (2) A point that is the same distance from the vertices of a regular polygon.

Centi A prefix meaning one hundredth in the metric system

Centigrade *See* Celsius

Central angle An angle formed by two radii of a circle.

Chord A straight line segment joining any two points on a circle.

Circle A set of points in a plane that are equidistant from a fixed point.

Circle graph A graph that shows the percentage of a whole represented by each pie-shaped sector, also known as a pie chart.

Circumference The distance around a circle or the perimeter of a circle.

Class interval The width of a class or the range of values of an interval.

Class frequency The number of items in a class interval.

Clearing fractions When working with fractional equations, multiplying both sides of the equation by a common denominator.

Combining like terms When working with an expression, combining terms that have the same variable in the same form.

Coefficient The constant or numerical factor in a term.

Common divisor A number that is a factor of two or more numbers.

Common multiple A number that is a multiple of two or more numbers.

Complementary angles Two angles whose sum is 90 degrees.

Complex fraction A fraction whose numerator and denominator contain one or more fractions.

Commission Earning based on a percent of total sales.

Commutative Property of Addition The sum remains the same regardless of the order of addition. $a + b = b + a$ when a and b are real numbers.

Commutative Property of Multiplication The product remains the same regardless of the order of multiplication. $a \times b = b \times a$ when a and b are real numbers.

Commutative Property of Addition and Multiplication The property that allows one to change the order in which operations are done.

Composite figure A figure made up of two or more different shapes.

Composite number A number that has more than two factors.

Compute To find a numerical result by adding, subtracting, multiplying, dividing or other arithmetic operations.

Cone A solid geometric figure with one circular base whose side tapers evenly to a point.

Congruent Having the exact same size and shape.

Consecutive integers In order, one after the other. Example: 7, 8, 9, . . .

Constant A number, letter, or symbol whose value remains fixed.

Coordinate system A 2-dimensional system where two perpendicular lines intersect and the coordinates of a point are the distances from the point of intersection.

Coordinates An ordered pair of numbers that identify a point on a coordinate plane.

Corresponding angles Angles formed by the intersection of a transversal with parallel lines.

Cosine ratio In a right triangle, the cosine of an angle is the ratio of the leg adjacent to that angle to the length of the hypotenuse.

Cross multiplication The product of one numerator and the opposite denominator in a pair of equivalent fractions. A method for solving proportion problems.

Cube (1) A regular solid having six congruent faces. (2) The third power of a number.

Cube root A number whose cube is equal to a given number.

Cubic unit A unit used in measuring volume or capacity.

Cylinder A solid geometric figure with equal, parallel, circular bases.

Data Information, generally numerical information, usually organized for analysis.

Decimal (1) A number written using base ten. (2) A number containing a decimal point.

Decimal fraction A fraction with a denominator of 10, 100, 1000 or a multiple of 10 written as a decimal.

Decimal point A dot separating the ones and tenths places in a decimal number.

Degree (angle measure) A unit of angular measure. One degree equals the central angle of a circle formed by rays that cut $^1/_{360}$ of its circumference.

Degree (temperature measure) Celsius is the metric unit of measurement of temperature. Fahrenheit is the customary unit of measurement for temperature.

Deka A prefix meaning 10 in the metric system.

Denominator The number below the line in a fraction. It tells the number of equal parts into which a whole is divided.

Dependent variable A variable whose value is determined from the choice of an independent variable in a function.

Descarte, René A French mathematician (1596–1650) who is credited with the development of the rectangular coordinate system.

Diagonal A line segment that joins two vertices of a polygon but is not a side of the polygon.

Difference The result after one quantity is subtracted from another.

Digit Any one of the ten symbols: 0, 1, 2, 3, 4, 5, 6, 7, 8, or 9.

Dimensions (1) The lengths of sides of a geometric figure. (2) The number of coordinates needed to locate a point.

Distributive Property $a \cdot (b + c) = (a \cdot b) + (a \cdot c)$ and $a \cdot (b - c) = (a \cdot b) - (a \cdot c)$ where a, b and c stand for any real numbers.

Dividend A quantity to be divided (the number inside the division sign).

Divisor The quantity by which another quantity is to be divided (the number outside the division sign).

Edge The line segment where two faces of a solid figure meet.

Endpoint Either of two points marking the end of a line segment.

Equal Quantities having the same value.

Equation A statement that two expressions are equivalent or equal.

Equilateral triangle A triangle in which all three sides and angles are equal.

Equivalent fractions Fractions that can be reduced to the same fraction.

Equivalent Naming the same value.

Estimate To find or "guess" a number close to an exact number.

Euclid The first known geometer, a Greek mathematician who lived about 300 BC, developed geometric principles, combined them with his own concepts and wrote the first geometry book.

Evaluate To find the value of a mathematical expression.

Even number A whole number that is divisible by 2.

Exact number A number obtained from counting.

Exponent The number that tells how many equal factors there are.

Exponential form A way of writing a number using exponents.

Expression A variable or combination of variables, numbers, and symbols that represents a mathematical relationship.

Face A plane figure that serves as one side of a solid figure.

Factors Numbers, variables, or expressions multiplied together.

Fibonacci sequence A special series of numbers in which each number is the sum of the two preceding numbers.

Figure A closed shape in 2 or 3 dimensions.

Formula A general mathematical statement or rule.

Fraction A number representing part of a whole; the bottom number represents the whole, and the top number represents the part of the whole.

Frequency The number of times something occurs in an interval.

Frequency distribution A table used to tabulate the number of occurrences of a particular measurement.

Frequency polygon A broken-line graph with the data value on the horizontal axis and the frequency of its occurrence on the vertical axis.

Function A relationship or association such that for each value of x there is exactly one value for y.

General form of the equation of a line $Ax + By = C$, where A, B, and C represent integers and A and B are not both zero.

Geometry The mathematics of properties and relationships of points, lines, angles, surfaces, and solids.

Graph (1) A visual presentation of numerical information. (2) A plot of ordered pairs that satisfy an equation.

Greatest common factor The largest factor of two or more numbers.

Height (altitude) The perpendicular distance from a vertex to the opposite side of a plane figure.

Hexagon A polygon with six sides.

Histogram A bar graph in which the data values are represented on the horizontal axis and the frequency is represented on the vertical axis.

Horizontal Parallel to or in the plane of the horizon.

Hypotenuse The side opposite the right angle in a right triangle.

Independent variable The variable in a function whose value is chosen arbitrarily.

Inequality A statement that a given quantity is not equal to another quantity.

Initial side The original position of the ray in an angle.

Inscribed angle An angle formed by two chords with one common endpoint inside a circle.

Integers The positive and negative integers and zero.

Interval A set containing all the numbers between its endpoints and including one endpoint, both endpoints, or neither endpoint.

Interpolation The process of estimating values between the data points on a graph.

Interquartile range The difference between the upper quartile and the lower quartile.

Irrational number A number that cannot be expressed as a rational number.

Isosceles triangle A triangle in which two sides and two angles are equal.

Joint variation Variation in which one variable varies directly with several other variables.

Least common denominator (LCD) The smallest common multiple of the denominators of two or more fractions.

Least common multiple (LCM) The smallest common multiple of a set of two or more fractions.

Leg In a right triangle, one of the two sides that form the right angle.

Length The distance along a line or figure from one point to another.

Like terms Terms that are identical except for their numerical coefficients.

Linear equation An equation that can be written in the form $ax + by = c$, where a and b are real numbers and are not zero.

Line segment A finite portion of a line bounded by and including two endpoints.

Literal number A letter or other symbol used to represent a real number.

Lower quartile The median of the lower half of an ordered set of data.

Lowest terms When the numerator and denominator of a fraction do not contain common factors.

Mean The sum of a set of numbers divided by the number of elements in the set.

Measures of central tendency Numbers that are representative of the distribution of data range, mean, median, and mode.

Median The middle number when data are arranged in ascending order, or the mean of two middle numbers when the set has two middle numbers.

Metric system A system of measurement based on tens. The basic unit of length is the meter. The basic unit of mass is the gram. The basic unit of capacity is the liter.

Midpoint The point on a line segment halfway between the endpoints.

Minute One-sixtieth of a degree.

Minuend In subtraction, the minuend is the number you subtract from.

Mixed number A number with an integer part and a fraction part.

Mode The data value that occurs most frequently in a set of data.

Monomial A polynomial of one term.

Multinomial An expression containing two or more terms.

Multiple The product of a whole number and any other whole number.

Multiplicand In multiplication, the multiplicand is the top factor being multiplied.

Multiplication The operation of repeated addition.

Multiplier In multiplication, the multiplier is the bottom factor being multiplied.

Natural numbers The numbers in the set 1, 2, 3, 4, . . .

Negative integers The set –1, –2, –3, . . .

***n*th root** A number which, when used as a factor n times, gives the original number.
Number line A diagram that represents numbers as points on a line.
Numeral A symbol used to represent a number.
Numerator The number written above the line in a fraction.
Numerical coefficient A numerical factor.

Oblique triangle A triangle that contains an oblique angle.
Obtuse angle An angle measuring between 90 degrees and 180 degrees.
Obtuse triangle A triangle whose largest angle measures greater than 90 degrees.
Octagon A polygon with 8 sides.
Odd number A whole number that ends in 1, 3, 5, 7, or 9.
Opposite (1) Directly across from. (2) Having a different sign but the same number.
Opposite side The side opposite a given angle in a triangle.
Order of operation Rules describing what sequence to use in evaluating expressions: (1) Evaluate within grouping symbols. (2) Evaluate powers or roots. (3) Multiply or divide left to right. (4) Add or subtract left to right.
Ordered pair A pair of numbers that gives the coordinates of a point on a rectangular coordinate system in the order (horizontal coordinate, vertical coordinate).
Origin In a rectangular coordinate system, the point of intersection of the two axes described by the ordered pair (0, 0).

Parallel lines Lines in a plane with no points in common.
Parallelogram A quadrilateral with two pairs of parallel sides.
Pentagon A polygon that has five sides.
Percent A special ratio that compares a number to 100, denoted by % and meaning per hundred.
Percentage A number that is a given percent of another number.
Percentile The numbers that divide a set of data into 100 equal parts.
Perimeter The distance around a figure.
Perpendicular Two lines that intersect to form a right angle.
Perpendicular bisector A line that divides a line segment in half and meets the segment at right angles.
Pi The ratio of the circumference of any circle to its diameter, approximately equal to 3.142.
Place value The value of the position of a digit in a number.
Plane A flat surface that extends infinitely in all directions.
Plane figure Any 2-dimensional figure.
Point An exact position in space.
Polynomial A collection of several terms.
Positive number Numbers that are greater than zero.
Power An exponent.
Precision A reference to the decimal position of the last significant digit in a number.
Prime number A number that has exactly two positive factors, itself and 1.
Prism A 3-dimensional figure that has two congruent and parallel faces that are polygons.
Principal The original or unpaid balance of the amount borrowed for a loan.
Polygon A closed figure in a plane with three or more straight sides. The remaining faces are parallelograms.
Product The result of multiplication.
Proper fraction A fraction whose numerator is an integer smaller than its integer denominator.
Proportion A statement of equality between two ratios.
Pyramid A polyhedron whose base is a polygon and whose other faces are triangles that share a common vertex.
Pythagorean theorem The sum of the squares of the lengths of the two legs of a right triangle is equal to the square of the length of the hypotenuse.

Quadrants The four equal regions into which the axes divide the plane in a rectangular coordinate system.
Quadrilateral A polygon with four sides.
Quartiles Along with the median, the quartiles divide an ordered set of data into four groups of about the same size.
Quotient The answer in a division problem.

Radian A central angle that intercepts an arc equal to the radius of a circle.
Radical sign A symbol that indicates the root to be taken.
Radicand The number expression under the radical.
Radii The plural of radius.
Radius The distance from the center to any point on the circle.
Range The difference between the greatest number and the least number in a set of data.
Rate A ratio comparing two different units.
Ratio A comparison of two or more quantities.
Rational number A number that can be expressed as a ratio of two integers.
Rationalizing the denominator Multiplying the numerator and denominator of a fraction by a quantity that results in a rational denominator.
Ray A line extending infinitely to one side of a given endpoint.
Rectangle A parallelogram with four right angles.
Rectangular coordinate system A system formed by two perpendicular number lines, or axes, used to graph ordered pairs of numbers.
Reduce Simplify a fraction into its simplest form.
Regular polygon A polygon in which the sides are equal and the interior angles are equal.
Remainder The amount left over in a division problem.
Repeating decimal A decimal that has an infinitely repeating sequence of digits.
Resultant The vector sum of two or more vectors.
Revolution One turn of 360 degrees about a point.
Rhombus A parallelogram with four equal sides.
Right angle An angle measuring exactly 90 degrees.
Right triangle A triangle with a right angle.
Rise The vertical change in a line.
Round To drop or zero-out digits in a number and change the digit in a specific place.
Run The horizontal change in a line.

Sample A number of events or objects chosen from a given population to represent the entire group.

Scale A ratio between two sets of measurements.
Scalene triangle A triangle with no equal sides.
Scientific notation A notation in which a number is written as the product of a number between 1 and 10 and a power of 10.
Secant A line that intersects a circle at two points.
Sector The region formed by two radii of a circle and the intercepted arc.
Segment A part of a circle bounded by a chord and the arc it creates.
Semicircle An arc that is exactly half of a circle. A diameter intersects a circle at the endpoints to two semicircles.
Sequence A collection of numbers in a certain order.
Set A collection of distinct elements or items.
Side A line segment connected to other segments to form a polygon.
Significant digit Any digit that denotes an actual physical amount.
Sign number Positive or negative number.
Similar figures Figures that have the same shape, but not necessarily the same size. Corresponding sides of similar figures are proportional.
Similar triangles Triangles that have equal corresponding angles and proportional corresponding sides.
Simplest form A fraction whose numerator and denominator have no common factor greater that 1.
Simplify Combine like terms and apply properties to an expression to make computation easier.
Sine ratio In a right triangle, the ratio of the length of the opposite leg to the length of the hypotenuse. The value of the sine of the angle depends upon the measure of the angle.
Slant height (1) The perpendicular distance from the vertex of a pyramid to one edge of its base. (2) The shortest distance from the vertex of a right circular cone to the edge of its base.
Slope In a line, the ratio of rise over run.
Slope form An equation of a line given in terms of the slope and a point on the line.
Slope-intercept form The equation of a line given in terms of the slope and the y intercept of the line.
Solid A geometric figure with 3 dimensions.
Solution A number (or ordered pair) that produces a true statement when substituted for the variable (or variables).
Sphere A 3-dimensional figure made up of all points that are equally distant from a point called the center.
Straight angle An angle with a measure of 180 degrees.
Subtraction An operation that gives the difference between two numbers.
Subtrahend In subtraction, the subtrahend is the number being subtracted.
Square A rectangle with four equal sides.
Straight line A line extending infinitely in both directions.
Substitution The replacement of variables with equivalent expressions or numbers.
Sum The answer to an addition problem.
Supplementary angles Two angles whose sum is 180 degrees.
Surface area The total area of the faces (including the bases) and curved surfaces of a solid figure.

Tangent (1) Touching at exactly one point. (2) In a right triangle, the ratio of the length of the leg opposite an angle to the length of the adjacent leg.
Term A number, variable, product, or quotient in an expression.
Terms of a proportion Each element of a proportion is a term.
Theorem A proven mathematical generalization.
Transposition A property stating that moving any term from one side of an equation to the other and reversing the sign does not change the equation.
Transversal A line that intersects two or more parallel lines.
Trapezoid A quadrilateral with one pair of parallel sides and one pair of nonparallel sides.
Triangle A polygon with three sides.
Trigonometric ratios Sine, cosine, and tangent are ratios among the sides of right triangles.
Trigonometry A branch of mathematics dealing with triangle measurement and related topics.
Trinomial A polynomial with three terms.

Unit rate A rate with a denominator of 1.
Upper quartile The median of the numbers greater than the median in an ordered set of numbers.

Variable A quantity that changes or can have different values.
Vector A quantity that has both magnitude and direction.
Vertex The point where two line segments, lines, or rays meet to form an angle.
Vertical angles Angles formed when two lines intersect.
Volume The number of cubic units it takes to fill a solid.

Weight Determined by the mass of an object and the effect of gravity on that object.
Whole number Any of the numbers 0, 1, 2, 3, . . .

***X*-axis** The horizontal number line in a rectangular coordinate system, representing the independent variable.
***X*-coordinate** The first element of the ordered pair (x, y). Also called the abscissa.
***X* intercept** The point where a graph intersects the x-axis.
***Y*-axis** The vertical number line in a rectangular coordinate system, representing the dependent variable.
***Y*-coordinate** The second element of the ordered pair (x, y). Also called the ordinate.
***Y* intercept** The point where a graph intersects the y-axis.

Zero A number that is neither positive nor negative.

INDEX